Pioneering British Women Chemists

Their Lives and Contributions

Other Related Titles from World Scientific

Women at Imperial College: Past, Present and Future
by Anne Barrett
ISBN: 978-1-78634-262-1

Women in Their Element: Selected Women's Contributions to the Periodic System
edited by Annette Lykknes and Brigitte Van Tiggelen
ISBN: 978-981-120-628-3
ISBN: 978-981-120-768-6(pbk)

The Enigma of Ferment: From the Philosopher's Stone to the First Biochemical Nobel Prize
by Ulf Lagerkvist
ISBN: 978-981-256-421-4(pbk)

The Periodic Table and a Missed Nobel Prize
by Ulf Lagerkvist
edited by Erling Norrby
ISBN: 978-981-4295-95-6(pbk)

Tu Youyou's Journey in the Search for Artemisinin
by Wenhu Zhang, Yiran Shao, Dan Li, Manyuan Wang
Translated by Junxian Yu
ISBN: 978-981-3207-63-9
ISBN: 978-981-3207-64-6(pbk)

Pioneering British Women Chemists

Their Lives and Contributions

Marelene Rayner-Canham
Geoff Rayner-Canham

Grenfell Campus, Memorial University of Newfoundland, Canada

NEW JERSEY · LONDON · SINGAPORE · BEIJING · SHANGHAI · HONG KONG · TAIPEI · CHENNAI · TOKYO

Published by

World Scientific Publishing Europe Ltd.

57 Shelton Street, Covent Garden, London WC2H 9HE

Head office: 5 Toh Tuck Link, Singapore 596224

USA office: 27 Warren Street, Suite 401-402, Hackensack, NJ 07601

Library of Congress Cataloging-in-Publication Data
Names: Rayner-Canham, Marelene F., author. | Rayner-Canham, Geoffrey, author. |
    Rayner-Canham, Marelene F. Chemistry was their life.
Title: Pioneering British women chemists : their lives and contributions /
    Marelene Rayner-Canham, Geoff Rayner-Canham.
Other titles: British women chemists
Description: London ; Hackensack, NJ : World Scientific, [2020] | Complements: Chemistry was
    their life : pioneering British women chemists, 1880–1949 (London : Imperial College Press,
    2008). | Includes bibliographical references and index.
Identifiers: LCCN 2019036783 | ISBN 9781786347688 (hardcover)
Subjects: LCSH: Women in chemistry--Great Britain--History. | Women chemists--Great
    Britain--Biography. | Chemists--Great Britain--Biography. | Chemistry--Great Britain--History.
Classification: LCC QD20 .R3855 2020 | DDC 540.92/520941--dc23
LC record available at https://lccn.loc.gov/2019036783

British Library Cataloguing-in-Publication Data
A catalogue record for this book is available from the British Library.

Copyright © 2020 by World Scientific Publishing Europe Ltd.

All rights reserved. This book, or parts thereof, may not be reproduced in any form or by any means, electronic or mechanical, including photocopying, recording or any information storage and retrieval system now known or to be invented, without written permission from the Publisher.

For photocopying of material in this volume, please pay a copying fee through the Copyright Clearance Center, Inc., 222 Rosewood Drive, Danvers, MA 01923, USA. In this case permission to photocopy is not required from the publisher.

For any available supplementary material, please visit
https://www.worldscientific.com/worldscibooks/10.1142/Q0228#t=suppl

Desk Editors: Herbert Moses/Jennifer Brough/Shi Ying Koe

Typeset by Stallion Press
Email: enquiries@stallionpress.com

About the Authors

Marelene Rayner-Canham and **Geoff Rayner-Canham** of the Grenfell Campus of Memorial University, Corner Brook, Newfoundland, Canada, have been long-time researchers in the history of women in chemistry. Their first book was *Harriet Brooks: Pioneer Nuclear Scientist* (1992), followed by *A Devotion to Their Science: Pioneer Women of Radioactivity* (1997), and then *Women in Chemistry: Their Changing Roles from Alchemical Times to the Mid-Twentieth Century* (1998). During their archival work, they identified an extraordinarily high number of forgotten British women chemists, resulting in *Chemistry Was Their Life: Pioneering British Women Chemists, 1880–1949* (2008) and then *A Chemical Passion: The Forgotten Saga of Chemistry at British Independent Girls' Schools, 1820–1940* (2018). Over the recent years, they have uncovered so much new material on the lives and contributions of pioneering British women chemists that they have now written this new book on these early, and mostly forgotten, pioneering British women chemists.

In addition to the books, the Rayner-Canhams have authored 42 research papers on the history of science, and they have been invited speakers at many conferences. They also authored invited chapters in *Women in Their Element: Selected Women's Contributions to the Periodic System* and *The Posthumous Nobel Prize in Chemistry. Volume 2: Ladies in Waiting for the Nobel Prize.*

Acknowledgements

Our research could never have been successful without the help of the archivists at each of the many, many institutions we have visited to undertake the research. They responded promptly to our e-mail enquiries and during our visits found the rarely accessed material needed for our investigations.

Through the many years that our research was in progress, the successive administrations at the Grenfell Campus of Memorial University (MUN), Corner Brook, NL, Canada, have been consistently and enthusiastically supportive. In particular, we have been grateful for a series of MUN Vice-President's/Social Sciences and Humanities Research Council Grants and Grenfell Campus Principal's Research Grants, both sources supporting our travel to archives across Britain.

Memorial University has a campus at Old Harlow, Essex, England, and we have been fortunate to sojourn there each time we have had the opportunity to visit Britain to hunt down primary material. In this context, we wish to thank the staff at the Harlow Campus, especially Ms. Sandra Wright and Ms. Dawn Bird, for their hospitality and helpfulness. Our travel needs, primarily the indispensable BritRail passes, were organized by the unflappable Ms. Raelene Parsons, of LeGrow's Travel, Corner Brook, NL, Canada.

viii *Acknowledgements*

Finally, we are grateful to Ms. Jennifer Brough, Editor, World Scientific Publishing (WSPC), for publishing what we believe is an important missing piece of the historical record in British Chemistry. Mr. Merlin Fox, WSPC, is thanked for his encouragement to undertake this book.

Contents

About the Authors		v
Acknowledgements		vii
Introduction		xi
Chapter 1	Pioneering Amateur British Women Chemists	1
Chapter 2	Chemistry at Pioneering Girls' Schools	21
Chapter 3	Cambridge and Oxford Women's Colleges	65
Chapter 4	London Co-educational Colleges	107
Chapter 5	London Women's Colleges	137
Chapter 6	English Provincial Universities	171
Chapter 7	Universities in Scotland and Wales	217
Chapter 8	Domestic or Household Chemistry	259
Chapter 9	The Professional Societies	285
Chapter 10	Women Chemistry Teachers	315
Chapter 11	Hoppy's "Biochemical Ladies"	355
Chapter 12	Women Crystallographers: The Bragg Descendants	381

x *Contents*

Chapter 13	Women in Pharmacy	425
Chapter 14	Roles of Chemist's Wives	459
Chapter 15	Women Chemists and the First World War	485
Chapter 16	Women Chemists and the Inter-War Period	509

Index 561

Introduction

In 1947, the Chemical Society published a book titled *British Chemists*.[1] This book contained the compiled biographies of prominent British chemists. None were women and the accounts read as if women were essentially absent from the chemical enterprise in the latter decades of the 19th century and the first half of the 20th century.

Nothing could be further from the truth: there were significant numbers of women chemistry students and women chemists; for example, we identified 896 who were members of the Royal Institute of Chemistry and/or the Chemical Society over the years 1880–1949. To remedy this inequity, following from our early research, we co-authored the book: *Chemistry Was Their Life: Pioneering British Women Chemists, 1880–1949*,[2] published in 2008.

Eleven years have elapsed since the publication of that book. During this time, we have continued our research into this fascinating and much-overlooked area of women's history and of the history of chemistry, including visiting many more archives. Access to the Rare Books Room of the Cambridge University Main Library helped us trace the interest of women in chemistry back farther than we had previously imagined. Also, researching the family trees of some of these pioneering women chemists has enabled us to fill in some missing background information.

---

[1] Findlay, A. and Mills, W. H. (eds.), (1947). *British Chemists*. Chemical Society, London.
[2] Rayner-Canham, M. and Rayner-Canham, G. (2008). *Chemistry Was Their Life: Pioneering British Women Chemists, 1880–1949*. Imperial College Press/World Scientific Publishing, London.

xii *Introduction*

We came to realize that *Chemistry Was Their Life* had become embarrassingly outdated. Not only had we acquired an enormous amount of additional information but also we discovered that the history of women in chemistry was richer and more diverse than we had indicated in the previous book. A chance encounter and conversation with Mr. Merlin Fox of World Scientific Publishing persuaded us that a successor book was essential if justice was to be truly done to the memory and legacy of these women chemists. Though obviously we have had to build on some of the content of that earlier book, we feel this truly is a new book and not simply a revised edition. We hope that you, the Reader, will concur.

Documentation

For the most part, these women chemists did not gain any recognition. It was unthinkable for a woman to occupy anything more than a supporting role. The majority of the women remained unmarried, thus when they died, no record of them survived. Any relative would probably have been unaware of the scientific contributions of their deceased aunt and hence destroyed her papers and correspondence. For this reason, many of our accounts raise more questions than provide answers. Some of the women simply "disappear" from the record, perhaps through marriage, perhaps death, or perhaps a change in career direction brought about by a lack of opportunity to practice their chosen profession.

Obituaries are usually a good source of information on an individual. For those women chemists who died in the early part of the 20th century, obituaries were sometimes published. These were usually in school or university student magazines or sometimes in professional journals. It is of note that in many instances, the obituary was written by another woman chemist. Paradoxically, the longevity of most of these women meant they did not die until the late decades of the 20th century when a one-line name and date of death in *Chemistry in Britain* was the most any of them received.

In Their Own Words

For the women chemists for whom we have biographical information, what comes through most strongly is their enthusiasm and dedication.

Introduction xiii

The early women chemists saw themselves as the pioneers: they had to succeed for the sake of the young women who followed them. At the same time, they truly found chemistry enthralling, it was indeed the centre of their lives, and for those in academia, they were determined to convey this belief to their students.

We have endeavoured to convey not just factual information but also this sense of excitement through the words of the women chemists themselves. Many of the quotations we have gleaned were from the student magazines from the 1880s to the 1920s. In the course of our researches, we visited archives of nearly every long-established British university and college together with the archives of over 50 independent girls' schools, including perusing the student magazine at each institution. Curiously, by the late 1920s, without exception, the school, college, and university magazines changed from the erudite, content-rich, and informative, to uninformative literary journals. We have obtained not only meaningful text quotes but also relevant content set in rhyming format. We have included several of the rhyming contributions in part (where lengthy) or in whole.

Life for Women Chemistry Students

In those early decades, women chemistry students in the women's colleges led sheltered lives. As we recount in Chapters 3, 5, and 8, the day's activities were regimented to a degree which would seem unbelievable to today's students.

Though the co-educational universities (see Chapters 4, 6, and 7) all claimed to welcome women students, their daily experience was often otherwise. We have included a selection of comments from student magazines by male correspondents that indicate significant hostility to the presence of women. Carol Dyhouse has summarised the situation:

> In many cases they [women students] were excluded from membership of existing societies and student unions, and found it necessary (or expedient) to form their own. The presence of a minority group of women frequently served to under*line* rather than under*mine* the norms of the dominant male culture, and male students often went in for exaggerated displays of masculinity, particularly in informal settings, where the presence of women

xiv *Introduction*

might supply both a target for aggression and an audience. Women students generally responded by keeping a very low profile.[3]

Under-Recognition of Women Chemists' Contributions

The quantity of research accomplished by these women was enormous. So why has their work been forgotten? Much was published under their names alone and was never retrieved when historians were looking under the names of the "great men." Alternatively, when the women's research was co-authored with one of the "great men," then they were overshadowed by their more famous supervisor. An example of the latter is Mollie Barr (see Chapter 16). Of the 36 publications of the biochemist Alexander Glenny, F.R.S., between 1932 and 1955, 25 were co-authored by Mollie Barr,[4] yet Barr's lifetime of contributions to the field was never recognized — until now.

The under-recognition of women's contributions in science has been well documented by Margaret Rossiter.[5] Rossiter was building on the contribution of the sociologist, Robert K. Merton. Merton identified the "Matthew Effect," the minimisation of the role of secondary contributors in science, naming the effect after the Biblical quotation of Matthew 13:12 "whosoever hath not, from him shall be taken away even that he hath." Rossiter argued that the effect was far more endemic and serious for women scientists, for whom she coined the term the "Matilda Effect" after the American suffragist, Matilda J. Gage, who both experienced it and articulated the phenomenon.

Birth Name or Married Name?

For consistency, all women chemists are discussed under their birth name (their married name is provided in parentheses). The change in name upon

---

[3] Dyhouse, C. (1995). The British Federation of University Women and the status of women in universities, 1907–1939. *Women's History Review* **4**: 465–484.

[4] Oakley, C. L. (1966). Alexander Thomas Glenny, 1882–1965. *Biographical Memoirs of Fellows of the Royal Society* **12**: 163–180.

[5] Rossiter, M. W. (1993). The Matthew Matilda effect in science. *Social Studies of Science* **23**: 325–341.

marriage has always been problematic for academic women. Sharon McGrayne remarked the change in name even caused confusion in the case of Dorothy Crowfoot (Mrs. Hodgkin):

> Dorothy published her penicillin studies under her maiden name 'Crowfoot' and announced vitamin B_{12} as 'Hodgkin.' Years later some scientists still did not know that the Crowfoot of penicillin fame was the Hodgkin of B_{12} fame.[6]

Supportive Male Chemists

Just as for the admission of women to Parliament, it was sympathetic males who had to fight the battles for women's admission to university and to professional societies. Sylvia Strauss has described the importance of sympathetic men to the suffrage campaign in Britain in her book *"Traitors to the Masculine Cause": The Men's Campaigns for Women's Rights.*[7] In our book, we have attempted to be meticulous and acknowledge the names and contributions of men who were "traitors to the masculine chemistry cause."

Those (male) chemists who took on women research students are also to be lauded. For example, F. Gowland Hopkins (see Chapter 11) hired a high proportion of women researchers in his biochemistry laboratory at Cambridge, despite the opposition from some of his colleagues. Jocelyn Thorpe, at Imperial College, even reserved spaces in his research laboratory specifically for women chemists.

Opening the Story

In our opinion, it is crucial to show that interest in chemistry by women extended back through the centuries. In Chapter 1, we trace that saga back into the 1600s. Yet, it is almost certain that there is much more to be

---

[6] McGrayne, S. B. (1993). *Nobel Prize Women in Science*. Birch Lane Press, New York, p. 248.

[7] Strauss, S. (1982). *Traitors to the Masculine Cause: The Men's Campaigns for Women's Rights*. Greenwood Press, Westport, Connecticut.

xvi *Introduction*

discovered of women's involvement with chemistry in those long-passed centuries. We also discuss the fascinating story of Marcet's *Conversations on Chemistry* whose impact on very early women's scientific education has still not been fully explored.

Then in Chapter 2, we investigate the role of the independent girls' schools and highlight the enthusiasm of the girls for chemistry. If it had not been for these academic schools which aimed to provide excellence in education, particularly in chemistry, there would have been no cohorts of chemistry-educated young women to enter colleges and universities to take chemistry degrees. This chapter provides a synopsis of our complete book on the topic, *A Chemical Passion: The Forgotten Saga of Chemistry at British Independent Girls' Schools, 1820–1940.*[8]

Colleges and Universities

It was a challenge to provide sufficient background about the institutions themselves without overwhelming the Reader with trivial details. We felt it was necessary to include some background, particularly on the women's colleges in order to make sense of how access to higher education for women took place. Only then could we describe, in context, the environment for women chemists at each institution and then provide individual life stories.

In Chapter 3, the Oxbridge women's colleges are the focus. Then Chapter 4 covers the co-educational London colleges, including the oft-forgotten Birkbeck College. We also place in this chapter, Battersea Polytechnic, which had a significant role to play in some women chemists' education. In Chapter 5, the importance of the London women's colleges and their chemistry graduates is explored. We have also included in that chapter, the women chemists of the London School of Medicine for Women.

The provincial universities and their women chemists are to be found in Chapter 6, followed by the women chemists at the universities of

---

[8] Rayner-Canham, M. and Rayner-Canham, G. (2018). *A Chemical Passion: The Forgotten Saga of Chemistry at British Independent Girls' Schools, 1820–1940.* Institute of Education Press, UCL, London.

Introduction xvii

Scotland and Wales in Chapter 7. As we show, higher chemistry education for women in Scotland can be traced back a long way, though the story was somewhat tarnished by the challenges faced by the "Edinburgh Seven."

The story of early 20th century chemistry for women in Britain would not be complete without an account of the controversy of academic chemistry versus domestic chemistry. In Chapter 8, we introduce the interesting institutions involved in the story, together with the women chemists who were involved in the debate, and those who taught at the colleges.

The Professional Societies

In Chapter 9, we summarise the access of women to professional societies. Each society handled the admission of women in a different way, with the long battle for admission to the Chemical Society taking up a substantial portion of the chapter. In addition to describing the long struggle, we highlight the women who signed the 1904 petition for admission. Most of those women, together with many of the signatories of a letter in 1909 to *Chemical News* seem to have been the "movers and shakers" of their time.

Women Chemistry Teachers

The teaching of chemistry to young women in secondary schools was a major source of employment. However, it was more than that. These pioneering women teachers enthralled their young protégées with a true enthusiasm for chemistry. The life stories of a selection of these women, to be found in Chapter 10, has to be an essential part of the narrative of this book.

Favoured Fields for Women Chemists

One of our previous studies indicated that women scientists tended to "cluster" in certain fields.[9] There were two areas in which women chemists

---

[9] Rayner-Canham, M. and Rayner-Canham, G. (1996). Women's fields of chemistry: 1900–1920. *Journal of Chemical Education* **73**: 136–138.

xviii *Introduction*

flourished: biochemistry (Chapter 11) and crystallography (Chapter 12). In those chapters, we focus not only on the women and their work but also on the environment in which they worked and the mentors who contributed to their advancement. In the early years, these were the only two fields in which pioneering women chemists received the accolade of election to Fellowship of the Royal Society.

In Chapter 13, we provide an account of the pioneering women pharmacists, some of whom had been active chemists. Even more than crystallography and biochemistry, pharmacy was a direction in which women chemists could find employment. In addition, the fight for the admission of women to the Pharmaceutical Society has interesting parallels to, and differences from, the admission of women to the Chemical Society.

Options for Women Chemists

Up to this point, we have used institutional and subject narratives in which to embed the biographical accounts. These encompass a relatively small number of women chemists of the period. In the last three chapters, we look at the roles of women chemists from other perspectives.

Chapter 14 focuses on the effects of marriage on women chemists. Though a significant proportion stayed single — particularly those in academia — many did marry, of whom most relinquished their career. A significant proportion of the women who did marry, married a chemist. Of the working chemistry couples, some wives continued with their original field while the majority joined their husband's research group. The degree of recognition of the women's work depended very much on their marriage partner. Some wives were cited as co-authors, while it would seem others received no credit for their collaboration.

The Great War and Its Aftermath

The First World War proved a turning point in the acceptance of women chemists, as it did for educated women in the wider society. Therefore, we devote the whole of Chapter 15 to the different roles which women chemists played in this War, especially in the production of fine chemicals in academic laboratories and in the many H. M. Factories around the country. With the destruction of many records after the First World War, it is unlikely we will ever determine the full role of women chemists during this period.

It may come as a surprise to many readers, the proportion, and even the total number, of women students at universities reached a maximum in the late 1920s and the trend applied equally to women chemists.[10] In Chapter 16, we examine this phenomenon and provide observations on the subsequent decline. We give an overview of the different employment avenues for women chemists in the inter-War period and choose exemplars for each.

The Time Frames

The book starts with the earliest British women involved in the chemistry enterprise who we could find. There is no specific "end point" to our narrative. Some of the women chemists lived incredibly long active lives, and their stories spill over into the 1940s and the 1950s. Nevertheless, we have tried to keep our emphasis upon the 1880s to 1930s range. We tell stories of hope and inspiration as the decades flow by, but as we show, the anticipated golden age for women chemists never comes, for as Evelyn Fox Keller has observed, the mid-20th century represented the "nadir of the history of women in science."[11]

We hope that this book will finally bring awareness of the forgotten role of British women chemists in the late-19th and the first half of the 20th centuries. At last, in addition to the HIStory of British chemists, we now have *HERstory*.

---

[10] Rayner-Canham, M. and Rayner-Canham, G. (1996). Women in Chemistry: Participation during the early twentieth century. *Journal of Chemical Education* **73**: 203–205.

[11] Fox Keller, E. (1987). The gender/science system: Or, is sex to gender as nature is to science? *Hypatia* **2**: 40.

Chapter 1

Pioneering Amateur British Women Chemists

Caroline Fox (24 May 1819–12 January 1871)

Chemistry held a fascination for some British women long before access to formal science education became possible. These were intelligent women of the upper and upper-middle classes who had the opportunity to socially encounter prominent scientists of their day.[1] Here, we will provide examples of such women over a 400-year time-period. In addition, we will illustrate the acceptability of chemistry as an interest for women.

---

[1] Rayner-Canham, M. and Rayner-Canham, G. (2009). British women and chemistry from the 16th to the mid-19th century. *Bulletin for the History of Chemistry* **34**(2): 117–123.

The Earliest Women Experimental Chemists

In our research, we found several society women of the 16th to 17th century period who had an interest in chemical changes. At the beginning of this period, it was not "modern chemistry" but more of a combination of alchemy and pharmacy that was used in order to develop medicinal remedies. These concoctions were sometimes herbal in origin, sometimes mineral, or sometimes combinations of both. Lynette Hunter has contended that, for these women researchers, the 1640–1660 era corresponded to a transformation from "Kitchin-Physick" to a more science-based "Ladies Chemistry."[2] Not only did it become a more scientific approach, but also, according to Hunter, it reflected an acceptance of chemistry as a field of study for women: "The practice of chemistry, general or experimental, was clearly seen as appropriate to the aristocratic lady, possibly even peculiar to her class in the 1640s."[3]

We can trace chemistry experimentation by women back even earlier to the late 1500s. The chemistry-related activities of Mary Sydney Herbert, Countess of Pembroke[4] were documented by her biographer, John Aubrey. Many upper-class women of the time had intellectual pursuits, but Herbert's interest in chemistry was exceptional. Though there are no surviving records of Herbert's experiments, Aubrey described how her research focussed on the extraction of substances from plants by chemical procedures:

> Her Honour's genius lay as much towards chymistrie as poetrie. … She was a great Chymist, and spent yearly a great deale in that study. … She kept for her Laborator in the house Adrian Gilbert (vulgarly called

---

[2] Hunter, L. (1997). Sisters of the Royal Society: The circle of Katherine Jones, Lady Ranelagh. In Hunter, L. and Hutton, S. (eds.), *Women, Science, and Medicine, 1500–1700: Mothers and Sisters of the Royal Society*. Sutton Publ., Thrupp, Stroud, Gloucestershire, p. 179.

[3] Note 2, Hunter, p. 186.

[4] Hannay, M. P. (1997). 'How I these studies prize': Countess of Pembroke and Elizabethan science. In Hunter, L. and Hutton, S. (eds.), *Women, Science and Medicine 1500–1700: Mothers and Sisters of the Royal Society*. Sutton Publ., Thrupp, Stroud, Gloucestershire, pp. 108–121.

Dr. Gilbert) half-brother to Sir Walter Raleigh, who was a great Chymist in those days.[5]

Chemical extractions were also the pursuit of Lady Margaret Clifford, daughter of Henry Clifford, 2nd Earl of Cumberland. Her own daughter, Anne, recorded:

> She was a lover of the Study and practice of Alchimy, by which she found out excellent Medicines, that did much good to many; she delighted in the Distilling of waters, and other Chymical extractions, for she had some knowledge in most kinds of Minerals, herbs, flowers and plants.[6]

A few decades later, there is documentation of two related women practicing chemistry: Dorothy Drury (née Moore) and her niece Katherine Jones, Lady Ranelagh (née Boyle). They were both members of an intellectual circle centred around Samuel Hartlib which also involved Katherine's brother, Robert Boyle.[7] There are many references to Moore's recipes and experiments in Hartlib's records.[8] In particular, Moore worked with Dr. (Arnold or Gerard) Boate on "Paris chemistry" in 1654. Founded 6 years previously, the Paris school espoused modernistic chemical ideas.[9] For the last 23 years of his life, Robert Boyle lived with his sister Katherine. Katherine had chemistry facilities constructed for him, and there is evidence that she worked alongside him in the laboratory.[10]

---

[5] Aubrey, J. (1962). Mary Herbert: Countess of Pembroke. In Dick, O. L. (ed.), *Aubrey's Brief Lives*. University of Michigan Press, Ann Arbor, MI, p. 138.

[6] Meads, D. (ed.), (1930). *The Diary of Lady Margaret Hoby 1599–1605*. Routledge, London, pp. 57–58; also cited in: Note 4, Hannay, p. 110.

[7] Maddison, R. E. W. (1963). Studies in the life of Robert Boyle, F.R.S.: Part VI, The Stalbridge period, 1645–1655, and the invisible college. *Notes and Records of the Royal Society of London* **18**: 104–124.

[8] Hunter, L. (2004). *The Letters of Dorothy Moore, 1612–1664: The Friendships, Marriage and Intellectual Life of a Seventeenth-Century Woman*. Ashgate, Hants., p. xxx.

[9] See, for example, Debus, A. G. (2003). The chemical philosophy and the scientific revolution. In Hellyer, M. (ed.), *The Scientific Revolution*. Blackwell, Oxford, p. 168.

[10] DiMeo, M. (2014). "Such a sister became such a brother": Lady Ranelagh's influence on Robert Boyle. *Intellectual History Review* **25**: 21–36.

4 *Pioneering British Women Chemists: Their Lives and Contributions*

Partially overlapping with Moore and Jones, the socialite, Margaret Cavendish, Duchess of Newcastle devoted many spare hours to working in the Cavendish family laboratory. Among her studies across the sciences, she learned the process of chemical distillation and the dissolving power of strong acids.[11] She insisted on attending demonstrations by famous scientists, such as Robert Boyle, at the most prestigious scientific institution of the day, the Royal Society.[12] In fact, in 1667, Cavendish was the first woman to be admitted to a meeting of the Royal Society, as Samuel Pepys commented: "after much debate, pro and con., it seems many being against it [her attendance]."[13] He continued: "Several fine experiments were shown her of colours, loadstones, microscopes, and of liquors among others, …"[13] However, her success was a reflection of her own social position and influence rather than a breakthrough for her gender.

Women and Chemistry in the 18th and Early 19th Centuries

Margaret Alic has written about the growing scientific interest by women as the 19th century progressed.[14] For these upper-middle class women, mathematics, biology, geology, and astronomy were relatively easy to practice for they required little in the way of facilities or expenditures. However, as Weldon commented in 1825, the pursuit of chemistry could only be accomplished by professionals and the very wealthy:

---

[11] (a) Grant, D. (1957). *Margaret the First: A Biography of Margaret Cavendish, Duchess of Newcastle, 1623–1673*. Toronto University Press, Toronto; (b) Merrens, R. (1996). A nature of 'infinite sense and reason': Margaret Cavendish's natural philosophy and the 'noise' of a feminized nature. *Women's Studies* **25**: 421–438; and (c) Whitaker, K. (2002). *Mad Madge: The Extraordinary Life of Margaret Cavendish, Duchess of Newcastle, The First Woman to Live by Her Pen*. Basic Books, New York.

[12] Mintz, S. I. (1952). The Duchess of Newcastle's visit to the Royal Society. *Journal of English and Germanic Philology* **51**: 168–176. For a more positive interpretation of Cavendish's science, see: Lewis, E. (2001). The legacy of Margaret Cavendish. *Perspectives on Science* **9**: 341–365.

[13] The diary of Samuel Pepys M.A. F.R.S., 30th May 1667 (accessed 11 January 2019).

[14] Alic, M. (1986). *Hypatia's Heritage: A History of Women in Science from Antiquity Through the Nineteenth Century*. Beacon Press, Boston.

[Chemistry] requires such an appropriation of time and property; such a variety of expensive and delicate instruments; such an acquisition of manual dexterity; and so much thought and attention to its successful prosecution, as will necessarily confine the *professed pursuit* of it to a few professors, and enthusiastic amateurs, whom fortune and opportunity favour.[15]

The lack of chemical laboratory facilities is probably why Mary Somerville, hailed at her death as "The Queen of Nineteenth Century Science," contributed little to chemistry.[16] Her sole venture, performed in collaboration with Michael Faraday, was a study of light absorption by different materials using the degree of darkening of silver chloride.[17]

Georgiana, Duchess of Devonshire

Related to Margaret Cavendish (see the preceding discussion) by marriage, Georgiana Cavendish (née Spencer), Duchess of Devonshire had an equal fascination for science, especially mineralogy and chemistry. One of her husband's relatives was the scientist, Henry Cavendish, and she met with him on numerous occasions, though the Duke forbad her from visiting Henry Cavendish in his laboratory.[18] Despite the disapproval of the Duke, she established and equipped her own small laboratory at Chatsworth House where she could study geology and experiment in chemistry. In 1792, while she was in Switzerland, it was commented that her days were spent: "Chemistry and mineralogy in the morning ..."[19]

---

[15]Weldon, W. (1825). *A Popular Explanation of the Elements and General Laws of Chemistry*. Weldon's Laboratory, New Bond-Street, London, p. iii.

[16]Badilescu, S. (1998). Jane Marcet and Mary Somerville — chroniclers of science in nineteenth-century England: In search of connections. *Chemical Intelligencer* **4**(4): 46–52.

[17]James, F. (ed.), (1993). *The Correspondence of Michael Faraday*, Vol. 2, Institution of Electrical Engineers, London, Letters 821 and 824.

[18]Bickley, F. (1911). *The Cavendish Family*. Constable & Co., London, p. 202.

[19]Letter, Henry Pelham to Earl of Chichester, 19 September 1792. BL Add. MSS 33,129, f88. Cited in: Foreman, A. (1998). *Georgiana, Duchess of Devonshire*. HarperCollins, London, p. 276.

6 *Pioneering British Women Chemists: Their Lives and Contributions*

The Duchess's interest in chemistry had been encouraged by a life-long friendship with the scientist, Sir Charles Blagden.[20] On the Duchess's return to London, her biographer, Amanda Foreman comments: "... she continued to pursue her new interests and filled her days with lectures at the Royal Academy, conducting chemistry experiments in a back room at Devonshire House, ..."[21] Later in 1793, Lady Sutherland described Georgiana's routine in a letter to Lady Stafford:

> ... the Duke has got the gout, & the Dss is "at home" every night at 12 o'clock, afterwards she sits with him till 3. She is busy studying Chemistry, and goes out a little, she is going this morning to a chemical lecture.[22]

In December 1793, Georgiana was staying in the Bristol area and she visited the laboratory of the scientist, Thomas Beddoes. Her seriousness and the extent of her knowledge in chemistry quickly became apparent to Beddoes, who commented in a letter to Erasmus Darwin: "... that she manifested upon this occasion a knowledge of modern chemistry superior to what he should have supposed 'that any duchess, or any lady in England was possessed of'."[23]

Elizabeth Fulhame

Elizabeth Fulhame (mid-1700s–1800s) was the one exception: she not only practiced chemistry but also received recognition among the chemists of her time. Regrettably, we know little about Fulhame's life except that she was born about the middle of the 18th century and that she was married to the physician, Dr. Thomas Fulhame.[24]

---

[20] Foreman, A. (1998). *Georgiana, Duchess of Devonshire*. HarperCollins, London, p. 277.

[21] Note 20, Foreman, p. 287.

[22] Letter, Lady Sutherland to Lady Stafford, 23 October 1793. PRO 30/29/5/5, f.49. Cited in: Note 20, Foreman, p. 287; See also p. 276 and p. 293.

[23] Stock J. E. (1811). *Memoirs of the Life of Thomas Beddoes, M.D. With an Analytical Account of his Writings*. John Murray, London, p. 100.

[24] Davenport, D. A. (2004). Fulhame, Elizabeth [known as Mrs. Fulhame] (fl. 1780–1794). *Oxford Dictionary of National Biography*. Oxford University Press, Oxford (accessed 11 January 2019).

Fulhame's contribution to chemistry was a book on combustion, *An Essay on Combustion with a View to a New Art of Dying and Painting, wherein the Phlogistic and Antiphlogistic Hypotheses are Proved Erroneous*, published in 1794.[25] Fortunately, the Preface of Fulhame's book explains how she developed her interest in the subject:

> The possibility of making cloths of gold, silver, and other metals by chymical processes, occurred to me in the year 1780; the project being mentioned to Doctor Fulhame and some friends, was deemed improbable. However, after some time, I had the satisfaction of realizing the idea in some degree by experiment.[26]

Fulhame decided to publish her research work in book form to ensure her claim to the discoveries and prevent "prowling plagiary." Also, she appeared hopeful that her technique of producing gold and silver cloth by metal deposition had commercial possibilities and that the book would solidify her claim to the process. She accepted that a chemistry tome authored by a woman was likely to have a negative reception in some quarters, explaining later in the Preface:

> It may appear presuming to *some*, that I should engage in pursuits of this nature; but averse from indolence, and having much leisure, my mind led me to this mode of amusement, which I found entertaining and will I hope be thought inoffensive by the liberal and the learned. But censure is perhaps inevitable; for some are so ignorant, that they grow sullen and silent, and are chilled with horror at the sight of anything that bears the semblance of learning, in whatever shape it may appear; and should the *spectre* appear in the shape of *woman*, the pangs which they suffer are truly dismal.[26]

Fulhame's interest was in reduction reactions that led to the deposition of metals. Over the years, she studied the reduction of metal salts of gold,

---

[25]Fulhame, Mrs. (1794). *An Essay on Combustion with a View to a New Art of Dying and Painting, Wherein the Phlogistic and Antiphlogistic Hypotheses are Proved Erroneous.* J. Cooper, London.

[26]Note 25, Fulhame, p. xix.

silver, platinum, mercury, copper, and tin using as reducing agents, hydrogen gas, phosphorus, potassium sulphide, hydrogen sulphide, phosphine, charcoal, and light. The realization that metals could be produced by aqueous chemical reduction at room temperature rather than by high-temperature smelting was probably her greatest contribution to chemistry. In addition, her use of light as a reducing agent for metal salts was the first recorded example of photochemical imaging — the chemical basis of the photographic process. This discovery resulted in the inclusion of her name and work in an 1839 review of the origins of photography by the British scientist, Sir John Herschel.[27]

From her references to the work of contemporary chemists, such as Lavoisier, Kirwan, and Scheele, Fulhame was obviously well educated in chemical principles. She disliked both the phlogiston and anti-phlogiston theories of oxidation and reduction, though her conclusions were fairly close to the anti-phlogistonist approach of Lavoisier.

Fulhame's particular concern was that reductions and oxidations seemed to occur only in the presence of water. In particular, she observed that water needed to be present for the combustion of carbon in air. Fulhame concluded that the combustion was a two-part process, the first involving the reaction of carbon with the oxygen of the water to give carbon dioxide and hydrogen gas, followed by reaction of the hydrogen gas with oxygen in the air to give water. Thus, she had proposed the novel idea that reactions could require more than one step. Fulhame was also the first to publish the concept of a catalytic process.[28] In the specific reaction of carbon and oxygen, the catalytic role of water has since been established.[29] However, Fulhame committed the error of extrapolating her findings and claiming that water was needed for all combustion processes.

---

[27]Schaaf, L. J. (1990). The first fifty years of British photography: 1994–1844. In Pritchard, M. (ed.), *Technology and Art: The Birth and Early Years of Photography*. Royal Photographic Society Historical Group, Bath, pp. 10–12; see also, Laidler, K. J. (April 1992). The story of photochemical imaging. *Chem 13 News*, 8.

[28]Laidler, K. J. (1986). The development of theories of catalysis. *Archive for History of Exact Sciences* **35**: 345–374.

[29]Hening, G. R. (1961). Surface reactions of single crystals of graphite. *Journal de Chimie Physique* **58**: 12–19.

Despite her trepidations, Fulhame's work was quite favourably received, one commentary being titled "An essay on *combustion*, by a *lady!*"[30] The prominent French chemist J.F. Coindet wrote a lengthy and positive review[31] while Benjamin Thompson, Count Rumford, repeated her work on the reduction of gold salts by light:

> This agrees perfectly with the results of similar experiments by the ingenious and lively Mrs. Fulhame. It was on reading her book that I was enduced to engage in these investigations; and it was by her experiments that most of the foregoing experiments were suggested.[32]

Fulhame's book was translated into German in 1798[33] while an American edition appeared in 1810.[34] In fact, Fulhame gained greatest recognition in the United States. The chemist, J. Woodhouse, mentioned: "The celebrated Mrs. Fulhame, a lady whom I am proud to quote on this occasion ... This distinguished lady, who is equally an example to her sex, and an ornament to science."[35] The Chemical Society of Philadelphia elected her a corresponding member[36] and in an oration stated that: "Mrs Fulhame has now laid such bold claims to chemistry that we can no longer deny the sex the privilege of participating in this science also."[37] Her work was then forgotten until 1903, when the famous British chemist, J.W. Mellor rediscovered it and devoted a whole paper to an appreciation of her contributions.[38]

---

[30] Anon. (1795). An essay on *Combustion*, by a *Lady*! *Gentleman's Magazine* **65**: 501.

[31] Coindet, J. F. (1797). *Annales de Chimie* **26**(series 1): 58.

[32] Davenport, D. A. and Ireland, K. M. (1989). The ingenious, lively and celebrated Mrs. Fulhame and the dyer's hand. *Bulletin for the History of Chemistry* **5**: 37.

[33] Fulhame Mrs., (1798). *Versuche über die Wiederherstellung der Metalle durch Wasserstoffgas, Phosphor, Schwefel, Schwefelleber, geschwefeltes Wasserstoffgas, gephosphortes Wasserstoffgas*. Translated by A.G.W. Lenten, Göttingen, Germany.

[34] Fulhame, Mrs. (1810). *An Essay on Combustion*. Humphries, Philadelphia, PA.

[35] Woodhouse, J. (1799). An answer to Dr. Joseph Priestley's considerations on the doctrine of phlogiston, and the decomposition of water; founded upon demonstrative experiments. *Proceedings of the American Philosophical Society* **4**: 465.

[36] Miles, W. (1959). Early American Chemical Societies. *Chymia* **3**: 95–113.

[37] Smith, E. F. (1972). *Chemistry in America: Chapters from the History of the Science in the United States*. Reprint Edition Arno Press, New York, p. 35.

[38] Mellor, J. W. (1903). History of the water problem (Mrs. Fulhame's theory of catalysis). *Journal of Physical Chemistry* **7**: 557–567.

10 *Pioneering British Women Chemists: Their Lives and Contributions*

Though Fulhame was wrong in her overemphasis on the role of water in oxidation, she deserves credit particularly for her discovery of photoreduction and the very concept of catalysis.[39] But above all, we would identify her as the first solo woman researcher of modern chemistry.

Public Lectures and the Scientific Lady

Even if women (with very few exceptions, such as Fulhame) were unable to practice chemistry, they were eager to learn about it, and the "scientific lady" became an accepted term in the vocabulary of the time.[40] R. Higgitt and Charles Withers have extensively reviewed the participation of women at the British Association for the Advancement of Science (BAAS) meetings.[41] However, the most important venue for women to learn about chemistry was the Royal Institution. Though the first public lecture at the Royal Institution took place in 1800, it was Humphry Davy's charismatic chemistry lectures during the 1802–1812 period that brought the affluent to the institution's premises on Albemarle Street.[42] Notably, about half the audience were women, which pleased Davy. However, Davy believed that women should absorb scientific knowledge and transmit it to their offspring but certainly not participate in science themselves.[43]

---

[39]Laidler, K. J. and Cornish-Bowden, A. (1997). Elizabeth Fulhame and the discovery of catalysis: 100 years before Buchner. In Cornish-Bowden, A. (ed.), *New Beer in an Old Bottle: Eduard Buchner and the Growth of Biochemical Knowledge.* Universitat de València, Valencia, Spain, pp. 123–126.

[40](a) Meyer, G. D. (1955). *The Scientific Lady in England 1650–1760: An Account of her Rise, with Emphasis on the Major Roles of the Telescope and Microscope.* University of California Press, Berkeley; and (b) Phillips, P. (1990). *The Scientific Lady: A Social History of Women's Scientific Interests 1520–1918.* Weidenfeld & Nicolson, London.

[41]Higgitt, R. and Withers, C. W. J. (2008). Science and sociability: Women as audience at the British Association for the Advancement of Science, 1831–1901. *Isis* **99**: 1–27.

[42]Foote, G. A. (1952). Sir Humphry Davy and his audience at the Royal Institution. *Isis* **43**: 6–12.

[43]Golinski, J. (1992). *Science as Public Culture: Chemistry and Enlightenment in Britain, 1760–1820.* Cambridge University Press, Cambridge, p. 194.

Caroline Fox

One of the women who attended a lecture at the Royal Institution during the time of Davy's successor, Michael Faraday, was Caroline Fox. Fox born 24 May 1819, daughter of the amateur scientist Robert Were Fox and Maria Barclay, kept a journal from 1835 until shortly before her death. She grew up in a household surrounded by her father's science as her mother's relation, Mary Anne Schimmelpenninck, commented:

> Imagine the back drawing-room strewn with reflectors, and magnets, and specimens of iron, and borax, cobalt, copper ore, blow-pipes, platina, &c., &c.; deflagrations, fusions, and detonations on every side; whilst we were deeply interested in watching the fusions of the ores, or their assaying; only that now and then I, having a house of my own, had a fellow feeling with Maria [wife of Robert Fox], at seeing a certain beautiful zebra-wood table splashed with melted lead or silver, and the chased Bury Hill candlestick deluged with acids.[44]

According to her journal, Caroline Fox attended many of the BAAS meetings. In addition, she visited the Royal Institution on 13 June 1851 to watch Faraday's chemical experiments. She reported:

> We went to Faraday's Lecture on "Ozone." He tried the various methods of making Ozone which Schönbein has already performed in the kitchen, and he did them brilliantly. He was entirely at his ease, both with his audience and his chemical apparatus; he spoke much and well of Schönbein, who now doubts whether Ozone is an element, and is disposed to view it simply as a condition of oxygen, in which Faraday apparently agrees with him.[45]

Fox died on 12 January 1871 at Budock, Cornwall.

---

[44] Hankin, C. (ed.), (1858). *The Life of Mary Anne Schimmelpenninck*. Longman, London, p. 114.

[45] Pym, H. N. (ed.), (1883). *Memories of Old Friends: Being Extracts from the Journals and Letters of Caroline Fox, of Penjerrick, Cornwall, from 1835 to 1871*. Smith, Elder & Co., London, pp. 269–270.

Women Writers on Chemistry

In addition to scientific lectures for women, some women's magazines in the 18th and early 19th centuries carried articles on science.[46] Ann Shteir has examined three of the journals: the *Lady's Magazine* (1770–1832), the *Lady's Monthly Museum* (1798–1828), and the first volume of the *Lady's Companion at Home and Abroad* (1849–1850).[47] At the beginning of the 19th century, the *Lady's Magazine* frequently carried articles on science as Shteir noted:

> For a short time, the *Lady's Magazine* included scientific excerpts drawn from contemporary publications, such as an essay on the "Progress and Utility of Chemistry" from the recently founded *Quarterly Journal of Science* and a portion of Sir Humphry Davy's 1821 address to the Royal Society on "the present State of Science."[48]

However, the ownership of the magazine changed in 1822 and the new editor eschewed "the abstruse mysteries and tedious details of science." When the topic of science for ladies was discussed in 1831, botany was considered the most suitable as: "… ladies will not, in pursuing botany, have to discolour their fingers in trying chemical experiments on substances which they may have previously risked their necks to obtain."[48]

The *Lady's Companion at Home and Abroad* too published articles on chemistry under its first editor Jane Loudon. Loudon believed her task was to educate her readers and provide mental stimulation — or "mental cultivation" as she called it. Included in this goal was a steady stream of scientific articles. Again, quoting Shteir:

---

[46] Benjamin, M. (1991). Elbow room; Women writers on science, 1790–1840. In Benjamin, M. (ed.), *Science and Sensibility: Gender and Scientific Enquiry, 1780–1945*. Blackwell, Oxford, pp. 27–59.

[47] Shteir, A. B. (2004). Green-stocking or blue? Science in three women's magazines, 1800–1850. In Henson, L. *et al.* (eds.), *Culture and Science in the Nineteenth Century Media*. Ashgate, Aldershot, pp. 3–13.

[48] Note 47, Shteir, p. 6.

In 1850 one of the male contributors presented a series of articles about fermentation and combustion under the title 'Chemistry and Everyday Life.' Edward Solly, a teacher and lecturer on chemistry who was associated with the Royal Institution and the Horticultural Society, joined Loudon's crusade to bring science into general female education. Applauding the increased 'desire for knowledge', he celebrated the importance of the sciences.[49]

After Loudon gave up her position as editor in June 1850, the direction of the magazine changed dramatically. When a long-standing subscriber expressed her dismay that drawings of flowers were now ornamental rather than botanical, the new (male) editor expressed his opinion that the designs of bonnets and sleeves were more important for women's minds than the mysteries of the botanical world.

Jane Marcet

Books were the primary means for women to learn about science.[50] Equally, writing about science enabled women to participate in the scientific enterprise without violating gender norms.[51] The first book on chemistry specifically for women was *La Chymie charitable et facile, en faveur des dames*[52] authored in France by Marie Meudrac[53] in 1666. This was sufficiently popular that it was reprinted in 1674 and again in 1711.

However, for British women, it was through Jane Marcet's book that they were able to comprehend the mysteries of chemistry. Titled *Conversations on Chemistry: In Which the Elements of that Science are*

---

[49] Note 47, Shteir, p. 10.

[50] Rossiter, M. W. (1986). Women and the history of scientific communication. *Journal of Library History* **21**: 39–59.

[51] Neeley, K. A. (1992). Woman as mediatrix: Women as writers on science and technology in the eighteenth and nineteenth centuries. *IEEE Transactions. Professional Communication* **35**: 208–216.

[52] Meudrac, M. (1666). *La Chymie charitable et facile, en faveur des dames*. Se vend rüe des Billettes et rüe du Plastre, Paris.

[53] (a) Bishop, L. O. and DeLoach, W. S. (1970). Marie Meudrac — First lady of chemistry? *Journal of Chemical Education* **47**: 448–449; (b) Tosi, L. (2001). Marie Meurdrac: Paracelsian chemist and feminist. *Ambix* **48**(2): 69–82.

14   *Pioneering British Women Chemists: Their Lives and Contributions*

Familiarly Explained and Illustrated by Experiments,[54] the first edition was published in 1806.

Jane Haldimand was born in London on 1 January 1769,[55] the daughter of Anthony Francis Haldimand, a Swiss merchant and banker, and Jane Haldimand. When Marcet was 15 years of age her mother died, and she had to take charge of the large family. In 1799, she married Alexander Marcet, a Swiss physician who, at the time, held an appointment at Guy's Hospital, London. Her husband was an amateur chemist and a prominent member of London's scientific society.

Having had difficulty understanding some of the chemistry lectures at the Royal Institution, Jane Marcet performed her own experiments, using chemicals and equipment procured by her husband. Finding this experience gave her a deeper understanding of chemistry, she decided to write an introductory textbook accompanied by experimental work so that others could understand the subject. Like Elizabeth Fulhame before her, Marcet was concerned about the reception of her efforts, stating in the Preface:

> In writing these pages, the author was more than once checked in her progress, by the apprehension that such an attempt might be considered by some, either as unsuited to the ordinary pursuits of her sex, or ill-justified by her own imperfect knowledge of the subject.[56]

To illustrate the concern for her reputation, all the early additions were published anonymously, only those from the 13th British edition (1837) onwards bearing her name as author. The book was extremely successful,

---

[54] Anon. [Marcet, J. H.] (1806). *Conversations on Chemistry: In Which the Elements of that Science are Familiarly Explained and Illustrated by Experiments*. 1st edn., Longman, Hurst, Rees and Orme, London.

[55] (a) Morse, E. J. (23 September 2004). Marcet, Jane Haldimand (1769–1858). *Oxford Dictionary of National Biography*, Oxford University Press (accessed 11 January 2019); (b) Derrick, E. M. (1985). What can a nineteenth century chemistry textbook teach twentieth century chemists? *Journal of Chemical Education* **62**: 749–751; (c) Anon. (1973) Jane Haldimand Marcet (1769–1858). In Thomson, D. L. *Adam Smith's daughters*. Exposition Press, New York, pp. 9–28.

[56] Marcet, J. H. (1817). *Conversations on Chemistry: In Which the Elements of that Science are Familiarly Explained and Illustrated by Experiments*. Volume 1: *On Simple Bodies*. 7th edn., Longman, Hurst, Rees, Orme, and Brown, London, p. viii.

18 editions appearing in Britain, the last in 1853. There were at least four French translations published in Paris together with one in Geneva,[57] an Italian translation of a French edition,[58] and a German translation.[59]

The book was written in a conversational style, as its name *Conversations on Chemistry* indicated. The style had been popularized by the biologist, Maria Jacson, in her book *Botanical Dialogues: Between Hortensia and her Four Children*[60] published in 1797. Jeremiah Payne had pioneered the format in the context of chemistry in his book, *Dialogues in Chemistry*, carried out between an anonymous tutor and two boys, Charles and James.[61]

In *Conversations on Chemistry*, Marcet explained that a conversational style was particularly appropriate for the female audience:

> Hence it was natural to infer, that familiar conversation was, in studies of this kind, a most useful auxiliary source of information; and more especially for the female sex, whose education is seldom calculated to prepare their minds for abstract ideas, or scientific language.[62]

The conversations themselves consisted of a series of discourses by a Mrs. B and two students, Emily, who is serious and hard-working, and Caroline, who is more spontaneous and inquisitive. The characters are sketched only superficially, the two students being wealthy, well-educated town-dwellers between 13 and 15 years of age. Jeffery Leigh and Alan Rocke have provided evidence that the names chosen were those of Emily

---

[57] Jacques, J. (1986). Une chemiste qui avait de la conversation. *Nouveau Journal de Chimie* **10**: 209–211.

[58] Cole, W. A. (1988). *Chemical Literature 1700–1860: A Bibliography with Annotations, Detailed Descriptions, Comparisons and Locations*. Mansell, London, p. 420.

[59] Leigh, G. J. (2017). The international publication history of Conversations on Chemistry: The correspondence of Jane and Alexander Marcet during its writing. *Bulletin for the History of Chemistry* **42**(2): 85–93.

[60] Shteir, A. B. (1990). Botanical dialogues: Maria Jacson and women's popular science writing in England. *Eighteenth-Century Studies* **23**(3): 301–317.

[61] Joyce, J. (1807). *Dialogues in Chemistry, Intended for the Instruction and Entertainment of Young People: In which the First Principles of that Science are Fully Explained*. Law and Gilbert, London.

[62] Note 56, Marcet, pp. vi–vii.

and Caroline Sebright. Sir John Sebright, an amateur chemist, was a close friend of the Marcets, and Sebright encouraged his daughters (especially the eldest, Frederica) to develop an interest in the subject.[63]

One of the great attractions of *Conversations on Chemistry* was the inclusion in subsequent editions of the latest chemical advances. Marcet learned of these discoveries from the famous British scientists of the period, through both correspondence with them and through Royal Institution lectures.[64] As an example, she incorporated the isolation of the alkali metals potassium and sodium by Sir Humphry Davy in the 7th edition, as this exchange illustrates:

> *Caroline*: I doubt, however, whether the metals will appear to us so interesting, and give us so much entertainment as those mysterious elements which conceal themselves from our view. Besides, they cannot afford so much novelty; they are bodies with which we are already so well acquainted.

> *Mrs. B.*: You are not aware, my dear, of the interesting discoveries which were a few years ago made by Sir H. Davy respecting this class of bodies. By the aid of a Voltaic battery, he has obtained from a variety of substances, metals before unknown, the properties of which are equally new and curious.[65]

Science education for daughters of affluent families had become very fashionable in antebellum America[66] and, as a result, Marcet's book was even more popular in the United States.[67] The first American edition, called *Conversations on Chemistry*, appeared in 1806, and, interestingly, it was printed by the same publisher as the American edition of Elizabeth Fulhame's *Essay on Combustion*. There were a total of 23 U.S. impressions that appeared between 1806 and 1850. According to some accounts,

---

[63] Leigh, G. J. and Rocke, A. J. (2016). Women and chemistry in Regency England: New light on the Marcet circle. *Ambix* **63**, 28–45.

[64] Leigh, G. J. (2017). The changing content of Conversations on Chemistry as a snapshot of the development of chemical science. *Bulletin for the History of Chemistry* **42**(1): 7–27.

[65] Note 56, Marcet, pp. 314–315.

[66] (a) Warner, D. J. (1978). Science education for women in antebellum America. *Isis* **69**: 58–67; and (b) Tolley, K. (2003). *The Science Education of American Girls: A Historical Perspective*. Routledge, New York.

[67] Lindee, S. M. (1991). The American career of Jane Marcet's Conversations on Chemistry, 1806–1853. *Isis* **82**: 8–23.

the total sales in the U.S. amounted to a phenomenal 160,000 copies.[68] Marcet died at Piccadilly, London, on 28 June 1858.

Readers of Conversations on Chemistry

Conversations on Chemistry had a major role in initiating Michael Faraday's career in science. Faraday, at the time, was a bookbinder's apprentice. It was reading the pages while he was binding the book that provided his initial chemical knowledge[69] and helped persuade him that science was his true vocation. As Faraday himself remarked:

> So when I questioned Mrs. Marcet's little book by such experiments as I could find means to perform, and found it true to the facts as I could understand them, I felt that I had got hold of an anchor in chemical knowledge, and clung fast to it. Hence my deep veneration for Mrs. Marcet.... You may imagine my delight when I came to know Mrs. Marcet personally; how often I cast my thoughts backwards, delighting to connect the past and the present; how often, when sending a paper to her as a thank-offering, I thought of my first instructress, and such like thoughts will remain with me.[70]

At age 71, Marcet wrote to Faraday for news of his latest work:

> Dear Mr. Faraday, I have this morning read in the "Athenaeum," some account of a discovery you announce ... respecting the identity of the imponderable agents, heat, light, and electricity; and as I am at this moment correcting the sheets of my "Conversations on Chemistry" for a new edition, might I take the liberty of begging you to inform me where I could obtain a current account of this discovery?[71]

---

[68] Miles, W. (1952). Books on chemistry printed in the United States, 1755–1900: A study of their origin. *The Library Chronicle (Philadelphia)* **18**: 51–62.

[69] James, A. J. L. (1991). Michael Faraday — the chemist. *Education in Chemistry* **28**: 128–130.

[70] Letter, Michael Faraday to Auguste de la Rive, n.d.. Cited in: Armstrong, E. V. (1938). Jane Marcet and her conversations on chemistry. *Journal of Chemical Education* **15**: 53–57.

[71] Letter, Jane Marcet to Michael Faraday, 24 November 1845. Cited in: Armstrong, E. V. (1938). Jane Marcet and her conversations on chemistry. *Journal of Chemical Education* **15**: 53–57.

Marcet's book was read by Lady Helen Hall (née Hamilton Douglas), wife of the Scottish geologist and chemist, Sir James Hall. Lady Hall wrote to Marcet explaining her reasons for reading *Conversations on Chemistry*: "I was at that time keen to improve myself by reading and attending lectures, keen to acquire knowledge, for the pleasure of conversing with my husband and communicating instruction to my young family."[72]

A reader of the French translation of *Conversations on Chemistry* was Anne Louise Germaine Necker, Madame de Staël. Madame de Staël wrote to Jane Marcet's husband, Alexander Marcet: "I have proposed the study of chemistry in the dialogues of Mrs. Marcet ... the beginning [is] most clever and the work admirably clear."[73]

The novelist Maria Edgeworth was another luminary of the time who was influenced by *Conversations on Chemistry*. In fact, her chemical knowledge, acquired by reading Marcet's book, possibly saved the life of Edgeworth's younger sister. The sister had swallowed acid and Maria recalled from the text that milk of magnesia was an effective antidote. Following the incident, Edgeworth wrote of the benefits of studying chemistry by women:

> Chemistry is a science particularly suited to women, suited to their talents and to their situation. Chemistry is not a science of parade, it affords occupation and infinite variety, it demands no bodily strength, it can be pursued in retirement; there is no danger of its inflaming the imagination, because the mind is intent upon realities. The knowledge that is acquired is exact; and the pleasure of the pursuit is a sufficient reward for the labour.[74]

Among Marcet's circle of friends was the Swiss geologist, Horace-Bénédict de Saussure. His daughter, Albertine Necker de Saussure,

---

[72]Letter, Lady Helen Hall, to Jane Marcet, 5 August 1829. Marcet Collection, Archive Guy de Pourtalès, Eoty, Switzerland. Cited in: Polkinghorn, B. (1993). *Jane Marcet: An Uncommon Woman*. Forestwood Publications, Aldermaston, Berkshire, p. 30.

[73]Letter, Mme. de Staël, to Alexander Marcet, 1816. Marcet Collection, Archive Guy de Pourtalès, Eoty, Switzerland. Cited in: Note 72, Polkingham, p. 30.

[74]Edgeworth, M. (1974). *Letters for Literary Ladies*. London, 1795. Reprint, Garland, New York, p. 66.

acknowledged that she had been influenced by *Conversations on Chemistry* in writing her own award-winning book, *L'Education Progressive*.[75] In de Saussure's book, she stated that science, and chemistry in particular, should have been an essential part of a girls' education.

Commentary

Here we have given examples of some of the earliest women who had, in one way or another, participated in some aspect of chemistry. Unfortunately, we will never know the total and complete extent of this involvement as women's history in this context is so fragmented. Likewise, we are unlikely to learn much more about the hundreds of women who viewed chemistry from the sidelines at the Royal Institution lectures or the thousands of women who purchased a copy of Marcet's book.

---

[75]Orr, C. C. (1995). Albertine Necker de Saussure, the mature woman author, and the scientific education of women. *Women's Writing* **2**: 141–153.

Chapter 2

Chemistry at Pioneering Girls' Schools

Frances Dove (27 June 1847–21 June 1942)

It is a common myth that chemistry was not taught at British girls' schools until well into the mid-20th century. For example, in Barry Turner's *Equality for Some: The Story of Girls' Education*, he remarks: "Even in schools where science was taught, it was unusual for a pupil to acquire an

22 *Pioneering British Women Chemists: Their Lives and Contributions*

adequate grounding for advanced study."[1] In Gillian Avery's *The Best Type of Girl: A History of Girls' Independent Schools*, the comment is made: "In the privately owned schools there was on the whole a marked absence of science before the 1950s,"[2] while subsequent discussion in the book supported that thesis. Thus, there is a clear implication that science (including chemistry) was of marginal relevance to English girls' schools until the mid-20th century.

Yet, as we show in subsequent chapters, there were many women chemists and their chemistry education started at the secondary school level. How chemistry came to be so widely taught at British independent girls' schools is covered in considerable detail in our book, *A Chemical Passion: The Forgotten Story of Chemistry at British Independent Girls' Schools, 1820s–1930s*.[3] Therefore, in this chapter, we will provide a selective study of key aspects.

The Early Quaker Schools

It was Quaker schools which led the way. The Quaker religious principles allowed women to play an active role in both the public and private spheres of the Religious Society of Friends.[4] There were two Quaker women in particular, Priscilla Wakefield and Maria Hack, who popularized science and promoted scientific literacy in the early 19th century.[5]

At least as early as the 1820s, chemistry was taught in Quaker girls' schools. As an example, chemistry was specifically part of the curriculum

---

[1] Turner, B. (1974). *Equality for Some: The Story of Girls' Education*. Ward Lock Educational, London, p. 145.

[2] Avery, G. (1991). *The Best Type of Girl: A History of Girls' Independent Schools*. André Deutsch, London, p. 254.

[3] Rayner-Canham, M. and Rayner-Canham, G. (2017). *A Chemical Passion: The Forgotten Story of Chemistry at British Independent Girls' Schools, 1820s–1930s*. Institute of Education Press, London.

[4] Allen, K. and MacKinnon, A. (1998). 'Allowed and expected to be educated and intelligent': The education of Quaker girls in nineteenth century England. *History of Education* **27**(4): 391.

[5] Leach, C. (2006). Religion and rationality: Quaker women and science education 1790–1850. *History of Education* **35**(1): 69–90.

at Sarah and Harriet Hoare's Quaker school in Frenchay, near Bristol. In 1820, Jane Heath, a student at the school, wrote to her mother:

> We rise a little before seven and study Geography till eight with dissected maps. After breakfast we make our beds and into school again by nine ... I have then to write a page of English and Natural History, lectures on Chemistry, Botany, etc., besides parsing and a slateful of exercises so that I cannot always finish before dinner.[6]

Newington Academy for Girls

The students at the (Quaker) Newington Academy for Girls, Newington Green, were also taught chemistry. Opened in 1824, the original school *Prospectus* stated:

> The course of Instruction shall comprehend a Grammatical knowledge of the English language, Writing, Arithmetic, Geography, Astronomy and the Use of Globes, Ancient and Modern History, Elements of Mathematics, of Physics or Experimental Philosophy, Chemistry, Natural History — The French Language and Needlework.[7]

Among the founders and benefactors was the Quaker scientist, William Allen. Allen's main occupation was as a pharmacist and manufacturing chemist, but from 1802, he assisted William Babbington with his chemistry lectures at Guy's Hospital. Allen also gave science lectures at the Royal Institution at Humphry Davy's request. It was Allen who taught chemistry at the school, as his biographer, James Sherman, noted:

> He [Allen] went through a course of lectures, which he annually repeated, on mechanics, chemistry, and natural and experimental philosophy, ... by familiar explanations, and by a variety of experiments with his extensive and valuable apparatus.[8]

---

[6]Heath, J. (1820). Letter to her mother, Hampshire Record Office, 12M58/2.

[7]Shirren, A. J. (1951). *The Chronicles of Fleetwood House*. Barnes Printers, London, p. 160.

[8]Sherman, J. (1851). *Memoir of William Allen F.R.S.*. Charles Gilpin, London, p. 346.

In her memoirs, *Three Score Years and Ten,* Sophia Elizabeth de Morgan (née Frend) described how she had attended some of Allen's classes:

> ... I made acquaintance with William Allen, who kindly allowed me to attend the lectures on chemistry which he gave, with experiments, to a class of young girls. From him I learned the meaning and importance of Dalton's discovery. The atomic theory, then beginning to be understood, was the first step in the raising of chemistry to the rank of a science. Mr. Allen's quick perception of facts was greater than his power of following out extensive inferences. He was a good observer and classifier, but stopped at facts and phenomena.[9]

De Morgan, raised a Unitarian, had received no formal education. However, in her teenage years, during the 1810s and 1820s, she would sit in the branches of an old oak tree and self-educate. She commented: "I used to climb this on fine mornings, and had read a very miscellaneous collection of books in the oak: ... books on natural philosophy and chemistry, ..."[10] Though it is unknown what she read, it is feasible that one of her chemistry books was Jane Marcet's *Conversations on Chemistry* (see Chapter 1).

The Mount School, York

The teaching of chemistry to girls at The Mount School, York, can be traced back to at least 1835.

The evidence comes from the letters written in that year by the schoolgirl, Anne White. These letters were reprinted in Winifred Sturge and Theodora Clark's School history, *The Mount School, York*:

> Well we went yesterday evening to the boys' school [Bootham Quaker boys' school]. H[annah] Brady [Head Mistress] came with us and we took tea with the brothers and cousins of the girls. ... After tea we looked

---

[9]De Morgan, S. E. (1895). *Three Score Years and Ten: Reminiscences of the Late Sophia Elizabeth De Morgan.* Edited by her daughter Mary A. de Morgan. Bentley, London, pp. 106–107.

[10]Note 9, De Morgan, p. 108.

at pictures until John Ford was ready to show us some chemical experiments. We then went to the dining-room where we found 10 or 12 more boys sitting and were shown a great many more experiments, most of which I saw before, but I was very glad to see them again. John Ford showed us phosphorus and sulphur burning in oxygen, potassium jumping about in water, phosphorus dissolved in water and boiled in a flask, several striking affinities, particularly making plaster of Paris, which I do not think I ever saw before. ... I spent the pleasantest evening I have had since I came.[11]

Chemistry was subsequently taught at the Mount School, itself. From 1882 until 1892, the teacher was Edward Grubb. However, his attempts at experiments were not always successful:

To his lectures [Edward Grubb's] on chemistry his audience came in a mood of prophetic sympathy, awaiting the experiment: "Will it? Won't it?" It generally wouldn't! Why should it? For before the laboratory was built in 1884 there was no science equipment worth the name.[12]

Poetic verse with a chemistry theme was a regular item in the high school magazines of each of the academic girls' schools, and in *A Chemical Passion* (see Footnote 3), we provided many examples. Here, we will include a poignant poem from the Mount School:

Oh charming lab oh pleasant lab
When I enter through thy door
I see thee smeared with many a dab
Of H_2SO_4

Solutions, solids, crystallines
Upon thy shelves do stand

---

[11] White, A. (1931). Letter to her mother, 5th day, 3rd month, 1835. Cited in: Sturge, H. W. and Clark, T. *The Mount School — York: 1785 to 1814; 1831–1931*. J. M. Dent and Sons, London, pp. 43–44.

[12] Sturge, H. W. and Clark, T. *The Mount School — York: 1785–1814; 1831–1931*. J. M. Dent and Sons, London, p. 131.

With acids, bases, alkalis
Forming a glorious band

Oh many are thy odours
And pleasant ones but few
But worse than all the others
Is that of SO_2

Often upon thy bench is seen
A kettle, cup, or spoon
And many a brew of tea I ween
Is made each afternoon

And in thy corner cupboard
Are things we can but guess
They are not so apparent
As fumes of H_2S

Oh charming lab! oh pleasant lab!
I oft shall think of thee
In the dim and distant future
And the days that are to be.[13]

Polam Hall, Darlington

Grubb moved to Darlington to teach at another Quaker girls' school, Polam Hall. The school had been founded in 1847 as *Jane, Barbara, and Elizabeth Procter's Boarding School for the Daughters of Friends*. In an advertisement for the school of 1850, and again of 1851, it is stated that: "A Lecture on some branch of Natural History, Natural Philosophy, Chemistry, &c., is delivered weekly to the Pupils by James Cooke."[14] An aged hand-written timetable of a student at the school has also survived. Chemistry is shown for Thursday, 10:50 a.m. to 11:30 a.m. and Friday

---

[13]"Vera" (May 1893). untitled. *The Mount Magazine* n. pag.

[14]Advertisements, *The British Friend Advertiser*, p. 3. In (1850). *The British Friend: A Monthly Journal Chiefly Devoted to the Interests of the Society of Friends*. **50**: between pp. 172–173.

from 5:30 p.m. to 6:50 p.m.[15] Though undated, it is most likely from the 1892–1895 period when Edward Grubb taught chemistry at Polam Hall.

Academic Independent Girls' Schools in England

The Quaker schools were isolated from the wider educational community. It was not until the latter half of the 19th century that a range of academic schools for middle-class and upper-middle class girls sprang into existence. Prior to that time, nearly all educational opportunities for girls were small private schools, catering to a very specific segment of society. As Joyce Senders Pedersen has described:

> Girls' private schools were, as the name suggests, the private property of the women who kept them. Such schools were for personal profit and were totally dependent for their income on student fees. Services in the schools were generally sold on a piecemeal basis. The basic instruction covered only instruction in English subjects (and in more fashionable schools French), while lessons in accomplishments such as music, German, and drawing were charged as "extras."[16]

The New Girls' Schools

The new generation of schools were totally different. They were much larger; they had a range of specialist teachers; the students came from a wider range of society; and most importantly, they stressed academic subjects. These girls' schools were truly revolutionary and their growth was not welcomed by all of society. The first generation of headmistresses and teachers at these schools met fierce opposition from several directions as Sara Delamont notes:

> The feminist pioneers who opened academic secondary schools for young girls in the second half of the nineteenth century did so against a

---

[15] Box of Miscellanea, Archives, Polam Hall School, Darlington.

[16] Pedersen, J. S. (1979). The reform of women's secondary and higher education: Institutional change and social values in mid and late Victorian England. *History of Education Quarterly* **19**(1): 64.

body of medical opinion, religious orthodoxy, and a widespread belief among their potential clientele (that is, middle- and upper-middle-class parents) that such institutions were dangerous. A pupil, or a member of staff, at an academic secondary school was held to be in physical danger (her health would suffer, she might become subfertile or die of brain fever); in moral danger (away from the control of her mother she might meet anybody) and liable to forfeit her marriage prospects for men who would not want a wife who knew algebra.[17]

The dedication and fearlessness of these pioneering headmistresses cannot be overemphasized. As a result of their efforts, and the girls' schools they founded, late-19th century girls were able to obtain an academic education, including chemistry, and hence had the requirements to enter university.[18] It was very young shoulders who bore this burden incredibly well, as Nonita Glenday and Mary Price commented: "Most of the headmistresses were in their early twenties, very young for the responsibilities they were carrying, and had led sheltered lives."[19]

The very first of the pioneering headmistresses were Frances Buss and Dorothea Beale. It is fitting that their names are often recalled together, as exemplified by the title of Josephine Kamm's book *How Different from Us: A Biography of Miss Buss and Miss Beale*, for the two of them fundamentally changed the nature of educational opportunities for middle- and upper-middle class girls.[20] Their goals of girls' academic education were similar, though the schools which they developed were significantly different. Buss founded a day school, North London Collegiate School (NLCS), while Beale took over the boarding school, Cheltenham Ladies' College (CLC). However, one striking commonality in backgrounds was their own educational experiences at Queen's College, Harley Street.

---

[17]Delamont, S. (1993). Distant dangers and forgotten standards: Pollution control strategies in the British Girls' School, 1860–1920. *Women's History Review* **2**(2): 234.

[18]Richardson, J. (1974). The great revolution: Women's education in Victorian times. *History Today* **24**: 420–427.

[19]Glenday, N. and Price, M. (1974). *Reluctant Revolutionaries: A Century of Headmistresses 1874–1974*. Pitman Publishing, London, p. 23.

[20]Kamm, J. (1958). *How Different from Us: A Biography of Miss Buss and Miss Beale*. Bodley Head, London.

Queen's College, Harley Street

Queen's College had been founded by Frederick Denison Maurice, a Christian Socialist. The Christian Socialists believed that religion and social progress went hand-in-hand, and that, in particular, women had rights to education. The college was run by men; thus, it was deemed essential that an adult woman be present in each classroom to chaperone the proceedings.[21] The title "Lady Visitors" was used for these individuals, one of whom was Jane Marcet (see Chapter 1). Maurice obviously thought very highly of Marcet as in his discussion of Lady Visitors for Queen's College, he footnoted: "Among these ladies we have the high honour and privilege of reckoning one whose life has been devoted to earnest and successful efforts for the instruction of both sexes and all ages — Mrs Marcet."[22]

In her analysis of the Queen's College curricula from 1848 to 1868, Shirley Gordon concluded:

> The scientific studies at Queen's College were, for the period, amongst the most remarkable in the curriculum.... the class was to hear of the leading facts and classifications of chemistry and to see experiments to demonstrate the points.[23]

One of the students of that time, Maud Beerbohn Tree, recalled: "At 12.30 I think we descend into cavernous depths of the earth to that quaint home of Natural Philosophy, Chemistry, and Mathematics known as Mr. Cock's room."[24] Later in the 19th century, chemistry was taught as a separate subject, first by William A. Miller, then by J. Millar Thomson, and followed

---

[21] Spencer, S. (2006). The lady visitors at Queen's College: From the back of the class to a seat on the council. *Journal of Educational Administration and History* **36**: 47–56.

[22] Maurice, F. D. (a reprint, 1848) (1898). Original objects and aims of Queen's College. In Tweedie, E. (ed.), *The First College Open to Women, Queen's College, London: Memories and Records of Work Done, 1848–1898*. Queen's College, London, p. 17.

[23] Gordon, S. C. (1955). Studies at Queen's College, Harley Street, 1848–1868. *British Journal of Educational Studies* **3**(2): 144–154.

[24] Tree, M. B. (1898) Quick, thy tablets, memory! In Tweedie, E. (ed.), *The First College Open to Women, Queen's College, London: Memories and Records of Work Done, 1848–1898*. Queen's College, London, p. 44.

30 *Pioneering British Women Chemists: Their Lives and Contributions*

by H. Forster Morley in 1897, all chemistry professors of King's College, London. The chemistry examiner for 1897 was listed as Richard T. Plimpton, a Chemistry Professor of University College, London.[25]

Girls' Day Schools

When Buss opened the NLCS, originally called the North London Collegiate School for Ladies in 1850, she was only 23 years of age.[26] Buss wanted her pupils to have equal opportunities to boys' and that included real science. The author of a history of NLCS, Nigel Watson stated:

> Robert Buss [Frances Buss's father] made a memorable [first] science teacher as Annie Martinelli, an early pupil later remembered: "His talents were simply wonderful. His Chemistry series was marvellous, especially for smells and explosions.[27]

NLCS became the model for the girls' schools financed by the Girls' Public Day School Company (GPDSC) — later the Girls' Public Day School Trust.[28] The formation of the GPDSC was the greatest success of the National Union for Improving the Education of Women of all Classes, widely known as the Women's Educational Union. The funds raised by the Company were used to found independent, affordable, academically selective girls' schools across England. Schools were only approved in towns and cities where there was local support for such ventures. Each of the GPDSC schools followed the lead of NLCS in promoting chemistry as a core subject, as we will show later in this chapter.

---

[25]Croudace, C. (1898). A short history of Queen's College. In Tweedie, E. (ed.), *The First College Open to Women, Queen's College, London: Memories and Records of Work Done, 1848–1898.* Queen's College, London, pp. 18–33.

[26]Scrimgeour, M.A. (ed.), (1950). *North London Collegiate School 1850–1950.* Oxford University Press, Oxford.

[27]Watson, N. (2000). *And Their Works Do Follow Them: The Story of North London Collegiate School.* James and James, London, p. 16.

[28]Kamm, J. (1971). *Indicative Past: A Hundred Years of the Girls' Public Day School Trust.* Allen and Unwin, London.

The importance of the NLCS as a role models for girls' academic day schools cannot be overemphasized. In addition to the GPDSC schools, many others used NLCS as a template, including Manchester High School for Girls (MHSG);[29] King Edward VI High School for Girls (KEVI), Birmingham;[30] Mary Datchelor School, London;[31] to name but a few, and even some boarding schools, such as Milton Mount College, Gravesend.[32]

To illustrate the recognition of the pioneering role of the NLCS, in 1900, there was a formal three-day Jubilee celebration of its founding, including a service at St. Paul's Cathedral. An account of the several events in *The Magazine of the Manchester High School* ended with:

> Thus was bought to a close one of the most interesting events of our time — the Jubilee of the first "High School for Girls," the success of which has meant so much for all of us; for to Miss Buss's efforts we owe the inception of the whole movement, and were it not for her valiant struggles against much opposition, the establishment of Girls' High Schools, and the opening of the Universities to women, might have been greatly retarded.[33]

Girls' Boarding Schools

Originally called the Young Ladies' College, Cheltenham, CLC was founded in 1854.[34] Over the first 4 years, the enrolment declined. Then the Governors appointed Dorothea Beale as the second Principal and everything changed. As part of initiating a strong academic focus for the

---

[29] Burstall, S. A. (1911). *The Story of the Manchester High School for Girls: 1871–1911*. Manchester University Press, Manchester.

[30] Vardy, W. I. (1928). *King Edward VI High School for Girls Birmingham 1883–1925*. Ernest Benn, London.

[31] Anon. (1957). *The Story of the Mary Datchelor School 1877–1957*. Hodder and Stoughton, London.

[32] Harwood, H. (1959). *The History of Milton Mount College*. Independent Press Ltd., London.

[33] Anon. (June 1900). The Jubilee of the Frances Mary Buss Schools, April 4th, 1900. *The Magazine of the Manchester High School*, 62–64.

[34] Clarke, A. K. (1953). *A History of Cheltenham Ladies' College*. Faber & Faber, London.

School, the teaching of science, particularly chemistry, and construction of science laboratories became a priority for Beale.

CLC was the role model for many girls' boarding schools. Among those to follow the CLC template were Downe House, Newbury;[35] Roedean School (initially Wimbledon House School), Brighton;[36] St. Swithun's School (initially Winchester High School), Winchester;[37] and Wycombe Abbey School, High Wycombe.[38]

Academic Independent Girls' Schools in Wales and Scotland

Though there was a clear narrative structure for the development of academic girls' schools in England, this was not the case for Wales or Scotland. Wales and Scotland each had distinct histories of education — especially for girls.

Girls' Schools in Wales

In Wales, the Association for Promoting the Education of Girls in Wales (APEGW), established in 1886, was the driving force for academic girls' schools.[39] One of those on the committee was Dilys Davies, formerly the first chemistry teacher at NLCS. Davies, under her married name of Dilys Glynne Jones, became the Vice-President of the APEGW. When academic girls' schools were established, it is of note that several of the science teachers had themselves attended chemistry-active English independent girls' schools.

The earliest academic girls' schools in Wales were the two Howell's Schools.[40] Founded in 1858, one in Llandaff, South Wales, and the other

---

[35]Ridler, A. (1967). *Olive Willis and Downe House: An Adventure in Education*. John Murray, London.

[36]Anon. (1985). *A History of Roedean School, 1885–1985*, privately printed.

[37]Bain, P. (1984). *St. Swithun's: A Centenary History*. Phillimore & Co., Sussex.

[38]Flint, L. (1989). *Wycombe Abbey School 1896–1986: A Partial History*, privately printed.

[39]Evans, W. G. (1990). *Education and Female Emancipation: The Welsh Experience, 1847–1914*. University of Wales Press, Cardiff, pp. 132–136.

[40]Sully, J. (2010). *Howell's School, Llandaff, 1860–2010: A Legacy Fulfilled*. Howell's School, Llandaff.

in Denbigh, North Wales, these schools came about as a result of accumulated funds from a bequest of Thomas Howell of Monmouthshire. Though the schools had an initial requirement of Anglican staff, girls of any denomination were to be accepted.

In the 1880s, there was an attempt to form a Welsh Girls' Day-School Company to parallel that of the GPDSC in England. One GPDSC school had opened, that of Swansea High School in 1888. However, the school had closed in 1895, having been unsuccessful in spreading the concept of the GPDSC into Wales (see Footnote 39). Instead, a number of academic girls' schools were established by local authorities, including the City of Cardiff High School for Girls[41] in 1895 and the County School for Girls, Pontypool,[42] in 1897.

Girls' Schools in Scotland

The beginning of the academic education of girls in Scotland dates back to the early 19th century. From about 1815 until 1835, specific subjects were taught by individual teachers at their home, as a Mrs. Leigh Furlong commented in *New Movement in Education*, published in 1855:

> In every street might, at the period to which I refer [1815-1835], be seen some delicate girl, hurrying from class to class (according as each teacher found their partizans), which were held at their respective houses. These young creatures were generally encumbered with a load of books, and the time lost in their endeavour to make the respective master's classes hours meet, being found at length an evil which called for some better plan ...[43]

---

[41] Carr, C. (1955). *The Spinning Wheel: City of Cardiff High School for Girls, 1895–1955.* Western Mail and Echo Ltd., Cardiff.

[42] Anon. (1947). *Jubilee: The History of the County School for Girls, Pontypool 1897–1947.* The Griffin Press, Pontypool.

[43] Furlong, L. (1855). *New Movement in Education. What It Is! "Oral" Education on the "Scottish" System explained in an introductory lecture on the re-opening of the Ladies "Oral" Educational Institute.* Thomas Hachard and William Gurner, London, p. 6.

34 *Pioneering British Women Chemists: Their Lives and Contributions*

Progress towards the formation of schools occurred largely, and independently, at the two metropolises of Edinburgh and Glasgow.[44] In the 1830s, several academic girls' schools opened in Scotland, chemistry being a common component of the curriculum. For example, at the Scottish Institution for the Education of Young Ladies, Dr. David Boswell Reid, physician and chemist, was the enthusiastic chemistry teacher and author of the textbook: *Rudiments of chemistry; with illustrations of chemical phenomena of daily life.*[45] In a report on the school, the diligence of the young women was noted: "…In the Chemical Class, the ladies answered many questions, wrote exercises, and practised, under Dr. Reid's superintendence, many useful experiments."[46]

The Scottish educational system was totally independent from that of the English. Nevertheless, it is very apparent that the independent Scottish girls' schools founded in the latter part of the 19th century emulated the English models. Most Scottish day schools, such as the Park School, Glasgow,[47] followed the GPDSC school model. Similarly, Scottish boarding schools, such as St. Leonards School, St. Andrew's,[48] often followed the CLC school model. In addition, many of the science teachers and headmistresses had experiences of the respective English school models either as former students or as former teachers.

At the Scottish girls' schools, too, relating chemistry incidents in poetic verse was to be found in school magazines. Here, we cite a poem recounting a laboratory accident at St. George's School, Edinburgh.

> *A troop of damsels hurried to the lab.*
> *To try to make some chlorinz (Cl_2).*

---

[44]Moore, L. (2003). Young ladies' institutions: The development of secondary schools for girls in Scotland, 1833–c.1870. *History of Education* **32**(3): 249–272.

[45]Reid, D. B. (1836). *Rudiments of Chemistry: With Illustrations of Chemical Phenomena of Daily Life*. William and Robert Chambers, Edinburgh.

[46]Anon. (1835). *Report of the Scottish Institution for the Education of Young Ladies; with an Appendix containing separate reports, by the different teachers, of the course of instruction, and the system pursued, in their respective classes*. Oliver & Boyd, Edinburgh, p. 8.

[47]Lightwood, J. (1981). *The Park School: 1880–1980*. University of Glasgow, Glasgow.

[48]Macaulay, J. S. A. (ed.), (1977). *St. Leonards School 1877–1977*. Blackie & Son, Glasgow.

The fair one with the golden locks was there,
Clad in a robe of purple casement-cloth,
Her lissom limbs chased in long black gym-hose,
A new St George's tie of flaming red;
Cheeks like wild rose, and eyes of speedwell blue.
A wondrous apparatus was set up –
Flask, bottles, tubes and H_2SO_4.
"The manganese dioxide, where is it?
Run! Fetch the jar of black MnO_2!
The fair one, like a streak of lightening, swooped,
To snatch the jar with dainty snow-white hand.
Another nymph swooped too, a buxom lass,
With auburn tresses glinting thick and bright.
Both seized the jar of fateful manganese,
And either nymph was loth to give it up.
A frightful crash resounded thro' the Lab –
(Cursed be the hand that pulled the stopper out!)
But, Oh! our fair one with the golden locks,
Was like a sweep before he's had a bath.
The wretched sooty-black MnO_2
Had made the wild-rose cheeks grimy black!
The speedwell eyes were piteous with dismay –
Alas! the lovely tie of flaming red
Was all besmirched, and grievous to behold.
The class proceeded with the fair one black.
The manganese dioxide did its work,
Till ghastly fumes of chorine filled the lab;
Then paled each check, and rolled each yellowing eye.
The fair one sat with wan and pensive mien,
With drooping lash, and golden hair all grey,
Received with silence every scornful gibe.
I wept to see this too pathetic sight!
After the class she vanished, in distress.
To cleanse her dainty face and change her dress.[49]

---

[49] "Scientist" (January 1927). A Disaster. *St George's Chronicle* (90): 14.

Chemistry Courses for Girls' Schools

If chemistry was to be a focus of the independent girls' schools, it had to be at least as rigorous as that at the boys' schools. Thus, defining content was a topic of specific concern. In this task, the GPDSC schools seem to have taken the lead.

To ensure a consistency of excellence in the teaching of chemistry in all GPDSC schools, a *Conference on the Teaching of Science with Especial Reference to Chemistry*[50] was organized at the Notting Hill High School in 1896 by the GPDSC Council. Chemistry teachers from all GPDSC schools were required to attend. Henry Armstrong (see Chapter 10) was one of the presenters, while the other was Ida Freund (see Chapter 3), the leading woman chemistry educator from Newnham College, Cambridge.[51] Following from the conference, a detailed syllabus for the teaching of chemistry, including laboratory work, was produced by W. W. Fisher, Aldrichian Demonstrator of Chemistry at the University of Oxford.[52]

Then in 1900, specifically for science teachers at GPDSC schools, a *Conference on the Teaching of Science* was held at Kensington High School.[53] The major presentation was by Wyndham Dunstan, Professor of Chemistry at the Royal Pharmaceutical Society and previously Lecturer in Chemistry at the University of Oxford. He spoke on "The Teaching of Science in High Schools for Girls" and following from his discourse, Dunstan was asked to devise a specific syllabus. His proposal was published in 1902 and reprinted in 1912.[54]

---

[50] Anon. (11 June 1896). *Conference on the Teaching of Science with Especial Reference to Chemistry*. Archives, Institute of Education.

[51] Palmer, B. (2012). Ida Freund: Teacher, Educator, Feminist, and Chemistry Textbook Writer (accessed 11 January 2019).

[52] Fisher, W. W. (October 1896). *Schedule of Teaching in Chemistry*. Girls' Public Day School Company, Archives, Institute of Education.

[53] Anon. (5 March, 1900). *Conference on the Teaching of Science*. Archives, Institute of Education.

[54] Dunstan, W. (July 1902). *Syllabus and Examination Schedule, Based on a Course of Instruction and Laboratory Work in Elementary Chemistry and Physics; together with suggestions for Teachers*. Girls' Public Day School Trust, reprinted January 1912, Archives, Institute of Education.

Importance of Chemistry Laboratories at Girls' Schools

In 1898, Clare de Brereton Evans (see Chapter 5) authored a chapter on *The Teaching of Chemistry* in a compilation titled *Work and Play in Girls' Schools by Three Head Mistresses* (the three Headmistresses were Dorothea Beale of CLC, Lucy Soulsby of Oxford High School (a GPDSC school), and Frances Dove of Wycombe Abbey). De Brereton Evans was convinced that junior, as well as senior, girls needed exposure to practical chemistry:

> For success in examinations it is now necessary to have a certain amount of practical knowledge of chemistry, and examination classes are therefore given some practical training, but this reform still remains to be extended universally to the junior classes, which need even more than the senior ones that the teaching should be objective: a child may learn and repeat correctly a dozen times that water is composed of oxygen and hydrogen, and the thirteenth time she will assure you that its constituents are oxygen and nitrogen; but let her make the gases for herself, test them and get to know them as individuals, and mistakes of this kind will become impossible.[55]

Between 1899 and 1904, the progressive girls' magazine, *Girls' Realm*, published a series of articles on famous girls' schools. In many of these articles, the author would comment upon the chemistry facilities. The most prolific author of such school reviews was Christina Gowans Whyte. Whyte waxed a lyrical about practical chemistry and its important role in a girl's education in her article in 1900 on Roedean School:

> Science has a laboratory, fully equipped in every respect for the course overtaken in the school curriculum. The mysteries, amounting almost to terrors, which the imagination of a girl sometimes conjures up at the idea of chemistry, may fade from her mind when she sees the white room and the dainty, spotless appearance of tubes and bulbs and blowers in that sanctum. People have a way of believing that such studies, as for instance the study of chemistry, are only useful to the bookworms of

---

[55] de Brereton Evans, C. (1898). The teaching of chemistry. In Beale, D., Soulsby, L. H. M., and Dove, J. F. (eds.), *Work and Play in Girls' Schools by Three Head Mistresses*. Longmans, Green, and Co., London, pp. 310–311.

38 *Pioneering British Women Chemists: Their Lives and Contributions*

women, who intend to take university degrees. The girl at Roedean School, no doubt, already discovers the great practical advantages to be derived from the study of chemistry.[56]

It was also believed that teaching rigorous science, especially chemistry, with its inherent dangers, was a fitting training for life. This view was eloquently expounded by Whyte in her article on Bedford High School also in *Girls' Realm*:

> Practical work in chemistry is included in the curriculum. Practical chemistry is one of the best among modern educational improvements. It teaches things which go much deeper into our consciousness than mere words could ever go. It teaches consequences-the stern, certain consequences of doing quite the right or the wrong thing. It never makes a mistake, or slurs over a little bit of carelessness, or pretends everything is right when everything is the reverse. And for girls who have to go through with life, it is not a bad thing to learn when young to expect the natural consequences of an action, even to the correct or incorrect testing of a compound or simple liquid.[57]

Some of the headmistresses regarded chemistry, and a chemistry laboratory in particular, as a defining feature of an academic girls' school. For example, Frances Dove, when she founded Wycombe Abbey School (following her years at St. Leonard's School, Scotland), considered a chemistry laboratory of utmost importance, as Lorna Flint recounted in a history of the School:

> But the dearest of all projects to Miss Dove was the provision of a laboratory. Her first endearingly wistful reference to it occurs in her Report to the Council in 1902, where she confesses that she is 'longing for the time when Practical Physics and Chemistry can find a place' among the School's subjects.... the Minutes of May 1907 record that a laboratory was 'urgently needed.'[58]

---

[56]Whyte, C. G. (1900). Famous girls' schools IV: Roedean. *Girls' Realm* **2**: 1063–1064.

[57]Whyte, C. G. (1900–1901). Famous girls' schools: The Bedford High School. *Girl's Realm* **3**(2): 872.

[58]Note 38, Flint, p. 19.

Construction of Chemistry Laboratories

To teach chemistry required a chemistry laboratory. The first British school to offer practical chemistry in a laboratory setting was the (boys') Manchester Grammar School in 1868.[59] The girls' schools were not far behind. The earliest we could find for independent girls' schools was the NLCS in 1875 and CLC in about 1876.

As a result of the networking among GPDSC schools, they were aware of each other's laboratory facilities. However, among the other independent girls' schools, many believed that their own school was one of the very rare schools to pioneer the teaching of chemistry and to have a chemistry laboratory. As an example, in the history of St. Paul's Girls' School, Hammersmith, it is noted that: "In 1904, St. Paul's was one of the few girls' schools in Great Britain to pay serious attention to Science education."[60] Similarly, in the history of Milton Mount College, it was stated that: "In many aspects Milton Mount was a pioneer among girls' schools. [In 1883] It was the first... to have a laboratory ..."[61] In fact, most of the independent girls' schools which we visited had installed chemistry laboratories in the 1880s, 1890s, or at the latest, in the early 1900s.

Some of the schools were unable to construct proper laboratory facilities until the early 1900s. To ensure that their students were not disadvantaged, at some of the schools, students were permitted to perform experimental chemistry work in classrooms. This was the case at Ipswich High School. In a contribution to a collection of school reminiscences, Louisa Frost, a student from 1898 to 1901, recalled:

> We were encouraged to "do it yourself" and, as General Elementary Science was compulsory in those days and we had no lab., we ... burned our hair (long) and nails, ... made oxygen and hydrogen, and discovered that Epsom Salts was only $MgSO_4 \, 7H_2O$![62]

---

[59]Perkin, W. H. (1900). The teaching of chemistry and its development. *School World* 1: 377–380.

[60]Bailes, H. (2000). *Once a Paulina. A History of St. Paul's Girls' School*. James & James Ltd., London, p. 47.

[61]Note 32, Harwood, p. 13.

[62]Anon. (1978). Frost, Louisa, 1898–1901. In *Ipswich High School G.P.D.S.T. 1878–1978*. n. pub. 7–8.

40 *Pioneering British Women Chemists: Their Lives and Contributions*

Similarly, Winchester High School (later St. Swithun's School) initially lacked proper laboratory facilities, but the students took matters into their own hands to encourage the authorities to add a chemistry laboratory (which they did in 1896). Edith Findley in her history of the school stated:

> In 1893, Miss Fletcher was appointed Science Mistress ... But still we had no [Chemistry] laboratory. Well do I remember these class-room chemistry lessons, where smelly experiments such as the preparation of chlorine had to be hastily pushed outside on the window-sill, and my joy when I was promoted to the honour of cleaning up the mess.... Apropos of our difficulties, I may mention that when in 1895 the question of a new addition to the building and possibly a laboratory was mooted, the Chemistry class prepared an energetic flask of chlorine mixture and left it where the Visiting Committee could not fail to be conscious of its objectionable odour — whether it influenced their decision I do not know, but to our joy we heard the laboratory was to be built.[63]

Basement and Attic Chemistry Laboratories

When the schools were founded, practical chemistry was undertaken in whatever space was available — in some, an attic room; in others, a basement room. The first chemistry laboratory at Wimbledon House School (later Roedean School) was an attic room as Joan Waller recalled in the *Roedean School Magazine*:

> Our first laboratory is now no more than a shadowy recollection of a small room at the top of one of the houses at Sussex Square, where, in our early days, to our great joy and pride, we were allowed to break test-tubes and generally to misconduct ourselves.[64]

Whereas attic rooms (or basements) as chemistry laboratories were a temporary expedient at most schools, at others they lasted into the 1920s

---

[63]Findley, E. (1934). *S. Swithun's School Winchester 1884–1934*. Warren and Son Ltd., Winchester, p. 20.

[64]Waller, J. (Summer 1899). The new science rooms. *Roedean School Magazine* **2**: 84–85.

and 1930s. In a history of Alice Ottley School, Worcester, Valentine Noake described how the attic laboratory was in use into the 1930s:

> A slightly older class would be instructed in setting up the apparatus for preparing oxygen or isolating ammonia gas, and would then go to the benches and fit flasks with rubber stoppers, bend glass tubing to appropriate shapes, pour acids cautiously into thistle funnels, and collect and test the results. It was sheer joy to be allowed to prepare the evil-smelling sulphuretted hydrogen [hydrogen sulphide] in the fume cupboard: it gave the budding chemist a sense of vast and noisome power.[65]

Custom-built Chemistry Laboratories of the 1890s to 1910s

The first widespread wave of construction of custom-built chemistry laboratories occurred in the 1890s and 1900s. For example, at Belvedere School (formerly the Liverpool High School), the new chemistry laboratory was opened in 1905. Alison MacKenzie recalled the excitement in the *Belvedere School Liverpool Chronicle*:

> By the time I was old enough to be trusted with a test tube and a bit of magnesium ribbon, which I dearly loved, the new Lab. over the gymnasium had been built. We were vastly thrilled when the workmen's dustsheets were removed and a new doorway disclosed, which led through to these two new Labs. I well recall the momentous day when I was appointed laboratory assistant to help each morning to put out the apparatus for the classes. Alas, I wasn't much of a scientist in those days and must have been a sore trial.[66]

In the preceding discussion, we noted that the first chemistry laboratory at Roedean had been an attic room. Upon moving the school to a new site in the late-1890s, a very spacious suite of science facilities were constructed in 1899, causing great excitement among the chemistry students (see Footnote 64). In addition to a large chemistry laboratory

---

[65]Noake, V. (1952). *History of the Alice Ottley School Worcester*. Trinity Press, Worcester. p. 133.
[66]MacKenzie, A. (1930). Reminiscences. *Belvedere School Liverpool Chronicle*. 46.

42 *Pioneering British Women Chemists: Their Lives and Contributions*

with a balance room, there was also a science lecture room and a science library.

Laboratories of the 1920s to 1930s

During the inter-War period, many schools built new large, spacious, and well-equipped laboratories. At some of the schools, not just chemistry laboratories, but entire new science wings or buildings were constructed. Reports of the provision of new laboratory facilities in the inter-War period were usually mere brief notes, in contrast with the detailed descriptions of the labs of the 1880s to 1910s. Chemistry laboratories were largely taken for granted. An exception was at Belvedere School, where two students, Barbara Gollin and Lois Buckley, enthusiastically reported in 1924 on the new laboratory (succeeding the 1905 laboratory described earlier) in the *Belvedere School Liverpool Chronicle*:

> We have had to wait patiently for a long time for the new lab., but now that it has materialised it has surpassed our wildest expectations. When the Modern Studies people beheld its shining glories, then were they truly grieved that they had forsaken the enlightened cause of Science for the sombre path of Modern Studies.[67]

Chemistry Laboratories in Wales and Scotland

A similar sequence of chemistry laboratory construction occurred in Wales. For example, at Howell's School, Llandaff, it was recounted by Janet Sully in a history of the school that the chemistry laboratory of 1885 was a bacon loft, with a custom-built laboratory not being constructed until 1904:

> Early in 1885, Miss Kendall [the Headmistress] asked for gas to be laid on in the Experimental Room for Chemistry. As there was no money for the building of a Laboratory, the former bacon loft above the stables or

---

[67]Gollin, B. and Buckley, L. (March 1924). The new chemistry laboratory. *Belvedere School Liverpool Chronicle* 33–34.

coach-house was fitted with gas and water and adapted for that purpose. ... the only means of entry was by a ladder.[68]

At Park School, Glasgow, it was only 4 years after the Park School opened in 1879 that the addition of chemistry to the syllabus was considered an urgent issue. Joan Lightwood described the events in her history of the School:

> ... discussions with Miss Kinnear upon an important matter which Professor Lindsay and Professor Young raised at the Board Meeting in December 1884. They both spoke on the advisability of introducing the study of Chemistry into the curriculum and submitted an estimate of the amount needed to fit out a Laboratory — £50 to £60. Dr. Young was asked to proceed in the matter.[69]

As mentioned earlier, St. Paul's Girls' School and Milton Mount College (and many others) in England proclaimed that they were pioneers of practical chemistry for girls. Park School made a similar claim. In the 1930 history of the school, the author noted that the 1885 chemistry laboratory was: "... surely one of the first laboratories to be introduced into a girls' school."[70] Likewise, in a compilation of reminiscences of the County Girls' School, Pontypool, Wales, the same claim was made of being among the first to install a chemistry laboratory:

> At first there was no opportunity for the study of Science, but in June 1898 the large room above the present dining-hall was equipped as a Chemistry laboratory. The fact that this came about at an unusually early date for girls' schools bears witness to Miss Dobell's determination to give to the girls the same opportunities as those enjoyed by their brothers.[71]

---

[68] Sully, J. (2010). *Howell's School, Llandaff, 1860–2010: A Legacy Fulfilled.* Howell's School, Llandaff, p. 62.

[69] Lightwood, J. (1981). *The Park School: 1880–1980.* University of Glasgow, Glasgow, p. 20.

[70] "M. M." (1930). The Girls' School Company, Glasgow. In Anon. The Park School Glasgow 1880–1930. William Hodge & Co. Ltd., Glasgow, p. 11.

[71] Anon. (1947). *Jubilee: The History of the County School for Girls, Pontypool 1897–1947.* The Griffin Press, Pontypool, p. 19.

Examiner's Reports

We have shown so far that chemistry was taught at many independent girls' schools, but of what quality? To establish credibility, there were external examiners to ensure the girls met the proper academic standards.

At several schools, one or more reports of the examiners over the years have survived. For example, at MHSG, the *School Reports*[72] of the 1880s and 1890s commented on the results of chemistry theory and practical examinations. The 1885 *School Report* noted that, for the chemistry examination, students were expected to have studied Miller's *Organic Chemistry*[73] and the introductory chapter of Tilden's *Chemical Philosophy*.[74]

The examiner's reports at Mary Datchelor School were summarised by the Headmistress, Florence Grimshaw, and included in the subsequent issue of the *Datchelor School Magazine*. For example, in 1900, the Cambridge Syndicate Examiners stated:

> Forms VI and Upper V were examined by papers restricted to *Chemistry* only. Here more attention should be paid to the experimental side of the science. There was a thorough knowledge of fundamental principles, laws, and facts, but some weakness was shown in applying this knowledge to the devising of experimental proofs or illustrations. Still, the work was very well done and some of it was excellent.[75]

For the GPDSC schools, a complete set of reports exist specifically on chemistry from 1884 through until the end of our period of study. In 1884, the Examiner, W. W. Fisher, visited seven GPDSC schools, of which only Croydon High School gained his praise. He gave some overall impressions of chemistry at the GPDSC schools, including:

---

[72] *School Reports*, Archives, Manchester High School for Girls.

[73] Miller, W. A. (1869). *Elements of Chemistry: Theoretical and Practical.* Part III, *Organic Chemistry*. Longmans, Green, Reader and Dyer, London.

[74] Tilden, W. A. (1884). *Introduction to the Study of Chemical Philosophy*. Longmans, Green, and Co., London.

[75] Grimshaw, F. E. (1900). Headmistress's summary of examiners reports. *Datchelor School Magazine* **13**(1): 16.

Most of the girls answered correctly questions about the gases entering into the composition of the air, and gave methods for preparing common gases, such as oxygen, hydrogen, carbonic acid, and the like; a majority of them successfully worked the problem set. But it appears evident from the answers, that the girls have little knowledge of the things they describe and write about — and are only reproducing book-work. The teaching does not appear to have been sufficiently experimental, or illustrated by actual experiments.[76]

In 1885, at St. Leonard's School, St. Andrews, Scotland, Miss Dove was headmistress and chemistry teacher. Dove was very concerned that the standard of chemistry education be of the utmost rigour and comparable to that of the best boys' schools. To do so, she had Thomas Purdie, Professor of Chemistry at St. Andrew's University, as an external examiner.[77]

At St. Swithun's, a student at the school recalled the university entrance practical chemistry examination:

In 1897, four of us entered for our first practical exams. in the Joint Board Certificate. In those days, a 'don' was in charge in cap and gown. An enterprising examiner had given red phosphorus as the unknown substance. About ten minutes after we had commenced, a nervous candidate dropped a glowing match on the 'unknown'— result, a wild flare and we all 'knew.' Hardly had the examiner extinguished this when it was discovered that a pile of dusters was on fire; this in turn was extinguished. Then suddenly the bottom came out of a medicine bottle improvised to contain sodium hydrate [sodium hydroxide], devastating a varnished table and all the candidate's papers. Wearily our friend came for the third time to the rescue, remarking, "My life is insured — I only hope yours are!" Possibly we had more thrills out of our Science lessons in those days than under modern conditions. To me at least they are a real joy to recall ... (see Footnote 63).

---

[76] Anon. (1884). *Minutes of the Council and Committees, Reports of Examiners, &c. for 1884*. Archives, Girls' Public Day School Company.

[77] Purdie, T. (22 July 1885). *Report on Science Teaching at St Leonard by Prof. Thomas Purdie, Prof of Chemistry, St Andrew's University at the Request of Miss Dove*. n. pag.

Academic Successes

To be judged as academically successful, it was necessary for the talented girls to be entered into nationally accredited examinations. There were five such bodies: The Oxford and Cambridge Schools Examination Board; The Oxford Local Examinations Delegacy; The London Matriculation; the Department of Science and Art, South Kensington; and the College of Perceptors. Here, we will mention two: the Department of Science and Art, South Kensington; and the College of Perceptors.

Established by the British Government to promote education in art, science, technology and design, the Science and Art Department, South Kensington, functioned from 1853 to 1899.[78] The department offered examinations in all sciences, including separate examinations for inorganic chemistry and organic chemistry. Of all the science examinations, inorganic chemistry had the greatest number of registrants.[79] From our own research, many girls' schools entered their students in the Science and Arts Department examinations. For example, in 1880, CLC had 15 students enter and pass the inorganic chemistry examination — compared with nine students for mathematics.[80]

Incorporated by Royal Charter in 1849, the College of Perceptors examined and provided certificates for secondary school students, both boys and girls, in a wide variety of subjects. B. S. Crane noted that:

> ... until the end of the [19th] century, there were more candidates for the examinations of the College [of Preceptors] than for the local examinations of Oxford and Cambridge combined.... a majority of the schools were for girls ...[81]

In these examinations, the girls proved to be exceedingly academically capable. Andrea Jacobs has documented the academic successes of girls in the 1860–1902 period:

---

[78]Edmonds, E. L. (1957). The science and art department: Inspection and/or examination? *The Vocational Aspect of Secondary and Further Education* **9**(19): 116–127.

[79]Crane, B. S. (1959). Scientific and technical subjects in the curriculum of English secondary schools at the turn of the century. *British Journal of Educational Studies* **8**(1): 52–64.

[80]Anon. (1880). Examinations 1880. *Cheltenham Ladies' College Magazine* 230–232.

[81]Note 79, Crane, p. 58.

By the end of the decade [the 1870s], the number of girls entering for the Schools' Examinations of the College of Perceptors exceeded boys. In 1879 there were 7645 candidates of whom 4428 were female and the following year there were 9148 candidates of whom 5121 candidates were female. Not only were there more female candidates but their success rate was often higher.[82]

In fact, as the 1880s progressed, the greater academic successes of girls than boys became an issue of concern. Jacobs commented: "By February 1884 the gap between the performance of the girls and that of the boys had widened to such an extent that it could no longer be dismissed."[83] However, the conclusion drawn at the time was that the problem lay with the girls: the girls were working too hard. By contrast, it was argued, boys had a healthier, more casual, attitude to their studies. This dedication to, and enthusiasm for, their studies by girls was seen as a cause for concern. For example, it was believed that the excessive use of the brain by girls could have led to lasting damage to their reproductive organs.[84] This "dedication and enthusiasm" certainly applied to girls' attitude to chemistry, as we will discuss subsequently.

Science Clubs and Societies

The enthusiasm of many of the girls for science — particularly for chemistry — can be found in their interest in science clubs and societies. Our source of information on the science clubs/societies was the student school magazines. Particularly in the early decades, the magazines of the respective schools contained lengthy accounts of their club/society activities. We had access to the early magazines of 59 independent girls' schools, of which about three-quarters had active science clubs or societies at some time during our timeframe, while four others had science activities without a formal club. In terms of chronology, of the schools

---

[82] Jacobs, A. (2001). The girls have done very decidedly better than the boys: Girls and examinations 1860–1902. *Journal of Educational Administration and History* **33**(2): 120–136.

[83] Note 82, Jacobs, p. 129.

[84] Note 82, Jacobs, p. 132.

Early Science Clubs and Societies

Among the schools we visited, the earliest science club of all was The Science Circle of Wimbledon High School. Founded in 1889,[85] Martha Whiteley (see Chapter 4) was reported in the *Wimbledon High School Magazine* as having given a lecture to the Science Circle on the Spectroscope in April 1892. However, only five attended the meeting and so it was decided to dissolve the group.[86]

Wimbledon House School (later Roedean School), too, was a pioneer, with a Natural Science Club founded in 1893 with much enthusiasm. According to the *Wimbledon House School News*: "The first meeting was also, as it were, a kind of house-warming for the new Laboratory.... There are between 30 and 40 members of this Club and the attendance on all occasions has been admirable."[87]

Inter-War Science Clubs

Then there was a second burst of enthusiasm for science societies in the inter-War period. At Bedford High School for Girls, they called their society members the "Ionians." They derived the name from the term the "Ionian Enlightenment" which was the appellation for the set of advances in scientific thought centred around Ionia of ancient Greece in 6th century B.C.[88] The Ionian Society at Bedford flourished through the 1930s, with a report in *The Aquila* in 1937 stating: "This year, there are more Ionians than there have been in previous years, and our membership numbers well over one hundred; ...."[89]

---

[85] Anon. (December 1889). Report of the natural science circle. *Wimbledon High School Magazine* 12.

[86] Anon. (April 1892). The science circle. *Wimbledon High School Magazine* 4–5.

[87] Anon. (Christmas term 1893). Natural science club. *Wimbledon House School News* 11.

[88] Freely, J. (2012). *The Flame of Miletus: The Birth of Science in Ancient Greece (and How It Changed the World)*. I. B. Tauris, London.

[89] Anon. (March 1937). The Ionians. *The Aquila, Magazine of the Bedford High School for Girls* 10.

The numbers of girls interested in science at many schools has to be considered as astounding. It was not just limited to a few of the schools but seems to have been a widespread phenomenon. At Wycombe Abbey School, the Science Club had been founded in 1918,[90] and by 1933, it was reported in the *Wycombe Abbey Gazette* that: "Numbers in the Science Club have increased greatly this term and now number over 100."[91]

At some schools, there were so many prospective members that a separate science club for the juniors was started. At Sutton High School, there was a rigorous admission standard even for the Junior Branch of their Science Club. A Club report in the *Sutton High School Magazine* noted:

> A Junior Branch of the Science Club has been formed for girls in the Lower IV and Upper III Forms. In order to qualify for membership, these girls must pass a short test on general scientific knowledge or demonstrate before a committee their interest in some scientific subject. The Club has, at present, a membership of thirty.[92]

Student Chemistry Presentations

The clubs catered to all the sciences, so there was a wide range of topics among the student presentations. Nevertheless, a significant number of presentations were on chemistry, and some of these were illustrated with demonstrations.

Presentations on the History of Chemistry

The history of chemistry, especially a discovery which could be interwoven with the life of a famous chemist, was always a popular topic at science club/society meetings. For example, it was reported in the *Malvern*

---

[90] Anon. (1918). Science club notes. *Wycombe Abbey Gazette* **6**(8): 116.

[91] Anon. (1933). Science club notes. *Wycombe Abbey Gazette* **10**(5): n. pag.

[92] "E.M.A." (1933). Junior science club. *Sutton High School Magazine* 30.

50   *Pioneering British Women Chemists: Their Lives and Contributions*

College for Girls Magazine that a student member of the Science Society spoke on the discovery of elemental oxygen:

> Hilda [Donald] gave a paper on the "Discovery of Oxygen." There are three names associated with the discovery of oxygen — Scheele, Priestley, and Lavoisier. They made experiments with air and discovered that one part supported combustion while the other did not.... Oxygen is essential to the support of animal life and there is only one element (fluorine) into combination with which it does not enter. Two common methods of preparing oxygen are (1) heating Barium Peroscide [sic: Peroxide], and (2) heating Potassium Chlorate with Manganese Dioxide.[93]

Presentations on Contemporary Topics

Contemporary events were the focus of some presentations. In the latter part of the First World War, some talks related to the war effort, including this paper at the Science Society described in the *Chronicle* of Winchester High School (later, St. Swithun's):

> In the Autumn Term, T. Rose read a most interesting paper on "Some Aspects of Chemistry During the Present War." She described the structure of a typical shell, giving illustrations on the blackboard, and then discussed the nature and properties of some of the high explosives employed in the manufacture of the shells. She also referred to the important part played by dye works during the war, explaining that they were the only works fully equipped for the manufacture of high explosives at its outbreak.[94]

Demonstrations by Students

In some cases, we do have details of experiments performed, such as this early account in the *The Paulina* by the Field Club of St. Paul's Girls' School:

> K. Grant then performed some very effective colour changes, after which M. Hooke showed the surprising results obtained by burning various substances in oxygen and in chlorine. An amusing diversion was

---

[93] Anon. (July 1916). The science society. *Malvern College for Girls Magazine* (20): 16–19.

[94] "J.O.E." (1917–1918). Science society report. *Winchester High School Chronicle* (22): 15.

created when too much sulphur dioxide was evolved, with the result that many of our guests were almost asphyxiated. After strenuous efforts to extinguish the flames, the Vice-President [of the Club] gallantly bore off the obnoxious liquid.[95]

The properties of liquid air held a particular fascination. Experimentation was usually undertaken by teachers, visiting lecturers, or at industrial plants. However, at Streatham Hill High School, the girls themselves demonstrated the properties to the Science Club, even the highly dangerous reaction of burning sulphur with liquid oxygen as described in the *Streatham Hill High School Magazine*:

As the subject of the paper promised to be very interesting, there were a hundred and four people present. Doris Mathews and Betty Boyd read a most interesting paper on "Liquid Air," which they supplemented with exciting experiments. Grapes, bananas, oranges, and tomatoes could only be broken with a hammer after immersion in liquid air. A rubber ball cracked like an egg; a frozen piece of beef was hard enough to crack a plate. Mercury made an excellent hammer with which nails were knocked into a board. The last experiment was perhaps the best. A hole had been made in a large block of ice, into which liquid oxygen was poured. A piece of burning sulphur was then dropped in, and it burnt with a glorious mauve flame that lit up the whole room. The meeting ended with a hearty vote of thanks to Doris and Betty.[96]

Professional Speakers

Career opportunities related to chemistry were sometimes among the lectures given. So there was an implicit assumption that chemistry was not being taught simply "to be as good as the boys" but with the possible intent of using it as a pathway into career opportunities. For example, at Wimbledon High School, it was reported in the *Wimbledon High School for Girls Magazine* that in 1924: "Miss Iva Lewis [chemistry teacher] read

---

[95] Seaton, D. (July 1909). Field club report. *The Paulina, Magazine of the St. Paul's School for Girls* (15): 8.

[96] Anon. (February 1928). Science club. *Streatham Hill High School Magazine* (29): 16–17.

52 *Pioneering British Women Chemists: Their Lives and Contributions*

a paper on 'Chemistry as a Career for Women'..."[97] At Roedean, a lecture was given in 1919 by Ida Smedley Maclean (see Chapter 9) of the Lister Institute. The report of the event in the *Roedean School Magazine* noted: "...Mrs McLean [sic] showed the wonderful possibilities of *Applied Chemistry as a Career for Girls...*"[98]

Lectures with Experiments

Lectures which involved experiments were a particular highlight of the Science Club programmes. Described in the *Science Club Minute Books*, the Science Club at Downe House enjoyed a lecture, with experiments, on plant chemistry:

> Dr. Druce M.Sc. of Battersea [John F. G. Druce, chemistry master at Battersea Grammar School] came down and gave us a lecture in the new lab on the "Chemistry of Plants." ... He did several very interesting experiments such as the extraction of chlorophyll from leaves and proving the presence of many of the common metals in the ash of hay.[99]

For some schools, particularly in the greater London area, science students were taken to lectures at other institutions. The *Blackheath School Record* reported that, over the years, several school parties had attended lectures at the Royal Institution, including the following commentary of the 1929 lecture:

> During the spring term some members of the Form VI were taken to three lectures at the Royal Institution, one given by Sir William Bragg on "Crystal Analysis," the second on "The Distribution of Chemical Elements" by Professor Goldschmidt, and the third on "Penetrating Radiations" by Sir Ernest Rutherford.[100]

---

[97] Eborall, M., and Kaye, M. (May 1924). The science club. *Wimbledon High School for Girls Magazine* (32): 32.

[98] Anon. (Lent 1919). 'Announcements: Lectures.' *Roedean School Magazine* **20**(2): 48.

[99] Anon. (1 June 1924). *Science Club Minute Books*. Archives, Downe House School.

[100] Anon. (1929). Science notes. *Blackheath School Record* (44): 6.

Radium Demonstrations

The discovery of radium and radioactivity was a major source of excitement during the 1910s and 1920s. In addition to the girls themselves speaking on the subject, many Science Clubs had visiting lecturers on the topic. The inherent dangers of radioactivity were not appreciated, and samples of radium compounds were widely available and handled openly. According to *The Magazine of the City of London School for Girls*, a group from that school was taken to King's College, London, to view a lecture with demonstrations:

> ... we attended the lecture on Radium and X-rays. ... The lecturer performed the following demonstration to illustrate this [the rays from radium causing other substances to become luminous]. He had a large flask the inside of which was coated in zinc sulphide. A closed glass tube was fitted into the cork of the flask which contained a very small quantity of radium emanation [radioactive radon gas]. The tube was opened, thus allowing the small quantity of the gas to pass into the flask, and immediately the bulb of the flask was filled with a green glow due to the fact that the small amount of radium emanation had caused the zinc sulphide to become luminous.[101]

While at Downe House, the *Downe House Magazine* described how it was a science teacher who performed the experiments with radium: "Miss Phelips gave a display of practical experiments including the decomposition of water by radium."[102]

Liquid Air Demonstrations

The properties of liquid air were a highlight of meetings at many of the schools. The following series of demonstrations at St. Martin-in-the-Fields School illustrates how liquid mercury was handled as an everyday substance. The hazard of burning sulphur in pure liquid oxygen and the

---

[101] Anon. (March 1924). The Scientific Novelties Exhibition at King's College. *The Magazine of the City of London School for Girls* **29**(4): 79–81.

[102] Anon. (1925). Science club. *Downe House Magazine* (46): 14.

54 *Pioneering British Women Chemists: Their Lives and Contributions*

toxicity of sulphur dioxide gas were not considered, either. Instead, the Science Club Secretary, Marguerite Dubois, described the "exciting experiments" in *St. Martin's Past and Present*:

> At our next meeting during the Spring Term, Miss Sutton [chemistry teacher] gave a lecture on "Liquid Air," and did many exciting experiments with it. After she had shown us a kettle containing liquid air boiling on ice some of us were inclined to be sceptical and think that liquid air must be very hot, but when we saw her bang nails into a block of wood, with a hammer made by freezing mercury in liquid air, we were convinced to the contrary. A very beautiful effect was obtained by dropping a piece of burning sulphur into a little pool of liquid air in a hole scooped out of an ice-block. The sulphur burned with a lovely blue flame, which appeared very beautiful when seen through the ice.[103]

Expeditions to Industrial Plants

When it came to visits to industrial plants, as the reader will see subsequently, such words as "thrilling" permeate the accounts. To that generation of girls, industrial chemistry was pure excitement. However, the authors of the reports were always careful to include the chemical aspects of their visit. The girls seemed fearless, for some of the accounts described red-hot metal, flames, toxic gases, and so on.

Local Gas Works

Without exception, every Science Club/Society visited at least one Gas Works. Urban Britain relied heavily on coal gas for heating, cooking, and — in the early decades — for lighting. The hydrocarbon gas mixture was produced by heating coal in the absence of air (a process known as pyrolysis). Thus, a local Gas Works was the one chemical industry which every large community in England possessed. These visits were not just to see the awe-inspiring buildings and processes but to understand the chemistry involved.

---

[103] Dubois, M. (September 1932). Science club. *St. Martin's Past and Present* (11): 43.

The most descriptive account of a gas works visit was given in the *Belvedere School Liverpool Chronicle* by Molly Baddeley. The Science Club members toured the Liverpool Gas Company's Works at Garston:

> After walking a few paces we turned a corner and saw, in the near distance, the thrilling sight of a huge mass of burning cinders shooting down the wall of a large building, while from the bottom of this wall banks of smoke and steam arose. Our excitement was at once aroused, and we pressed eagerly forward toward this scene of splendour.... The series of furnaces is surrounded by a passage. We entered this passage and found ourselves in what might almost have been part of the infernal regions, for it was intensely hot and dark except for one spot, where, from an opening in the wall, huge flames were belching forth.... In order to show us the intense heat of the furnaces a small plug was taken from the wall of one. We gazed in awe at the almost white-hot interior, until the heat from the opening drove us away.... [Later] we saw the tests for sulphuretted hydrogen being made, by holding over the gas a paper dipped in solution of lead acetate, which any sulphuretted hydrogen quickly turns brown. We realised, from the care taken in these tests, why all our efforts to find traces of sulphuretted hydrogen in our gas supply have been in vain.[104]

Sulphuric Acid Plants

There were factories in several locations in Britain producing sulphuric acid and for nearby girls' schools, tours of such plants were also popular. A group from Clapham High School commented in the *Clapham High School Magazine* on inhaling the sulphur dioxide, and their tour seemed somewhat hazardous:

> In November, 1923, a party of the Upper and Lower VI Science girls was taken over the sulphuric acid factory at Greenwich. They were shown the pyrites burners where the sulphur dioxide is obtained, and the preparation of the oxides of nitrogen, used in the Chamber Process. They then passed on to the actual leaden chambers, and when later they mounted

---

[104]Baddeley, M. (March 1925). Visit to the gas works. *Belvedere School Liverpool Chronicle* 56–57.

56 *Pioneering British Women Chemists: Their Lives and Contributions*

a ladder, and came on the roofs of the chambers, the sulphur dioxide was everywhere in their eyes and in their mouths. They also saw the finished sulphuric acid flowing out from the absorption towers.[105]

According to the account in *Laurel Leaves*, the magazine of Edgbaston High School for Girls, Birmingham, ladder-climbing was also part of the experience for the chemistry students when they toured a combined sulphuric acid and hydrochloric acid plant:

> We had been advised to take old mackintoshes, and were very glad of this as we had a lovely time climbing dirty ladders to view various parts of the workings.... We were very thrilled at being allowed to climb up a little ladder in pairs to look into the furnace in which the sodium hydrogen sulphate was being converted to sodium sulphate, which could then be sold.[106]

Soap Manufacturing Plant

Soap manufacturers were another favourite industry for Science Club visits. Marjorie Davis of the Science Club of Colston's Girls' School, Bristol, described in the *Colston's Girls' School Magazine*, their tour of Messrs. Christopher Thomas' Soap Works:

> It was interesting to see all the stages in the process of soap-making from the beginning, when the fats were melted by steam, then pumped from the ground floor of the factory to the top floor, and there poured into vast cauldrons where, with water and caustic soda, they are gradually boiled into soap.... We also saw the vacuum chambers where the valuable by-product, glycerine, is distilled and purified.[107]

Steel Works

Though Sheffield High School did not have a science club, the girls of the Senior Science Division were taken out annually during the late 1920s and

---

[105] Anon. (1925). School reports. *Clapham High School Magazine* (25): 54.

[106] "H.K.L. M.A.C." (December 1937). A visit to a sulphuric and hydrochloric acid works. *Laurel Leaves. Edgbaston High School Magazine* 31–32.

[107] Davis, M. (1929). Science club. *Colston's Girls' School Magazine* (2): 37.

early 1930s. In 1933, they toured the Steel Works at Templeborough, reporting in the *Sheffield High School Magazine* that:

> We saw one of the furnaces being tapped, which proved a most thrilling sight. During this spectacular process the molten steel is run out of the furnace into a massive iron ladle, amidst showers of sparks; the heat was so terrific we had to stand some distance away in order to watch in comfort.[108]

Annual Science Exhibitions

Another common activity was an annual science show produced for the school and often for the parents as well. For example, at St. Paul's School for Girls, it was reported in *The Paulina* that the Science Club held a scientific exhibition for all the members of the school. The hazardous, mercury vapour-producing, Pharaoh's serpent reaction was included in their repertoire: "Among the attractions were the making of silver mirrors and Pharaoh's serpents in the Chemistry Room."[109]

The Science Club of Wimbledon High School started an annual science exhibition in 1924, and these continued to the end of the 1920s. The report for 1929 in the *Wimbledon High School Magazine* described how their exhibition was serious science:

> At the end of the Summer Term, 1929, the Chemical, Physical and Botanical Laboratories were full of excited girls, dressed in black overalls, looking very worried but very learned. It was the Science Exhibition. I am sure every parent must have been most impressed and most interested by the various demonstrations.... Lower V showed the solubility of ammonia gas and hydrochloric acid gas in water. "Burning Air" (Upper IV), Invisible Inks (Upper IVB), "Supersaturation" (Lower V) and "Surface Tension" were other experiments which were much enjoyed.[110]

---

[108] Anon. (1933–1934). Visit to the steel works. *Sheffield High School Magazine* **12**(1): 8.

[109] Anon. (December 1923). Science club. *The Paulina, Magazine of the St. Paul's School for Girls* (57): 4.

[110] Stevenson, W. (April 1930). The science exhibition, 1929. *Wimbledon High School Magazine* (33): 33.

Science Club Magazines and Science Libraries

The NLCS Science Club had its own magazine, *Our Science Club Magazine*, started in 1891. In 1912, the Magazine was re-named *The Searchlight, NLCS Student Magazine for Science*, and the issues of 1912, 1913, and 1914 have survived in the NLCS Archives.

An Experiment

The 1912 issue of *The Searchlight* contained the description of an experiment which some of the senior students performed, titled "Another Experiment" much to the distress of the chemistry teacher, Miss Stern:

Object. To do something exciting.

Experiment. We placed a considerable quantity of potassium chlorate & crushed sugar in a trough & poured onto it a little concentrated sulphuric acid.

Results. 1. The mass burst into flames.

2. Miss Stern was much alarmed & required a lengthy explanation, but even after she had received this, would only concede that the brilliant experiment <u>might</u> have been a success if we had used smaller quantities.

Conclusion. Experiments which may be thought in any way exciting should <u>not</u> be performed when nervous people are about.[111]

Several Science Clubs founded their own science libraries. The Science Club at Roedean School regularly added chemistry books to their science library over the period from 1913 to 1928.

Demise of Many of the Science Clubs

The peak of interest in the science clubs occurred during the inter-War years. Whether the inter-War Club flourished more in the 1920s or in the

---

[111] "Chief Experimenter" (1912). Another experiment. *The Searchlight: NLCS Science Club Magazine* 26.

1930s seems to largely relate to the enthusiasm of the chemistry teacher of the period.

The Science Society of Notting Hill High School was active in the 1930s. It started with great promise in 1928: "During the Science Society's first year it has shown itself both active and versatile in its pursuits."[112] But by 1939, it was no more: "At the end of the Summer Term it was decided that the Society should not continue unless the girls were keener."[113]

At some schools, science clubs faded away, then years later, a new Science Club was formed, without any school recollection that there had been one previously. The students at Ipswich High School had founded a Science Club in 1907, which disappeared. Upon the arrival of a dynamic new science teacher, Miss Virgo, in 1921, the school instituted a Science Club again: "A Science Club for the girls of the Upper School was started in the Spring Term."[114] The Club was active throughout the 1920s, with an 'expedition' to a nearby sulphuric acid works in 1929 being described in enthusiastic detail in the *Ipswich High School Magazine*.[115]

From School to University

With the graduation from these schools of the first cohorts of academically educated young women, the demand arose for access to a university education. The headmistresses at many of the schools urged their students on to this new goal. Sturge and Clark described how, following the opening of Girton College, Cambridge in 1864, Lydia Rouse, Headmistress of The Mount School, York: "... set constantly before her schoolgirls the prospect of college life as a practical ambition, and they strained to look through her window, standing not without effort on tiptoe."[116]

---

[112]Anon. (1928). Science society. *School Magazine, Notting Hill High School for Girls* (43): 9.

[113]"M.T." (1939). Science society. *School Magazine, Notting Hill High School for Girls* (54): 10.

[114]Virgo, Miss. (1921). Science club. *Ipswich High School Magazine* 16.

[115]Pipe, A. (1929). Expedition to the Sulphuric Acid Works, Bramford. *Ipswich High School Magazine* 18.

[116]Note 12, Sturge and Clark, pp. 108–109.

Even the supporters of women's admission to university, such as Lilian Faithfull, Beale's successor as Headmistress at the CLC, looked back from 1900 at the speed of change. Faithfull wrote in *King's College Magazine, Ladies' Department*:

> Few people probably realize the courage and independence of those pioneers, who were certain to be dubbed "blue," and looked askance at as alarmingly intellectual and "advanced." Now-a-days all that is changed, it is hardly too much to say that throughout England college education is regarded as a desirable continuation of a girls school education.[117]

The First Generation

For those pioneers, to decide to go to college was a brave act in itself. It required a strong self-image, particularly if the family held to the conventional view that a well-brought-up Victorian or Edwardian girl should stay quietly at home until a suitor appeared on the horizon. It was an avenue open only to the daughters of the expanding middle class, the rapidly growing business and professional sector of society, who were already populating the independent girls' schools.[118] The daughters of the poor were simply financially unable to attend university, while the daughters of the upper classes were, for the most part, given an education that would prepare them for their intended life of leisure rather than one that might promote intellectual development.

Being the eldest daughter (or particularly, an only child) also seems to have favoured the pursuit of a university education.[119] Daughters of clergymen seemed to be overrepresented among these early cohorts.[120] As an

---

[117] Faithfull, L. M. (1900). *King's College Magazine, Ladies' Department* 4(2): 6–10.

[118] Howarth, J., and Curthoys, M. (June 1987). The political economy of women's higher education in late nineteenth and early twentieth-century Britain. *Historical Research* **60**: 208–231.

[119] Sulloway has argued that birth order is of vital importance in determining traits, in particular, he argues that the first-born child has unique behavioural characteristics. See: Sulloway, F. J. (1996). *Born to Rebel: Birth Order, Family Dynamics and Creative Lives*. Random House of Canada, Toronto.

[120] Wade, G. A. (1908–1909). Where do the cleverest girls come from? Is it from the Rectory, the Vicarage, the Parsonage, and the Manse? *Girl's Realm* **11**: 343–348.

additional driving force, to go to college was one of the few avenues (nursing and missionary work being among the others) for a young woman to escape the family home without the necessity of marriage.[121]

Going away to college was an exhilarating experience for this first generation. As one student remarked, she and her fellow co-eds were happy: "... in the glorious conviction that at last, *at last*, we were afloat on a stream that had a real destination, even though we hardly knew what that destination was."[122]

The Second Generation

For the second generation, there were different factors. In particular, by the end of the First World War, with so many young men having been killed, marriage prospects were minimal. Muriel Glyn-Jones recounted her own experience in *The History of Royal Holloway College 1886–1986*:

> My father decided to discuss my future with me. He said he would like to see me happily married, but after what had gone on in Flanders for the past four years that could only be doubtful. He would never be able to leave enough money for me to live on, but he was prepared to spend any money I could use for my education. It was later decided that I should go to Royal Holloway College ...[123]

At co-educational facilities, women students were often constrained as to where they were allowed to go. In several institutions, women had to enter lecture rooms through a different door from men, and it was common for the lecture room itself to have a separate ladies row or section. Yet, to these women students of the Victorian and Edwardian eras, the slights and insults were a small price to pay for the excitement of being in the first

---

[121]Vicinus, M. (1985). *Independent Women: Work and Community for Single Women 1850–1920*. University of Chicago Press, Chicago, Illinois, p. 138.

[122]Vicinus, M. (1982). 'One life to stand beside me': Emotional conflicts in first-generation college women in England. *Feminist Studies* **8**: 603–628.

[123]Bingham, C. (1987). *The History of Royal Holloway College 1886–1986*, Constable, London, p. 134.

62 *Pioneering British Women Chemists: Their Lives and Contributions*

assault on the bastions of learning: to be in those "hallowed halls," where studying philosophy or physics was a joy, an end in itself.

By our standards, the activities of women students were severely circumscribed by the university rules, such as the need for chaperones. However, compared to the societal restrictions at home, universities were a haven of freedom for women. In *Late Victorian Britain 1875–1901*, the historian, J. F. C. Harrison, described this feeling:

> For middle class girls the opportunity to have a room of one's own, to be able to organize one's life free from patriarchal dominance, to have cocoa, tea or coffee parties unsupervised, to discuss what one liked with friends, to play games of hockey, and cycle around town — all this was immensely liberating, despite many restrictions and controls imposed by the college authorities.[124]

Choice of University

For the 898 women chemists who became Associates or Fellows of the Royal Institute of Chemistry and/or Fellows of the Chemical Society between 1880 and 1949, we found information on the college or university attended for 841 of them. For London, Bedford College was the most popular (100 women chemistry students), followed by University College (57), Royal Holloway College (36), King's College for Women (24), Imperial College (23), Queen Mary College (19), Birkbeck College (15), and Battersea Polytechnic (12).

Over 60% of the numbers attending Cambridge (45) resided at Newnham College, while for Oxford (28), Somerville and St. Hugh's Colleges were the predominant choice. Among the provincial universities, the women chemists' preferences — usually based upon nearness to home — were Manchester (38), Birmingham (27), Liverpool (22), Leeds (21), Newcastle (13), and Nottingham (10).

In Scotland, Glasgow was the most popular (48), then Edinburgh (19), followed by Aberdeen (10) and St. Andrew's (10). The dominance of

---

[124]Harrison, J. F. C. (1990). *Late Victorian Britain 1875–1901*. Fontana Press, London, p. 171.

Glasgow was possibly due to the existence of a separate women's college, Queen Margaret College, while the total for Edinburgh includes Heriot-Watt College (later Heriot-Watt University). The combined number for the Welsh constituent university colleges (Aberystwyth, Bangor, and Cardiff) was 15.

As students could obtain external degrees from the University of London, we find Battersea Polytechnic and Nottingham in the list, both offering London external B.Sc. (Chemistry) degrees during this time (as did Exeter, Northern Polytechnic, and others). In addition to those women chemists identifiable with a particular college, an additional 39 women had London chemistry degrees with no indication of their affiliated institution.

School–University Links

Some independent girls' schools had links with a specific college or university. The most apparent link was between MHSG and the University of Manchester: most MHSG graduates went to their neighbouring university, as a former MHSG student, Mary McNicol, commented in 1902:

> Owens College [later, Manchester University] has always seemed to me the natural place at which [Manchester] High School girls should study after having finished their school course, and, judging by the number of Old Girls there, other people think so too. No High School girl needs ever feel lonely at Owens, for she will always come across girls she has known at school. A considerable number will be found in the Arts Classes, and several in the Science Department.[125]

As other examples, nearly all NLCS graduates planning chemistry degrees entered Bedford College or Royal Holloway College, with about half going to each. For the chemistry-focussed graduates of South Hampstead High School, nearly all went to Bedford College, according to our dataset,

---

[125]McNicol, M. (December 1902). College letters, *Magazine of the Manchester High School* 115.

as did a significant proportion of CLC graduates planning to follow a chemistry career.

Commentary

In historical accounts of the first cohorts of young women entering the universities, it is sometimes overlooked that the women had to have received academic schooling first. The story of academic girls' education, including chemistry, can be traced back to those oft-forgotten Quaker schools of the early 19th century who were the real pioneers. However, it was the founding of the many academic girls' schools across Britain in the later 19th century which truly opened the doors of educational opportunities for young women. For our story, the crucial factor was the role of the teaching of chemistry at these schools as a means of illustrating their academic rigour. It was the largely chemistry-obsessed matriculants of these schools whose stories and accomplishments fill the pages of the rest of this book.

Chapter 3

Cambridge and Oxford Women's Colleges

Ida Freund (15 April 1863–15 May 1914)

"Newnham and Girton *par excellence* are synonymous everywhere with the University education of women,"[1] proclaimed Millicent Morrison in the pages of *Girl's Realm*. *Girl's Realm* (published 1898–1915), which was one of the three influential girl's magazines of the late 19th century, the others being *Atalanta* (published 1887–1898) and the *Girl's Own*

---

[1] Morrison, M. H. (1915). Newnham College: A pioneer foundation for the higher education of women. *Girl's Realm* **17**(203): 719.

Paper (published 1880–1927).[2] Each of the magazines promoted the Oxbridge women's colleges, for example, Atalanta had articles: "Oxford and Cambridge Colleges for Women,"[3] "A Letter from Cambridge,"[4] "Girton College,"[5] "Newnham College,"[6] and "Women Students at Oxford."[7] As a result, admission to one of these colleges represented the pinnacle to which young academically gifted students at independent girls' schools could aspire. But we will begin with a brief overview of their founding, starting with the women's colleges of Cambridge.[8]

Cambridge Women's Colleges

In fact, the story of the Cambridge women's colleges began not at Cambridge, but at Hitchin in Hertfordshire. It was in Hitchin that Emily Davies co-founded the first college for women. Born in 1830, the daughter of an Anglican clergyman, Davis became frustrated with the fact her brothers had followed careers while she was expected to remain at home and undertake parish visiting with her father.[9] In 1858, Davies met Barbara Leigh Smith Bodichon,[10] daughter of the Unitarian, Benjamin

---

[2] Dixon, D. (1998–1999). Deprived and oppressed: Victorian and Edwardian magazines for girls. *British Library Newspaper, Library News*, 25.

[3] Knatchbull-Hugessen, E. (1889–1890). Oxford and Cambridge colleges for women. *Atalanta* **3**: 421–423.

[4] Anon. (1892–1893). A letter from Cambridge. *Atalanta* **6**: 461–462.

[5] Meade, L. T. (1893–1894). Girton College. *Atalanta* **7**: 325–331.

[6] Meade, L. T. (1893–1894). Newnham College. *Atalanta* **7**: 525–529.

[7] Carr, K. (1896–1997). Women students at Oxford. *Atalanta* **10**: 43–48.

[8] Sutherland, G. (1994). Emily Davies, the Sidgwicks and the education of women in Cambridge. In Mason, R. (ed.), *Cambridge Minds*. Cambridge University Press, Cambridge, pp. 34–47.

[9] Bennett, D. (1990). *Emily Davies and the Liberation of Women 1830–1921*. Andre Deutsch, London.

[10] Hirsch, P. (2000). Barbara Leigh Smith Bodichon: Feminist leader and founder of the first University College for women. In Hilton, M. and Hirsch, P. (eds.), *Practical Visionaries: Women, Education and Social Progress 1790–1930*. Longman/Pearson Education, Harlow, pp. 84–100.

Leigh Smith. Davies and Bodichon formed an inspiring duo: Davies, the practical organizer; and Bodichon, the charismatic socialite.

In 1866, Davies authored a book expounding her views, *The Higher Education of Women*,[11] and following from this, with the help of a committee of well-wishers and fund-raisers, she opened Benslow House, Hitchin, in 1869. Davies believed the location of Hitchin between London and Cambridge would enable lecturers from both universities to visit and teach at her college. The impracticability of this plan soon became apparent as neither Cambridge nor London lecturers would travel the distances involved, so the decision was made to move the college nearer to Cambridge. A site was chosen at Girton, then 2 miles from Cambridge, and with fund-raising by Bodichon, the building was completed and opened in 1873.[12]

Parallel with the development of Girton, a second college was being established much closer to Cambridge. This college owed its origin to Henry Sidgwick,[13] a professor of Physics at Cambridge, and Millicent Garrett Fawcett, women's rights activist. In 1869, Sidgwick had proposed that informal lectures be offered for women, and with their success, it became apparent that the time had come for formal instruction for ladies. He chose Anne Jemima Clough[14] to be the organizer of this endeavour.

Clough was a very different individual to Davies. Though Clough's family was socially middle-class, a series of business disasters had left them in a precarious financial position. She had always been interested in the cause of education for women. One of Clough's greatest triumphs had been the organization of lectures on astronomy to women and girls in

---

[11] Davies, E. (1866). *The Higher Education of Women*. Alexander Strahan, London and New York.

[12] Bradbrook, M. C. (1969). *'That Infidel Place': A Short History of Girton College 1869–1969*. Chatto and Windus, London.

[13] Collini, S. (2004). Sidgwick, Henry, 1838–1900. *Oxford Dictionary of National Biography*. Oxford University Press (accessed 13 July 2018).

[14] (a) Gallant, M. P. (1997). Against the odds: Anne Jemima Clough and women's education in England. *History of Education* **26**: 145–164; and (b) Sutherland, G. (2000). Anne Jemima Clough and Blanche Athena Clough: Creating educational institutions for women. In Hilton, M. and Hirsch, P. (eds.), *Practical Visionaries: Women, Education and Social Progress 1790–1930*. Longman/Pearson Education, Harlow, pp. 101–114.

68 *Pioneering British Women Chemists: Their Lives and Contributions*

Liverpool, Manchester, Sheffield, and Leeds, which had attracted a total of over 550 enthusiasts. In 1872, Sidgwick rented and furnished a house at 74 Regent Street for Clough's college, but as the institution flourished, it became necessary to find their own premises. As a result, in 1875, Newnham Hall was constructed at Newnham, then a hamlet on the edge of the city of Cambridge.

The First Victory

From the founding of Newnham and Girton up until 1879, women were admitted to the Tripos (Final) examinations only by permission of the individual examiners. At issue that year was the proposal for the formal admission of women to the examinations and for the issuance of a certificate noting their accomplishments but with no formal degree. The Cambridge University Senate overwhelmingly supported the proposal. Eleanor Andrews, a student at Newnham, in a letter to her sisters at home, anticipated the next step: "When women get the Degrees (for this is only the thin end of the wedge) it will be nothing on this. We all feel it is the great crisis in the history of women's colleges."[15] She had no way of knowing another 69 years would elapse before the University of Cambridge allowed this.

The First Defeat

The first official attempt to give women degree status occurred in May 1897.[16] The modest proposal was to grant women "titular degrees." In this way, women could use the degree letters after their name and have unrestricted access to the university library, but would receive none of the rights, such as membership of the university, which accompanied degree status for men. The opponents mobilized voter support, going to the

---

[15]Andrews, E. A. (1988). Hurrah! we have won! In Phillips, A. (ed.), *A Newnham Anthology.* 2nd edn., Newnham College, Cambridge, p. 19.

[16]Shils, E. and Blacker, C. (1996). Preface. In Shils, E. and Blacker, C. (eds.), *Cambridge Women: Twelve Portraits.* Cambridge University Press, Cambridge, pp. xii–xiv.

lengths of chartering a special train on 21 May to carry non-resident M.A.s (mainly clergymen) from London Kings Cross to Cambridge. In addition, free lunches were provided by the men's colleges for those opposed to the motion.

The Senate of the university held the official ballot, the results being 662 votes in favour and 1,713 opposed. Gillian Sutherland described what ensued:

> Once the result of the ballot was announced, triumphant [male] under-graduates marched to Newnham [the closer woman's college], there to be faced down by the grim-faced women dons, assembled behind barred gates. They turned back to the construction of a huge celebratory bonfire in the Market Square, doing hundreds of pounds worth of damage.[17]

Trinity College and the "Steamboat Ladies"

Though the cause of women's rights had been thwarted at Cambridge, for a brief period, namely 1904–1907, women Oxbridge graduates were able to obtain a degree from the Trinity College, Dublin (TCD). It was pre-dominantly graduates of Girton and Newnham Colleges, Cambridge, and Somerville College, Oxford, who took advantage of the opportunity.

The first women students were admitted to TCD in January 1904.[18] For many years, Trinity had enjoyed close relationships with Cambridge and Oxford, in particular, the reciprocal recognition of examinations. Thus, it seemed unfair that Irish women who had, in the past, successfully completed the degree requirements at Cambridge and Oxford, should not have formal title as the women soon to be graduating from TCD would have. To right this wrong, the new Provost of Trinity College, Anthony Traill, a strong supporter of women's higher education, proposed that past Irish women who had completed their Oxbridge qualifications be granted TCD degrees. In June 1904, the Senate of TCD approved such a move

---

[17] Note 8, Sutherland, pp. 44–45.
[18] McDowell, R. B. and Webb, D. A. (1982). *Trinity College Dublin 1592–1952: An Academic History*. Cambridge University Press, Cambridge, pp. 344–353.

70 *Pioneering British Women Chemists: Their Lives and Contributions*

with a termination date of December 1907, by which time TCD would produce its first class of women graduates.[19]

Traill and the Senate had assumed that, among the past women Oxbridge graduates, only those of Irish ancestry would take advantage of the opportunity; but events proved otherwise. Word spread rapidly among British women. All the women's colleges at both Cambridge and Oxford encouraged their graduates to take advantage of the situation. The Clothmakers' Company, which had provided scholarships for students at Girton, Newnham, and Somerville, announced it would pay the Commencement (Graduation ceremony) fees of any Oxbridge women who wished to obtain a degree from TCD.

The whole expedition could be accomplished in as little as 34 hours using the fast mailboats. Women graduates often arranged to travel as a group of friends from the same graduation year in a college or from the same high school. A former student of Manchester High School for Girls (and later teacher at the school), Caroline Coignou reported: "It was a merry gathering; old friends who had not met since college days greeted each other."[20]

With this flood of women arriving at the gates of Trinity, attempts were made to abrogate the agreement, but Traill insisted the Senate could not renege on its commitment. He mollified the Senate by promising that all fees collected would be committed to the construction of a women's residence at Trinity.

During the near-4 year period this dispensation was in effect, approximately 700 Oxbridge women availed themselves of the opportunity to purchase a formal degree. Following the expiration of the agreement, Girton and Newnham students jointly petitioned the Trinity Senate to renew the privilege, but this was denied. Subsequent petitions, even one by two Indian princesses, were also dismissed.[21] The "window of opportunity" had closed.

---

[19]Parkes, S. M. (1996). Trinity College, Dublin and the 'steamboat ladies' 1904–1907. In Masson, M. R. and Simonton, D. (eds.), *Women and Higher Education: Past, Present and Future*. Aberdeen University Press, Aberdeen, 1996, pp. 244–250.
[20]"C.C." [Coignou, C.] (July 1905). Commencement at Trinity College, Dublin. *The Magazine of the Manchester High School* 50–51.
[21]Bailey, K. C. (1947). *A History of Trinity College Dublin 1892–1945*. The University Press, Dublin, p. 12.

The 1920s Battle

In 1920, two options were proposed: full University of Cambridge membership for women or a separate university for women. A joint Newnham and Girton Committee distributed leaflets that explained why the women's colleges deserved equal status and why a separate women's university was unacceptable.[22]

The debate became public in the Correspondence column of *The Times*. Support of women's rights came from many sources including two letters from Cambridge professors: one from F. Gowland Hopkins in Biochemistry (see Chapter 11) and the other being a stirring statement signed jointly by Ernest Rutherford in Physics and William Pope in Chemistry:

> For our part, we welcome the presence of women in our laboratories on the ground that residence in this University is intended to fit the rising generation to take its proper place in the outside world, where, to an ever increasing extent, men and women are being called upon to work harmoniously side by side in every department of human affairs.[23]

The Second Defeat

Despite the women's efforts, the first motion was defeated (as was the second). Embarrassed by Cambridge having now become the only university to bar women from degrees (Oxford having succumbed that same year, see the subsequent discussion), the Council of the Senate put forward a pair of linked motions, the first having provisos to appease some of the less rigid opponents. The first motion was that women be admitted to formal membership, provided that the number of women students was limited to 500 and that women would not be members of Senate. In the event this motion failed, then women would be admitted to the title of degree, but they would have no privileges. When the vote was taken in 1921, the first motion was defeated, while the second passed.

---

[22]Strachey, J. P. (1921). Admission to Membership. *Newnham College Roll Letter* 33–36.
[23]Rutherford, E. and Pope, W. (8 December 1920). Letter to the Editor. *The Times* (London) p. 8.

72 *Pioneering British Women Chemists: Their Lives and Contributions*

Following the defeat of the women's cause by the result of the first vote, the Reverend Pussy Hart aroused the assembled jubilant male undergraduates, over 1,000 strong, to head for Newnham College. Though the authorities expected some sort of demonstration after the vote, the size and speed overwhelmed them. The Memorial Gates of Newnham College were shut and barred, but the mob, using a hand cart, smashed the lower part of the gates.[24]

The aftermath was very different, as Mary Roberts (Mrs. Henn), a Newnham student, reported:

> Next day the University was steeped in gloom and guilt, to an extent I imagine seen neither before nor since. Apart from the official and collective apology of the University, Miss Clough was inundated with contrite statements from every conceivable group of [male] undergraduates, athletic and otherwise.[25]

Thus, another attempt at equal rights for women students had failed.

Success at Last

Twenty-seven years passed before the issue arose again. A Second World War had come and gone, while women at Newnham and Girton were behaving in every other way as if they were full members of the university. The issue had become an even greater embarrassment to the university authorities. Another report was drafted in 1947, and to the relief of all, its recommendations giving women equal rights passed without dissent.

With the arrival of 1948 and after 67 years of controversy, women finally had formal degree status with all the rights and privileges thereto attached. Dorothy Dale (Mrs. Thacker) described the new reality:

> For examinations we sat with the men students and our names were in the same alphabetical list as theirs. Degrees were conferred on us in the Senate House by the Vice-Chancellor and we became eligible for all

---

[24]Clough, B. A. (October 29, 1922). *Newnham College Roll Letter* 48–53.

[25]Henn, M. E. (1988). The breaking of the gates. In Phillips, A. (ed.), *A Newnham Anthology*. 2nd edn., Newnham College, Cambridge, pp. 150–151.

Cambridge and Oxford Women's Colleges 73

University posts and prizes; we were now full members of the University on an equal footing with men.[26]

Life of Women Students

Life in the women's colleges, particularly in the earliest years, was very circumscribed.[27] A former student, "M.L." described the regularity of life at Girton in 1888:

> The daily routine is — Prayers at eight, breakfast from eight to nine, study hours from nine to one, lunch when you like between twelve and three, noise hours from one to three, dinner at six, study hours again from half-past seven to nine, and then noise hours again from nine until half-past ten.[28]

The pioneering women students were often taught separately. To instruct Girton students, the lecturers had a considerable distance to travel with such lectures being given in the afternoon as their classes at the University in Cambridge itself were held in the morning. The Girton women therefore spent their mornings reading. Some lecturers already allowed mixed classes, another former student, "J.W.B" reporting:

> Notably among these are the Natural Science students, who have the reputation of never being in their rooms except during the evening, the afternoon generally being spent on practical work, either in Cambridge or the laboratory here.[29]

The publication of the Tripos results was the highlight of the year. There was a specific set of examinations for science students: the Natural

---

[26]Thacker, D. (1988). Women in laboratories. In Phillips, A. (ed.), *A Newnham Anthology*. 2nd edn., Newnham College, Cambridge, p. 80.

[27]Platt, H. (July 1952). Open letters and articles: Three old girls. *Magazine of the Manchester High School* 17.

[28]"M.L." (March 1888). Life at Girton. *Magazine of the Manchester High School*, 219; For a full account of daily life at Newnham at that period, see: Field, E. E. (20 June 1889). Newnham College, Cambridge. *The College, Dundee* 1(5): 133–138.

[29]"J.W.B." (1883). Girton College. *Magazine of the Manchester High School* 72.

Science Tripos, with a strong chemistry component.[30] In 1882, the success of women students in these science examinations provided especial excitement:

> The scene last year, when news was brought out that we had obtained *three* first classes in Natural Science, was a never-to-be-forgotten one; the wildest excitement ensued, bells were set ringing, and many other frivolous things were done in the heat of our enthusiasm.[31]

Even in the early 20th century, chaperone rules were enforced. The issue of modifying these rules arose in 1917, provoking a dispute between the "traditionalists" and "moderate reformers":

> We were discussing whether a Newnham student could be allowed to take an afternoon walk with an [male] undergraduate, and a senior member said, 'Well ... yes ... if she is engaged to him.' To which a friend of mine on the committee replied that it seemed a heavy price to pay for a walk.[32]

Women Chemistry Students

At Girton and Newnham, women students could learn of the principles of science, but the laboratory experience was another question altogether. There was already a shortage of laboratory space at Cambridge, and the intrusion of women only exacerbated the situation. During the 1870s, Philip Main generously organized early morning practical chemistry sessions for women students at his laboratory in St. John's College.[33] His classes ceased in 1879 with the completion of chemical laboratories at Newnham and Girton.

With their own facilities within the walls of the women's colleges, women chemistry students no longer needed chaperones to attend the

---

[30]Roberts, G. K. (1980). The liberally-educated chemist: Chemistry in the Cambridge Natural Science Tripos, 1851–1914. *Historical Studies in the Physical Sciences* **11**: 157–163.

[31]Note 29, "J.W.B.", p. 73.

[32]Wallas, M. G. (1988). A restless generation. In Phillips, A. (ed.), *A Newnham Anthology*. 2nd edn., Newnham College, Cambridge, p. 118.

[33]Gardner, A. (1921). *A Short History of Newnham College*. Cambridge University Press, Cambridge, p. 37.

laboratory sessions. In a letter to her mother in 1889, a Newnham student, Catherine Holt, reported on one day of chemistry:

> ... we got arsenic and phosphorus fumes at the [chemistry] lecture yesterday morning and a frightful smell of ammonia at the Laboratory afterwards. It was perfectly disgusting. All the same Chemistry is great fun and I did some splendid experiments yesterday.[34]

However, after 1913, upon the retirement of Ida Freund (see the next section), women chemistry students from Newnham had to undertake the experimental component of their courses in the University Chemistry Laboratories. The women at Girton were able to keep their own laboratory until Beatrice Thomas (see the respective section) retired in 1935.

In the University Chemistry Laboratories, an entirely different atmosphere prevailed as the arrival of women students was against the wishes of the male laboratory staff. They made life difficult for the women pioneers, as M. D. Ball remembered:

> ... the lab boys took a delight in leaving some essential bit of apparatus out of our lists so that we had to walk the whole length of the lab to the store to ask for it. An ordeal for some of us, especially as they appeared to be too busy to attend to us for several minutes while we waited at the door.[35]

Ida Freund

Though smaller, Newnham was regarded as the more science-oriented of the two Cambridge women's colleges. It was Newnham's own Chemistry Laboratory, in the middle of the college grounds, where Ida Freund[36] "reigned supreme." Born 5 April 1863 in Austria, Freund was raised by her maternal grandparents following the death of her mother. She studied at the State School and the State Training College for Teachers in

---

[34]Letter, Holt, C. D. to Holt, L. (31 October 1889). In Cockburn, E. O. (ed.), (1987). *Letters from Newnham College 1889–1892.* Newnham College, Cambridge, p. 13.

[35]Ball, M. D. (1988). Newnham scientists. In Phillips, A. (ed.), *A Newnham Anthology.* 2nd edn., Newnham College, Cambridge, p. 77.

[36]Wilson, H. (1988). Miss Freund. In Phillips, A. (ed.), *A Newnham Anthology.* 2nd edn., Newnham College, Cambridge, pp. 71–72.

76 *Pioneering British Women Chemists: Their Lives and Contributions*

Vienna.[37] Following her grandmother's death in 1881, Freund moved to Britain to join her uncle and guardian, the violinist, Ludwig Strauss.

It was arranged that Freund would be sent to Girton to complete her education, a decision that she bitterly opposed at the time. Completing her undergraduate studies in 1886, Freund accepted a 1-year Lectureship at the Cambridge Training College for Women (later Hughes Hall, see Chapter 10). The next year, she was offered a position as Demonstrator at Newnham and she was promoted to Lecturer in Chemistry in 1890.

Freund was mentioned by Holt in another letter to her mother:

> I attended my first lecture today; it was Chemistry; there were about 8 students from this college and three from Girton; ... Afterwards we adjourned for a couple of hours to the laboratory here; Miss Freund is the presiding genius, a jolly, stout German, whose clothes are falling in rags off her back. We made lots of horrible smells and got back here for lunch at a quarter past one.[38]

Hilda Wilson, who entered Newnham in 1905, recalled how every year, just before the examinations, Freund would summon her chemistry students for a study session.[39] For the 1907 session on the lives of important chemists, each student was provided with a large box of chocolates containing a written biography of a famous chemist. The following year when the Periodic Table was the focus, a large Periodic Table was provided with each element location consisting of an iced cake showing its name and atomic weight in icing, while the group numbers were made of chocolate and the dividing lines were rows of candy sticks.

During Freund's youth, a cycling accident necessitated the amputation of a leg, but this did not affect her mobility, as Ball recalled:

> Miss Freund was a terror to the first-year student, with her sharp rebuke for thoughtless mistakes. One grew to love her as time went on, though

---

[37](a) "E.W." [Walsh, E.] (1914). Ida Freund. *Girton Review*, 9–13; and (b) Ogilvie, M. B. (2004). Freund, Ida (1863–1914), *Oxford Dictionary of National Biography*. Oxford University Press (accessed 13 July 2018).

[38]Letter, Holt, C. D. to Holt, L. (12 October 1889). In Cockburn, E. O. (ed.), (1987). *Letters from Newnham College 1889–1892*. Newnham College, Cambridge, p. 11.

[39]Note 36, Wilson, p. 72.

we laughed at her emphatic and odd use of English. Yet how brave she was trundling her crippled and, I am sure, often painful body about in her invalid chair smiling, urging, scolding us along to "zat goal to which we are all travelling which is ze Tripos."[40]

Freund's most renowned work was a chemistry text, *The Study of Chemical Composition*,[41] which remained popular for many years.[42] The historian of chemistry, M. M. Pattison Muir, commented that her text "is to be classed among the really great works of chemical literature."[43] The book itself was reprinted in 1968 as a classic in the history of chemistry.[44] Freund also wrote a manual of laboratory procedures, *The Experimental Basis of Chemistry*, to be used to illustrate chemical concepts.[45]

Despite her mobility challenges, Freund became a fervent traveller, "wheelchairing" her way around England, Scotland, Germany, Austria, Switzerland, and Italy. She died on 15 May 1914 and, following her passing, the Ida Freund Memorial Fund was instituted at the University of Cambridge to further train women teachers in the physical sciences.

Dorothy Marshall

During the 1897–1906 period, Dorothy Blanche Louisa Marshall[46] was the Demonstrator in the Girton Chemistry Laboratory. Marshall, born

---

[40] Note 35, Ball, p.76.

[41] Freund, I. (1904). *The Study of Chemical Composition. An Account of its Method and Historical Development, with Illustrative Quotations*. The University Press, Cambridge.

[42] Hill, M. and Dronsfield, A. (2004). Ida Freund — Pioneer of women's education in chemistry. *Education in Chemistry* **41**(5): 136–137.

[43] Cited in Gardiner, M. I. (1914). In Memoriam: Ida Freund. *Newnham College Roll Letter* 34–38.

[44] Note 41, Freund, I. (1968). Reprint edition, Dover Publications, New York.

[45] Freund, I., Hutchinson, A. and Thomas, M. B. (eds.), (1904). *The Experimental Basis of Chemistry; Suggestions for a Series of Experiments illustrative of the Fundamental Principles of Chemistry*. The University Press, Cambridge.

[46] (a) Butler, K. T. and McMorran, H. I. (eds.), (1948). *Girton College Register, 1869–1946*. Girton College, Cambridge, p. 638; and (b) Creese, M. R. S. (1998). *Ladies in the Laboratory: American and British Women in Science 1800–1900*. Scarecrow Press, Lanham, Maryland, p. 265.

78 *Pioneering British Women Chemists: Their Lives and Contributions*

12 December 1868 in Mayfair, London, was the daughter of Julian Marshall, connoisseur and collector, and Florence Ashton Thomas, musician and author. Marshall was educated at King Edward VI High School for Girls, Birmingham (KEVI), and then went to Bedford College, London, in 1886. From there, she transferred to University College, London (UCL), studying chemistry, physics, and electrical technology and graduating with a B.Sc. (Hons.) in 1891.

Staying at UCL as a postgraduate researcher until 1894, Marshall studied heats of vaporisation of liquids, her work resulting in three lengthy publications, one of which was co-authored with William Ramsay. Following a 1-year Demonstratorship at Newnham, in 1896, she was appointed Demonstrator in Chemistry at Girton, and then promoted to Resident Lecturer in 1897.

Marshall left Girton in 1906 to become Senior Science Lecturer at Avery Hill Training College. Avery Hill did not, as it happened, teach science until 1930.[47] Soon after being appointed Acting Principal in February 1907, Marshall became ill (as both her predecessor and successor as Principal resigned due to illness, it is possible that "illness" was a way out of a thankless position). Following her resignation, Marshall became Senior Science Mistress of Huddersfield Municipal High School in 1908. She moved south in 1913, taking a position as Chemistry Mistress at Clapham High School. Like so many other women chemists (see Chapter 15), Marshall started war work in 1916. She was appointed Scientific Research Assistant at the National Physical Laboratory (NPL), where she stayed until she was required to retire in 1930, aged 62. Marshall died on 26 February 1956 at Teddington Middlesex.

Beatrice Thomas

In the Girton Laboratory, Marshall was followed by another former KEVI student, Mary Beatrice Thomas.[48] Just as Freund had "reigned supreme"

---

[47] Shorney, D. (1989). *Teachers in Training, 1906–1985: A History of Avery Hill College.* Thames Polytechnic, London, p. 53.

[48] (a) "D.M.P." (Michaelmas Term 1954). In Memoriam: Mary Beatrice Thomas, 1873–1954. *Girton Review*, 14–24; and (b) "D.M.A." [Adcock, D. M.] (1955). Mary Beatrice Thomas, 1873–1954. *Newnham College Roll Letter* 37–38.

at Newnham, so her close friend Thomas was a dominant figure in chemistry at Girton from 1906 to 1935. Thomas was born on 15 October 1873 in Aston, Birmingham, daughter of "Wild William" Thomas, a surgeon and later Professor of Anatomy at Mason College, Birmingham and Mary Thomas.

With her father's support, but opposition from her mother, Thomas followed the well-trod trail from KEVI to Newnham in 1894 where she studied chemistry and physiology, completing the degree requirements in 1898. After graduation, she was a Demonstrator in Chemistry for 2 years at the Royal Holloway College (RHC); the subsequent year, she held a Priestley Scholarship at the University of Birmingham. From 1902 to 1906, she was Demonstrator in Chemistry at Girton College, under Marshall. Following Marshall's departure, Thomas was appointed Lecturer and Director of Studies in Natural Science at Girton, a post that she held until retirement in 1935.

Initially, Thomas was able to undertake some research, but in later years, with the pressure of teaching and a hostility to women in the University Chemistry Research Laboratory, she focussed her life on her students. A student, "D.M.P." described how Thomas was a scientist and a feminist:

> ... an eager devotee of the gospel of dedication to that search for truth and for scientific knowledge which, at the end of the Victorian age, had for many ardent spirits taken the place of orthodox religion. We found her unyielding in her demands that we should put academic tasks before all other interests, that we should maintain the tradition of the earnest pioneers, who never for one moment allowed themselves to forget their goal of proving and demonstrating that no distinction can be drawn between the intellectual powers of men and women, once they are offered similar educational opportunities.[49]

Thomas, who had been a chemistry student under Freund, became close friends with her former Demonstrator. Upon Freund's death, it was Thomas who organized the Ida Freund Memorial Fund and co-edited Freund's unfinished *Experimental Basis of Chemistry* (see Footnote 45).

---

[49] Note 48(a), "D.M.P.", p. 17.

80 *Pioneering British Women Chemists: Their Lives and Contributions*

Though she taught at larger classics-oriented Girton, her sympathies lay more with the smaller, science-oriented Newnham. There must have been some sharing between the two women's colleges for one of her obituarists, a former Newnham student, "D.M.A.", commented: "Fortunately for Newnham science students, Miss Thomas was able to undertake our supervision in Chemistry so that we were able to appreciate her qualities as a teacher."[50]

Known as "Tommy" by her students, Thomas demanded the highest standards. She was never perturbed by any laboratory incident. In a touching posthumous letter to Thomas, Dorothy Russell remembered:

> I recall how, on one occasion, a moth elected to commit suicide in my flask of standardised acid. This naturally did not escape your eye! You regarded the corpse with considerable disfavour and merely said: "Of course that acid will have to be re-standardised."[51]

Thomas was an ardent supporter of women's rights. However, women at Cambridge had to be careful not to let their sympathies become too obvious as "D.M.P." noted in her obituary of Thomas:

> The position of women in the University was uncertainly poised, and it was still very necessary to avoid adverse criticism, so no member of the women's colleges could join the W.S.P.U. [Women's Social and Political Union] — the 'Militant Suffragettes' led by Christabel Pankhurst — but most of us belonged to the N.U.W.S.S. [National Union of Women's Suffrage Societies] which pursued the same aims by non-violent and more legitimate methods.[52]

When Thomas purchased a house, she found an avenue to make a personal protest. The "passive resisters," including Thomas, refused to pay the education rates levied on householders, on the basis of non-representation. As "D.M.P." described: "Time after time, ... she submitted to the arrival

---

[50] Note 48(b), "D.M.A.", p. 38.

[51] Russell, D. S. (Michaelmas Term 1954). Posthumous Letter to Miss M. B. Thomas. *Girton Review* 25.

[52] Note 48(a), "D.M.P.", p. 21.

of the bailiffs and ensuing sale of her goods, without offering obstruction, but with signal success in making the authorities appear ridiculous."[53]

Cambridge had been Thomas's life and even in retirement she welcomed visits from her former students and from the daughter of Ida Smedley (see Chapter 9), who called her "Aunt Bea." Also, she attended meetings of the Faculty of Physics and Chemistry regularly until her health declined further. Her rebelliousness continued in altercations with nationalised gas services, doctors, and any other figure of authority who crossed her path, until her death on 14 June 1954.

Mary Johnson (Mrs. Clark)

Mary Johnson,[54] a graduate of Newnham, became one of the first women researchers in the Cambridge Chemistry Laboratories. Born on 14 January 1895 at Durham, she was the daughter of John Wright Johnson, coal merchant and quarry owner, and Betsy Eleanor Burdis. Educated at Bede College School, Sunderland, Johnson entered Newnham in 1913. She had actually wanted to study mathematics, but inadequate preparation at school resulted in a change of plans and she took chemistry instead. When she completed Part II of the Natural Science Tripos in 1917, she was placed above all the men in her year.

Johnson was awarded a Newnham Bathurst Research Scholarship in 1917–1918 for research in organic chemistry, but she could not be registered for a Ph.D. as, at the time, women were ineligible for higher degrees at Cambridge. From 1920 to 1925, she was a researcher in the Cambridge Chemistry Laboratories holding various research Fellowships. Her work on the dyestuffs of the pyrazolone series, undertaken with Charles Stanley Gibson,[55] was published in 1921.

According to family lore, being a senior in the laboratory, Johnson was asked who she wanted to work across the bench from her. She

---

[53] Note 48(a), "D.M.P.", pp. 21–22.

[54] White, A. B. (ed.), (1979). *Newnham College Register, 1871–1971*, Vol. I, *1871–1923*, 2nd edn., Newnham College, Cambridge, p. 37.

[55] Simonsen. J. L. (1950). Charles Stanley Gibson, 1884–1950. *Obituary Notices of Fellows of the Royal Society* **7**: 114–137.

82 *Pioneering British Women Chemists: Their Lives and Contributions*

responded that she didn't care: "so long as he is a good tidy worker,"[56] and in due course, Leslie Marshall Clark was assigned to the bench. In 1925, they announced their engagement and Johnson submitted her resignation. They were married on 26 February 1926 and had two daughters, both of whom followed in her footsteps to Newnham.

Clark was running a research laboratory for Imperial Chemical Industries at the beginning of the Second World War. According to family history, he was quoted as saying that: "since she was a better chemist than he was, he would have her in his laboratory like a shot, if it wasn't for the children."[56] Later in the War, Johnson briefly taught chemistry at Howell's School, Bangor, in North Wales. Unfortunately, she had a massive stroke in 1945 which severely incapacitated her until her death on 26 April 1969 at Cambridge.

Sosheila Ram

Of all the early chemistry graduates of the Cambridge women's colleges, one of the most remarkable was Sosheila Ram.[57] Ram, born 29 November 1899 in Edinburgh, was the daughter of Labbhu (Murgai) Ram, a doctor, and Sukhda Ram. She was educated at Mexborough Secondary School, Yorkshire, and entered Newnham in 1918. Completing the first part of the Natural Science Tripos in 1921, she spent 1921–1922 at St. Mary's Training College, London, from which she received a Certificate in Education from the University of London. Then she travelled for her first time to India where she was appointed Lecturer in the Department of Science at the Lady Hardinge Medical College for Women.

The college had been named after Lady Winifred Hardinge, wife of the then-Viceroy of India. In 1912, she had proposed that a college for the medical education of women should be founded in Delhi, the new capital of India.[58] Sadly, Lady Hardinge died before completion of the building,

---

[56]Curriculum vitae and family history supplied by Margaret Spufford (daughter), (1998). Newnham College Archives.

[57]White, A. B. (ed.), (1979). *Newnham College Register, 1871–1971*, Vol. I, *1871–1923*, 2nd edn., Newnham College, Cambridge, p. 292.

[58]Mathur, N. N. (1998). Indian Medical Colleges: Lady Hardinge College, New Delhi. *National Medical Journal of India* **11**(2): 97–100.

Cambridge and Oxford Women's Colleges 83

and upon its opening in 1916, it was named after her as the Lady Hardinge Medical College (LHMC) and Hospital, with an associated Nursing School.

Alice Bain (see Chapter 7) was appointed the first Professor of Chemistry at LHMC, then when Bain left in 1923, Ram was awarded the position. Ram returned on two occasions to Newnham, where she continued her studies, being granted an M.A. in 1925. In 1939, Ram became a Consulting and Analytical Chemist with Gindlay & Co., Delhi, the first woman of Indian heritage to establish and run her own laboratory. Her research in analytical chemistry, some of which was published in *The Analyst*, resulted in her election to Fellowship of the Royal Institute of Chemistry in 1945. In addition to her new career, she resumed teaching at the LHMC as a Lecturer in Biochemistry for the 1942–1944 period.

Throughout her time in India, Ram had endeavoured to further the education of women and to spread knowledge of modern dietetics. Thus, in 1953, she decided to devote the rest of her life to social work at remote foothill communities in the Himalayas, using Dagshai as a base. This outreach work was to lead to her death on 28 March 1957 as the following account describes:

> Her social work won for her the love and gratitude of many, both men and women, but her interest in prohibition brought her the animosity of a few individuals. One moonlight night, on her return from a visit of mercy to a village home, amidst the glorious scenery of the Simla Hills, she was ambushed and most cruelly molested. Within a few weeks she was dead.[59]

Delia Simpson (Mrs. Agar)

The most prominent woman researcher in the Cambridge Chemistry Laboratories was the spectroscopist, Delia Margaret Simpson.[60] Born on

---

[59] Anon. (1958). Sosheila Ram, 1899–1957. *Newnham College Roll Letter* 45–46.

[60] (a) White, A. B. (ed.) (1981). *Newnham College Register, 1871–1971*, Vol. II, *1924–1950*, 2nd edn., Newnham College, Cambridge, p. 83; (b) Newton, A. A. (1999). Delia Agar, 1912–1998. *Newnham College Roll Letter* 101–103; (c) Lynden-Bell, R. M. and Sheppard, N. (1999). The scientific work of Delia Agar. *Newnham College Roll Letter* 103–105; and (d) Mason, J. (1999). Delia Agar. *Newnham College Roll Letter* 105.

84 *Pioneering British Women Chemists: Their Lives and Contributions*

5 February 1912 in Greenwich, the daughter of Robert Simpson, sanitary inspector, and Delia Maud Pope, teacher, she was educated at the Haberdashers' Aske's Hatcham Girls' School. Simpson entered Newnham in 1930, completing her studies in 1934.

Simpson was a Bathurst Research Student from 1934 to 1936, following which, with a Travelling Scholarship, she spent the years 1937–1939 undertaking research at the University of Vienna. Her initial research had been in the field of optical rotary dispersion and circular dichroism, but she discovered that she found spectroscopy much more interesting.

Before the Second World War, Simpson had studied the spectra of small molecules in the vacuum ultraviolet as part of the spectroscopy group. A feature of the pre-war days was a stimulating weekly meeting of spectroscopists organized by Gordon Sutherland in Sutherland's rooms in Pembroke College, Cambridge. Norman Sheppard,[61] Sutherland's obituarist, recalled:

> For this purpose the 'group' included also Dr (later Professor) W. C. Price and Dr Delia M. Simpson (later Mrs Agar) who were carrying out research in vacuum ultraviolet spectroscopy. Talks were given about work in progress and significant recent spectroscopic publications were reviewed.[62]

With the onset of war, Simpson shifted her focus to infrared spectroscopy. This technique was used to analyse samples of enemy fuels for the Air Ministry, in particular to identify the use of synthetic oils in place of natural oil, which, because of the naval blockades was in very short supply in Germany. This war work was undertaken with the women chemists of Bedford. As we will mention in Chapter 5, Bedford personnel and students had been evacuated to Cambridge, and for the duration of their stay, Simpson held the position of Demonstrator and Assistant Lecturer in Chemistry for Bedford College. Following the return of the Bedford

---

[61] Griffiths, P. (2015). In Memoriam: Norman Sheppard. *Applied Spectroscopy* **69**: 160A–162A.

[62] Sheppard, N. (1982). Gordon Brims Black McIvor Sutherland. 8 April 1907–27 June 1980. *Biographical Memoirs of Fellows of the Royal Society* **28**: 588–626.

students to London in 1944, she was appointed College Lecturer in Chemistry at Newnham and elected to a Teaching Fellowship.

Simpson continued research after the War, working for a time with biologists on fluorescence spectroscopy. Her major focus continued to be infrared and Raman spectroscopy, resulting in a total of 20 papers over her career including a two-part review of the spectra of hydrocarbons in *Quarterly Reviews* of the Chemical Society, co-authored with Sheppard.[63] However, in 1954, she was appointed Director of Studies in Natural Sciences, which required her to spend most of her time teaching at Newnham and Downing Colleges. In 1952, Simpson had married the electrochemist John Newton Agar, a Fellow of Sydney Sussex College, Cambridge. After retirement in 1977, Simpson continued to attend spectroscopy seminars until close to her death on 29 March 1998, aged 86.

Rosemary Murray

Murray's span covers the latter part of our time frame and her story shows how far women chemists had come since the opening of universities to the first women students. Though her university education was at Oxford, her subsequent career flourished upon her move to Cambridge, hence the inclusion of her life story in this section.

Alice Rosemary Murray[64] was born on 28 July 1913 at Havant, Hampshire, daughter of Admiral Arthur John Layard Murray and Ellen Maxwell Spooner, and grand-daughter of William Spooner, Warden of New College, Oxford. She was educated at Downe House, Newbury, and entered Lady Margaret Hall, Oxford, to study chemistry, completing a B.Sc. in 1936 and a D.Phil. in 1938.

Murray was appointed Assistant Lecturer at RHC in 1939, but early in 1940 she was granted leave of absence to join the Admiralty Research

---

[63] Sheppard, N. and Simpson, D. M. (1952). The infrared and Raman spectra of hydrocarbons I: Acetylenes and olefins. *Quarterly Reviews London* **6**: 1–33; Sheppard, N. and Simpson, D. M. (1953). The infrared and Raman spectra of hydrocarbons II: Paraffins. *Quarterly Reviews London* **7**: 19–55.

[64] Wilson, A. (2014). *Changing Women's Lives: A Biography of Dame Rosemary Murray*. Unicorn Press, London.

86 *Pioneering British Women Chemists: Their Lives and Contributions*

Department at Portsmouth.[65] During that summer, she obtained a post as Lecturer in the Chemistry Department at the University of Sheffield, where she also carried out research in organic chemistry for the Ministry of Supply. In 1942, finding life at Sheffield frustrating, Murray joined the Women's Royal Naval Service, undertaking a variety of roles and rising to the rank of Chief Officer.

In 1946, Murray received a telegram from the Mistress of Girton College asking if Murray would like to be interviewed for a position as chemist. As Murray told her biographer, Karen Walton:

> Although I had not considered what I would do after the war, this seemed too good an opportunity to miss, so off I went in Wren uniform to Cambridge and had various interviews and was shown the chemistry laboratories by the professor. I gathered later that I caused quite a stir among the technicians — a wren officer in uniform going round the chemistry lab.[66]

Initially appointed Lecturer in Chemistry, Murray was promoted in 1949 to a Fellowship, and Tutor and Demonstrator in Organic Chemistry. In that period, concerns were raised that there was an 11:1 ratio of men to women undergraduates at Cambridge. To help redress the balance, the suggestion was circulated that there was a need for another women's college. The University gave permission and Murray was chosen as Tutor in Charge. Starting New Hall in 1954 with 16 students in a converted guesthouse, her obituarist, Alison Wilson, commented:

> Murray's aim was to widen access to all bright girls, regardless of back-ground and schooling. Some students found her austere, but those who penetrated her reserve found her the kindest of women, and a wonderful teacher. ... Despite financial difficulties and the climate of student unrest in the late Sixties, Murray did not buckle under pressure, but maintained a positive outlook. Students and Fellows endured Spartan conditions

---

[65]Anon. (November 1940). *College Letter, Royal Holloway College Association*, 12.

[66]Walton, K. D. (1996). In retrospect: Dame Rosemary Murray. In Walton, K. D. (ed.), *Against the Tide: Career Paths of Women Leaders in American and British Higher Education*. Phi Delta Kappa Educational Foundation, Bloomington, Indiana, pp. 167–175.

because heating and lighting were kept to the minimum to save expense, but how could they complain when their leader (designated President in 1964) was living in the same circumstances?[67]

In 1972, New Hall was formally recognized as a College of the University of Cambridge.[68] Murray held the post of Vice-Chancellor of the University of Cambridge from 1975 until 1977, the first woman to hold this position. She was made Dame Commander of the Order of the British Empire in 1977. Murray remained President of New Hall until 1981. She died on 7 October 2004.

Oxford Women's Colleges

Though the authorities at Oxford were almost as recalcitrant as those at Cambridge, the progress of women's rights were not accompanied by the sort of demonstrations of overt hostility that occurred twice at the gates of Newnham. The first event in the Oxford saga was initiated in 1865 by Mark Pattison, Rector of Lincoln College, a forgotten early advocate of women's education. As Josephine Kamm described:

> ... [he] had informed the Schools' Inquiry Commission that he and several other men and women had started classes in Oxford for girls of seventeen and eighteen and upwards who had left school, and he also mentioned an earlier experiment of the same kind which had been started by the Midland Institute, Birmingham. His aim, he said, was the establishment of regular classes — or a college — for girls in every large town in the country, with women in the majority on the governing body.[69]

Such an eventuality never came to be and it was not until the 1870s that the "women issue" first arose at Oxford. Balliol and Worcester Colleges of the

---

[67]Wilson, A. (18 October 2004). Obituaries: Dame Rosemary Murray, First woman to be Vice-Chancellor of Cambridge University. *The Independent* (accessed 17 July 2018).

[68]Murray, M. (1980). *New Hall, 1954–1972: The Making of a College*. New Hall, Cambridge.

[69]Kamm, J. (2010). *Hope Deferred (Routledge Revivals): Girls' Education in English History*. Taylor & Francis, London, p. 181.

88 *Pioneering British Women Chemists: Their Lives and Contributions*

University of Oxford were offering scholarships to school students based on the results of the Senior Local examinations. A student by the name of A. M. A. H. Rogers was placed in the First Class of Junior candidates in 1871, and then headed the Senior candidate list in 1873 (see Footnote 69). Henry Daniel of Worcester College wrote to Rogers' father offering a scholarship to Worcester College for the successful student. Informed that the student in question was Annie Mary Anne Henley Rogers, the offer was hastily withdrawn. As Vera Brittain commented: "Thus, for the first time in history, the question of admitting women to Oxford University was raised. It was, of course, dropped as hurriedly as a red-hot coal ..."[70]

Signing herself always, "A. M. A. H. R.," Rogers was to subsequently triumph and to continue her role as a thorn in the flesh of the powers-that-were at Oxford, becoming an expert in the University Statutes, particularly as they pertained to women's rights.[71] Brittain observed: "If the women at Oxford could be said to owe their triumph to any one individual, the credit is hers. She was their forerunner, their expert, their champion, and the symbol of their struggle."[72]

There was an interesting parallel between the cause of women's education at Cambridge and at Oxford. At Cambridge, as we have seen, Henry Sidgwick had been a key "agent of change." At Oxford, it was Henry's younger brother, Arthur Sidgwick,[73] and his wife, Charlotte Wilson, who were at the centre of the battle.[74]

Founding of the Colleges

The first of the Oxford women's colleges, Lady Margaret Hall (LMH), opened its doors in 1879. It was founded by Elizabeth Wordsworth,

---

[70] Brittain, V. (1960). *The Women at Oxford: A Fragment of History.* George G. Harrap, London, p. 18.

[71] Rogers, A. M. A. H. (1938). *Degrees by Degrees: The Story of the Admission of Oxford Women Students to Membership of the University.* Oxford University Press, Oxford.

[72] Note 70, Brittain, p. 19.

[73] Howarth, J. (2004). Sidgwick, Arthur (1840–1920). *Oxford Dictionary of National Biography.* Oxford University Press (accessed 1 February 2019).

[74] Howarth, J. (1984). 'In Oxford but ... not of Oxford': The Women's Colleges. In Brock, M. G. and Curthoys, M. C. (eds.), *The History of the University of Oxford,* Vol. VII, *Nineteenth-Century Oxford,* Part 2, Clarendon Press, Oxford, pp. 237–311.

great-niece of the poet William Wordsworth, and it was named after Lady Margaret Beaufort, a learned medieval noblewoman and mother of King Henry VII. LMH was the first venture into higher education by Anglican high churchmen, and all students admitted were required to be Anglicans.[75] Ethel Harvey, a student in 1903, described how every morning there were mandatory prayers at 8.05 a.m. in the chapel and then prayers again at 8:30 p.m. She added: "Everybody is expected to attend Divine Service at least once on Sundays, and Miss Wordsworth's Bible Class after dinner."[76]

St. Hugh's Hall (named after Hugh of Avalon, Bishop of Lincoln in the 12th century) was founded in 1886 also by Elizabeth Wordsworth.[77] Wordsworth was concerned that daughters of impecunious families, such as clergy, could not afford the fees of LMH, and therefore had to attend a non-religious based college: Somerville, Newnham, or Girton. She hoped the college would provide a steady supply of church workers and teachers. Students at St. Hugh's attended the same classes and programs as those at Lady Margaret Hall and underwent equivalent religious instruction.

St. Hilda's was opened in 1893 as another Anglican College. The founder was Dorothea Beale of Cheltenham Ladies' College (CLC) fame (see Chapter 2), who chose the name in recognition of the 7th century St. Hilda of Whitby, head of the most important place of learning for women of its time. There were close ties between the institutions, with CLC having representatives on the St. Hilda's Council until 1955 and, up to 1962, about 10% of all students ever at St. Hilda's were Cheltonians.[78]

In the late 19th century, many Anglicans had deep reservations about the higher education of women, and this was reflected in an early emphasis on training to women for teaching, social, or missionary work. All three of the Anglican Colleges suffered financially, as Janet Howarth commented:

---

[75]Howarth, J. (1994). Women. In Harrison, B. (ed.), *The History of the University of Oxford,* Vol. VIII, *The Twentieth Century.* Clarendon Press, Oxford, pp. 345–377.

[76]Harvey, E. (July 1903). College letter. *Magazine of the Manchester High School* 70–73.

[77]Schwartz, L. (2011). *A Serious Endeavour: Gender, Education and Community at St. Hugh's.* Profile Books, London.

[78]Rayner, M. E. (1993). *The Centenary History of St. Hilda's College, Oxford.* St. Hilda's College, Oxford, p. 110.

Anglican ambivalence towards women's higher education had various consequences for the Church of England halls at Oxford. One was that they were much poorer than Girton, Newnham or Somerville: church people, although they had given generously to Keble, were slow to respond to appeals for money to bring women students to Oxford.[79]

When LMH was founded, the religious requirement was a deterrent to potential non-Anglican women students. To provide an alternative, Somerville College was opened in 1879, being named after the pioneering woman scholar and scientist, Mary Somerville.[80] The supporters of Somerville, which included William Sidgwick (the eldest of the Sidgwick brothers) and the chemist A. G. Vernon Harcourt, insisted that there be no religious tests or obligations for Somerville women students. In fact, it transpired that over one-third of the early students at Somerville were nonconformists.[81] This made Somerville unique among the Oxford women's colleges. In fact, Somerville seemed closer philosophically to Newnham and Girton at Cambridge than to Somerville's sister Oxford colleges. For example, Newnham, Girton, and Somerville were the three colleges from which nearly all the "steamboat ladies" came. Somerville, with its links to university liberalism, was the first to offer opportunities for research to Oxford women students.

The Oxford colleges did not serve the same role as those at Cambridge, as A. Mary Baylay reported back in 1901 to students at North London Collegiate School (NLCS) in *Our Magazine*:

> At Oxford, unlike Cambridge, most of the lectures attended are in the men's Colleges, we are coached by tutors of the University, we read in the Bodleian and Taylorian; in the final dread week we assemble in the lobby of the Schools with pallid men in black coats and white ties...[82]

---

[79]Howarth, J. (1986). Anglican perspectives on gender: Some reflections on the centenary of St. Hugh's College, Oxford. *Oxford Review of Education* **12**(3): 299–304.

[80]Adams, P. (1996). *Somerville for Women: An Oxford College 1879–1993*. Oxford University Press, Oxford, pp. 38–40.

[81]Note 75, Howarth, p. 346.

[82]Baylay, A. M. (July 1901). Some aspects of life in the women's colleges at Oxford. *Our Magazine: North London Collegiate School for Girls* **26**: 42.

The necessity to travel to the men's colleges for lectures and laboratories put a particular burden on students at Lady Margaret Hall, which, like Girton at Cambridge, was sited about a half-hour outside of the city. As a result, the "new women" with their bicycles had a distinct advantage, as Isobel Sargeant recalled in the *Magazine of the Manchester High School*: "One's morning energy in cycling grimly to all the lectures recommended by an optimistic tutor is only equalled by one's evening energy in cycling even more grimly to all the 'right' societies."[83]

University Degrees for Women

On Trinity Sunday, 1884, Dean Burgon preached a sermon in New College Chapel which closed with a message for women: "Inferior to us God made you, and inferior to the end of time you will remain. But you are not the worse off for that."[84] Nevertheless, there were no dramatic incidents on the pathway towards the acceptance of women students at Oxford as there was at Cambridge. Brittain contended that the Oxford method of publishing graduation lists in alphabetical order rather than the Cambridge method of descending marks meant that women who excelled at Oxford were not so obvious.[85]

By the end of 1894, university examinations had been opened to women. However, the offering of a formal B.A. degree to women was far more contentious. When the matter came to a vote on 3 March 1896, the women's cause was defeated in Congregation, the governing body, by 215 votes to 140.[86] Though the vote took place in a civil manner, Rogers, by then Tutor in Classics at St. Hugh's, observed: "... we have more actual enemies to the firm establishment of women's education in Oxford than we thought."[87] A second unsuccessful attempt was made in 1908. However, a small victory occurred on 1 November 1910 when the women's colleges

---

[83] Sargeant, I. (July 1952). Open letters and articles: Three old girls. *Magazine of the Manchester High School* 17–18.

[84] Note 70, Brittain, p. 69.

[85] Note 70, Brittain, p. 70.

[86] Note 74, Howarth, p. 266.

[87] Note 74, Howarth, p. 267.

92 *Pioneering British Women Chemists: Their Lives and Contributions*

were granted the title of Recognized Societies, and the previously invisible women students were accepted as Registered Women.[88]

It was not until May 1920 that, without opposition, the final statute passed to permit women to graduate with full membership of the university. That Parliament had passed the Sex Disqualification (Removal) Act in 1919 was certainly no coincidence, just as it had been in 1920 when the Chemical Society finally admitted women chemists (see Chapter 9).

But this was not the end of the story. On 14 June 1927, fearful of the rising proportions of women, Congregation debated and passed a motion by 229 votes to 164 that: "The University has a right to remain predominantly a men's university."[89] This motion limited the number of women students in residence to 840. The quota for women was raised to 970 in 1948 but not abolished until 1957. In addition, the motion prohibited the foundation of any new women's college if it would make the proportion of women greater than 20%.

The vote seemed to be an example of the anti-feminist wave of the late 1920s and early 1930s, which we will examine in Chapter 16. As cited by Howarth: "the 'fundamental question', as one friendly male don had put it, was 'feminine influence and the *monstrous regiment of women*'."[90] In other words, many Oxford men saw the increase in numbers of women students as destroying the essential masculine identity of the university.

Women Chemistry Students

It was Harold Baily Dixon[91] who, in the 1880s, first raised the issue of admission of women chemistry students to his regular lectures. It is likely that his concern originated from being raised by a feminist mother. Dixon's obituarist commented:

---

[88]Rogers, A. M. A. H. (October 1911). The present position of women at the University of Oxford. *Journal of Education* **33** (new series): 677.

[89]Note 75, Howarth, p. 356.

[90]Letter, Tod, M. N. to Craig, E. S. 9 December 1926. Cited in Howarth, Note 73, p. 357.

[91]"H.B.B. and W.A.B." (1931). Obituary Notices: Harold Baily Dixon. *Journal of the Chemical Society* 3349–3368.

His mother was a woman of innate good taste and manners, with advanced views upon the subject of women's suffrage; she went by herself to Ibsen's plays, when they were produced in London, and called in a lady doctor when her youngest son was born.[92]

Dixon's request for women's admission was at first rejected by the Oxford college administrators. Then, chemist Vernon Harcourt, a powerful figure in the Oxford hierarchy, intervened on Dixon's behalf. The authorities relented in 1886, though stating that: "... but only by special permission in each case, and with the accompaniment of some elderly person."[93] Dixon had no problem with this decree (apart from the 'elderly') as his wife, Olive Beechey Hopkins, attended all his lectures and she acted as chaperone to the women students in his class.[92]

There were a significantly lower proportion of women students in the early years studying the Natural Sciences at the four residential Oxford colleges than at the two Cambridge colleges. For example, between 1881 and 1913, the proportion of women students studying science at Cambridge was in the 12%–14% range, while at Oxford, it was more generally in the range of 3%–4%. Howarth has shown that the reasons for the weakness of science at Oxford were complex, though she noted that chemistry at Oxford was one of the few stronger sciences.[94]

In the early years, Somerville was the home of most of the Oxford women chemists. This may have been due to students with non-traditional goals being more likely to have come from nonconformist backgrounds. In addition, the influence of co-founder Vernon Harcourt and the presence on the Somerville Council of a series of science professors' wives[95] would have encouraged a scientific focus. Somerville was unique at Oxford in having a women's Scientific and Philosophical Society, which flourished from 1892 until 1897.[96] The Somerville students set up a small

---

[92] Note 91, "H.B.B. and W.A.B.", p. 3350.

[93] Bowen, E. J. (1970). The Balliol–Trinity Laboratories, Oxford 1853–1940. *Notes and Records of the Royal Society of London* **25**(2): 227–236.

[94] Howarth, J. (April 1987). Science education in Late-Victorian Oxford: A curious case of failure? *English Historical Review* **102**: 334–371.

[95] Note 74, Howarth, p. 283.

[96] Note 80, Adams, pp. 128–129.

94 Pioneering British Women Chemists: Their Lives and Contributions

science museum and held meetings with a speaker followed by appropriate experiments or practical demonstrations. The Society even published an annual record of its learned proceedings.

Mary Watson

The first two women students at Oxford studying chemistry were Margaret Seward (Mrs. McKillop) (see Chapter 4) and Mary Watson[97] both Somervilleans. Watson, born in October 1856 at Thame, Oxfordshire, was the daughter of John Watson, land agent and farmer, and Anne Bruce. She was educated at home and at St. John's Wood High School. Watson entered Somerville in 1879 on a Clothworkers' Scholarship, completing a first in geology in 1882 and a second in chemistry in 1883. Following graduation, she was appointed Science Mistress at CLC. Watson held that position until 1886 when she had to resign following her marriage to John Styles, Headmaster at Cheltenham Grammar School. They retired to Hampshire, where she died on 20 February 1933.

Mary Rich

A couple of years behind Seward and Watson was Mary Florence Rich. Rich[98] was born on 18 October 1865 in Weston-super-Mare, the daughter of Thomas Rich, chemist (pharmacist) and later cocoa fibre manufacturer, and Elizabeth Cuddeford. She was educated at Haberdashers' Aske's Hatcham School for Girls and entered Somerville as a Clothworkers' Scholar in 1884. Seward coached Rich in physics and acted as chaperone when necessary. Pauline Adams described Rich's experiences:

> In her second year, year Miss Rich was the only woman in a class of eight studying quantitative analysis in Mr. Harcourt's [A. G. Vernon Harcourt] laboratory at Christ Church. Mr. Harcourt, whose refusal to lecture separately to women students had hastened their admission to university lectures, had equally strong objections to allowing chaperones

---

[97]P. Adams, Librarian and Archivist, Somerville College, is thanked for providing biographical notes on Watson.

[98]Note 80, Adams, pp. 39–41; P. Adams, Librarian and Archivist, Somerville College, is thanked for providing additional biographical notes on Rich.

in his laboratory; an arrangement was therefore arrived at by which, at times when Miss Rich was in the laboratory, Miss Seward carried out research with Mr. W. H. Pendlebury on the rate of chemical change.[99]

In Rich's third year at Somerville, life was not much easier, as Adams noted:

> When in her third year Miss Rich was at last able to devote herself exclusively to chemistry she found herself 'frightfully handicapped by being the only woman; always having to depend on someone to accompany me to lectures or library.'[99]

Completing the Oxford Honours Chemistry degree requirements in 1887, Rich then became a "steamboat lady" to acquire an M.A. from Dublin. That same year, she accepted a position as Science Mistress at Howell's School, Llandaff, Wales. Rich taught at Howell's only until 1888, teaching at Grantham Ladies' College, Grantham, from 1889 to 1892, and then at Wimbledon House School (later Roedean School) from 1892 to 1905.

Rich left Roedean in 1905 to set up her own girls' school, Granville School, Leicester, of which she was Principal from its opening in 1906. In Rich's obituary in the journal *Nature*, it was noted that the school: "... acquired a considerable reputation ..."[100] Due to ill-health, Rich had to resign in 1923. However, she then became an Honorary Assistant in the Botanical Department of Queen Mary College, London. Previously, her hobby had been fresh-water algae, and the academic position now enabled her to pursue her interest as serious research, which led to recognition of her contributions.[101] Rich died on 20 April 1939.

The Alembic Club

Most universities had a chemistry club or society, but they differed in their attitude to women chemists. The Chemical Club of Cambridge University seems to have accepted women members without comment. In fact,

---

[99] Note 80, Adams, p. 40.
[100] Fritsch, F. E. (1939). Miss M. F. Rich. *Nature* **143**: 845.
[101] Child, M. D. (Michaelmas 1939). Miss Florence Rich. *Roedean School Magazine* **41**: 8.

two of the early women members, Ida Freund and M. Beatrice Thomas, presented research papers to a meeting of the Club in 1904.[102]

The attitude at Oxford could not have been more different from that at Cambridge. From 1882, there had been a male-only Junior [undergraduate] Scientific Club.[103] There was also a Chemical Club, which in 1901 became the prestigious male-only Alembic Club.[104] The Scientific Club was divided into a Senior Club for graduates and faculty, and a Junior Club for undergraduates: both Clubs held occasional open meetings but in addition, each week there were members-only seminars. These seminars were a focus of the life of the chemistry department.

In 1932, during the fourth year of her undergraduate degree at Oxford, Dorothy Crowfoot (see Chapter 12) discovered the existence of these meetings and that, being a woman, she was barred from them.[105] The situation was no better when Crowfoot later returned from Cambridge to Oxford as a Fellow and Tutor. The Senior Alembic Club ignored her existence. On one occasion, she arrived early for an open session of the Club and entered the room while the closed session was still in progress. One of the members lifted her off the ground and forcibly ejected her from the room.

It was not until 1950 that the Club voted to admit women as members, following an acrimonious debate in which some of the older members threatened to resign, though only Dalziel Hammick[106] actually did resign, later to rejoin.[107] A decade later, the Club had a woman President, Muriel Tomlinson, of St. Hilda's College (see the respective section).

---

[102] Berry, J. A. and Moelwyn-Hughes, E. A. (December 1963). Chemistry at Cambridge from 1901 to 1910. *Proceedings of the Chemical Society* 361.

[103] Rowlinson, P. J. (1983). Student participation in science teaching: The early years of the Oxford University Junior Scientific Club. *Oxford Review of Education* 9: 133–136.

[104] As mentioned in: Pickard, R. H. (1935). Obituary Notices: George William Fraser Holroyd. 1871–1934. *Journal of the Chemical Society* 407.

[105] Ferry, G. (1998). *Dorothy Hodgkin: A Life*. Granta Books, London, pp. 132–134.

[106] Bowen, E. J. (1967). Dalziel Llewellyn Hammick, 1887–1966. *Biographical Memoirs of Fellows of the Royal Society* 13: 107–124.

[107] Smith, J. C. (n.d.). *The Development of Organic Chemistry at Oxford*. Part II, *the Robinson Era 1930–1955*. unpublished, p. 54. H. Anderson, Memorial University, is thanked for a copy of this manuscript.

Women Chemists and the Dyson Perrin Laboratory

Despite the misogyny of the Alembic Club, a significant number of women chemists worked in the Dyson Perrins Laboratory (DP) in the 1930s. Oxford had hoped the arrival in 1919 of Frederick Soddy[108] from Aberdeen (subsequently to be rejoined by his Research Assistant, Ada Hitchins, see Chapter 7) would boost the university's research profile in physical and inorganic chemistry. However, Soddy's time there was not a happy one with promises of support not forthcoming and considerable friction with the constituent colleges.[109] Soddy then turned his attentions to problems of economics instead of chemistry.

It was to be organic chemistry that mainly provided Oxford with its chemical claim to fame. The impetus was the construction of the DP "for long the envy of all other university chemistry departments."[110] It was the DP which attracted brilliant young organic chemists to Oxford,[111] including Robert Robinson.[112] Robinson had arrived at Oxford in 1930[113] with his chemist-spouse, Gertrude Walsh (see Chapter 14) who worked with him in the laboratory. However, she was not the first woman to undertake research in the DP, as Elinor Ewbank had arrived in 1920.

---

[108]Fleck, A. (1957). Frederick Soddy. Born Eastbourne 2 September 1877. Died Brighton 26 September 1956. *Biographical Memoirs of Fellows of the Royal Society* **3**: 203–216.

[109]Morrell, J. (2009). Research as the thing: Oxford Chemistry 1912–1939. In Williams, R. J. P., Chapman, A., and Rowlinson, J. S. (eds.), *Chemistry at Oxford: A History from 1600 to 2005*. Royal Society for Chemistry, London, pp. 145–147.

[110]Hartley, H. (1955). Schools of chemistry in Great Britain and Ireland–XVI The University of Oxford, Part II. *Journal of the Royal Institute of Chemistry* **79**: 176–184. See also: Knowles, J. (2003). The Dyson Perrins Laboratory at Oxford. *Organic and Biomolecular Chemistry* **1**: 3625–3627.

[111]Bleaney, B. (1994). The physical sciences in Oxford, 1918–1939 and earlier. *Notes and Records of the Royal Society, London* **48**(2): 247–261.

[112]Todd, Lord, and Cornforth, J. W. (1976). Robert Robinson. 13 September 1886–8 February 1975. *Biographical Memoirs of Fellows of the Royal Society* **22**: 414–527.

[113]Tomlinson, M. L. (1987). The Dyson-Perrins Laboratory in Robinson's Time. *Natural Products Chemistry* **4**: 73–75.

98 *Pioneering British Women Chemists: Their Lives and Contributions*

Elinor Ewbank

Elinor Katherine Ewbank[114] was born on 19 October 1880 at Ryde, Isle-of-Wight, her parents being Henry Ewbank, vicar, and Louisa Caroline Wollaston. Educated at Highfield School, Hendon, Ewbank entered LMH in 1899. She obtained a B.Sc. in Chemistry in 1903 and then undertook research with Edward Charles Cyril Baly at the Spectroscopy Laboratory, University College, London, co-authoring two papers with him in 1905. With the commencement of the First World War, Ewbank volunteered as a nurse, being sent in 1915 to aid Russian troops, and then later transferred to aid Italian troops.[115]

Returning to England in 1917, from 1919 to 1920, Ewbank worked in the organic chemistry research division of the Department of Scientific and Industrial Research (DSIR). Then from 1921 until about 1930, Ewbank undertook research at the DP, five of her seven publications being co-authored with Nevil Sidgwick[116] and the last one being co-authored with Hammick. In the 1930s, she co-authored two publications on the soils of Cyprus, her affiliation being given as the Department of Biochemistry and Colloidal Chemistry, Hebrew University, Jerusalem. Ewbank died on 21 January 1958 at Oxford.

Kathleen Rogers (Mrs. Penfold)

Kathleen Margaret Rogers[117] was a student with another DP researcher, Sydney Plant.[118] Rogers, daughter of Gilbert Pearson Rogers and Kathleen Hudson, was born on 30 December 1911 at Bexley, Kent. She was educated at St. Swithun's School before entering St. Hilda's in 1932,

---

[114]R. Staples, Archivist, Lady Margaret Hall College, is thanked for providing brief biographical notes on Ewbank.

[115]Avert, C. and Pipe, H. (1991). *Lady Margaret Hall Register 1879–1990*. n.pub. p. 47.

[116]Tizard, H. T. (1954). Nevil Vincent Sidgwick, 1873–1952. *Obituary Notices of Fellows of the Royal Society* **9**(1): 236–258.

[117]Verity, G., Brown, B. and Rayner, M. E. (1993). 1931. *St. Hilda's College, Oxford Centenary Register 1893–1993*. St. Hilda's College, Oxford, n. pag.

[118]Tomlinson, M. (1956). Obituary Notices: Sydney Glenn Preston Plant, 1896–1955. *Journal of the Chemical Society* 1920–1922.

completing her B.Sc. in 1935 and M.A. in 1939. Rogers' research at the DP was published in 1935 and 1936. Her marriage to D. Penfold in 1938 terminated her academic life. However, after her two sons had grown up, she took a position as a teacher at The Croft House School, a girls' boarding school in Shillingstone, Dorset. Rogers died in 1989 at Poole in Dorset.

Katherine Ross (Mrs. Wilson)

Katherine Isobel Ross[119] was one of the two women chemists of the DP who found subsequent employment at RHC. Born on 7 February 1907 at Durban, Natal, South Africa, her father was Edward R. Ross, a railway consultant. Ross was educated at Birkhamstead School for Girls and completed a B.Sc. at St. Hilda's College in 1930. Though her publications from the DP (with J. M. Gulland) are dated 1931, the actual research must have been accomplished the year earlier, for she was appointed Assistant Lecturer and Demonstrator in Chemistry at the RHC in 1930.[120] In 1934, Ross resigned from RHC and joined Imperial Chemical Industries at Billingham-on-Tees where she worked until 1936. She resigned in order to marry Charles Wilson, a technical chemist, and had one son. Ross died in February 1984 in Surrey.

Elizabeth Lavington (Mrs. Hedley)

Elizabeth Mary Lavington[121] was the other woman chemist at the DP who went to RHC. She was born on 6 May 1910 at Barnet, Hertfordshire, the daughter of Henry Wykeham Lavington, builder and contractor, and Mary Ann Drury. Lavington was educated at Benenden School, Kent, and then entered St. Hugh's, completing a B.A. (Hons.) in Chemistry in 1932.

---

[119] (a) Verity, G., Brown, B. and Rayner, M. E. (1993). *St Hilda's College, Oxford Centenary Register 1893–1993*, (1926). St. Hilda's College, Oxford, n. pag.; and (b) Anon. (November 1930). Katherine Ross. *College Letter*. Royal Holloway College Association, 16.

[120] Anon. (March 1932). News. *Chronicle of the Berkhamsted School for Girls*.

[121] (a) Soutter, A. M. and Clapinson, M. (2011). *St. Hugh's College Register 1886–1959*. St. Hugh's College, Oxford, 123; and (b) *Calendars*, University of Oxford.

Her research with T. W. J. Taylor at the DP was published in 1934, the same year that she received a B.Sc. In 1933, Lavington undertook research work at the Lister Institute of Preventative Medicine in London and then took an office position at the Imperial Chemical Industries in 1935. The following year, Lavington replaced Ross as Demonstrator and Assistant Lecturer in Chemistry at RHC. In 1938, she married Rev. Henry Hedley. No further information could be found on Lavington until her death at Whitehaven, Cumbria in 2000.

Chika Kuroda

Chika Kuroda,[122] a Japanese chemist, was one of several overseas women attracted to DP for postgraduate studies. Kuroda, born on 24 March 1884 in Saga Prefecture, Japan, entered the Division of Science at Rika Women's Higher Normal School in 1902 and graduated in 1906. She then taught at Fukui Normal School for a year before enrolling in the graduate program at Kenkyuka Women's Higher Normal School in 1907. Finishing the course in 1909, Kuroda was appointed as an Assistant Professor at Tokyo Women's Higher Normal School.

Then, in 1913, when Tohoku Imperial University became the first of Japan's Imperial Universities to accept women students, Kuroda was admitted to the Chemistry Department of the College of Science. It was Riko Majima who inspired Kuroda's interest in organic chemistry, particularly natural pigments. She completed her Bachelor of Science in 1916, becoming the first woman in Japan to do so. Following graduation, Kuroda was appointed an Assistant Professor at the Tohoku Imperial University in 1916 and then Professor at the Tokyo Women's Higher Normal School in 1918.

Kuroda arrived at Oxford in 1921 as a Japan Minister of Education overseas student. She spent 2 years working with W. H. Perkin, Jr.,[123] returning to Japan in 1923 to resume her role as Professor at the Tokyo Women's Higher Normal School. In 1924, Kuroda was commissioned by

---

[122] Wikipedia: Chika Kuroda (accessed 1 February 2019).

[123] "J.F.T." (1931). Obituary Notices: William Henry Perkin, 1860–1929. *Proceedings of the Royal Society of London, Series A* **130**: i–xii.

the RIKEN Institute to research the structure of carthamin, the pigment of safflower plants. Her thesis earned her a doctorate in science in 1929. Throughout the 1930s and 1940s, Kuroda's research examined the composition of plant pigments, resulting in the award of the Majima Prize by the Chemical Society of Japan in 1936. She was appointed Professor at Ochanomizu University in 1949. Kuroda retired in 1952, but continued to lecture at Ochanomizu University as a Professor Emeritus until shortly before her death on 8 November 1968 in Fukuoka, aged 84.

Earliest Women Chemistry Staff

The "official" history of the Oxford University Chemistry Department contains a paragraph titled: "A Note on Women Fellows in Chemistry" in which only Dorothy Crowfoot Hodgkin (see Chapter 12) and Muriel Tomlinson (see the respective section) are identified as early women Fellows.[124] In fact, at least two individuals preceded them: Elizabeth Farrow and Margaret Leishman.

Elizabeth Farrow (Mrs. Cutliffe)

Elizabeth Monica Openshaw Farrow[125] was born on 23 August 1899 at Swinton, Lancashire, the daughter of Rev. J. W. Farrow and Elizabeth Openshaw. Educated at Oxford High School for Girls, Farrow entered St. Hugh's College in 1918, receiving her B.A. (Hons.) in Chemistry, in 1922, followed by an M.A. in 1925. In 1923, Farrow was appointed Tutor in Natural Science at St. Hugh's, then in 1923, she was appointed Lecturer and Demonstrator, Old Chemistry Department, being made a Senior Demonstrator. Two years later, she was promoted to University Demonstrator in Chemistry and elected Fellow of St. Hugh's College.

---

[124]Williams, R. J. P. (2009). Recent times, 1945–2005: A school of world renown. In Williams, R. J. P., Chapman, A. and Rowlinson, J. S. (eds.), *Chemistry at Oxford: A History from 1600 to 2005*. Royal Society for Chemistry, London, pp. 195–291.

[125](a) Anon. (1954–1955). Elizabeth Monica Openshaw Cutliffe (née Farrow). *St. Hugh's College Chronicle* (27): 19–20; and (b) Soutter, A. M. and Clapinson, M. (2011). *St. Hugh's College Register 1886–1959*. St. Hugh's College, Oxford, 50.

Farrow resigned her positions in 1930 upon her marriage to Eric Francis Cutliffe, following which she had a son and a daughter (the daughter following in her mother's footsteps, entering St. Hugh's in 1949 to study chemistry). Probably because of the wartime shortage of science teachers, Farrow accepted a temporary appointment to teach at Oundle School, Northamptonshire. Then in 1945, she was listed as being on the staff at Epsom College, Surrey. Farrow died on 12 December 1954.

Margaret Leishman

Margaret Augusta Leishman[126] was born on 3 June 1903 in East Twickenham, Middlesex, the daughter of General Sir William Boog Leishman and Maud Elizabeth Gunter. She was educated at Croham Hurst School, Croydon, then graduated from St. Hilda's College with a B.Sc. in Chemistry in 1926.

In 1926, Leishman was appointed Demonstrator in Inorganic and Physical Chemistry at Bedford College under James Spencer (see Chapter 5). Four years later, in 1930, she was invited back to Oxford. In addition to becoming Demonstrator in Chemistry at Oxford and Lecturer at St. Hugh's College, she joined the research group of M. P. Applebey at the DP. Leishman's research was published in 1932, the same year that she was appointed as Fellow and Tutor at St. Hugh's. After all these years of connection with Oxford, in 1937, she accepted a position as Assistant Mistress at St. Swithun's School, Winchester. Leishman remained at St. Swithun's until retirement, dying in 1976.

Muriel Tomlinson

The one women chemist who, more than any other, gained recognition in the Oxford chemistry laboratories, was Muriel Louise Tomlinson.[127]

---

[126]Verity, G., Brown, B. and Rayner, M. E. (1993). *St. Hilda's College, Oxford Centenary Register 1893–1993, (1922)*. St. Hilda's College, Oxford, n. pag.; and *Staff Records*, Archives, Bedford College.

[127]Christie, M. (1990–1991). Obituaries: Muriel Louise Tomlinson. *St. Hilda's College Report and Chronicle* 30–32.

Cambridge and Oxford Women's Colleges   103

She was born on 17 July 1909 in Birmingham, daughter of William Seckerson Tomlinson and Annie Wall. Her initial interest in chemistry began when she was at King's High School, Warwick, and it had a very unusual cause, as Tomlinson recounted:

> When I was at school I was first attracted to chemistry because of the delightful blue colour the word conjured up for me. To me all words have colour: the colours are perhaps less vivid than they were when I was a child and I remember being amazed when I discovered that to most people words are not coloured at all and that this 'colour-thinking' is experienced by relatively few people.[128]

This "colour-thinking" as Tomlinson called it, is a rare condition known as synesthesia, in which stimulation of one sense causes an unusual response in another.[129]

Entering St. Hugh's College in 1928, she was advised by her tutor to take the Part I of her examinations at the end of her second year. She accepted the advice, then discovered that none of the men were taking the examination until their third year. Tomlinson recalled: "I was very frightened but it was too late to change & I carried on."[130] She obtained a first class. Despite the hard work, she still found time to master the art of punting, and as her friend Margaret Christie observed: "... forty years later she could still put young colleagues to shame on the river."[131]

Tomlinson graduated in 1931, staying at Oxford as Senior Scholar and Gilchrist Student at LMH and undertaking research at the DP with Sydney Plant on indoles and carbazoles. After obtaining a D. Phil. in 1933, she spent 2 years collaborating with Robinson, the last year holding a Mary Somerville Research Fellowship at Somerville.

---

[128] Hand-written draft of a speech given by Muriel Tomlinson at the St. Hilda's College Gaudy, 25 June 1977, copy provided by E. Boardman, Archivist, St. Hilda's College.

[129] Hubbard, E. M. and Ramachandran, V. S. (2005). Neurocognitive mechanisms of synesthesia. *Neuron* **48**(3): 509–520.

[130] E-mail, Muriel Tomlinson to Margaret Goodgame, 7 December 2006. M. Goodgame is thanked for this quote.

[131] Note 127, Christie, p. 30.

104   *Pioneering British Women Chemists: Their Lives and Contributions*

In 1935, Tomlinson obtained an appointment at Girton College, Cambridge, first as Lecturer, then as Fellow, teaching the main organic chemistry general course and organizing the University Practical Classes in organic chemistry. She continued to do this through the years of the Second World War. This was an especially difficult task with the arrival of the students from Queen Mary College, Bedford College, and Guy's Hospital when these London colleges were evacuated to Cambridge.

Tomlinson was invited back to Oxford as Fellow at St. Hilda's in 1946 and then became university lecturer in 1948. One of her students in the 1950s, Margaret Goodgame, recalled:

> She was keen on practical work, and expected us to work in the labs most of the hours they were open — 11 am to 7 pm Monday to Friday plus Saturday morning. Lectures were usually 9–11 am. Essays and other tutorial work were to be done at other times. She was an assiduous demonstrator and a stickler for good practice in the lab, coming down hard on anything she considered sloppy. Consequently when she was demonstrating many of the male students used to slip off to the library (or wherever!). But for the girls, who all worked together at one end of the large laboratory, it was better to be there, even if doing something badly, than not to be there.[132]

At the same time, Tomlinson resumed organic chemistry research with Robinson and Plant at the DP. Goodgame commented that, during her time at Oxford, Tomlinson was the only women in the DP, in "what was very much a man's world."[132] Tomlinson's name appears on a total of 29 research papers: 4 on her own, 5 with Robinson, and 8 with her lifelong friend Plant, with Tomlinson also writing Plant's obituary for the *Journal of the Chemical Society*. In addition, she authored a text: *An Introduction to the Chemistry of Benzenoid Compounds*.[133]

A great traveller, she was usually accompanied by either of her friends, Mary Cartwright or Helen Gardner. Christie recalled: "She was a good, if aggressive driver with a liking for large powerful Volvos."[134]

---

[132] Personal communication, e-mail, M. Goodgame, 7 December 2006.

[133] Tomlinson, M. L. (1971). *An Introduction to the Chemistry of Benzenoid Compounds*. Pergamon Press, Oxford.

[134] Note 127, Christie, p. 31.

After retirement in 1975, she returned to Warwick, where she had spent her childhood, and she became Chairman of the Board of Governors of the Junior Division of King's High School. She died on 17 May 1991.

Commentary

Though we traditionally consider Oxford and Cambridge as similar institutions, for pioneer women chemists, they offered very different experiences. At Cambridge, Newnham and Girton provided a self-contained "haven" in which the women had their own classes, and more importantly, their own chemistry laboratories with women staff, particularly Freund, Marshall, and Thomas. At Oxford, undergraduate women students experienced the male environment and culture for both lectures and laboratory. Women were not even admitted to the Balliol–Trinity Chemistry Laboratories at Oxford until 1904 (see Footnote 128), and, in addition, exclusion from the Alembic Club must have added to the negative atmosphere for women chemistry students.

The situation seemed to be reversed for the research laboratories. The Oxford Dyson Perrins Research Laboratory had women researchers throughout the period of our study, while in the Cambridge Chemistry Research Laboratories there did not appear to be any mentors for women students: for example, Mary Johnson published her research as sole author while Delia Simpson undertook her research in the Physics Laboratory. For Cambridge, it was biochemistry which provided the supportive research atmosphere, as we show in Chapter 11.

In the "official" history of the Oxford Chemistry Department, one of the authors expresses puzzlement at the lack of women chemistry Fellows. They postulate that: "A question arises as to the wish of women to be involved in the very demanding lifestyle of a chemistry fellow/lecturer or parallel professions."[135] We will show in the following chapters that the "very demanding lifestyle" was not a discouragement to women chemists at other academic institutions.

---

[135]Williams, R. J. P. (2008). Recent Times, 1945–2005: A School of World Renown, Section 7.2.1: A Note on Women Fellows in Chemistry. In: Williams, R. J. P., Chapman, A. and Rowlinson, J. S. (eds.), *Chemistry at Oxford: A History from 1600 to 2005*. RSC Publishing, Cambridge, p. 213.

Chapter 4

London Co-educational Colleges

Frances Micklethwait (7 March 1867–25 March 1950)

In south-east England particularly, a momentous date was 1878. In that year, the University of London opened its doors to women for the granting of degrees. The students of North London Collegiate School for Girls celebrated in verse:

> *The time is come," Minerva said,*
> *For you, the school's elite,*
> *To study Greek and Chemistry,*
> *Acoustics, Light and Heat;*

108 *Pioneering British Women Chemists: Their Lives and Contributions*

And talk of that mysterious sphere
Where parallels all meet.

And when you've passed the London [Examinations], dears,
Go in for a degree;
A glorious field lies open now —
B.A. and B.Sc.!
With these as aids you soon may climb
Ambition's loftiest tree![1]

The University of London became an umbrella organization for a number of constituent colleges. It was a unique institution in that women, once they were admitted, had a choice of both co-educational colleges and women-only colleges. In this chapter, we will look at the contributions of women chemists who attended the five co-educational London colleges of the time: Birkbeck College, University College, King's College, Imperial College, and East London College (Queen Mary College). This chapter also includes the polytechnics, particularly Battersea Polytechnic, which provided a different pathway for the education of women chemists.

Admission of Women to the University of London Examinations

The University of London was founded in 1836 as an official body to administer the examinations for the two rival colleges of the time: University College and King's College. However, it was not until 1856 that the first woman, Jessie Meriton White,[2] petitioned to be allowed to sit the university's examinations. White had come from an affluent noncon-formist family, and she had studied philosophy at the Sorbonne between 1852 and 1854. During her continental travels, she met the Italian revolutionary leader, Giuseppe Garibaldi. Inspired by his vision for Italy, White decided to devote her life to the campaign for the unification of Italy. She returned to England in 1855 with the intent of gaining a medical degree

---

[1]"Two Little Oysters" (April 1879). A Tale of 1879. *Our Magazine, North London Collegiate School for Girls* 12.

[2]Daniels, E. A. (1972). *Jessie White Mario Risorgimento Revolutionary*. Ohio University Press, Athens, Ohio.

as this would make her more useful to the revolutionary cause. However, on 9 July 1856, The University Senate, on the advice of legal counsel, rejected White's application to sit for London University examinations because Senate did not consider themselves "empowered to admit Females as candidates for degrees."[3]

The attempt by the second applicant, Elizabeth Garrett[4] (later Mrs. Garrett Anderson), though equally unsuccessful, followed a different path. Garrett had decided in 1860, at the age of 24, to become a doctor. The case for admitting Garrett was put to the Senate meeting of 9 April 1862. The vote was close: by seven votes to six, the resolution was adopted that the Senate found: "no reason to doubt the validity of Counsel's opinion given in the case of Miss Jessie Meriton White in 1856."[5]

The Senate of the university was not the only challenge. The Senate had been established in 1837, but following agitation by graduates for representation in decision-making, Convocation was established by the Charter of 1858. Convocation consisted of all doctors, all Bachelors of Law of 2 years' standing, and all Bachelors of Arts of 3 years' standing who paid a registration fee. Its consent was required for any changes in the University Charter. In fact, William Shaen,[6] one of the most forthright male proponents for women's rights of the period, believed Convocation to be the more powerful body and therefore key to women's admission:

> ... though the Senate, and not Convocation, is the governing body, still the Senate would not force female graduates upon the others, in opposition to the wishes of Convocation; and, on the other hand, there could be no doubt, that when Convocation makes up its mind that degrees shall be open to women, the Senate would be of the same opinion.[7]

It was Convocation that first declared itself in favour of permitting women to take degrees. This occurred on 12 May 1874. Senate followed suit on

---

[3]Willson, F. M. G. (2004). *The University of London, 1858–1900: The Politics of Senate and Convocation*. Boydell Press, Suffolk, pp. 85–144.

[4]Manton, J. (1965). *Elizabeth Garrett Anderson*. Methuen, London.

[5]Note 3, Willson, pp. 92–93.

[6]Shaen, M. J. (ed.), (1912). *William Shaen: A Brief Sketch*. Longmans, Green & Co., London.

[7]Note 3, Willson, p. 98.

110 *Pioneering British Women Chemists: Their Lives and Contributions*

1 July of the same year, but there seemed to be no enthusiasm for seeking the necessary statutory changes, so exclusion remained the practice. The division of powers between Senate and Convocation of the university bedevilled further progress on the issue. In fact, it was not until 14 May 1878 that a new Supplemental Charter was finally approved admitting women to all degrees of the university. Finally, in 1880, the first four women graduated with B.A. degrees.[8]

Florence Eves

The very next year (1881), the first woman graduated with a London University B.Sc. in Chemistry. This was Florence Elizabeth Eves. Eves was born on 31 March 1858, daughter of George Eves, architect and surveyor, and Fanny Wyatt Murray[9]. Educated at North London Collegiate School, she entered Newnham College, Cambridge in 1878. Eves completed the Natural Science Tripos in 1881, and in the same year, she was awarded the B.Sc. conjoint in Chemistry and Botany from University College, London. From 1881 to 1887, she was a Demonstrator at Newnham College. Eves was considered for the position of Director of the Balfour Biological Laboratory in 1884, but was not chosen.[10]

For the next 3 years, Eves held the position of Assistant Mistress at Manchester High School for Girls, followed by 1 year as a Natural Science teacher at a school in Brighton, then 2 years as a teacher at St. Leonard's School, St. Andrews. In 1892, Eves career took a completely different direction, becoming a worker for social reform. She became Head of Women's House, Christian Socialist Union in Hoxton, dying on 11 February 1911 at Uxbridge.

---

[8]Murray, J. H. and Stark, M. (1880). Record of Events: London University, *The Englishwoman's Review of Social and Industrial Questions 1880*. Routledge Reprint. n. pag.

[9]White, A. B. (ed.), (1979). *Newnham College Register 1871–1971*, Vol. 1, *1871–1923*. Newnham College, Cambridge, p. 65.

[10]Richmond, M. L. (1997). 'A Lab of one's own': The Balfour Biological Laboratory for women at Cambridge University, 1884–1914. *Isis* **88**: 422–455.

Birkbeck Literary and Scientific Institution (Later Birkbeck College)

We begin our coverage of the co-educational colleges of London University with Birkbeck. It was in the evening of 11 November 1823 that a meeting was convened at the Crown & Anchor Tavern on the Strand, London. About 2000 people attended to hear Dr. George Birkbeck speak on the importance of educating working people. At a subsequent meeting at the Tavern on 2 December 1823, the formation of the London Mechanics' Institute was proposed to accomplish this goal of providing courses in science (particularly chemistry), arts, and economics.[11]

In 1830, it was agreed to admit women students to lectures, though they could not become members of the Institute. An added complication of women's participation was raised at a meeting of the governing committee in December 1833, specifically: "the propriety of admitting females attending lectures through the front entrance."[12] However, female enrolment was low in the early decades, for example, of the 1081 students in 1839, only 11 were women.[13]

Then in 1858, with a change in University of London regulations, the students of the Institute became eligible to sit University of London examinations and granted University of London degrees. The name of the Institution was changed to that of the Birkbeck Literary and Scientific Institution in 1866 and then to Birkbeck College in 1907. However, it was not until 1920 that the college was admitted as a constituent college of London University.

University College

University College (UCL) had been founded in 1826 to provide a secular alternative to the religious universities of Oxford and Cambridge. In 1868,

---

[11]Burns, C. D. (1924). *A Short History of Birkbeck College*. London University, pp. 17–26.

[12]Note 11, Burns, p. 43.

[13]Note 11, Burns, p. 56.

112 *Pioneering British Women Chemists: Their Lives and Contributions*

a committee was established to provide classes for "ladies," parallel to those offered to men. As Gillian Sutherland has commented:

> This Committee spent a considerable amount of time on devices to keep male and female students separate, not only in the lecture rooms but in their comings and goings. Eventually it was agreed that lectures for men should begin and end on the hour, while those for women should begin and end on the half-hour. Practical considerations, such as the use of equipment, gradually brought a handful of shared classes. But there had to be a separate entrance to these; and in the autumn of 1870 a whole new door was knocked into the Chemistry Laboratory, the Ladies' Association having given a guarantee to cover the cost of blocking it off again, if the female demand for chemistry were to fall off.[14]

Emily Aston

Emily Alicia Aston,[15] the first woman chemistry researcher at UCL, was born in Paddington on 4 January 1866 to Joseph Keech Aston, a barrister, and Sarah Alice Eccles. Like Frances Buss and Dorothea Beale (see Chapter 2), Aston was educated at Queen's College, Harley Street. She was admitted to Bedford College (see Chapter 5), to study Chemistry. It was while at Bedford that Aston undertook her first research work: the investigation on "multiple sulphates" under the direction of Spencer Pickering.[16]

In 1885, Aston entered UCL, receiving her B.Sc. in Chemistry and Geology in 1889. Aston remained at UCL after graduation, taking additional courses, particularly analytical chemistry. However, most of her time was spent undertaking research. It was in 1887 that William Ramsay[17] had been offered and accepted the Chair of Chemistry of UCL. During his previous position at Bristol College, Ramsay had hired a

---

[14] Sutherland, G. (1990). The plainest principles of justice: The University of London and the higher education of women. In Thompson, F. M. L. (ed.), *The University of London and the World of Learning, 1836–1986*. The Hambledon Press, London, p. 39.

[15] Creese, M. R. S. (1998). *Ladies in the Laboratory? American and British Women in Science, 1800–1900*. Scarecrow Press, Lanham, Md., p. 265.

[16] "A.H." (1926). Obituary Notices of Fellows Deceased. Percival Spencer Umfreville Pickering, 1858–1920. *Proceedings of the Royal Society, Series A* 111: viii–xii.

[17] "J.N.C." (1917). Obituary Notices of Fellows Deceased. Sir William Ramsay, 1852–1916. *Proceedings of the Royal Society Series A* **93**: vlii–liv.

woman research student: Katherine Williams (see Chapter 6). After Aston graduated, she joined Ramsey's research group, co-authoring five publications in the field of physical chemistry. In Morris Travers's biography of Ramsay, it is noted that:

> Ramsay carried out further experimental investigations in the same direction with John Shields and with Miss Emily Aston, and made attempts to deduce from the results the degree of complexity of associating liquids.[18]

Aston's research versatility was remarkable. For example, she performed chemical analyses under the direction of the geologist, Thomas George Bonney. These analyses were on water and mineral samples sent by explorers in North Africa and the Middle East, while geologist Catherine Raisin[19] undertook the mineral classification. In addition, Aston co-authored a publication in the field of organic chemistry with J. Norman Collie and one on physical chemistry with James Walker.

During the late 1890s, Aston spent some time at the Sorbonne in Paris where she worked with Paul Dutoit on electrolytic conductivity and molecular association resulting in two publications; then she undertook research with Philippe Auguste Guye at the University of Geneva, Switzerland on optical rotation.

In all, Aston co-authored 18 research papers, certainly the most of any woman chemist of the period. Her time in Geneva seems to have been the end of her research career. According to census information of 1901 and 1911, her occupation was noted as "private means." No subsequent information could be discovered, except her death on 18 March 1948 at Uckfield, Sussex.

Lucy Newton

One of Aston's published research notes[20] is particularly intriguing as it was not co-authored with a university professor, but with another woman

---

[18] Travers, M. W. (1956). *A Life of Sir William Ramsay*. Edward Arnold, London, p. 96.

[19] Burek, C. V. and Higgs, B. (eds.), (2007). *The Role of Women in the History of Geology*. Geological Society, London, pp. 27–29.

[20] Aston, E. and Newton, L. (1897). A note on the estimation of zinc oxide. *Chemistry News* **75**: 133–134.

114 *Pioneering British Women Chemists: Their Lives and Contributions*

chemist, Lucy Newton[21]. Newton was born on 15 September 1867 at Scarborough, Yorkshire, her father was Edward Hotham Newton, bank director and brewer, and her mother, Elizabeth Blundus Taylor. Educated privately in Scarborough and Nottingham, she entered Newnham College, Cambridge, in 1886, completing the Natural Science Tripos in 1889. Newton stayed on at Newnham for an additional year, and then held the position of Assistant Demonstrator in the Newnham chemistry laboratory from 1891 to 1893. It was then that she moved to UCL to undertake analytical chemistry research with Aston. No more is known of her life. Newton died on 29 March 1903 in Malta, aged 35.

Katherine (Kate) Burke

Another researcher in Ramsay's group at UCL was Katherine (Kate) Alice Burke.[22] Born in 1866 in Marylebone, London, her father was William Henry Burke, marble manufacturer, and her mother, Mary Anne Keith. Burke obtained her B.Sc. (London) in 1899 following her studies at Bedford and Birkbeck Colleges.

That same year, Burke entered UCL to work in Ramsay's laboratory under Frederick Donnan.[23] Burke had two publications with Donnan and one with Edward Charles Cyril Baly.[24] In addition, she acted as a private Research Assistant to Ramsay, part of her task being to translate into English a book by the Danish chemist, Julius Thomsen, on systematic researches in thermochemistry.[25] The translation appeared in print in 1905.

In 1906, she was appointed as Assistant in the Department of Chemistry, being promoted to Assistant Lecturer in 1921. Donnan noted in Burke's obituary in the *Journal of the Chemical Society*:

---

[21] Note 15, Creese, p. 268.

[22] Personal communication, letter, Keith Austin, 12 June 1998. Archivist, Senate House, University of London.

[23] Freeth, F. A. (1957). Frederick George Donnan (1870–1956). *Biographical Memoirs of Fellows of the Royal Society* **3**: 23–39.

[24] Donnan, F. G. (1948/49). Edward Charles Cyril Baly, 1871–1948. *Obituary Notices of Fellows of the Royal Society* **6**: 7–21.

[25] Donnan, F. G. (1926). Obituary Notices: Katherine A. Burke. *Journal of the Chemical Society* 3244.

In 1906 she was appointed a member of the Chemical Staff at University College, and from that time until her death on July 6th, 1924, she continued her teaching work, having charge of the practical laboratory work for students of the Intermediate Science class, and giving courses of lectures to more advanced students on the chemical aspects of radioactive transformations (see Footnote 25).

Burke was obviously well-regarded at UCL. In the 1925 Report of the University College [Academic] Committee, it was stated that:

> Deceased. Miss K. A. Burke. Assistant Lecturer in the Dept. of Chemistry. That this meeting of the Academic Staff desire to record their sense of loss that has come to the College through the death of Miss Katherine A. Burke, who by her devoted work for the welfare of the students had won the esteem of the College community.[26]

Ida Homfray

Ida Frances Homfray[27] was also a researcher with Ramsay. Born on 17 December 1869 at Newcastle-under-Lyme, Staffordshire, she was the daughter of George Homfray, engineer, Coal and Iron Mines, and Marion Sarah Block. Homfray worked with Ramsay between about 1900 and 1910, authoring seven papers on surface properties some, like Aston's, involving collaboration with Guye at the University of Geneva. During her time at UCL, she obtained a B.Sc. by research in 1905, followed by a D.Sc. in 1910. Her research focussed on the absorption of gases by charcoal, studies that were very relevant to the gas masks of the First World War. No later records of her could be found except her death in Hampstead on 8 February 1948.

---

[26]Anon. (1925). Report of the University College Committee, University of London, University College, p. 22.

[27]Creese, M. R. S. (1991). British women of the nineteenth and early twentieth centuries who contributed to research in the chemical sciences. *British Journal for the History of Science* **24**(3): 275–305.

Effie Marsden (Mrs. Solomon)

Effie Josephine Gwendoline Marsden[28] was one of the two women researchers who worked with the physical organic chemist, Edward Baly. Between 1905 and 1910, she co-authored five papers with Baly Marsden, daughter of Algernon Moses Marsden, fine arts dealer, and Louise Frances Hyam, completed her secondary education at Kensington High School for Girls in 1899.

Marsden then entered UCL, joining Baly's research group. Using absorption spectroscopy, she made major contributions to the study of keto-enol tautomerism. To illustrate the overlap of the women chemists, of Marsden's seven publications, one was also co-authored by Katherine Burke (see the respective section), while another was part co-authored by Maud Gazdar (see the following section). In 1913, she married Reginald Saul Solomon and they had one daughter, Elizabeth Annette Solomon. Marsden died on 20 December 1946 in Hendon, London.

Maud Gazdar (Mrs. Taylor)

Maud Gazdar[29] was the other woman researcher with Baly. She was born in January 1883 in Edmonton, Middlesex, daughter of Jamshidi (James) Gazdar and Mary Gazdar. Gazdar was educated at South Hampstead High School for Girls and completed a B.Sc. (Hons.) at UCL in 1908. After graduation, she was appointed demonstrator in Chemistry while undertaking research with Baly and with Samuel Smiles,[30] having a total of three publications. In 1911, Gazdar left UCL to become a Research Assistant in the Biochemistry Department at Trinity College, Dublin. She completed an M.B. (B. Medicine) and a B.S. (B. Surgery) at the London School of Medicine for Women in 1918 (her sister, Gertrude, having obtained the same qualifications in 1908).

Upon graduation, Gazdar was appointed House Physician at the Royal Free Hospital, London (RFH), becoming Clinical and Obstetric Assistant

---

[28] Note 27, Creese, p. 289; and *Student Records*, Archives, University College, London.

[29] *Calendars*, London School of Medicine for Women; and *Staff Records*, Archives, Bedford College for Women.

[30] Benett, G. M. (1953). Samuel Smiles, 1877–1953. *Obituary Notices of Fellows of the Royal Society* **8**: 583–600.

in the Skins Department, 1919 to 1920, and then Clinical Assistant in the Gynaecological Department. She was awarded an M.D. in 1921. In 1924, Gazdar went into private practice as physician and surgeon in Bishop Stortford, Hertfordshire. She married John F. Taylor in 1928, retiring the following year to Godfreys Farm, Radwinter, Essex.

King's College

King's College was founded shortly after University College in 1831. King's was a male-only college, as Negley Harte described: "Female persons, with extremely rare exceptions on Saturday mornings [for laboratory work], were not admitted to the sacred precincts of King's itself until 1915."[31] Thus, prior to that date, women students at King's were to be found not on the Strand, but in Kensington.

The Women's Department of King's College

The foundation of the Women's Department (originally the Ladies' Department) of King's College was primarily the effort of Rev. Canon Alfred Barry.[32] In 1877, the Women's Education Union was in search of an institution willing to organize lectures for ladies in the west end of London. Barry proposed to King's College Council that classes for ladies might be offered under King's auspices. The lectures would take place in Kensington, not the Strand campus, except chemistry and physics, which were to be offered at the Strand on Saturday mornings using existing King's laboratories and equipment. The Council approved the former, but balked at the latter, only agreeing in 1880 to women visiting the Strand for scientific laboratory work.

With the overwhelming success of the ladies' lectures, a formal Women's Department of King's College was established in 1881, with official recognition in 1885. The department expanded until 1910 when it was incorporated as a separate constituent college of the University of

---

[31] Harte, N. (1979). *The Admission of Women to University College, London, A Centenary Lecture.* University College, London, p. 20.

[32] Marsh, N. (1986). *The History of Queen Elizabeth College: One Hundred Years of Education in Kensington.* King's College, London, p. 6.

London: King's College for Women. As discussed by Christina Bremner in an article in the *Journal of Education* in 1913, the type of students had changed from those of the early years:

> In the early days a majority of students were drawn from the leisured classes, and sought culture largely for its own sake.... The majority are now being prepared for work, to take up positions in life where their services are used and paid.[33]

As with the other women's colleges, the rules pertaining to residence life, and even academic life, were close to jail-like, as Neville Marsh commented:

> Life in the lecture theatres and laboratories was as ordered as life in Hall. Degree students were expected to wear undergraduate gowns in lectures and the institutionalisation was completed by students having to wear coloured overalls in laboratories according to their course.[34]

First-year students wore brown; second-year, orange; and third-year, navy; while laboratory assistants wore green. However, the limited number of possible colours led to the scheme later being abandoned.

In 1915, the arts and science departments of King's College for Women were transferred to the Strand, at last making King's College a co-educational institution — in theory. In reality, the absorbed women students and staff had to adapt to functioning in a traditional male environment (see, for example, Rosalind Franklin's experiences at King's, Chapter 12).

Margaret Seward (Mrs. McKillop)

Margaret Seward[35] was the earliest chemist on staff at the Women's College, being there from 1896 to 1915. Seward, daughter of James

---

[33]Bremner, C. S. (January 1913). King's College for Women: The Department of Home Science and Economics. *Journal of Education* (35 new series): 72–74.

[34]Note 32, Marsh, pp. 147–148.

[35]Pottle, M. (23 September 2004). McKillop [née Seward], Margaret (1864–1929), scientist and university teacher. *Oxford Dictionary of National Biography*. Oxford University Press (accessed 15 January 2019).

London Co-educational Colleges 119

Seward, master at the Liverpool Institute, and Sarah Jane Woodgates was born on 22 January 1864 at Wigan and educated at Blackborne House, Liverpool. She entered Somerville College, Oxford, in 1881,[36] and in 1884, she was the first Oxford woman student to be enrolled for the honour school of Mathematics. Seward then changed her focus to chemistry and in 1885 became the first woman to obtain a First Class in the Honour School of Natural Science. Upon graduation, Seward was immediately appointed Natural Sciences Tutor at Somerville, in addition undertaking research with the Oxford chemist, W. H. Pendlebury. Two publications on chemical reactions resulted from her work, one of which was read to the Royal Society.[37]

In 1887, Seward accepted a position as the first Lecturer in Chemistry at the Royal Holloway College (RHC). This episode in her life will be described in the RHC section in Chapter 5. Resigning her position at RHC in 1891, Seward travelled to Singapore to marry John McKillop, a civil engineer. Seward, her husband, and son (Alasdair) returned to England in 1893, where she taught at the Bradford Girl's Grammar School, and then at Roedean School. Seward was appointed to King's College, Women's Department, in 1896 to teach Elementary Science and Chemistry, the Chemistry Laboratory having opened in 1895.

Seward was mentioned in an 1898 article on the college in *Girl's Realm*:

> As regards science, there are two laboratories; the larger is for chemistry, the smaller serves various scientific courses: zoology, biology, botany, geology etc. In both the laboratories Princess Alice of Albany works, under the supervision of Mrs McKillop, one of our foremost women science-lecturers, and lately lecturer on Chemistry at the Royal Holloway College.[38]

---

[36](a) Adams, P. (1996). *Somerville for Women: An Oxford College 1879–1993*. Oxford University Press, Oxford, pp. 35–40, 53; and (b) Personal communication, e-mail, B. Ager, 3 July 1998. Archivist, King's College, London.

[37]Pendlebury, W. H. and Seward, M. (1889). An investigation of a case of gradual chemical change. The interaction of hydrogen chloride and chlorate in the presence of potassium chlorate. *Proceedings of the Royal Society of London* **45**: 396–423.

[38]Rawson, Mrs. S. (1898–1899). Where London girls may study, II. King's College, Kensington Square. *Girl's Realm* **1**: 1201–1207.

120   *Pioneering British Women Chemists: Their Lives and Contributions*

The year 1912 was the turning point for chemistry and for Seward herself. A report in the *King's College Magazine, Women's Department* suggests that the King's College administration wanted a male Lecturer in Chemistry; as a result, Seward became marginalized:

> Mr. H. L. Smith has been appointed full-time lecturer in Chemistry, and as at present there is a very definite majority of Home Science over B.Sc. students in the Chemical Laboratory, he has taken charge of it, with Miss Masters [see Chapter 8] as Demonstrator. Mrs. McKillop has thus been set free to undertake in addition to her tutorial work, the organization of the Library, a business which has been pressing for a little time.[39]

The Home Science Committee terminated Seward's position as of Spring 1914, perhaps in preparation for the transfer of the science programs to the existing male-faculty departments of King's on the Strand. In 1915, Seward was back teaching economics at Bradford Girl's Grammar School and giving food lectures to old girls. She then worked in the Ministry of Food, being awarded an M.B.E. in 1919 for her wartime studies on nutrition and human health, which included authoring a book, *Food Values*.[40] From 1920 until her sudden death on 29 May 1929 at Brentford, she acted as Librarian of the Sociological Society at Lepay House.

The Laboratory of Professor Huntington

Though at the time, women were not permitted to attend King's College, The Strand, as students, women chemists were hired as research assistants for King's leading metallurgist, Alfred Kirby Huntington.[41] His most prominent student was Elison Anne Macadam. For most of the

---

[39] Anon. (1912). College notes. *King's College Magazine Women's Department* (46): 4, K/SER1/170.

[40] McKillop, M. (1916). *Food Values*. E. P. Dutton, New York; expanded and re-published as: McKillop, M. (1922). *Food Values: What They Are, and How to Calculate Them*. Routledge, London.

[41] Desch, C. H. (15 May 1920). A. K. Huntington, *Journal of the Society of Chemical Industry* 162.

biographical information on Macadam, we have to rely on the obituaries of her chemist husband, Cecil Henry Desch.[42]

Elison Macadam (Mrs. Desch)

Born in 1883 in Midlothian, Elison Anne Macadam (Mrs. Desch) was the daughter of William Iveson Macadam, Professor of Chemistry at the College of Surgeons, Edinburgh and Sarah MacDonald. She had wished to study for a degree in chemistry at the University of Edinburgh, but at that time women were excluded. Curiously, despite the formal ban of women from the Strand campus, she was able to study chemistry with F. C. Thompson and Herbert Jackson at King's College in order to sit the Institute of Chemistry examinations. After she successfully passed the examinations,[43] Thompson and Jackson recommended her to Huntington.

It was noted in McCance's obituary of Desch that: "Desch did not find it easy to work with Professor Huntington who was quick tempered and exacting."[44] Macadam undertook accurate analysis of metal samples which were then examined metallographically by Desch. In January 1909, Macadam and Desch were married. McCance quotes Huntington that: "Cecil Desch had robbed him of his best assistant."[44] The Desch's moved shortly afterwards to Glasgow, where Desch had been appointed as Lecturer in Metallurgical Chemistry. They had two daughters, Rosalind and Marjory. The only other reference to her was at a later appointment of Desch at the University of Sheffield, that: "In promoting the social activities of the University, Mrs Desch took an active part."[45] Macadam died on 6 February 1965 in Wandsworth, London.

---

[42] (a) McCance, A. (1959). Cecil Henry Desch, 1874–1958. *Obituary Notices of Fellows of the Royal Society.* **5**: 49–68; and (b) Thompson, F. C. (1958). Prof. C. H. Desch, FRS. *Nature*, **182**: 223–224.

[43] (a) Anon. (1906). New Associates, *Proceedings of the Institute of Chemistry of Great Britain and Ireland* 62; and (b) (1909). New Fellows, *Proceedings of the Institute of Chemistry of Great Britain and Ireland* (pt. i): 30.

[44] Note 42(a), McCance, p. 57.

[45] Note 42(a), McCance, p. 60.

Emily Forster

Emily L. B. Forster was another of Huntington's research students. She was born about 1870 at Chester, daughter of Robert Cochrane Forester, civil engineer, and Lydia Brougham Vaughan. Nothing could be found on Forster's early education, but we know that she was working with Huntington in 1909.

Forster must have then trained as a pharmacist, for she later published in the field giving her position in 1920 as late-Lecturer at the Westminster College of Pharmacy. In an article in the *Pharmaceutical Journal* in 1916, she makes it clear that women pharmacists faced opposition:

> Where to settle! ... The woman pharmacist has something else to weigh besides expense: it is the question of her sex, and the fact that at present she is a pioneer in her profession, and must naturally turn to where she thinks an enterprising woman will be respected and her ability made use of rather than to a locality that appears very "Early Victorian" ... The places to avoid are centres, such as cathedral towns, where anything new is looked upon with suspicion, and must stand the test of time before it can be trusted.[46]

The following year, she authored a book on how a woman could become a dispenser: *How to become a Dispenser: A New Profession for Women*.[47] Then in 1918, she authored *How to Become a Woman Doctor*[48] and in 1920, *Analytical Chemistry as a Profession for Women*.[49] Her next book was *Vegetarian Cookery*,[50] published in 1930, and re-published in 1942 as: *Everybody's Vegetarian Cookery Book*.[51] Forester died in 1939, aged 69.

---

[46]Forster, E. L. B. (12 August 1916). The ideal neighbourhood for the woman pharmacist. *Pharmaceutical Journal* **97**: 158.

[47]Forster, E. L. B. (1917). *How to Become a Dispenser: The New Profession for Women.* T. Fisher Unwin, London.

[48]Forster, E. L. B. (1918). *How to Become a Woman Doctor.* Charles Griffin & Co Ltd., London.

[49]Forster, E. L. B. (1920). *Analytical Chemistry as a Profession for Women.* Charles Griffin & Co., London.

[50]Forster, E. L. B. (1930). *Vegetarian Cookery.* W. Foulsham, U.K.

[51]Forster, E. (1942). *Everybody's Vegetarian Cookery Book.* W. Foulsham, U.K.

Imperial College

The Chemistry Department of Imperial College had two precursors.[52] The first was the Royal College of Chemistry which had been established in Hanover Square, London in 1845, the result of a private enterprise to found a college to aid industry.[53] In 1853, the Royal College of Chemistry affiliated with the Government School of Mines Applied to the Arts, effectively becoming its chemistry department. The name was changed to the Royal School of Mines (RSM) in 1863, and new buildings were constructed in South Kensington in 1872.

Some of the courses, including chemistry, were merged with other science subjects, to form the Normal School of Science (named after the École Normale in Paris). The Normal School of Science was renamed the Royal College of Science (RCS) in 1890. At first, the RCS saw itself primarily as a teacher training college in practical science and, as such, attracted women students including some from overseas.

The second precursor originated with the Cowper Street Schools which then became Finsbury Technical College.[54] Finsbury College was intended as the first of a number of "feeder" colleges for the Central Institution (renamed the Central Technical College) but was almost certainly the only one founded. Needing a large new location, one was found in South Kensington adjacent to the RSM and RCS.

In 1907, the RSM and RCS were incorporated as constituent colleges under the name of Imperial College (IC). The same year, the Central Technical College was renamed the City and Guilds College (CGC), and it was then incorporated as a third constituent College of IC. Wishing to avoid the conflicts within the University of London, primarily between King's College and University College, it was only in 1929 that IC was finally cajoled into joining the University of London.[55]

---

[52] Gay, H. and Griffith, W. P. (2017). *The Chemistry Department of Imperial College London: A History 1845–2000*. World Scientific Publishing, London.

[53] Note 52, Gay and Griffith, pp. 9–25.

[54] Note 52, Gay and Griffith, pp. 92–102.

[55] Gay, H. (2007). *The History of Imperial College London 1907–2007*. Imperial College Press, London, pp. 84–88.

Women Chemistry Students

At IC Chemical Society meetings, the women chemistry students were expected to serve tea. In a 1904 issue of the college newspaper, *The Phoenix*, a male student reported back to his parents that: "You buy a six-penny ticket, and grab what you can get, whilst you ask one of the lady students, who are quite a jolly lot, for a cup of tea."[56]

There was a sentiment among some male students that women did not belong in their midst but in the biological sciences. The following rebuttal titled "Some Fallacies about US" was published in *The Phoenix* in 1905:

> In patronizing fashion masculine voices have enunciated that *they* are the students for Mathematics, Mechanics, Chemistry and Physics (that is, everything they consider worth knowing), but that we, *par excellence*, are of the right mettle for Biology, and that in attempting to work at any other science we must surely feel sadly out of our element. ... It seems scarcely in accordance with the theorem under discussion that neither Girton nor Newnham is overrun by ardent girl biologists, that Holloway and Bedford Colleges possess but a few lonely votaries of the science...[57]

The article concluded with a note that 14 of the 15 women students then at IC agreed totally with the sentiments expressed in the article.

This was the period of the suffragette movement, and an article in a 1907 issue of *The Phoenix*, promoting the right of women to vote,[58] provoked an ominous response:

> From the article [in the preceding issue] and also from a certain incident that took place in the College, I understand that all the lady students have turned suffragettes ... My real object in writing this article is to warn our

---

[56]"D.M.L." (1905–1906). The letters of a self-made student to his parents. *The Phoenix* **18**: 13–14.

[57]"C.E.A.S." [Speed, C.]. (1904–1905). Some fallacies about US. *The Phoenix* **17**: 32–35. Clarisse Speed completed an Associateship of the Royal College of Science (A.R.C.S.) from 1902–1906 and then became a Lecturer in Music (Extra-Mural) at Cambridge University.

[58]"Suffragist." (1907–1908). The question of the hour. *The Phoenix* **20**: 64–65.

own Little Band to be very careful and take great care of themselves. I think that perhaps they had better wait until they are older; for the present they must be good little girls and study their Chemistry.[59]

Lucy Alcock

Lucy Alcock[60] was one of the pioneering women chemistry students at the RCS. She was a student from 1904 to 1907 gaining a B.Sc. (Hons.) in 1907 (followed by her sister, Mary, in 1908). Alcock, then became a Staff Member at IC for the 1907–1908 year. Born in Stoke-on-Trent in 1884, she was the daughter of Joseph Alcock, a tile merchant, and Barbara Bowen. A glowing account of a 1908 presentation by Alcock to the IC Chemical Society was reported in *The Phoenix*:

> Miss Alcock was then called upon to disclose the wonders of Colloids. Undismayed by the mysteries of a meaningless nomenclature, Miss Alcock laid bare all the secrets of science with a lucidity and clearness that would have done credit to Minerva herself. With a winning women's way we were conducted through the intricacies of one of the obscurest domains of chemistry, and the low murmur of pleasure which had been gradually growing through the meeting became a roar of delirious and deafening delight when the fair lecturer showed a colloidal solution of barium sulphate that had the consistency, appearance, and properties of condensed milk.[61]

Alcock then became a science teacher. In 1926, together with her mother, she emigrated to Canada,[62] where her brother Frank had become a farmer in Kinsella, Alberta, having himself emigrated in 1910. Alcock died there on 17 June 1936.

---

[59] "R.M.H." (1907–1908). Suffrage. *The Phoenix* **20**: 90.
[60] *Register of Old Students and Staff of the Royal College of Science*, (1951). 6th edn., Royal College of Science Association, London.
[61] Carvel, H. (1908–1909). Chemical Society. *The Phoenix* **21**: 120.
[62] Anon. (May 1927). Royal College of Science Association (Old Students and Staff). *The Phoenix* (new ser.) **12**(5): 81.

Martha Whiteley

In the first half of the 20th century, there were two women whose names became recognised across the British chemical community. These were the chemist, Martha Whiteley, and the biochemist, Ida Smedley Maclean (see Chapter 9).[63] Martha Annie Whiteley,[64] born on 11 November 1866 at Hammersmith, London, was the daughter of William Sedgwick Whiteley and Mary Bargh. She was educated at Kensington High School and then she entered RHC from which she graduated with a B.Sc. in Chemistry in 1890. Between 1891 and 1900, Whiteley was Science Mistress at Wimbledon High School and for the next 2 years, science Lecturer at St. Gabriel's Training College, Camberwell. During the 1898–1902 period, she was also undertaking research at the RCS.

In 1902, Whiteley was awarded a D.Sc. from the University of London, and the following year, she was invited to join the staff at the RCS as an assistant under William Tilden. Her appointment was reported in *The Phoenix*: "At the Prize Distribution, we had the very great pleasure of watching Dr. Whiteley take her place with the Staff, the first lady to occupy that position at the Royal College of Science."[65] Whiteley was promoted to demonstrator in 1908 and, with the drafting of male scientists for war work in 1914, she was appointed to the rank of lecturer.

Whiteley founded the Imperial College Women's Association in 1912, upon recommendation from then Rector Sir Alfred Keogh. Keogh remarked: "Here you have no ordinary woman. I know of no one more likely to inspire women students to great things in science than Dr. Whiteley."[66] This pioneering association provided Imperial College women science students with a sense of community and mutual support. Whiteley was elected as first President, a position she held for many years,

---

[63]Rayner-Canham, M. and Rayner-Canham, G. (2011). Forgotten pioneers. *Chemistry World* **8**(12): 41.

[64](a) Barrett, A. (2017). *Women at Imperial College: Past, Present and Future.* World Scientific, London, pp. 69–78; (b) Nicholson, R. M. and Nicholson, J. W. (2012). Martha Whiteley of Imperial College, London: A pioneering woman chemist. *Journal of Chemical Education* **89**: 598–601.

[65]"C.E.A.S." [Speed, C.]. (1904–1905). From our lady correspondent. *The Phoenix* **17**, 12.

[66]Quoted in note 64(a), Barrett, p. 73.

the admiring female students secretly giving her the affectionate title of *Queen Bee.*[67]

As will be described in Chapter 15, during the First World War, Whiteley became actively involved in war-related duties. Her contributions resulted in her being awarded the OBE in 1920.

Retaining her academic position after the end of the First World War, Whiteley was made an Assistant Professor (a senior post at Imperial, later designated as Reader) in 1920. As T. S. Moore, one of her obituarists noted:

> A former R.H.C. student who went on to the Imperial College in 1920 writes: "She was in charge of the undergraduates' organic laboratory while I was there. They were nearly all men and a very lively lot — being mainly ex-Service men, but she had them completely under control. She managed to turn them into fairly tidy, efficient practical workers.[68]

Whiteley maintained an active research program, authoring or co-authoring at least 15 publications. Mary Creese, in her biography of Whiteley, commented:

> Although several British women from about Whiteley's time were productive researcher workers, most of them made their contributions as assistants to male chemists. ... Whiteley, however, was probably the only one who found a place as an independent worker in an established area of chemistry and remained active in research, teaching and technical writing throughout a long career at a major educational institution — a notable achievement for a woman chemist of her generation.[69]

Whiteley co-authored a manual on organic chemical analysis,[70] but her greatest contribution was that on the *Dictionary of Applied Chemistry.*

---

[67] Note 55, Gay, p. 421.

[68] "T.S.M." [Moore, T. S.]. (December 1956). Martha Annie Whiteley, O.B.E., D.Sc., F.R.I.C.. *College Letter, Royal Holloway College Association* 60–61, RHC AS/902/99.

[69] Creese, M. R. S. (1997). Martha Annie Whiteley (1866–1956): Chemist and editor. *Bulletin for the History of Chemistry* **20**: 44.

[70] Thorpe, J. F. and Whiteley, M. A. (1925). *A Students Manual of Organic Chemical Analysis, Qualitative and Quantitative.* 2nd edn., 1927, Longman Green & Co., London.

128 *Pioneering British Women Chemists: Their Lives and Contributions*

She had helped extensively in the preparation of the second edition of (Edward) *Thorpe's Dictionary of Applied Chemistry*. Thus, when a third edition was contemplated, she was asked to co-edit it with Jocelyn Thorpe.[71] She retired in 1934, but continued to work on the volumes as editor-in-chief following the death of her co-editor.[72] This task was a labour of love that continued until 1954, when the last volume appeared. Whiteley was still editing and proofing other manuscripts until shortly before her death, just prior to her 90th birthday, on 24 May 1956.

Margaret Carlton

The second woman to teach in the Chemistry Department of IC was Margaret (Maggie) Carlton.[73] Carlton was born on 19 December 1894, daughter of Thomas Carlton, engineer surveyor, and Wilhelmina Carlton. She first attended Croydon High School and transferred to Sutton High School in 1908. Carlton then attended Birkbeck College, completing a B.Sc. (Hons.) in Chemistry in 1919.

Following graduation, Carlton moved to IC to become a research student with the Inorganic Chemistry Professor of the time, Herbert Brereton Baker,[74] completing her Ph.D. in 1925. As Hannah Gay noted:

> Baker, like many of the early professors, had a gifted woman assistant. Margaret Carlton carried out much of the research in his laboratory and was acknowledged when Baker won the Davy Medal from the Royal Society.[75]

Baker was very supportive of Carlton, looking after her interests including seeking salary raises for her.[76]

---

[71] Thorpe, J. F. and Whiteley, M. A. (1934, 1936, 1936). *Dictionary of Applied Chemistry*. Supplement 3 Vols, Longmans, Green & Co., London.

[72] Anon. (1942). Thorpe's Dictionary of Applied Chemistry. *Nature* **149**: 665.

[73] (1951). *Register of Old Students and Staff of the Royal College of Science*. 6th edn., Royal College of Science Association, London.

[74] Thorpe, J. F. (1935). Herbert Brereton Baker, 1862–1935. *Obituary Notices of Fellows of the Royal Society* 1: 522–526.

[75] Note 55, Gay, p. 155.

[76] Note 52, Gay and Griffith, p. 192n43.

London Co-educational Colleges 129

Upon Baker's retirement, Carlton was appointed Assistant Lecturer while assisting Henry Briscoe in his research. She was not promoted to Lecturer until 1946. Gay added:

> This limited career progress was typical for women of her generation. While several of the women working at the college in this period were acknowledged as gifted scientists, they were not seen as serious candidates for professional advancement (see Footnote 75).

Carlton retired in 1960.

Frances Micklethwait

One of Whiteley's assistants (another being Edith Usherwood, see Chapter 14) was Frances Mary Gore Micklethwait.[77] She was born on 7 March 1867, daughter of John P. Micklethwait, barrister and later farmer, and Mary Gore, farmer, of Chepstow, Monmouthshire. After a private education, she lived at home until 1897. That year, Micklethwait entered the Swanley Horticultural College, which had only just opened its doors to women,[78] and it was at Swanley where she gained her love of chemistry. In 1898, she transferred to the RCS (with which Swanley had links), obtaining an associateship in 1901.

Micklethwait joined the RCS/IC research group of Gilbert T. Morgan,[79] becoming one of the most prolific women authors of chemistry publications of her time,[80] co-authoring at least 22 papers between 1902 and 1914. In his personal reminiscences, Morgan commented:

> In 1904 I was joined by Miss Micklethwait, a member of this [Chemical] Society, with whom I collaborated for nine years. During this period we succeeded in arousing the interest of many senior students, about 14 of

---

[77] Note 64(a), Barrett, pp. 353–355.

[78] Opitz, D. L. (2013). "A triumph of brains over brute": Women and science at the Horticultural College, Swanley, 1890–1910. *Isis* **104**: 30–62.

[79] Irvine, J. C. (1941). Gilbert Thomas Morgan, 1872–1940. *Obituary Notices of Fellows of the Royal Society* **3**: 354–362.

[80] Burstall, F. H. (1952). Frances Mary Gore Micklethwait (1868–1950). *Journal of the Chemical Society* 2946–2947.

130   *Pioneering British Women Chemists: Their Lives and Contributions*

whom were included in our joint publications. The work, which covered a wide field, included various studies of the diazo reaction, the preparation of organic arsenicals and antimonials, and the examination of certain coordination compounds of coumarin.[81]

At the outbreak of the First World War, Micklethwait came under the wing of Whiteley (work which will be discussed in Chapter 15). For her war services, Micklethwait was awarded the MBE in 1918. George Kon, one of Micklethwait's students during the First World War (and later Professor of Chemistry), commented that Micklethwait had high expectations: "I soon realized that my ideas of what work meant had to be overhauled."[82]

After the War, Micklethwait worked briefly in the research laboratory of Boots Pure Drug Company and then returned to Swanley Horticultural College, where she taught until 1921, being principal for her last year there. From then until 1927, she compiled the index for the second edition of *Thorpe's Dictionary of Applied Chemistry*, the series co-edited by her friend, Whiteley. She died on 25 March 1950, aged 83.

Helen Archbold (Mrs. Porter)

As we will show in Chapter 11, most women biochemists of our era were part of the Hopkins' group at Cambridge. Archbold was one of the few renowned biochemists outside of that group, though she did interact with them. Helen Kemp Archbold[83] was born on 10 November 1899 at Farnham, Surrey. Her father, George Kemp Archbold, was a school headmaster, and her mother, Emily Broughton Whitehead, was a Belgian-trained professional singer. Educated at Clifton High School, Bristol, Archbold entered Bedford College in 1917, graduating with a B.Sc. in 1921.

---

[81] Morgan, G. T. (1939). Personal reminiscences of chemical research. *Chemistry and Industry* 665–673.

[82] Linstead, R. P. (1952–1953). George Armand Robert Kon, 1892–1951. *Obituary Notices of Fellows of the Royal Society* **8**: 172.

[83] (a) Northcote, D. H. (1991). Helen Kemp Porter. 10 November 1899–7 December 1987. *Biographical Memoirs of Fellows of the Royal Society* **37**: 400–409; (b) Note 64(a), Barrett, pp. 95–104.

London Co-educational Colleges 131

For her graduate studies, Archbold obtained one of the places reserved for women in Jocelyn Thorpe's organic chemistry research group at IC as her biographer, Donald Northcote noted:

> She [Helen] obtained one of the two or three places in Professor Thorpe's organic chemistry department that were made available to women by the persuasion of Dr. Martha Whiteley.... under the supervision of the strict and assiduous Dr. Whiteley set the high standards of Helen's future work.[84]

During the large-scale importation of apples, whole cargoes in the refrigerated holds of ships were found to have turned brown and unfit to eat. A joint Cambridge–Imperial research team was set up in 1918 to investigate this problem. At IC, it was V. H. Blackman who led their group and, needing a chemist, Archbold was invited to join. The other members at IC were Dorothy Haynes and Elsie Widdowson (see Chapter 16). As she lacked a background in biology, Archbold took evening classes in the subject at Birkbeck College. Over time, her research shifted from the chemical analysis of apples to the study of the metabolic processes of the maturing apple. This involved liaising with the Biochemistry Department at Cambridge, including working with F. Gowland Hopkins and Muriel Wheldale (see Chapter 11).

In 1931, she was transferred to the plant physiology team at IC to work on the origin of the starch in barley. This study continued into the 1940s and laid the foundation for her subsequent research on polysaccharide synthesis. The work with barley led to a lasting interest in the enzymes responsible for starch formation and breakdown. In 1947, she spent a year in St. Louis, United States, with the Nobel Laureates, Gerti and Carl Cori, the discoverers of the Cori cycle.

Returning to IC, she received her first major grant, enabling her to set up a research group and equip a laboratory. The original intention was to study enzyme systems by conventional means, but in the 1950s, her group focussed on the newly developed radioactive tracer methods. In particular, they were among the first to prepare radioactive biochemicals and use

---

[84] Note 83(a), Northcote, p. 404.

132 *Pioneering British Women Chemists: Their Lives and Contributions*

them to study the intermediate metabolism of plants. The work was so innovative that she was elected a Fellow of the Royal Society in 1956, and in 1959, she became the first woman professor ever at the Imperial College.

Archbold combined her academic brilliance with a taste for adventure and travel. Unfortunately, her personal life was dogged by tragedies. Both her first husband, William George Porter, and her second, Arthur Huggett, each died after only a few years of marriage. Archbold, herself, died in 1987 at Goring, Oxfordshire, aged 88.

East London College (Later Queen Mary College)

At the other end of the social spectrum from the Oxbridge Colleges (see Chapter 3) was East London College.[85] When it opened in 1888, it was titled the People's Palace Technical Schools, changing its name to East London Technical College in 1897. Then in 1915, the institution was incorporated in the University of London as East London College (ELC) and renamed as Queen Mary College (QMC) in 1934.

Kathleen Balls (Mrs. Stratton)

Kathleen Balls,[86] the daughter of Edwin Balls, a Clapham carpenter, and Catherine Anne Wood, was born on 2 May 1890 at Hackney, London. Educated at the City of London School for Girls, she entered East London Technical College in 1908 and completed a B.Sc. (Hons.) Chemistry degree in 1911. After graduation, Balls was appointed Science Mistress at Swansea Municipal School.[87]

At the outbreak of war, Balls was the science teacher at the County School, Enfield. With the greater urgency for university academic staff, she was released from her school teaching duties and hired by the

---

[85] Hickinbottom, W. J. (1956). Schools of chemistry in Great Britain and Ireland–XXVI Queen Mary College, London. *Journal of the Royal Institute of Chemistry* **80**: 457–465.

[86] *Staff and Student Records,* Archives, Queen Mary College, London.

[87] Anon. (December 1912). News of old girls. *Magazine of the City of London School for Girls* **16**(3): 50.

Chemistry Department of East London College.[88] This college had a specific need, for in 1915, John Theodore Hewitt,[89] the Head of Chemistry at the time, was commissioned in the Royal Engineers. The only remaining chemist, F. G. Pope, was placed in charge. Balls was hired to assist Pope as Lecturer and Demonstrator in Chemistry, enabling courses to be run through the War period. In 1917, Balls married Septimus Stratton in Hackney.

Balls' appointment at the college stipulated the occupancy of the position as being for the duration of hostilities only. Nevertheless, she actually continued at ELC into the inter-War era, obtaining an M.Sc. in 1919. Balls submitted her resignation in 1924. Her resignation was accepted, but she was asked to continue as lady superintendent, which she agreed to do.[90]

Though Balls had lost academic status, she remained active in chemistry at QMC, working with James Riddick Partington,[91] Hewitt's successor as senior physical and inorganic chemist. Balls and Partington co-authored three research publications between 1922 and 1936. In addition, Balls, under her married name of Mrs. Stratton, and Partington co-authored a book on chemical calculations.[92] For the Easter Term 1932, Balls replaced Mary Boyle (see Chapter 5) at the Royal Holloway College as Lecturer in Chemistry.[93] At least by 1939, Balls was back at QMC as Lecturer in Chemistry. In that year, she wrote a letter (as Lecturer) to the

---

[88] *Minutes, East London College Council.* (1 February 1916). A. Nye, Archivist, Queen Mary College Library, is thanked for this information and for copies of the two related letters between QMC and Balls.

[89] Turner, E. E. (1955). John Theodore Hewitt, 1868–1954. *Biographical Memoirs of Fellows of the Royal Society* **1**: 79–99.

[90] *Minutes, East London College Standing Committee.* (29 May 1924). See note 88.

[91] Partington is best known for his multi-volume series on the history of chemistry. See: Butler, F. H. C. (1966/1967). Obituary: James Riddick Partington. *British Journal for the History of Science* **3**: 70–72.

[92] Partington, J. R. and Stratton, Mrs. K. (1939). *Intermediate Chemical Calculations.* Macmillan, London.

[93] *Calendar, 1931–1932.* Royal Holloway College, p. xxxix. "Dr. Boyle, staff lecturer in chemistry was granted leave of absence for the Easter Term, and her place was taken by Mrs. K. Stratton, B.Sc."

134 *Pioneering British Women Chemists: Their Lives and Contributions*

administration requesting replacement for a lab coat and stockings following a chemical accident.[94] Balls died in June 1976 in Hastings.

Battersea Polytechnic

The role of women in polytechnics has been almost totally overlooked, and yet, they were an avenue for predominantly lower-social-class women to acquire chemical education, either pure or applied[95] and for women to find teaching positions in chemistry.

The great polytechnic movement in London was designed to:

> Promote the education of the poorer inhabitants of the metropolis by technical instruction, secondary education, art education, evening lectures, or otherwise, and generally to improve their physical, social and moral condition.[96]

The first polytechnic, the Polytechnic at Regent Street (now the University of Westminster), dated back to 1839, and though many of the offerings were gender-specific, some women did enter the co-educational Department of Science.

Another polytechnic, Battersea Polytechnic, attained considerable academic success. This success led to an application for recognition as a School of the University of London in 1911, though nothing came of the application.[97] Battersea Polytechnic was organized into six main departments: mechanical engineering and building trades, electrical engineering and physics, chemistry, women's subjects (specifically domestic science, to be discussed in Chapter 8), art, and music.

---

[94]Letter, (7 December 1939), Stratton, K., Archives, Queen Mary College. See Note 88.

[95]Stevenson, J. (1997). 'Among the qualifications of a good wife, a knowledge of cookery certainly is not the least desirable' (Quentin Hogg): Women and the curriculum at the Polytechnic at Regent Street, 1888–1913. *History of Education* **26**(3): 267–286.

[96]Chandler, A. R. (1998). The funding of higher education in London. In Floud, R. and Glynn, S. (eds.), *London Higher: The Establishment of Higher Education in London.* Athlone Press, London, pp. 178–198.

[97]Arrowsmith, H. (1966). *Pioneering in Education for the Technologies: The Story of Battersea College of Technology 1891–1962.* University of Surrey, Surrey, p. 30.

Mary Corner

Mary Corner[98] was one of those to obtain a University of London B.Sc. in Chemistry while studying at Battersea Polytechnic. She was born on 25 March 1899, the only child of John Corner, architect and sanitary engineer, and Elizabeth Mary Corner. Corner was educated at Beulah House High School, Balham, London, matriculating in 1915.[99] During the First World War, she worked in a pharmacy, entering Battersea Polytechnic in 1922 and graduating in 1927 with an Honours B.Sc. (London). It was during this time that she must have become acquainted with Robert Pickard,[100] an organic chemist at Battersea Polytechnic.

After graduation, Corner's career took a different direction, as her obituarist G. R. Davies described:

> After spending a further year at London University, she took a post in 1928 at the British Cotton Industry Research Association [in Manchester], thus following Dr. (later Sir) Robert Pickard, whom she greatly admired and who, in 1927, had relinquished the Principalship of the Polytechnic to become Director of the Research Association. Miss Corner first served in the rayon department and one of the papers published with her colleagues dealt with the microdetermination of metals in commercial rayon yarns. This work probably triggered off her interest in microanalysis, and it was not long before she was made head of the microanalytical department.[99]

In 1945, Corner obtained a similar post with the British Leather Manufacturers Research Association. Two years later, she accepted an invitation to become head of the newly formed Microanalytical Section of the Chemical Research Laboratory (later the National Chemical Laboratory).

As noted in her obituary, Corner had an "unfortunate accident" early in life and, "Burdened with a severe disability, she had, in addition, more

---

[98] Anon. (1963). Obituary: Mary Corner. *Journal of the Royal Institute of Chemistry* **87**: 147.

[99] Davies, G. R. (1963). Obituary: Mary Corner. *Analyst* **88**: 155.

[100] Kenyon, J. (1950). Robert Howson Pickard, 1874–1949. *Obituary Notices of Fellows of the Royal Society* **7**: 252–263.

than the usual share of suffering and trouble" (see Footnote 99). In the 1930s, she became a founder member of the Microchemical Club (to be later joined by Isabel Hatfield, see Chapter 9), and at the time of her death on 4 November 1962, she was Vice-Chairman of the Microchemistry Group of the Society for Analytical Chemistry.

Commentary

At both University College and Imperial College, women chemists were readily accepted. It was Ramsay's group that was welcoming at UCL; however, women remained as laboratory assistants. In those early years, many of the chemists at Imperial College were particularly supportive. For example, Tilden had been one of the women chemists' most vociferous advocates for admission to the Chemical Society (see Chapter 9). Of particular note, Whiteley was able to advance to the rank of Assistant Professor, a senior position at the time. Even more exceptional, Jocelyn Thorpe agreed (at Whiteley's suggestion) to assign a number of the research places in his laboratory to women students.

Religious affiliation played a significant role in how women chemists were treated. As we saw in Chapter 3, Unitarians and Nonconformists were advocates of the equality of women. In this chapter, we have shown that Anglican-affiliated King's College preferred to "banish" women students to Kensington so that they would not "contaminate" the male-exclusive college on the Strand. When co-education was imposed, the women-dominated Domestic Science program was excluded (see Chapter 8).

As a final point, at the time of the First World War, small East London College more closely resembled the English Provincial universities (see Chapter 6) in that a woman chemist (Balls) was hired to do most of the teaching during that War. In general, as we show in later chapters, it appears that women chemists played more important roles at smaller institutions.

Chapter 5

London Women's Colleges

Sibyl Widdows (27 May 1876–4 January 1960)

The account of the London co-educational colleges in the previous chapter provides only part of the story of the education of women chemists in the capital city. The two women's colleges, Bedford College and Royal Holloway College (RHC), of the University of London may have been small, but they graduated a disproportionately large number of women chemists. In fact, oft-forgotten Bedford College was more important in

138 *Pioneering British Women Chemists: Their Lives and Contributions*

terms of women chemists graduated than the Oxbridge women's colleges.[1]

Up to now, we have focussed on the traditional university pathway for women chemistry students. There were other avenues by which young women could obtain a tertiary-level education that involved chemistry studies. One of these pathways was the London School of Medicine for Women (LSMW),[2] which we consider also "belongs" in this chapter.

The chemistry departments at these three institutions shared another factor: all three hired significant numbers of women chemists as faculty and staff. Thus, not only did women chemistry students gain an education at these three institutions but they also provided havens of employment for qualified women chemists.

Bedford College

Bedford College had evolved from a set of classes first held in 1849 at the house of Mrs. Reid in Bedford Square. Mrs. Reid, a Unitarian, wanted women to be part of the governance of the college and for the college, then named the Ladies' College, Bedford Square, to be non-denominational.[3] The early history was full of turmoil: to Reid's disappointment, she was not flooded with women students. In addition, the lecturers who came from King's College, an Anglican institution, were forced by their Principal to resign from part-time teaching at Nonconformist Bedford.[4]

Three factors led to a change in fortune for the better. First, the formation of the "academic" girls' secondary schools under the Girls Public Day

---

[1] Rayner-Canham, M. and Rayner-Canham, G. (2006). The pioneering women chemists of Bedford College, London. *Education in Chemistry* **43**(3): 77–79.

[2] Rayner-Canham, M. and Rayner-Canham, G. (2017). Women chemists of the London School of Medicine for Women, 1874–1947. *Bulletin for the History of Chemistry* **42**: 126–132.

[3] Watts, R. E. (1980). The Unitarian contribution to the development of female education, 1790–1850. *History of Education* **9**(4): 273–286.

[4] Pakenham-Walsh, M. (2001). Bedford College 1849–1985. In Crook, J. M. (ed.), *Bedford College: Memories of 150 Years*. Royal Holloway and Bedford College, pp. 13–45.

School Company (GPDSC) (see Chapter 2) provided cohorts of educated and enthusiastic young women. Second, the college was moved to a more spacious facility on Baker Street (now the Sherlock Holmes Hotel). Third, the University of London examinations were opened to women.

In 1880, the first women at Bedford graduated with external University of London degrees. However, the future of Bedford College was not assured until its formal inclusion as a constituent college of the University of London in 1900. One of the people who had strongly pressed the case of Bedford as a constituent college was a Bedford alumna, Sophie Bryant, by then Headmistress of North London Collegiate School (NLCS) (see Chapter 2).

Foundation of the Chemistry Department

Though lectures in chemistry had been given as early as 1877, it was the hiring of Holland Crompton[5] in 1888 as Head that really marked the founding of the Department of Chemistry.[6] The University of London chemistry and physics programs required practical components, but Bedford initially lacked any labs. One of Crompton's first tasks, then, was the design of the new laboratory facilities — the first university-level ones for women in London.[7] An article in *Girl's Realm* reported: "… Further up the stairs you come upon the chemistry laboratory, where Mr. Crompton is supervising a large class. Each girl wears a smock-pinafore of coloured linen."[8]

Having more students than it could handle, Bedford College moved again, this time to Regent's Park, a site which it was to occupy from 1913 until the College's merger with Royal Holloway College in 1985. In 1906,

---

[5]Spencer, J. F. (1932). Obituary Notices: Holland Crompton. *Journal of the Chemical Society* 2987–2988.

[6]Turner, E. E. (1953). Schools of Chemistry in Great Britain and Ireland–XVII Bedford College, London. *Journal of the Royal Institute of Chemistry* **79**: 236–238.

[7]Harris, M. M. (2001). Chemistry. In Crook, J. M. (ed.), *Bedford College: Memories of 150 Years*. Royal Holloway and Bedford College, pp. 81–94.

[8]Rawson, S. (1898–1899). Where London girls may study: I. Bedford College. *Girl's Realm* **I**: 925–929.

140 *Pioneering British Women Chemists: Their Lives and Contributions*

Crompton was joined by a Demonstrator, James F. Spencer,[9] a physical chemist. The Department was split in 1919, Crompton becoming Head of the Department of Organic Chemistry, while Spencer became Head of the Department of Inorganic and Physical Chemistry. Having suffered from bad health, Crompton retired in 1927, to be replaced by Eustace E. Turner[10] in 1928. It was these three individuals who filled the senior positions of the chemistry department through the first part of the 20th century. However, they were assisted by a series of young women chemistry graduates who held Demonstratorships and Lectureships (see the subsequent sections).

Life for Women Chemistry Students

Life at Bedford in the early decades was unlike the universities of today. A dress code was enforced: for example, about 1915, during weekdays, students were required to wear "day dress," a jacket-suit or a dress and jacket in a dark colour, or in summer, a dark skirt and light blouse (ties were commonly worn). Dinner was a specific ceremonial occasion, as a student in 1915, Margaret McDonald, recalled:

> Staff and students forgathered in the Common Room and in order of precedence each chose a partner, senior offering an arm to juniors. Any fresher left over bought up the rear of the procession, sometimes with relief at not having to make conversational efforts.[11]

The early volumes of the *Bedford College Magazine* provide an insight into the enthusiasm for chemistry that prevailed at the institution. In particular, some of the articles highlight the differences from today's chemistry. During those early years, danger was regarded as an intrinsic part of laboratory life, as the following commentary from an 1898 issue shows:

---

[9] Anon. (1951). Obituary: James Frederick Spencer. *Journal of the Royal Institute of Chemistry* **75**: 127.

[10] Harris, M. M. (1966). Obituary: Eustace Ebenezer Turner, 1893–1966. *Chemistry and Industry* 1953–1955.

[11] McDonald, M. (1991). Cited in: Bentley, L. *Educating Women: A Pictorial History of Bedford College, University of London, 1849–1985*. RHC and Bedford New College, London, p. 40.

London Women's Colleges 141

A certain amount of excitement was caused one afternoon by the fact that one student was suddenly seen to be in flames. However, she lay down quite calmly, and was immediately knelt upon by her nearest neighbours, so that all danger was over before most people knew what had happened and they only caught a glimpse of her as she lay 'smiling and smouldering' on the floor. We consider the behaviour of those concerned a credit to the College and to the cause of women's education.[12]

Explosions also seemed commonplace according to a subsequent remark in 1900: "At present, one's life is a series of adventures in the Chemistry Laboratory, for bits of flying glass are as plentiful as the smuts on the window-sill"[13]; while drinking and eating in the lab was taken for granted:

B.Sc.'s on the whole, seem to enjoy life in all places and at all times, be it at their work, or having tea-fights in the Chemistry laboratory, when beakers take the place of cups, and flasks do for kettles. Some say that tea-cakes cooked on asbestos have a very savoury taste.[13]

Chemical spillage was also prevalent:

The final chemistry students have been performing original research on the action of saturated solution of caustic potash on their books and garments. It has been found that the result is the destruction of the parts acted on.[14]

The unique appearance of chemistry students was observed:

This science [chemistry] indeed leaves a hall-mark upon its devotees. They can usually be distinguished by the moth-eaten appearance of their clothes and the peculiar colour of their hands.[14]

Qualitative work was accomplished in part by Fresenius's analysis scheme using hydrogen sulphide, or sulphuretted hydrogen, as it was then called.

---

[12] "H.A.B." (June 1898). College notes: Science. *Bedford College London Magazine*, BC AS/200/1/2 (36): 16.

[13] Anon. (June 1900). College notes: Science. *Bedford College London Magazine*, BC AS/200/1/2 (42): 3.

[14] Anon. (June 1914). College notes: Science. *Bedford College London Magazine*, BC AS/200/1/4 (84): 5.

142 *Pioneering British Women Chemists: Their Lives and Contributions*

In the annual report for 1906, the purchase of a new Kipp's apparatus[15] was announced:

> We have, I think, somewhat diminished the odour of sulphuretted hydrogen by the substitution of a new generating apparatus instead of the old one, which always seemed to be out of order; though, poor thing, with so many calls on its attention, it is perhaps no wonder that it broke down, when old age made it feeble.[16]

Chemistry had the largest enrolment among the sciences.[17] In addition to the classes and laboratory sessions, there were weekly Chemical Society meetings. A student by the pseudonym of "C=M" was obviously critical of a 5 p.m. meeting held in 1905, at which 18 papers were presented:

> The victims of the evening invariably retire from the laboratory to remove traces of the day's labour from their hands five minutes before five.... There is no lack of information in the papers, but the authors generally consider it incumbent upon them to deliver it at such a rate that the audience is left almost too exhausted to applaud.[18]

As we found elsewhere, chemistry-based rhymes were popular in the early 20th century, and we provide one here from the chemistry class of 1917.

> *We Chemists are met with our overalls on*
> *For the Work that is before us —*
> *With acids and alkalis, Kipps and corks,*
> *And a dreadful odour around us!*
> *You ask, why we favour H_2S?*
> *Whatever in "Stinks" entrances?*
> *We're making every possible thing*

---

[15] Sella, A. (November 2007). Kipp's apparatus. *Chemistry World* 81.

[16] "μχ" (June 1906). College notes: Science. *Bedford College London Magazine* (60): 4, BC AS/200/1/3.

[17] Anon. (July 1918). Science letter. *Bedford College London Magazine* (93): 3, BC AS/200/1/5.

[18] "C=M" (March 1905). College notes: Science. *Bedford College London Magazine* (56): 11, BC AS/200/1/3.

Of interest and Importance
(chorus)
Chemistry! O Chemistry
Hark to the sounds of explosions!
Electric shocks to the Physicists leave –
But give to the Chemists their odours![19]

Edith Humphrey

Among the Bedford College chemistry alumna was the forgotten pioneer of coordination chemistry, Edith Ellen Humphrey.[20] Born on 11 September 1875, in Kentish Town, Middlesex, she was the daughter of John Charles Humphrey, clerk, and Louisa Frost. Humphrey attended NLCS, and it was there that she first took chemistry: "At the North London we did quite a bit of it. We had a good teacher who had no degree but who was very good."[21] Her father encouraged all his children, even the girls, to obtain a good education, and Humphrey entered Bedford College in 1893. Humphrey continued with chemistry, as she considered it to be "where things were happening."

Following completion of her B.Sc. degree in 1897, Humphrey applied and was accepted to do a Ph.D. at the University of Zürich. Zürich first admitted women in the mid-1860s, and it had become a haven for women students from all over Europe.[22] The anonymous author of the College Notes: Science in the *Bedford College London Magazine* wrote:

> At the beginning of this term [in 1898] the familiar face of Miss Humphrey was much missed, especially in the Chemical Laboratory, but all will be glad to hear of her well-being at Zürich. She is engaged upon a "Doktor-Arbeit" on a compound of Cobalt.[23]

---

[19] Anon. (January 1917). Departmental ditties and laboratory lilts: Chemistry. *Bedford College London Magazine* (90): 17, BC AS/200/1/5.

[20] Bernal, I. (1999). Edith Humphrey. *The Chemical Intelligencer* **5**(2): 28–31.

[21] Brandon, R. (11 September 1975). Going to meet Mendeleev. *New Scientist* 593.

[22] Bridges, F. (1890). Coeducation in Swiss universities. *Popular Science Monthly* **38**: 524.

[23] Anon. (December 1898). College notes: Science. *Bedford College London Magazine* (37): 16, BC AS/200/1/2.

144 *Pioneering British Women Chemists: Their Lives and Contributions*

Humphrey wrote a long essay on her life at Zürich, which was published in the *Bedford College Magazine*.[24] Her doctoral program extended for 4 years, and for the research, she had to provide her own apparatus and chemicals.

It was her thesis work with Alfred Werner that was to enter her in the annals of the history of chemistry.[25] Among the compounds she made was *cis*-bis(ethylenediamine)dinitrocobalt(III) bromide. This was the very first synthesis of a chiral octahedral cobalt complex, though at the time, the significance of her synthesis was overlooked. Werner was so impressed with her work that, for her last year, he took her on as his personal assistant, the first woman to be chosen for this prestigious post. More important for the impoverished Humphrey, she at last had some income in very expensive Switzerland.[26]

Following completion of her Ph.D. in 1901, it was recommended that Humphrey continue her studies with Wilhelm Ostwald[27] in Germany. There was a problem, as Humphrey herself commented: "But they wouldn't have me in Germany. They said I could go to lectures but not practical's because the men wouldn't do any work" (see Footnote 21).

Thwarted in her plans, Humphrey returned to England, spending the rest of her working life as a Research Chemist with the company Arthur Sanderson & Sons, who specialized in such products as fabrics and wallpapers. It was during her time with Sanderson that she became a signatory of the 1904 petition for admission of women to the Chemical Society and of the 1909 letter to *Chemical News* (see Chapter 9). She died in 1977 at the age of 102. The chiral crystals, synthesized by Humphrey, are now in Burlington House, Piccadilly, London, as they were donated in 1991 to the Royal Society of Chemistry on the occasion of its 150th anniversary by the Swiss Committee on Chemistry.

---

[24] Humphrey, E. (June 1900). The University of Zurich, *Bedford College London Magazine* (42): 25.

[25] Kauffman, G. B. (1966). *Alfred Werner: Founder of Coordination Chemistry*. Springer-Verlag, Berlin.

[26] Anon. (March 1901). College notes. *Bedford College London Magazine* (44): 18.

[27] Fleck, G. (1993). Wilhelm Ostwald 1853–1932, in Laylin, J. K. (ed.), *Nobel Laurates in Chemistry, 1901–1992*. American Chemical Society, Washington, D.C., pp. 61–68.

Women Chemistry Staff

Though men held the senior ranks in chemistry at Bedford College, women played key roles as junior staff. The first woman appointee occurred in 1898, and this was Barbara Tchaykovsky.

Barbara Tchaykovsky

Barbara "Ally" Tchaykovsky[28] was born on 26 September 1875 in New York, the daughter of Nikolay Vasilievitch Tchaykovsky, a Russian university professor, and Barbara Alexandrovna Tchaykovsky. The Tchaykovskys had fled to New York to escape the Tsarist regime. Shortly after her birth, the family moved again, this time to London. She was educated at NLCS and then completed a B.Sc. in Chemistry at Bedford College in 1897. Tchaykovsky was hired by Crompton the following year with the rank of Assistant Lecturer in Chemistry.

After working 2 years at Bedford, Tchaykovsky was offered a Reid Fellowship which she used to study medicine at the LSMW, completing an M.B. (B. Medicine) and B.S. (B. Surgery) in 1906 the following year, she was awarded the Diploma of Public Health from Cambridge University and then, in 1908, completed an M.D. from the University of London, being the recipient of the University Gold Medal in State Medicine.[29] That same year, Tchaykovsky was appealing for support for the release of her father who had been arrested in Russia.[30]

Tchaykovsky's first medical appointment was that of house-physician and assistant anaesthetist at the Royal Free Hospital; then in 1909, she accepted a part-time position as School Medical Officer with the London County Council. Tchaykovsky refused offers of a full-time position, as she believed it would interfere with her many social causes. For example, during the First World War, she became active with the East London

---

[28] Anon. (n.d.). Pioneer of the child welfare service: Death of Dr. Barbara Tchaykovsky. Unidentified newspaper clipping, Archives, North London Collegiate School.

[29] Anon. (3 March 1956). *British Medical Journal* 524.

[30] Price, J. M. (16 November 1909). Letter to the Editor: Russian Justice, Daughter of Tchaykovsky fears her father will be tried secretly. *The New York Times*, p. 8.

146 *Pioneering British Women Chemists: Their Lives and Contributions*

Federation of Suffragettes.[31] The horrifying rates of infant mortality in London was another of her areas of activism.[32] In a stirring appeal for action, Tchaykovsky reported that, for example, during 6 months in 1914–1915, 50,209 children's deaths occurred just in London alone.[33]

As a contributor to an obituary for Tchaykovsky, C.J.T. wrote: "Outside her official work she devoted herself to the furtherance of child welfare, and with the late Miss Margaret McMillan she was a pioneer in the establishment of nursery schools and school clinics."[34] Tchaykovsky was forced to retire from the position of School Medical Officer at the age of 65, but kept active with her voluntary work until her death in Watford on 4 February 1956, aged 80.

Paule Vanderstichele

The incumbents of the Demonstrator and Lecturer positions changed periodically, the next one of note being Paule Laure Vanderstichele.[35] Vanderstichele had obtained a university education in Belgium, but had fled to Britain following the German invasion of 1914. She entered Bedford College the same year and after completing a B.Sc. in 1917, she commenced work on an M.Sc. with Crompton, her research resulting in two publications, the second under her name alone. Concurrently with her studies, Vanderstichele was appointed as Demonstrator in organic chemistry. Following Nellie Walker's resignation (see Chapter 7), Vanderstichele was promoted to Senior Demonstrator. For the 1922 and 1923 years, she was a recognized Teacher of the University of London.

---

[31] (a) Jackson, S. and Taylor, R. (2014). *East London Suffragettes*. The History Press, London; (b) Winslow, B. (1996). *Sylvia Pankhurst: Sexual Politics and Political Activism*. UCL Press, London, pp. 77, 85; see also (March 1915). Appeal for Help, *Our Magazine: North London Collegiate School for Girls*, 16.

[32] "B.K.W." (Michaelmas 1916). Miss Tchaykovsky's Lecture. *Roedean School Magazine* (21): 11.

[33] Anon. (20 November 1915). An S.O.S. call to save the babies of England. *The Survey* **35**: 177.

[34] Note 29, Anon, 524.

[35] *Staff and Student Records*, Archives, Bedford College.

During the 1920s and 1930s, Vanderstichele taught at Bolton School, Bolton; Collegiate School for Girls, Leicester; and Southampton Girls' Grammar School. It was at Southampton that, during an experiment to collect hydrogen gas, there was an explosion and glass shards lodged in one of her eyes. She also taught at the High School for Girls, Pretoria, South Africa, most probably between 1930 and 1932. Vanderstichele must have returned again to South Africa, for it was noted that in 1945, she was teaching at Huguenot University College, Wellington, Cape Province.

Mary Crewdson

Mary Sumner Crewdson[36] was hired at Bedford College in 1919. Crewdson was born on 23 April 1889, daughter of Moses F. Crewdson, a Wesleyan Methodist minister, and Alice Sumner. She was educated at the High School, Wallasey; Derby Municipal Technical College; and NLCS and entered Bedford College in 1907. Completing her B.Sc. (Hons.) in 1910, Crewdson was a science teacher at Saltburn High School for Girls between 1910 and 1918. Returning to Bedford in January 1919, she was appointed as Demonstrator in inorganic and physical chemistry, following Spencer's advancement to Professor. Crewdson was promoted to Assistant Lecturer later in 1919, then Junior Lecturer in 1920, and finally Lecturer in 1922.

While teaching, Crewdson worked with Spencer on an M.Sc., which she completed in 1923. In 1926, her nomination for recognition as Teacher of the University of London was turned down on the grounds of insufficient research. She resigned due to ill-health before her appointment was terminated and that same year became Warden of Northcutt House and then Lindsell Hall, both residences of the College. Retiring in 1954, Crewdson died on 18 February 1966 at Burnham, Buckinghamshire, aged 76.

---

[36](a) *Staff and Student Records*, Archives, Bedford College; (b) *Student Records*, Archives, North London Collegiate School (Frances Mary Buss Schools for Girls).

Ivy Rogers

The most tragic story was that of Ivy Rogers.[37] Ivy Winifred Elizabeth Rogers was born on 14 February 1900, daughter of Ernest W. T. Rogers, public works contractor, and Sarah Ann Darby. From Notting Hill High School, Rogers came to Bedford in 1918 as a Junior Laboratory Assistant, being promoted to Senior Laboratory Assistant in 1925. In her spare time, she studied for a Special B.Sc. in Chemistry, completing it in 1928. It was Spencer who conveyed the news of her death to the chemical community in 1934:

> Miss Rogers died as a result of an accident on June 20th [1934] in the Chemistry laboratories of Bedford College.... Bedford College, by her tragic and untimely death, has lost a very willing, faithful and efficient servant and those who knew her have lost a loyal and staunch friend.[37]

The exact nature of the accident was not noted in the department files.

Mary Lesslie

The long-time staff members at Bedford were Mary Lesslie, an organic chemist, who started in 1927, and Violet Trew, a physical chemist, in 1930. The pair were the anchors of the Bedford Chemistry Department for the next 40 years.

Mary Stephen Lesslie,[38] daughter of Andrew J. W. Lesslie, hairdresser, and Mary Anne Stephen of Dundee, was born in 1901 and educated at the Morgan Academy, Dundee. She was only 16 years old when she entered University College, Dundee (now the University of Dundee), graduating with an M.A. (St. Andrews) in 1922, and a B.Sc. (St. Andrews) in 1924. She then completed a Ph.D. (St. Andrews) with Alexander

---

[37] Spencer, J. F. (1934). Ivy Winifred Elizabeth Rogers, 1900–1934. *Journal of the Chemical Society* 2016.

[38] Anon. (May 1988). Obituary: Mary Lesslie. *Royal Holloway and Bedford New College Association, College Journal* 17–19.

McKenzie[39] on stereochemistry in 1927, her results being published in two papers.

Lesslie immediately left Scotland to take up an appointment as Demonstrator at Bedford College. She carried with her from Dundee: "a trunk containing not only her personal effects and books but a vacuum desiccator, the first seen in the Bedford Department" (see Footnote 38). A significant proportion of the organic chemistry teaching fell on Lesslie's shoulders, particularly in her first year when Crompton was in failing health. With the arrival of Turner, the load was eased slightly and, in addition, Lesslie and Turner formed a research partnership that blossomed over the following decades. During the period 1928–1956, Lesslie co-authored 21 research publications with Turner, nearly all on organic stereochemistry involving the resolution of isomers.

Lesslie's initial appointment at Bedford terminated in 1931 and was non-renewable. With her research flourishing, she continued to work unpaid during the day at Bedford, taking evening employment elsewhere in order to survive. Fortunately, in 1932, she obtained a re-appointment as a Junior Lecturer, and subsequently rose to the rank of Senior Lecturer in 1947. Appointed a Recognised Teacher of the University of London, Lesslie was awarded a D.Sc. (London) in 1950.

During the Second World War, the staff and students of Bedford were evacuated to the University of Cambridge. Teaching duties became even more onerous for Lesslie as the degrees had to be completed within 2 years. In addition, she and Turner, together with Margaret Jamison (later Mrs. Harris), undertook a program of syntheses of pure hydrocarbon isomers, for the Ministry of Aircraft Production.[40]

Lesslie's speciality was getting recalcitrant Grignard reactions to work. In her obituary, this focus of her life was noted:

> As ever, it was in the laboratory that the Organic staff gathered for coffee: on one bench there would be a boiling kettle, and on the bench

---

[39] Read, J. (1952–1953). Alexander McKenzie, 2 December 1869–11 June 1951. *Obituary Notices of Fellow of the Royal Society* **8**: 207–228.

[40] Note 7, Harris, p. 84.

150 *Pioneering British Women Chemists: Their Lives and Contributions*

opposite a bubbling Grignard reaction, and something very pure and clean crystallising in slow perfection.[41]

The infrared spectra of each isomer was then obtained, much of the analysis being undertaken by Delia Simpson (Agar) (see Chapter 3). The purpose was to take samples of fuel from a captured German aircraft and, comparing the infrared spectra with Lesslie's reference samples, endeavour to identify the German fuel source by determining isomer ratios.

In 1954, Lesslie was appointed Dean of Lindsell Hall, following Crewdson's retirement. Her new duties were in addition to those of lecturing, running laboratory sessions, and supervising a Ph.D. student. During vacations, Lesslie would hurry north to Dundee to care for her elderly widowed father and to see her nephews and nieces. She was periodically asked by students why she, herself, had never married and her response was always: "It was not for want of asking!"[41] Her attitude is quite typical for the period in that she saw herself as dedicating her life to the cause of chemistry and to her students.

Lesslie retired to a small house in Dundee in 1968 where she survived for another 20 years, dying on 24 February 1987 at age 86. Her obituarist quoted Lesslie's former Ph.D. student, Yvonne Bernstein, as saying:

> Her knowledge was encyclopaedic and she shared it with her students unstintingly. She taught me patience when my instinct was to take short cuts or give up too easily. She had high principles as befitted her Calvinistic upbringing but combined these with tolerance and a marvellous sense of humour.[42]

Violet Trew

Violet Corona Gwynne Trew[43] born on 26 June 1902 to Harold George Gwynne Trew, chartered secretary of a petroleum company, and Ada

---

[41] Note 38, Anon, p. 18.

[42] Note 38, Anon, p. 19.

[43] (a) *Staff Register*, Archives, St. Martin-in-the-Fields High School for Girls; (b) *Staff and Student Records*, Archives, Bedford College for Women.

Nellie Blamey, was educated first at St. Olaf's, Beckenham, and then at James Allen's Girls' School, Dulwich. She completed her B.Sc. at Bedford in 1926 and a Ph.D. in 1928, becoming Bedford's first internal doctoral degree in chemistry. During 1927, she was also a part-time Demonstrator in Physical and Inorganic Chemistry.

After graduation, Trew was employed for 2 years in part-time positions, including Assistant Science Mistress at St. Martin-in-the-Fields High School for Girls from September 1929 to August 1930. Trew's fortune improved in 1930 with a full-time appointment as Junior Lecturer at Bedford College. In 1933, she was promoted to Lecturer, and then to Senior Lecturer in 1949.

The teaching of inorganic and physical chemistry largely fell to Trew. When Margaret Jamison arrived at Bedford as an undergraduate, she commented:

> On our first morning in 1932 we assembled in the inorganic laboratory, each to be allotted a cupboard (personal territory) and given, by Dr V C G Trew, a tube containing a powdered mixture whose composition we had to identify. We set about a regular series of tests; qualitative investigations such as fusion on a charcoal block using a blowpipe to get maximum heat, or bending platinum wire and forming a borax bead whose colour, after fusing with our mystery mixture, could (with other diagnostic tricks) lead to the identification of four or five components.[44]

The physical chemistry lab was also under Trew's command, the experiments being performed on Saturday mornings.

The 1930s were a period when spiritualism and all manner of occult sects thrived. Trew became involved with the Theosophical Society and the occult chemistry of Annie Besant.[45] Besant and her group believed that in a semi-trance, it was possible to make oneself infinitesimally small and visit atoms on a voyage of discovery. In fact, she and her group

---

[44] Note 7, Harris, p. 83.

[45] Butler, A. (1991). An extraordinary excursion into atomic structure. *Chemistry in Britain* **27**(1): 40–42.

152 *Pioneering British Women Chemists: Their Lives and Contributions*

reported the shapes of each atom as they saw them. As Jamison commented:

> Dr. Trew had another interest which proved pastorally helpful later. She was an early member of the Theosophical Society (she even inherited Annie Besant's typewriter). When cults began to capture the minds of innocent students who were diverted from their work, Miss Trew was able to produce a useful folder of relevant information.[46]

Trew's research area was magnetochemistry leading to 16 publications. The early research papers were co-authored with Spencer, while much of her later work was published with her as sole or senior author. In 1937, Trew was granted the honour of Recognised Teacher of the University of London, and, in 1955, she was awarded a D.Sc. (London). Trew retired in 1969 and died in 1995 at the age of 92.

Royal Holloway College

Whereas Bedford College was located in the middle of London, Holloway College, later Royal Holloway College, was built on 95 acres of Surrey countryside at Egham, though with easy rail access to London. It was Jane, the wife of Thomas Holloway, who suggested that part of Holloway's fortune be used to construct a college for women. This college would be: "... founded on those studies and sciences which the experience of modern times has shown to be the most valuable, and best adapted to meet the intellectual and social requirements of the students."[47] Modelled on the Chateau at Chambord in the Loire Valley, Holloway College was formally opened in 1886.

The difference between the two women's colleges was that Bedford was partially a day college, while RHC was entirely residential. Bedfordians thought of themselves as sophisticated "girls about town" and considered their "sisters" at RHC to be mere "girls," "protected" in a boarding-school

---

[46] Note 7, Harris, p. 85.

[47] Finch, A. (1963). Royal Holloway College. *Chemistry and Industry* 1132–1135.

atmosphere.[48] In reality, the students at both colleges came from a similar range of middle-class backgrounds.

Life for Women Chemistry Students

The chemistry laboratories, designed by the Oxford University chemist, Vernon Harcourt, did not open until 1889. Martha Whiteley (see Chapter 4), a chemistry student at RHC at the time, noted:

> ... there were only four science students, and no laboratories. They used three rooms in the North Tower, which to this day have sinks and taps in them, and did Chemistry, Physics or Botany according to the way the wind blew, because one chimney always smoked.[49]

Student life at RHC was very regimented in the early years, as one student, a Miss Dabis, recounted in 1888:

> The order of the day is as follows: A bell is rung at 7 a.m., when a servant brings around hot water. At 7.55 a short service is held in the chapel, which is absolutely compulsory, and is begun most punctually... Breakfast follows at 8.20 in the large dining hall, where three long tables are spread with good and plentiful provisions. Lectures begin at 9, and last till 1, five minutes being allowed after each lecture, and at 11 there is an interval of ten minutes for letters and lunch.[50]

As at Bedford, dinner was an important ritual, though at RHC it was organized slightly differently, as Marion Pick recalled from the 1902–1907 period:

> Each night at 6.55 p.m., we assembled in lines down the Library in order of our years and pair by pair, subsequently joined the procession led by the

---

[48] Mackinnon, A. (1990). Male heads on female shoulders? New questions for the history of women's higher education. *History of Education Review* **19**(2): 36–47.

[49] Whiteley, M. Cited in: Salt, C. and Bennett, L. (eds.), (1986). *College Lives: and Oral Panorama Celebrating the Past, Present and Future of the Royal Holloway and Bedford New College.* Royal Holloway and Bedford Colleges, London, p. 17.

[50] Dabis, A. (June 1888). Holloway College. *Magazine of the Manchester High School* 236–238.

154　*Pioneering British Women Chemists: Their Lives and Contributions*

Principal or Senior Resident, which passed from the Drawing Room.... The staff with their partners took the heads of the tables and a student marshal then saw the rank and file filled the tables in due order. After grace, an ample domestic staff waited on us and our business was to converse, which we did.[51]

Women Chemistry Staff

When RHC opened its doors in 1886, Margaret Seward (see Chapter 4) was appointed the first Lecturer in chemistry. Martha Whiteley, a student at the RHC at the time, recalled in the *College Letter, Royal Holloway College*:

Probably no one shouldered a heavier load than the Science Lecturer, Margaret Seward, for the College then possessed no science laboratories or equipment, and yet, during the four years during which she held that post, the first science building, comprising Chemistry, Physics, Botany and Zoology laboratories, was built and equipped under her direction; and, with occasional help in Physics and Zoology, she was responsible for all the science teaching in Chemistry, Physics and Zoology required to carry successfully the first group of science students through the Intermediate and Final B.Sc. Examinations.[52]

In 1891, Seaward resigned in order to marry and the Board of Governors appointed Mary W. Robertson as her successor.[53] According to the Minutes of the Meeting of the Board of Governors, Robertson had a B.A. (Hons.) and M.A. in Chemistry and Physics from the Royal University of Ireland. At the time of her appointment, she held the position of Lecturer at Alexandra College, Dublin.

---

[51]Pick, M. Cited in: Salt, C. and Bennett, L. (eds.), (1986). *College Lives: and Oral Panorama Celebrating the Past, Present and Future of the Royal Holloway and Bedford New College*. Royal Holloway and Bedford Colleges, London, p. 31.

[52]Whiteley, M. A. (November 1929). Margaret McKillop, M.B.E., M.A., 1864–1929. *College Letter, Royal Holloway College Association* 51–52, RHC AS/902/72.

[53]Minutes, Meeting of Governors, 27 June 1891, Board of Governors.

London Women's Colleges 155

However, in 1893, Robertson fell ill as was described in the Board Minutes:

> The Principal reported that Miss Robertson had been unable to return to College on account of ill-health & that she had therefore engaged as a temporary Lecturer this Term Miss Dorothy Marshall B.Sc. (Honours in Chemistry) [see Chapter 3] who has been well recommended by Professor Ramsay of University College.... The Principal further reported that she had requested Miss Whiteley, B.Sc., a former student to come to give assistance for a week.[54]

Robertson was not to return and in the summer of 1893, Eleanor Field was appointed as Robertson's replacement. Field was to dominate the department for the next two decades.

Eleanor Field

Elizabeth Eleanor Field[55] was born on January 1863 in Liverpool, Lancashire, her father was William Field, a purser. She was educated at Newnham College, graduating in 1887 and staying from 1889 to 1890 as an Assistant Demonstrator in Chemistry. For the following 2 years, Field held a Bathurst studentship at Newnham, carrying out research under Matthew Moncrieff Pattison Muir.[56] After leaving Cambridge, she held a post as Assistant Mistress for 2 years (1893–1895) at the Liverpool College for Girls. Then in 1895, Field was appointed Lecturer and Head of Chemistry at RHC.[57] Women Heads were titled "Senior Staff Lecturer," while the male faculty of the same rank were titled "Professors."

---

[54]Minutes, Meeting of Governors, 29 April 1893, Board of Governors, RHC GB/110/2.

[55](a) White, A. B. (ed.), (1979). *Newnham College Register, 1871–1971*, Vol. I, *1871–1923*, 2nd edn., Newnham College, Cambridge, p. 83; and (b) Personal communication, letter, S. Badham, 12 June 1998. Archivist, Royal Holloway College.

[56]Morrell, R. R. (1932). Obituary Notices: M. M. Pattison Muir, 1848–1931. *Journal of the Chemical Society* 1330–1334.

[57](a) Anon. (November 1932). Eleanor Field (1893–1913). *College Letter, Royal Holloway College Association*, 57–58, RHC AS/902/75; and (b) Fortey, I. C. (1933). Elizabeth Eleanor Field. *Newnham College Roll Letter* 52.

156 *Pioneering British Women Chemists: Their Lives and Contributions*

In addition to Field, there were a series of women chemistry demonstrators, including Mildred Gostling (see Chapter 14).

Field remained at RHC for 19 years. Her RHC obituary noted:

> In her early, young and vigorous days she proved herself an able teacher, but the teaching of chemistry was only one of her activities here: she was, in addition, a keen student of politics, art, life and human nature.... She was not herself a great conversationalist; her thought perhaps moved too slowly for sparkling repartee, but to listen to her slowly expressed but lucid exposition of some problem in Chemistry, or even to hear her tell a story, was an intense pleasure and a liberal education.[58]

After her retirement in 1913, Field lived near RHC caring for her father at Egham Hill. Then in 1922, she: "... elected to follow the fortunes of her cousin and ward to [Brno] Czechoslovakia."[58] It was there that she died on 17 November 1932.

Mary Boyle

Mary Boyle,[59] like Field, was a stalwart of the Chemistry Department at RHC. She was born in 1874 at Leeds to James Boyle, hemp rope and twine merchant and manufacturer, and Mary Boyle. Boyle entered the Royal College of Science in 1898. She took the full 3-year course for chemistry, receiving the Frank Hatton Prize, which was awarded to the student heading the list in the final examination. Boyle transferred to RHC in 1901, where she gained her B.Sc. after 1 year of additional study. She remained at RHC, first as a Demonstrator, then Senior Assistant Lecturer and Demonstrator in 1906, and subsequently as Staff Lecturer.

---

[58] Note 57(a), Anon, p. 58.

[59] (a) Moore, T. S. and Whiteley, M. A. (1945). Obituary: Mary Boyle, 1874–1944. *Journal of the Chemical Society* 719; and (b) Anon. (November 1930). IV. College news. 1. Staff news. *College Letter, Royal Holloway College*, 15.

In 1910, only 4 years after being appointed as Lecturer, Boyle received her D.Sc., the subject of her research being the iodosulphonic acids of benzene. One of her obituarists noted:

> This extensive research, the work of many years, was published in four large instalments from 1910 to 1919 [all under her name alone]. Five of the six theoretically possible diiodosulphonic acids, three of the six theoretically possible triiodosulphonic acids, and one tetraiodosulphonic acid, were first prepared by her, and incidentally she prepared a large number of new nitro- and amino-sulphonic acids. ... Our knowledge of this group of substances still rests mainly on Dr. Boyle's work. After 1920, the expansion of the teaching work in the Chemistry Department left her little time for research.[60]

Boyle retired in 1933, moving to Leeds to be near her family, particularly her many nephews and nieces and their children, and she became active in community service. She died in November 1944.

The Succession Controversy

Up until Field's resignation, the Chemistry Department had always been in the hands of women chemists. However, her up-graded replacement position of a University Professorship of Chemistry was advertised in the 15 March 1913 issue of *The Times* as open to men and women on equivalent terms. The opening of the position to men caused a flurry of correspondence in the pages of the RHC *College Letter*.[61]

Sibyl Widdows (see the respective section) sent a letter to many of the former RHC chemistry students expressing her opposition:

> I and several other Old Students feel rather strongly that this post which has been in the hands of women ever since the opening of the College, and which has been markedly efficiently run, ought not to pass out of

---

[60] Anon. (November 1944). Dr. Boyle. *College Letter, Royal Holloway College* 52–54, RHC AS/902/87.

[61] Multiple authors (July 1913). Correspondence. *College Letter, Royal Holloway College* 14–26, RHC AS/902/51.

158 *Pioneering British Women Chemists: Their Lives and Contributions*

their hands without some strong reason. Of course we do not want *any* woman to be given the post, but we think that if a woman of sufficient standing and ability applies it should be offered to her. It is becoming a very serious thing for science women the way in which the science posts in women's Colleges are gradually being placed in the hands of men when there are quite good and efficient women to fill them. It means that a woman can never obtain those opportunities for research and association with other scientists which are so necessary for their work.[62]

Accompanying this letter was a petition, pointing out that of the nine members of the Selection Committee for the post, only one was a woman. Among the members of the Vigilance Committee elected by the Royal Holloway College Association to promote the hiring of a woman Professor, was Margaret Seward, the first Lecturer in Chemistry at RHC.

A counter-petition was then sent by other RHC chemistry graduates, arguing that so long as there was a reasonable proportion of women on the staff at RHC, it was undesirable to reserve any specific post for a woman. In response to Widdow's original petition, 30 expressed the opinion that no gender preference should be given; 207 considered that preference should be given to a woman, of whom 156 added the caveat, "subject to equal qualifications"; while 48 stated that the position had to be given to a woman.

The Governors of the College continued with their original plan. George Barger,[63] who had been Head of Chemistry at Goldsmith's College, London, was appointed as the first Professor of Chemistry at RHC. Barger held the position for only 1 year before being appointed Research Chemist under the wartime Medical Research Committee. Barger was succeeded as Head by T. S. Moore,[64] who held the post from 1914 until 1946. One of Moore's obituaries was written by a former student, Rosa Augustin, and she added: "Teaching in a woman's college and with a daughter of his own (as well as a son), T.S. was immensely

---

[62]Widdows, S. (July 1913). *College Letter, Royal Holloway College* 15–16, RHC AS/902/51.

[63]Dale, H. H. (1941). George Barger, 1878–1939. *Obituary Notices of Fellow of the Royal Society* **3**: 63–82.

[64]Bourne, E. J. (1967). Obituaries: Professor T. S. Moore. *Nature* **214**: 1063.

London Women's Colleges 159

interested in the education of women and was delighted when we did well in our careers."[65]

Millicent Plant (Mrs. Georg-Plant)

In 1933, Moore was joined by Millicent Mary Theodosia Plant.[66] Born on 17 December 1906 in Bilston, Staffordshire, to George Harry Plant, architect's assistant, and Mary Ann Edwards, she received her B.Sc. from the University of Birmingham. Over the next few years, Plant undertook research at Birmingham with Norman Haworth[67] on carbohydrates. After arriving at RHC, while Moore taught the inorganic and physical courses, Plant took over responsibility for teaching organic chemistry. These two constituted the entire chemistry staff at RHC until the end of the Second World War.

Plant embarked on a research programme at RHC, her major publication being on aldol condensation products. However, her interests and influence spread far beyond the chemistry laboratory, as an article on her retirement noted:

> There are few sides of college life which did not feel the shaping pressure of her serious interest, for she thought of life in a resident college as a whole and was concerned for the total well-being of the students. We remember her against many backgrounds as well as the familiar one of the Chemistry Laboratory; in Choral, in Chapel; on the tennis-courts and in the swimming-bath; at Staff Meetings; on her potato-patch at Highfield — this unsorted and far from exhaustive list suggests the variousness of her gifts and interests.[66]

In 1947, Plant married Alfred Georg, resigned her position, and moved with him to Geneva. It would seem possible that she returned to England,

---

[65] Augustin, R. (1967). T. S. Moore 1881–1966. *Chemistry in Britain* **3**: 494.

[66] Anon. (December 1947). Resignation: Millicent Plant. *College Letter, Royal Holloway College Association* 16, RHC AS/902/90.

[67] Hirst, E. L. (1951). Walter Norman Haworth, 1883–1950. *Obituary Notices of Fellows of the Royal Society* **7**: 372–404.

160   *Pioneering British Women Chemists: Their Lives and Contributions*

as in 1954 she co-authored a publication on an algal polysaccharide under the name of M. M. T. Georg-Plant. Plant died in Switzerland.

London School of Medicine for Women

During a period when British Medical Schools were a male preserve, with male sports and male culture emphasized,[68] the London School of Medicine for Women, part of University of London, provided a collegial and non-aggressive learning environment for women medical students. As part of their education, the students were required to take inorganic and analytical chemistry and organic chemistry courses. All the courses were taught by women chemists who provided a welcoming and encouraging yet rigorous learning environment. Here, we will bring to light the forgotten women chemistry staff of the LSMW.

With women barred from admission to British medical schools, Sophia Jex-Blake, Elizabeth Garrett Anderson, Emily Blackwell, Elizabeth Blackwell, and Thomas Henry Huxley established LSMW in 1874.[69] The school began in a small house on Henrietta Street (now Handel Street). Sympathetic male faculty from other London teaching medical schools offered their services to lecture part-time at LSMW, and the school opened its doors in the same year.[70] The initial enrolment was 14, increasing the following year to 23. The radical lawyer William Shaen (see Chapter 4) agreed to act as solicitor for the school. Shaen is another of the forgotten heroes of women's educational rights, having also assisted in the foundation of Bedford College and having been connected with the campaign for women's entry to Cambridge University.

As an independent school, LSMW lacked any opportunity for the students to undertake clinical studies. Fortunately, in 1877 an agreement was reached with the Royal Free Hospital (RFH) to allow the women

---

[68] Dyhouse, C. (1998). Women students and the London Medical Schools, 1914–1939: The anatomy of a masculine culture. *Gender and History* **10**(1): 110–132.

[69] (a) Bell, E. M. (1953). *Storming the Citadel: The Rise of the Woman Doctor*. Constable, London; and (b) Blake, C. (1990). *The Charge of the Parasols: Women's Entry to the Medical Profession*. The Women's Press, London, pp. 167–171.

[70] Thorne, I. (1905). *Sketch of the Foundation and Development of the London School of Medicine for Women*. G. Sharrow, London.

London Women's Colleges   161

students access to the RFH wards. In the same year, the University of London admitted women to its medical examinations. Over the following decades, many of the women students of the LSMW received recognition from the University of London for outstanding performance. For example, in 1881, Mrs. Scharlieb took First Class Honours in Materia Medica and Pharmaceutical Chemistry while Miss Tomlinson was awarded Second Class Honours in Organic Chemistry. In 1886, over half of all the honours medical students of the University of London in Anatomy, Physiology, and Materia Medica were the women students of the LSMW.

In 1896, to indicate the increasingly close relationship with the RFH, the name was changed to the London Royal Free Hospital School of Medicine for Women (LRFHSMW). The number of women students increased, and by 1914, there were over 300, resulting in an urgent need for additional facilities. Construction was completed in 1916, including new science laboratories. It was reported in the LSMW *Magazine*: "On this (ground) floor is the *Maude du Cros* Organic Chemistry Laboratory and a Chemical Research Laboratory connecting with the Inorganic Department."[71]

Chemistry at the LSMW

Chemistry was part of the program of studies from the beginning, with Charles William Heaton[72] of Charing Cross Hospital as the first Lecturer. It was noted in the school's history that "... the Lectures on Chemistry were given in a room on the left-hand side of the garden entrance to the old building, which was also used for Practical Chemistry in the summer, ..."[73]

The chemistry students of the LSMW were particularly prolific in their production of chemistry-related poetic works. The ensuing example is only the first part of a complete inorganic chemical analysis procedure in rhyme.

---

[71] Anon. (1916). Opening of the extension of the school by Her Majesty the Queen, *Magazine of the London (Royal Free Hospital) School of Medicine for Women* **11**: 86–87.
[72] "S.A.V." (1894). Obituary Notices: Charles William Heaton. *Journal of the Chemical Society, Transactions* 386–388.
[73] Note 70, Thorne, p. 19.

162 *Pioneering British Women Chemists: Their Lives and Contributions*

> **Group IA**
> *Add HCl, and then you'll get*
> *The metals of Group I., you bet.*
> *Add water, and you'll find the lead*
> *Without a word, has softly fled.*
> *(If on this point your partner wrangles*
> *KI will give you golden spangles).*
> *You then proceed to add ammonia,*
> *The silver, you perceive, has flown-i-a!*
> *The blackened mercury will lead yer*
> *(To add to this some aqua regia*
> *If stannous chloride's added here*
> *On warming, Hg will appear).*[74]

Our particular interest in the Chemistry Department of the LSMW was that, apart from the initial appointment of Heaton, it was exclusively women-staffed until the LSMW ceased to be women-only in 1947.

Lucy Boole

The first woman to hold a position in the Chemistry Department of the LSMW was Lucy Everest Boole.[75] Boole, born in Cork on 5 August 1862, was one of five daughters of the mathematician, George Boole[76] and Mary Everest. Each daughter was highly talented; in particular, Alicia (Alice) Boole (Stott) followed in her father's footsteps to become a distinguished mathematician.[77]

---

[74]"EDF" (1916). A mnemonic of inorganic analysis. *Magazine of the London (Royal Free Hospital) School of Medicine for Women* **11**: 71–74.

[75](a) Anon. (1905). Obituary. *Proceedings of the Institute of Chemistry of Great Britain and Ireland*, Part II, **29**: 26; and (b) Anon. (October 1905). Obituary: Miss Boole. *Magazine of the London (Royal Free Hospital) School of Medicine for Women*, 454–455.

[76]MacHale, D. (1985). *George Boole: His Life and His Work*. Boole Press, Dublin.

[77]Coxeter, H. S. M. (1987). Alicia Boole Stott (1860–1940). In Grinstein, L. S. and Campbell, P. J. (eds.), *Women of Mathematics: A Biobibliographic Sourcebook*. Greenwood Press, Westport, CT., pp. 220–224.

Though having had little formal education, Lucy Boole obtained admittance to the School of the Pharmaceutical Society in 1883.[78] After passing the examinations, only the second woman to do so, she became the first woman researcher in pharmaceutical chemistry. Working with Wyndham Rowland Dunstan at the Pharmaceutical Society Laboratory, in 1889, Boole developed a procedure for the analysis of tartar emetic. The procedure, published in the *Pharmaceutical Journal*,[79] was in use as the official assay method until 1963.

In 1891, Boole was appointed Demonstrator in Chemistry at the LSMW under Heaton. Shortly after, as a result of Heaton's ill-health, she took over his duties. Upon his resignation in 1893, Boole was appointed as Lecturer in Chemistry. Unfortunately, later that year, deteriorating health caused Boole to submit her own resignation. Wishing to keep her, the Council of the school divided her duties, assigning her as Teacher of Practical Chemistry in 1893,[80] while hiring Clare de Brereton Evans (see the following section) to be Lecturer in Chemistry.

In 1894, Boole was elected the first woman Fellow of the Institute of Chemistry. Boole died in December 1904 at the age of 42. Included in her obituary was a comment from one of her former students:

> Miss Boole was no believer in 'cram-work,' it was the real deeper meaning of her science that she cared about; and while she taught us with conscientious care the facts necessary for us to know for our examinations, those who knew her well realised that to her that part of the subject was only the threshold to an inner world of knowledge untouched by examination requirements.[81]

---

[78] Shellard, E. J. (unpublished work). Chapter II. Some outstanding women pharmacists of the late 19th and early 20th century in *Women in Pharmacy*. We thank the Archivist of the Pharmaceutical Society for a copy of the manuscript.

[79] Dunstan, W. R. and Boole L. E. (1889). Chemical observations on tartar emetic. *Pharmaceutical Journal* **19**: 385–387.

[80] Note 75(b), Anon, p 454.

[81] Note 75(b), Anon, p. 455.

164 *Pioneering British Women Chemists: Their Lives and Contributions*

Clare de Brereton Evans

Clare de Brereton Evans was born in about 1866 in Bath to William de Brereton Evans, surgeon and retired Deputy Inspector General of Hospitals in Madras (India), and Emma Soames. She was educated at the Royal School for Daughters of Officers, Bath, and then entered Cheltenham Ladies College (CLC) as a boarding student continuing her studies there until she obtained a B.Sc. (London).[82] De Brereton Evans then moved to London to undertake research with Henry Armstrong at the Central Technical College. Her research with Armstrong resulted in her being awarded a D.Sc. in 1897, the first woman to receive that degree from the University of London.

As described earlier, Evans was appointed Lecturer in Chemistry at LSMW in 1893 to take over part of the duties of Lucy Boole. Evans combined her LSMW teaching with part-time research at University College, London, under Sir William Ramsay. Ramsay had many of his research group, including Evans, searching for new chemical elements. In 1908, she claimed to have isolated an unknown metal from the mineral thorianite.[83] Unfortunately, this was not the case.[84] In 1912, she resigned her position at LSMW so that she could devote herself full-time to her research.[85] Nothing is known of Evans after this period, except that she continued living at the same address of 47 Campden Hill Court until her death on 10 August, 1935, aged 69.

Sibyl Widdows

Sibyl Taite Widdows dominated the LSMW Chemistry Department for 40 years.[86] Born on 27 May 1876 at Lewisham, Kent, her parents were

---

[82] Creese, M. R. S. (1998). *Ladies in the Laboratory: American and British Women in Science, 1800–1900*. Scarecrow Press, Lanham, MD., pp. 265–266.

[83] Evans, C. de B. (1908). Traces of a new tin group element in thorianite. *Journal of the Chemical Society, Transactions* **93**: 666.

[84] Fontani, M., Costa, M and Orna, M. V. (2015). *The Lost Elements: The Periodic Table's Shadow Side*. Oxford University Press, Oxford, pp. 105–107.

[85] Anon. (1912). Hospital and school news. *Magazine of the London (Royal Free Hospital) School of Medicine for Women* **8**: 77.

[86] Anon. (1960). Sibyl Taite Widdows. *Journal of the Royal Institute of Chemistry* **84**: 233; and Personal communication, letter, K. Austin, 12 June 1998. Archivist, Senate House, University of London.

London Women's Colleges   165

Thomas Widdows, solicitor, and Elizabeth Shoosmith. Widdows was educated at Dulwich High School, London, and then obtained a B.Sc. (Hons.) degree in Chemistry at RHC in 1900.

In 1901, Widdows was appointed Demonstrator in Chemistry at LSMW under Boole and in 1904, took over Boole's position of Teacher of Practical Chemistry following Boole's death. When Evans resigned in 1912, Widdows was promoted to Lecturer in Inorganic Chemistry. Then in 1935, she was appointed Head of the Chemistry Department, though she had been acting Head for many years previously.

Her obituarist, Phyllis Sanderson, described Widdows as follows:

> Of miniature stature, alert and sprightly, Miss Widdows possessed such vitality and drive that it seemed a store of dynamite must be housed within her small frame.... Practical classes, certainly no play time, held an element of excitement (possibly mixed with terror) that kept everyone on their toes; for S.T.W. would systematically work her way down the laboratory, visiting student after student to ensure that each in turn was fully understanding what they were doing. Suddenly a loud scream of dismay would ring out and all would shudder, knowing full well that some unfortunate student had uttered an appalling chemical howler or had committed some dangerous crime such as heating an inflammable liquid with a naked flame.[87]

In addition to teaching, Widdows was an active researcher, authoring at least 12 publications involving analytical chemistry of biological relevance. For example, in the LSMW *Magazine* of 1921, it was commented that she had been determining the calcium content of the blood under various conditions, to see what may be the limit of physiological variation during menstruation and pregnancy.[88]

Widdows subsequently turned her attention to breast milk as this letter to the Editor of the *British Medical Journal* indicates:

> For some time at this school a group of workers has been investigating breast milk, from both biochemical and chemical aspects. ... It has now

---

[87] Sanderson, P. M. (1960). Obituary. *Royal Free Hospital Journal* **23**: 81–82.
[88] Anon. (1921). Research work at the LSMW. *Magazine of the London (Royal Free Hospital) School of Medicine for Women* **16**: 78.

166   *Pioneering British Women Chemists: Their Lives and Contributions*

been decided that this investigation should be extended to include secretions occurring before parturition, during menstruation, and other instances of mammary activity. As such cases are infrequent, may we ask the help of your readers in giving us the opportunity of getting into touch with women in whom the breasts become active before parturition, or independently of pregnancy?[89]

Retiring in 1942, Widdows died on 4 January, 1960 at Finchley, London. Sanderson remarked:

> As so many of her contemporaries, she was an ardent feminist and willingly sacrificed her own career as a chemist for the cause most dear to her heart, the training of women doctors at Hunter Street (LSMW), the only training ground in Medicine open to women in England at the time.[90]

Phyllis Sanderson

Widdow's successor as Head of Chemistry in 1942 was Phyllis M. Sanderson.[91] Born in 1901 at Hove, Sussex, daughter of Robert Sanderson, surgeon, and Agnes Mary Cooke, she was educated at Brighton and Hove High School. Sanderson completed her B.Sc. (Hons.) in Chemistry at University College, London, in 1924. After 1 year of postgraduate study at the Children's Hospital, London, she was appointed Demonstrator in Chemistry at the LSMW.

In addition to teaching, during the 1930s, Sanderson undertook research with Vincent Briscoe[92] at Imperial College, London. The study of industrial dusts, especially chemical aspects of silicosis in miners, resulted in 10 co-authored publications in specialist mining journals plus

---

[89] Widdows, S. T. (May 1931). Aberrant Mammary Secretion. *British Medical Journal* 16.

[90] Note 87, Sanderson, p. 81.

[91] "M. H. & F. K." (1966). Dr. Phyllis M. Sanderson. *Royal Free Hospital Journal* **27**: 190–191.

[92] Anon. (1961). Obituary: Henry Vincent Aird Briscoe. *Journal of the Royal Institute of Chemistry* **85**: 425.

a summary in the journal *Nature*.[93] Briscoe was awarded a medal by the Institute of Mining and Metallurgy for the method of dust analysis.[94] Sanderson received a Diploma of Imperial College for this work and then in 1939, a Ph.D. from University College for research on solubility of silica and silicates.

At LSMW, Sanderson was promoted to Senior Demonstrator in 1933; Assistant Lecturer in 1934; and Lecturer in 1946. From 1957 to 1963, Sanderson, giving her affiliation as Imperial College, was co-author of a series of research papers on heterocyclic organic compounds with Frederick Kurze, a Lecturer at the RFH. However, in these later years, Sanderson's major research occupation became the history of chemistry. In her obituary it was noted:

> It was typical of her sense of justice that in one of these studies she should have rescued from oblivion a hitherto obscure 18th Century scientist, William Cruickshank, by re-establishing his claims to several important discoveries that had been erroneously ascribed to another investigator.[95]

Sanderson died on 7 September 1965.

Anne Ratcliffe

Following behind Sanderson career-wise was Anne Ratcliffe, the last of the women lecturers at the LSMW.[96] Born in 1896, she, obtained her qualifications at University College, London. Ratcliffe completed a B.Sc. (Hons.) in Chemistry in 1924, then in 1939, an M.Sc. based on her research into sterols and carbohydrates in certain fungi.[97] Her initial

---

[93] Briscoe, H. V. A., Matthews, J. W., Holte, P. F. and Sanderson, P. M. (1937). A note on some new characteristic properties of certain industrial dusts. *Nature* **139**: 753–754.

[94] Gay, H. and Griffith, W. P. (2017). *The Chemistry Department of Imperial College London: A History 1845–2000*. World Scientific Publishing, London, pp. 207–208.

[95] Note 91, "M. H. & F. K." p. 191.

[96] "P. M. S." (1961). Anne Ratcliffe. *Royal Free Hospital Journal* **24**: 15.

[97] Ratcliffe, A. (1937). The sterols and carbohydrates in fungi I. Boletus edulis. *Biochemistry Journal* **31**: 240–243.

appointment at LSMW in 1929 was followed by promotion to Senior Demonstrator in 1940; Assistant Lecturer in 1945; Lecturer in 1947; and finally Senior Lecturer in 1949.

Upon her retirement in 1961, Sanderson wrote of Ratcliffe's character:

> That she is an inspired and tireless teacher was quickly realised by students ... Patient and kind though she is, however, Miss Ratcliffe would not tolerate shoddy work or bad manners ... She is one of those rare beings possessed of extreme intellectual honesty. Rather than risk passing on often erroneous textbook information to a student she would take infinite trouble reading original papers on the subject, and never would she say she understood anything unless she had probed to the depths and considered it from every possible angle (See Footnote 96).

Other Women Chemistry Staff

Because the Chemistry Department was, in many ways, an add-on to the LSMW, it was poorly documented, except for the Lecturers. During Evans' Lectureship, it was *Norah Ellen Laycock* who held the position of Demonstrator from 1906 to 1916.[98] Laycock was born on 21 January 1877 at Keithley, Yorkshire to Arthur Laycock, estate agent, and Ellen Elizabeth Scott. She obtained her B.Sc. degree from RHC in 1901. In 1904, she was appointed to the position of Demonstrator in Chemistry. In 1916, her appointment became a joint one shared between the Biology and Chemistry Departments and the following year changed again to solely a Demonstrator in Biology. In 1918, Laycock was promoted to Assistant Lecturer in Biology, a position which she held until her departure in 1929 to become Headmistress of Mayertorne School, Buckinghamshire. She died on 18 November 1951, at Wendover, Bucks, aged 74.

The Demonstrator in Chemistry who succeeded Laycock in 1916 was *Yvonne M. D. Cooper,*[99] who had obtained a B.Sc. in 1917 at RHC.

---

[98] Anon. (1916). School Notes. *Magazine of the London (Royal Free Hospital) School of Medicine for Women* **11**: 119.

[99] Anon. (1936). School Notes. *Magazine of the London (Royal Free Hospital) School of Medicine for Women* **31**: 76.

However, no other information could be found about her. About the same year, *Mrs. Effie Isobel Stirling-Taylor* was appointed to the Chemistry Department, though the position was not specified, and she retired in 1936.[100]

The only other individual for whom we have any information is *May Williams*.[100] Williams was born on 7 May 1886, daughter of Ralph Williams, Clerk in Holy Orders of Maida Vale, London, and Lucy Anne Williams. Educated at Notting Hill High School, she entered RHC in 1905, completing a B.Sc. (Hons.) Chemistry in 1909. She was appointed as Demonstrator in Chemistry at LSMW in the same year, and promoted to Senior Demonstrator in 1920, and to Assistant Lecturer in 1921. In 1921, Williams received an M.Sc. in Chemistry from RHC and LSMW based on her research on quinoline derivatives with John Addyman Gardner.[101] Gardner was a researcher at the Biochemical Laboratories at St. George's Hospital Medical School and also Lecturer in Organic Chemistry at LSMW. At her retirement in 1946, it was commented: "Miss Williams' brilliant gifts as a teacher, her renowned patience with the students to whom chemistry was no easy subject ... will be greatly missed."[102]

Commentary

Bedford College, a pioneering London magnet for aspiring women chemists, is no more. Absorbed into Royal Holloway College to initially form Royal Holloway and Bedford New College, the word "Bedford" was subsequently dropped from the name. All that remains of the name is the Bedford Library on the Royal Holloway Campus. The history of Bedford College as a pioneering institution in the field of women's education has been forgotten.

---

[100](a) *Royal Free Hospital School of Medicine for Women, University of London Report,* 1945/1946. Archives, Royal Holloway College; and (b) *Student Records,* Archives, Royal Holloway College.

[101]Ellis, G. W. (1947). Obituary Notice: John Addyman Gardner, 1867–1946. *Biochemistry Journal* **41**: 321–324.

[102]Anon. *Royal Free Hospital for Women, University of London Report,* 1945/1946.

Equally — if not more regrettable — the name of the London School of Medicine for Women, later the London Royal Free Hospital School of Medicine for Women (LRFHSMW), has vanished. In 1947, the LRFHSMW was required to become co-educational and the "for Women" dropped from its name. Then, in 1996, it was absorbed into the Medical School of University College, London. Finally, in 2008, after another merger, it disappeared into the UCL Medical School. The pioneering LSMW, and the enthusiastic women chemists who taught there were long forgotten.

Chapter 6
English Provincial Universities

May Sybil Leslie (14 August 1887–3 July 1937)

In England, up until 1850, there were four universities: the historic Cambridge and Oxford; Durham; and London. The English provincial, or civic, universities which came into existence over the subsequent decades were to change the availability and range of university education tremendously.[1] The provincial universities saw themselves as serving their specific cities and regions, thus most of the students commuted from home. Each institution started from College status and was initially only able to prepare students for University of London degrees. From the

---

[1] (a) Robertson, C. G. (1939). The provincial universities. *Sociological Review* **31**: 248–259; and (b) Morse, E. J. (1992). English civic universities and the myth of decline. *History of Universities* **11**: 177–204.

opening of the doors of each provincial college, women had equal rights of graduation. However, women's presence at university was not always accepted, as we will see in the following.

Pre-First World War, the distribution of the backgrounds of women students at provincial universities also differed substantially from those at the four long-established institutions, as Julie Gibert reported:

> The upper and professional class students who represented over half of the students at Oxford's women's colleges represented less than one-fifth of the female student body at the civic universities, while working-class occupational groups, entirely unrepresented at both Oxford and London, accounted for 8% of the students at the civics. A third notable discrepancy is the much higher proportion of 'semi-professional' rank students at the civic universities. At the civic universities 27% of students fell in this group, while at Oxford and London the figures were 4% and 7%, respectively.[2]

The Northern Universities

Before covering each individual institution, we need to introduce the reader to the complexity of the name changes among the northern universities. A (university) college was opened in each of the major northern cities: Owens College (Manchester); University College, Liverpool; and Yorkshire College (Leeds).[3] These Colleges shared a common philosophy as David Jones has explained:

> Middle class support and clientele, non-sectarianism, moderate fees, non-residence, the importance of evening classes and of part-time and non-degree students, and a strong desire to adapt to local needs were basic characteristics shared by all of the civic colleges.[4]

---

[2] Gibert, J. S. (1994). Women students and student life at England's civic universities before the first world war. *History of Education* **23**(4): 405–422.

[3] Jones, D. R. (1988). *The Origins of the Civic Universities: Manchester, Leeds & Liverpool*, Routledge, London, pp. 50–63.

[4] Note 3, Jones, p. 57.

In 1880, a Royal Charter established Victoria University, a University for the North of England to which Colleges could affiliate. Owens College did so immediately, followed by University College, Liverpool, in 1884, and Yorkshire College in 1887. However, each city aspired to its own university. University College, Liverpool, broke away in 1903 to become the University of Liverpool; then Yorkshire College was given its own Charter in 1904 as the University of Leeds. Owens College then became the Victoria University of Manchester.

Owens College (Victoria University of Manchester)

In his Will, John Owens, a Manchester merchant, left a substantial sum for the founding of a university-level institution and, as a result, Owens College was established in 1850. At the time, the primary concern was that the College be non-sectarian and no thought was given to the possibility of women wanting to attend.[5] By 1875, pressure was building for the admission of women. Manchester High School for Girls (MHSG) had opened its doors in 1874, thus it could no longer be argued that women students were ill-equipped to face a university education. Heated debate ensued in the correspondence columns of the *Manchester Examiner*, but as Mabel Tylecote observed: "The men poo-poohed, often with little taste, the claims of women as a sort of bad joke. The women often began with serious arguments and then ended by losing their tempers."[6]

A defining step was taken in 1877 with the opening of the Manchester and Salford College for Women at 223 Brunswick Street, near the Owens campus.[7] Most of the teaching at the College for Women was given by professors and lecturers of Owens College though there was no legal connection between the two institutions. The Women's College was absorbed into Owens College in 1882 as the Department for Women. Initially, women students were only permitted onto the Owens campus in

---

[5]Tylecote, M. (1941). *The Education of Women at Manchester University 1883 to 1933.* Manchester University Press, Manchester, p. 1.

[6]Note 3, Tylecote, pp. 8–9.

[7]The Editor. (December 1887). Editorial notes. *Iris: The Magazine of the Department of Women, The Owens College,* 5.

174   *Pioneering British Women Chemists: Their Lives and Contributions*

preparation for their final examinations and it was not until 1892 that they were first allowed to attend physics and chemistry instruction and laboratory work towards a B.Sc.[8]

Women Students

Even the Owens College Library was initially "off-limits," as a woman student in 1901 later recalled:

> It would have been the height of impropriety [for a woman student] to enter the library and demand a book in the hardened manner now usual … No, we had to "fill up a voucher," and a dear little maid-of-all-work, aged about 13, went to the library with it. If we were not quite sure of the volume required, she might have to make the journey ten times, but it was never suggested that she should be chaperoned.[9]

As at many co-educational Colleges, it was understood that certain rows in the lecture room were unofficially reserved for women students. A student, J. Harold Bailey, described in the *Owens College Union Magazine* of 1892 how this fact had not been imparted to a new male arrival:

> A tale is told of an Arts man commencing his courses late in the session, who came in some little time before the lecture began and took his seat in the middle of the row usually occupied by the ladies. Presently the ladies trooped in a body, but on catching sight of the intruder they shrank back in dismay, and crowded round the doorway undecided as to the proper thing to do under the circumstances. Just as the Professor entered the room one of the men students, thinking that the fun had gone far enough, went up to the innocent offender and explained matters,

---

[8] Anon. (April 1892). The department for women. *Owens College Magazine* **24**(3): 92–93.
[9] Purvis, J. (1995). Student life. In Purvis, J. (ed.), *Women's History: Britain, 1850–1945*. St. Martin's Press, New York, p. 193; see also: LaPierre, J. (1990). The academic life of co-eds, 1880–1900. *Historical Studies in Education* **2**: 225–245.

whereupon he meekly took another seat amidst the suppressed titters of his fellow men and the ladies were able to occupy their accustomed seats without running the risk of being contaminated by having a man sit amongst them.[10]

The women students were also harassed in the chemistry laboratory sessions, with the men seizing the lab stools, as was expressed in rhyme "A Lady's Lament" in the *Owens College Union Magazine* of 1902:

And we, in weak appeal, our voices raise,
If we want merely to retain a stool.
The only way that we can hope to keep
That "more primeval" beast our seats from filling,
Is to inflict (small wonder that I weep!)
A fine on each offender of a shilling.[11]

The Chemical Society

The Chemical Society, formed in 1877, was originally one of the male bastions of the University. The first attempt to admit women came in 1906 as the Society Minutes recorded:

It was proposed by Mr. E. W. Smith and seconded by Mr. Slade that "The Committee be asked to enquire into and report on the advisability of admitting women students into this society." On being put before the meeting the proposition was lost.[12]

---

[10]Bailey, J. H. (October 1892). Types of college men-and women: VIII The lady student. *Owens College Magazine* **25**(1): 24–25.

[11]"Mis Pickel." (December 1902). From the chemical side. *Owens College Union Magazine* **10**(81): 52.

[12]*Minutes 1905–1915 meeting.* (October 25, 1906). Manchester University Chemical Society.

176　*Pioneering British Women Chemists: Their Lives and Contributions*

Sentiment seemed to shift with the Chemical Society members according to the Minutes of 5 February 1907 as the Society had invited Ida Smedley (see Chapter 9) as a guest speaker:

> It was then proposed by Dr. Hutton and seconded by Prof. Carpenter that "The Union Committee be asked to grant permission for Dr. Ida Smedley to attend the meetings in the Union." The proposal was carried. It was proposed by Dr. N. Smith, seconded by Mr. S. R. Best & carried that "The committee shall at some special meeting or at the general meeting present a report on the question of admitting women students to membership of this society."[13]

However, in November 1907, the bar on women members was sustained: "… It was decided that women students be not admitted as members of the Society as it would spoil the social nature of the meetings."[14] Though it was agreed that Smedley could address the Chemical Society, the venue had also been a problem. The Student Union building was under control of the male-only Union Committee: "The secretary [of the Chemical Society] read a letter from the Union secretaries stating that the Union Committee could not see their way to allow Dr. Ida Smedley to use the Union Rooms for any reason whatsoever."[15]

Something must have happened over the following 12-month period as the Society reversed itself in a motion of November 1908:

> An Extraordinary General Meeting was held on Friday, November 13th, to consider the following resolution: "That the words (in Rule VI) 'This is to be interpreted as applying to men students only' be rescinded." Professor Perkin was in the chair. The resolution was proposed by Dr. Meldrum seconded by Mr. Wood, and after a long discussion in which several members took part, the resolution was carried by 48 votes to 19.[16]

---

[13] *Minutes 1905–1915 meeting.* (February 5, 1907). Manchester University Chemical Society.
[14] Anon. (November 1907). Chemical Society notes. *Manchester University Magazine,* 23.
[15] *Minutes 1905–1915 meeting.* (February 22, 1907). Manchester University Chemical Society.
[16] *Minutes 1905–1915 meeting.* (November 13, 1908). Manchester University Chemical Society. The 'Rule VI' itself was not given. The Society Rules changed periodically. Unfortunately, the Minute Book for 1892–1897 is missing from the Manchester University

English Provincial Universities   177

At the 26 October 1909 meeting, 12 of the 35 new student members were women. The *Minutes* noted that business began: "After welcoming women students as new members of the Society ..."[17] It would seem that the barrier of location had been overcome by holding the Chemical Society meetings in the Women's Union instead, and on 7 December 1909, Smedley presented her paper to the Society.

Leonore Kletz (Mrs. Kletz Pearson)

There were two women chemists who were on staff in the early years; one was May Badger (see Chapter 14) and the other was Leonore Kletz[18] (later Mrs. Pearson). She was born in Bolton on 18 June 1891, her parents being Louis Kletz, a furniture dealer, and Lena Kletz. Kletz attended North Manchester High School and Pendleton High School. She completed a B.Sc. in Chemistry at Victoria University in 1912, and an M.Sc. in 1913 in Organic Chemistry. Kletz worked for a year with Arthur Lapworth[19] as Schlunk Research Assistant and from March to June 1915 held a part-time temporary position teaching chemistry at MHSG.[20] She married Bernard Pearson on 24 June 1915, the day she left her position at the school.

Later in 1915, Kletz was appointed Assistant Lecturer and Demonstrator in the Chemistry Department back at Victoria University.[21] It is not clear for how long she held the position, but her seven research

---

Archives and it is probably during this period that the Rule was proposed. There was no mention of 'men only' in the original Rules of 1878 and the addition might well have occurred when women were first admitted to the chemistry laboratories in 1892.

[17] *Minutes 1905–1915 meeting.* (October 26, 1909). Manchester University Chemical Society.

[18] Anon. (1947). Obituary: Mrs. Leonore Pearson. *Journal and Proceedings of the Royal Institute of Chemistry of Great Britain and Ireland* **71**: 217.

[19] Robinson, R. (1947). Arthur Lapworth, 1872–1941. *Obituary Notices of Fellows of the Royal Society* **5**: 554–572.

[20] *Staff Records*, Leonore Kletz, Archives, Manchester High School for Girls.

[21] Burkhardt, G. N. (1954). Schools of Chemistry in Great Britain and Ireland–XIII The University of Manchester (Faculty of Science). *Journal of the Royal Institute of Chemistry* **78**: 448–460.

178 *Pioneering British Women Chemists: Their Lives and Contributions*

publications (all under the name of L. K. Pearson) are spread from 1915 to 1927. It was noted in her obituary that: "For several years she had given up the practice of chemistry but had recently renewed her activities in this direction" (see Footnote 18). Kletz died on 23 July 1947, aged 57.

Lucy Higginbotham

Among the women chemistry graduates from Victoria University was Lucy Higginbotham.[22] Higginbotham was born on 6 November 1896, the daughter of James Higginbotham, yarn merchant, and Martha Hannah Stones in Chorlton-cum-Hardy. She attended Penrhos College, Colwyn Bay, before completing her schooling at MHSG. Entering Victoria University in 1915, she graduated with a B.Sc. (Hons.) in Chemistry in 1919 and was the recipient of the Leblanc Medal for that year.

The following year, Higginbotham worked on her M.Sc. with Henry Stephen, a Research Associate with Lapworth. From 1920 until 1922, she was a Research Assistant with Lapworth, contributing to five of his publications. Higginbotham then joined the staff of the British Cotton Industry Research Association at the Shirley Institute, Manchester. Her task was to investigate the minor constituents of cotton, particularly the complex mixture of substances present in cotton wax and their reaction products formed during the bleaching and finishing of cotton fibre. Higginbotham's results were published in a series of papers in the *Textile Industry Journal* and in the *Memoirs of the Shirley Institute*.

In 1926, Higginbotham became ill and went for a medical operation to Berne, Switzerland, where she died on 22 November 1927. Her obituarist, Robert Fargher, observed:

> As a research worker she possessed unusual energy and a pronounced flair for the rapid exploration of a field of inquiry; outside her work, she was keenly interested in athletics, particularly tennis, golf, and motor cycling. She had many of the best characteristics of her native country,

---

[22] *Student Records*, Archives, University of Manchester.

and her frank outspoken genial personality endeared her to her colleagues at the University and the Shirley Institute.[23]

University College, Liverpool (University of Liverpool)

Chemistry was taught in Liverpool well before the founding of University College, Liverpool (LU) in 1881. An article in *Chemistry News* of 1874 reported on the Institutions in Liverpool which offered chemistry.[24] Among them were the College of Chemistry with Martin Murphy, F.C.S. as Professor of Chemistry; and the Analytical Laboratory and School of Technical Chemistry, with Lecturer Norman Tate. Tate was also Lecturer on Chemistry and Laboratory Practice at the Liverpool Operatives' Science Classes. In addition, a 100-lecture course on Chemistry and Practical Chemistry was offered at the Liverpool Royal Infirmary School of Medicine by J. Campbell Brown, D.Sc. London, F.C.S. All of these programs contained practical laboratory work.

Women Chemistry Students

In 1928, the *Liverpool University Chemical Society Magazine* (LU-Chem. Soc.) carried an article on "Women and Chemistry." In the article, the anonymous author muses that:

> I often wonder why women take up chemistry. Can it be that they imagine they will become chemists? I shudder at the thought. ... Women in the right setting are delightful creatures. A chemistry laboratory is not the right setting. A woman in a lab is as incongruous as a man at an afternoon tea party. ... If it is impossible to have a special "female" lab, then let the flapper vote give England a women's University.[25]

---

[23]Fargher, R. G. (1928). Obituary notices: Lucy Higginbotham. *Journal of the Chemical Society* 1056.

[24]Anon. (1874). Schools of Chemistry. *Chemistry News* 29, 132.

[25]Anon. (1928). Women and chemistry. *Liverpool University Chemical Society Magazine* **9**(1): 12–13.

180 *Pioneering British Women Chemists: Their Lives and Contributions*

This article provoked an immediate response from a woman chemistry student, defending the presence of women in chemistry:

> Life at a University offers many attractions, not the least of which is, that should she find after many years that she is a superfluous woman she will always have a university training, and perhaps a degree, which are useful sort of things to have when one is thinking of earning one's living. .... Besides, Chemistry offers so many more possibilities than Arts. Engineering would, of course, be the ideal faculty for this attractive woman, but–it simply isn't done!![26]

In the closing remarks, she commented about men "... who would label their doors 'No Admittance to Women.'"[26]

That writer seemed to accept that a degree was a "back-up plan" in the event of failure to marry. However, the next issue carried a rebuttal with a more strongly feminist stance:

> The author [of the attack on women chemists] seems to forget that we are now living in the 20th century, when that which used to be a "man's job" is a man's job no longer. In almost every occupation women are equalling and have equalled men.... He evidently does not know that darning socks and rocking cradles went out with crinolines. ...[27]

Then the author pointedly added: 'Women and men meet on equal terms and work on equal terms. At night, the man goes home to be waited on, while a woman goes home to do a "woman's job'."[27] This third contribution seemed to end the correspondence, but the exchange clearly indicates the degree of hostility facing women students from some of their male chemistry colleagues.

---

[26] "F.M.E." (1929). In defence of women. *Liverpool University Chemical Society Magazine* **10**(2): 15–16.

[27] "A Woman Chemist." (1929). Women and chemistry, Part II. *Liverpool University Chemical Society Magazine* **9**(3): 9–11.

English Provincial Universities 181

The Chemical Society

The arrival of women chemistry students at LU had a considerable effect on the student chemistry culture. The Chemical Society (LU-Chem. Soc.) had been founded in 1892,[28] and the social life of the Society focussed on the men-only Annual Dinner and Annual Kneipe (Beer Party). The latter event was an evening spent in drinking beer, smoking, singing songs, and telling stories.

When the women petitioned to join the Society in 1902, they requested a reduced subscription due to their exclusion from the male-only social functions. Their petition was rejected and they were barred from membership in the Society. In response, the women promptly organized their own Women's Chemical Society. The admission of women to the LU-Chem. Soc. was raised in a subsequent year (probably 1908), but again they were turned down. It was not until 1912 that women chemistry students were finally admitted and a Society dance was instituted. In 1914, the LU-Chem. Soc. had their first woman speaker, Dorothy Baylis (see below), one of the graduating class. The same year, the men-only Kneipe was dropped and a Smoking Concert took its place. For those males who still abhorred the presence of women, there was the refuge of the Research Men's Club.[29]

Membership did not result in equality for women. As at Imperial College (see Chapter 4), the women chemistry students were expected to serve tea to their male colleagues. A cutting letter to the *LU-Chem. Soc. Magazine* in 1923 commented: "Lady Chemists are overwhelmed by the extreme courtesy paid to them at Chem. Soc. teas. To the Victorian male mind, they still serve as Hewers of Bread and Drawers of Tea."[30]

Edith Morrison (Mrs. Corran)

Amongst the women pioneers at Liverpool was the first woman to obtain a Ph.D. and as it happened, the Ph.D. was in chemistry.[31] The recipient,

---

[28] Anon. (1929). The evolution of the Chemical Society. *Liverpool University Chemical Society Magazine* **10**(3): 7–8.

[29] Anon. (1922). *Liverpool University Chemical Society Magazine* **3**(1): 30.

[30] Anon. (1923). Rip-Raps. *Liverpool University Chemical Society Magazine* **4**(1): 18.

[31] (a) Edwards, L. P. (1999). *Women Students of the University of Liverpool: their Lives Career and Postgraduate Lives 1883–1937* Ph.D. Thesis, University of Liverpool, (Ed. Lib Thesis 2153); and (b) *Student Records*, Archives, University of Liverpool.

Edith Morrison, was born on 17 August 1902 in Liverpool, completing her B.Sc. at LU in 1922 and her Ph.D. in 1924. The *LU-Chem. Soc. Magazine* announced the event with pride:

> An event of a nature quite unprecedented in the annuls of the Chemical Society took place in the large Lecture Theatre on January 16th. A degree of Ph.D. was introduced by Liverpool University in 1922 but not until December 1924 was the degree conferred on a woman. The Faculty of Science and especially the Department of Chemistry, is very proud that the first successful woman was a chemist — Dr. Edith Morrison, who after two years brilliant research work on Photo-synthesis, under Prof. Baly, obtained a well-deserved degree.[32]

Morrison subsequently worked at Colman's Mustard Co. and married George W. Corran in 1927. In the 1935–1936 session, it was noted that "Dr. Corran" was asked to return to LU to give a public lecture to the LU-Chem. Soc.[33]

Eileen Sadler (Mrs. Doran)

The career of Eileen Sybella Sadler[34] illustrated the difficulty of a woman chemist in finding an academic post except outside of chemistry or at a non-mainstream institution. Sadler was born in Wallasey on 10 November 1901, daughter of Harold Sadler, a chartered accountant, and Clarabell Sadler. She was educated at Wallasey High School for Girls and then Malvern Girls' College. Entering LU in 1919, she graduated with a B.Sc. in Chemistry in 1924, followed by an M.Sc. in 1926, and a Diploma in Education in 1928.

Sadler's career plan was to become a high school teacher. However, in 1928, she was offered and accepted a position as Research Assistant in the LU Department of Pharmacology of the Faculty of Medicine.[35] In

---

[32] Anon. (1925). *Liverpool University Chemical Society Magazine* **6**(2): 2.

[33] Anon. (1935). *Liverpool University Chemical Society Magazine* **16**: n. pag.

[34] Personal communication, e-mail, A. Allen, 20 September 2007. Archivist, University of Liverpool.

[35] Anon. (December 1928). *Malvern Girls' College Magazine* 35.

1929, she was promoted to Demonstrator, then teacher of Pharmacy in 1933, and Assistant Lecturer in 1934. Sadler resigned her position in 1935 and accepted an appointment as Science Lecturer at the F. L. Calder College of Domestic Science, which later became a constituent college of the Liverpool Institute of Education, then Liverpool Polytechnic. Sadler married William F. Doran, who was a Lecturer in Organic Chemistry at LU, though like Sadler, he held only an M.Sc. qualification. She died in 1996 at Ryedale, Yorkshire, aged 94.

Women Analytical Chemists

Analytical chemistry was an area favoured by some women chemists during the inter-War period. In fact, a remarkable number of the women chemists who graduated from LU chose this direction. We have selected three individuals to exemplify this: Dorothy Baylis, Muriel Roberts, and Gertrude Andrew.

Dorothy Baylis

Dorothy Baylis[36] was born on 1 July 1891 at Hull, Yorkshire to Alfred Baylis, commercial traveller, and Fanny Freeborough. She completed her B.Sc. at LU in 1914 and from then until 1916, was an analyst with Lever Bros., Port Sunlight. Then from 1916 to 1917, Baylis worked as an Analyst to the West Riding of Yorkshire Rivers Board, Wakefield, followed by a period as Research Chemist to Brookes Chemicals, Ltd., Lightcliffe, from 1917 to 1919. During 1920 and 1921, she was a Research Chemist at British Dyestuffs Corporation, Manchester. Then Baylis changed her career path, accepting an appointment as Lecturer in Pure and Applied Chemistry at Leicester College of Technology. At the College, she simultaneously undertook independent research on the composition of wool grease. This research contributed to her M.Sc., which was granted in 1926. Baylis died on 17 January 1978 at Bexhill, Sussex.

---

[36](a) (1926). Certificates of candidates for election at the ballot to be held at the ordinary scientific meeting on Thursday, December 2nd, *Proceedings of the Chemical Society*, 114; (b) *Student Records*, Archives, University of Liverpool; and (c) Anon. (February 1914). *Liverpool University Chemical Society Magazine* 19.

Muriel Roberts

Whereas Baylis led a peripatetic life, Muriel Roberts[37] remained in one position for most of her career. Roberts was born on 28 December 1893 at Everton, Liverpool, to Robert Roberts, accountant and estate agent, and Ann Ellen Roberts. She completed her B.Sc. at LU in 1915, becoming a member of the Society for Chemical Industry in 1924 and of the Analytical Society in 1931. In 1932, Roberts name appears as Senior Analyst at the Liverpool City Laboratories, where she was the first woman Public Analyst. Her name was listed in 1934 as an Additional Public Analyst for the County Borough of Barrow-in-Furness. Roberts received certification under the Food and Drug Adulteration Act in 1938. She continued to be a Public Analyst in the Liverpool City Laboratories until her retirement. She served as a member of Convocation's Standing Committee for the University of Liverpool from 1932 to 1950. By 1947, Roberts was also President of the Liverpool Soroptimist Club. She died on 29 June 1985 at Holylake, Wirral, aged 91.

Gertrude Garland Andrew

Gertrude Garland Andrew,[38] daughter of William Andrew, clerk of New Brighton, and Katherine Bartle Clemo, was born on 9 October 1895 and educated at Wallasey High School, entering LU in 1913, studying botany, zoology, and chemistry. After graduation in 1916, she worked as a Chemist for Cow & Gate, Winchester, Somerset, later becoming a Public Analyst. Andrew joined the Analytical Society in 1926. By 1948, she was a Senior Assistant Chemist with the County Laboratories, Staffordshire. After retirement, Andrew was re-hired at the Laboratories as an Additional Public Analyst. She died on 28 February 1978, at Bath, aged 82.

---

[37] (a) University of Liverpool, student records; (b) Personal communication, e-mail, A. Allen, July 2011. Archivist, University of Liverpool; and (c) Anon. (1985). Personal news, deaths. *Chemistry in Britain* **21**: 800.

[38] (a) (1948). *Register* Royal Institute of Chemistry; (b) University of Liverpool, student records; (c) (1913–1914). *Liverpool University Chemical Society Magazine*; and (d) Anon. (1978). News review. *Chemistry in Britain* **14**: 267.

English Provincial Universities   185

Yorkshire College (University of Leeds)

The Yorkshire College of Science was founded in 1875 to provide: "… instruction in such sciences and arts are applicable or ancillary to the manufacturing, mining, engineering and agricultural industries of the County of York"[39] The journal *Nature* expressed surprise that it had not been founded long before:

> It is the largest county in England, carries on a greater variety of industries all more or less dependent for success on the results of scientific research, and boasts of a larger number of local scientific societies and field-clubs than any other county in the three kingdoms …[40]

With the support of the Yorkshire Ladies Education Association, but against some opposition, arts courses were added in 1877 and the "of Science" was deleted from the name. Yorkshire College was admitted to the Federation in 1887 and, like Manchester, initially had a separate Women's Department, but it seems to have become defunct early in the College's history.

Arthur Smithells

Of all the university chemistry professors, it was Arthur Smithells,[41] Chair of Chemistry at Leeds from 1885 until 1923,[42] who had been most active in promoting the science education of girls (see Chapter 8). The explanation for his interest dates back to his last days at Owens College, Manchester, before his appointment at the Yorkshire College:

> Part of his last session (1882) at the Owens College was spent in conducting a course of lectures and practical work on chemistry at the

---

[39] Shimmin, A. N. (1954). *The University of Leeds-the First Half Century.* Cambridge University Press, Cambridge, p. 16.

[40] Anon. (1874). The Yorkshire College of Science. *Nature* **9**: 157–158.

[41] Raper, H. S. (1940). Arthur Smithells, 1860–1939. *Obituary Notices of Fellows of the Royal Society* **3**: 96–107.

[42] Challenger, F. (1953). Schools of Chemistry in Great Britain and Ireland–IV The Chemistry Department of the University of Leeds. *Journal of the Royal Institute of Chemistry* **77**: 161–171.

186 *Pioneering British Women Chemists: Their Lives and Contributions*

neighbouring [Manchester] Girls' High School where he gained his first experience of teaching, an experience which may have directed his attention to the gaps in a girl's education, and implanted the germ of his present schemes for the scientific training of women.[43]

Smithells was obviously deeply influenced by his experience teaching high school girls, commenting: "… the first teaching I ever attempted was in a girls' high school, and I have at least a first-hand knowledge of a wrong way of doing it."[44]

Women Chemistry Students

The complaint was made in a letter of 1898 to the student magazine *The Gryphon* that the "industrious" and "enthusiastic" women students were showing up their laggardly male colleagues,[45] whereas at Owen's College, it was the men who grabbed the laboratory stools. The presence of women in the chemistry laboratory was equally unpopular as a correspondent in 1902 reported:

> Well might the new Hiawatha lament the depredations of the stool-snatcher, for the worst offenders are — tell it not to an Amazon — *snatcheresses*. And there is no redress — man must submit and stand. Look where you will, woman is in possession; aye, even of the very stink-cupboard.[46]

Leeds University Cavendish Society

The Yorkshire College had a scientific society to which women were welcome, the first woman speaker noted being a Miss Findlay who gave a presentation on "Recent Attempts in Colour Photography" on

---

[43]"J.B.C." (February 1910). Our photograph. *The Gryphon: Journal of the University of Leeds* **13**(3): 37, LUA/PUB/002/GRY/73.

[44]Smithells, A. (December 1912). Science in girls' schools. *School World* **14**: 460.

[45]"F.C.G." (March 1898). Students: Lady students. *The Gryphon: Journal of the Yorkshire College* **1**(3): 44, LUA/PUB/002/GRY/3.

[46]"Mel da Kahnt." (May 1902). The position of man in the laboratory. *The Gryphon: Journal of the Yorkshire College* **5**(5): 86, LUA/PUB/002/GRY/27.

English Provincial Universities 187

16 December 1897.[47] The society changed its name to Leeds University Cavendish Society in 1910, with discussions solely on chemistry and physics topics. A report on a meeting of early 1916 indicates that women chemistry students had begun to play a more active role: "... The meeting was indeed a remarkable one, for the women students turned up in great numbers and joined in a little discussion."[48]

Though the Cavendish Society was the recognised body for discourses on the physical sciences, some male members decided they wanted a women-free social venue, the Organic Lab "club." This "club" appears to have organized male-only dinners with songs, smoking, and musical solos, similar to the activities of the Manchester and Liverpool student chemistry societies. The "club" was first reported in 1898[49] and was still active in 1905,[50] while in 1911 a "Chemists' Dinner" seems to have been organized in a similar male-only format.[51]

May Leslie (Mrs. Burr)

Leeds' most famous chemistry alumna was May Sybil Leslie.[52] Leslie was born on 14 August 1887 in Woodlesford, Yorkshire, daughter of Frederick Leslie, school board attendance officer, and Elizabeth Dickinson. She studied chemistry at the University of Leeds and graduated with a B.Sc. (Hons.) in 1908. The following year, she was awarded an M.Sc. for

---

[47] Anon. (February 1898). Scientific society. *The Gryphon: Journal of the Yorkshire College* **1**(2): 30, LUA/PUB/002/GRY/2.

[48] "C.A.M." (March 1916). Cavendish Society. *The Gryphon: Journal of the University of Leeds* **19**(4): 64, LUA/PUB/002/GRY/1112.

[49] Anon. (March 1898). College news: The organic lab. *The Gryphon: Journal of the Yorkshire College* **1**(3): 48, LUA/PUB/002/GRY/3.

[50] Anon. (February 1905). Convivial chemists. *The Gryphon: Journal of the University of Leeds* **8**(3): 46, LUA/PUB/002/GRY/43.

[51] "H.E.W." (May 1911). Chemists' dinner. *The Gryphon: Journal of the University of Leeds* **14**(5): 75, LUA/PUB/002/GRY/82.

[52] (a) Rayner-Canham, M. (23 September 2004). Leslie [*married name* Burr], May Sybil (1887–1937) chemist. *Oxford Dictionary of National Biography* Oxford University Press (accessed 16 January 2019); and (b) Rayner-Canham, M. and Rayner-Canham, G. (1993). A chemist of some repute. *Chemistry in Britain* **29**: 206–208.

188 *Pioneering British Women Chemists: Their Lives and Contributions*

research with Harry M. Dawson[53] on the kinetics of the iodination of acetone, work that has since become a classic in its field.[54]

In that same year, 1909, Leslie was awarded a scholarship which she decided to use to work with Marie Curie[55] and her research group[56] in Paris. Her letters from Paris to Smithells are among the few accounts of life in the early Curie laboratory. Leslie spent 1909 to 1911 with Curie, and as she told Smithells, the only English woman in Curie's group.

> There are only two ladies besides myself, Norwegian Mlle. Gleditsch,[57] and French, Mlle. Blanquies.[58] Of the French lady I see very little because she does not spend all her time here, but of Mlle. Gleditsch I see much since she lives in the same pension. She has been exceedingly good to me and has prevented me from feeling lonely.[59]

Leslie's work involved the extraction of new elements from thorium, and for a chemist used to working with grams of pure chemicals in beakers, the manipulation of kilogram quantities of radioactive minerals in huge jars and earthenware bowls must have been a completely new experience.

> A number of people seem to be employing radium emanation [radon] at present and my electroscope is disgraceably sensitive to the influence of

---

[53]Whytlaw Gray, R. and Smith, G. F. (1940). Harry Medforth Dawson, 1876–1939. *Obituary Notices of Fellows of the Royal Society* **3**: 139–154.

[54](a) Personal communication, e-mail, J. Sichel, 31 August 1993; and (b) Dainton, F. S. (1993). Reputable memories. *Chemistry in Britain* **29**: 573.

[55](a) Quinn, S. (1995). *Marie Curie: A Life*. Simon and Schuster, New York; and (b) Pflaum, R. (1989). *Grand Obsession: Marie Curie and Her World*. Doubleday, NY.

[56]Davis, J. L. (1995). The research school of Marie Curie in the Paris faculty, 1907–1914. *Annals of Science* **52**: 321–355.

[57](a) Kubanek, A-M. W. (2010). *Nothing Less Than an Adventure: Ellen Gleditsch and Her Life in Science*. CreateSpace Independent Publishing Platform, Montreal, Canada; (b) Lykknes, A., Kragh, H. and Kvittingen, L. (2002). Ellen Gleditsch: Pioneer woman in radioactivity. *Physics in Perspective* **6**: 126–155.

[58]Micault, N. P. (2013). The Curie's lab and its women (1906–1934). *Annals of Science* **70**: 71–100.

[59]Letter, M. S. Leslie to A. Smithalls. (30 November 1909). Smithalls Collection, Archives, University of Leeds, MS/416/1A/219.

English Provincial Universities   189

anyone entering from the salle active so that I spend half my time in keeping dangerous people out & in airing the room. Formerly more care was taken to prevent the distribution of activity all over the laboratory, but as the foundations for a new Institute of Radioactivity for Mme Curie are now laid, all precautions seem to have been abandoned.[60]

Returning to England, Leslie took a position with Ernest Rutherford at the Physical Laboratory of Victoria University. There she continued with her work on thorium and extended her studies to actinium during the 1911 to 1912 period. Leslie spent the next 2 years as a science teacher at the Municipal High School for Girls in West Hartlepool. During this time, she managed to resume research with Dawson, this work being on ionization in non-aqueous solvents. From 1914 to 1915, Leslie held a position as Assistant Lecturer and Demonstrator in Chemistry at the University College in Bangor, Wales.

In 1915, Leslie entered the world of industrial chemistry, being hired to work at His Majesty's Factory in Litherland, Liverpool, a position that she obtained as a result of the call-up for military duty of the male research chemists. Her initial rank was that of Research Chemist, but in 1916, she was promoted to Chemist in Charge of Laboratory, a very high position for a woman at that time. Leslie's research involved the elucidation of the pathway in the formation of nitric acid and the determination of the optimum industrial conditions for the process. This work was vital for the munitions industry, which required massive quantities of nitric acid for explosives production. In June 1917, the Litherland factory closed[61] and Leslie was transferred with the same rank to the H. M. Factory in Penrhyndeudraeth, North Wales. She was awarded the D.Sc. degree in 1918 by the University of Leeds, mainly in recognition of her contribution to the war effort.

With the return of the surviving male chemists at the end of the First World War, Leslie lost her government position. She returned to the University of Leeds as Demonstrator in the Department of Chemistry in 1920, being promoted in the following year to Assistant Lecturer. Leslie

---

[60]Letter, M.S. Leslie to A. Smithalls. (8 June 1911). Smithalls Collection, Archives, University of Leeds, MS/416/1A/234.
[61]Rogans, E. S. F. (1993). Reputable memories. *Chemistry in Britain* **29**: 573.

190   *Pioneering British Women Chemists: Their Lives and Contributions*

then moved to the Department of Physical Chemistry in 1924 and was promoted to Lecturer in 1928. In 1923, Leslie had married Alfred Hamilton Burr,[62] a Lecturer in Chemistry at the Royal Technical College, Salford. She had first met Burr at the H. M. Factory in Litherland where he, too, had worked in 1916. Continuing active research at Leeds after marriage, Leslie was also invited by the famous British chemist, J. Newton Friend, to author one volume of the classic series *A Textbook of Inorganic Chemistry* and to co-author another.[63]

According to university records, Leslie resigned her position at the University in 1929 due to health reasons. In 1931, when Burr was appointed head of the Chemistry Department at Coatbridge Technical College, Scotland, she moved there with him. After Burr died in 1933, Leslie returned to Leeds, resuming research work at the University. Her first project was the completion of Burr's unfinished research on wool dyes, then she returned to her own interest in the mechanisms of reactions. In addition to performing research work, she was employed as sub-Warden of a woman's hall of residence (Weetwood Hall) at the university from 1935 to 1937.

Leslie died at Bardsey, near Leeds, on 3 July 1937, having given up research only a month earlier. No cause of death was recorded, but it was quite possibly radiation-related considering her exposure to high levels of radioactivity during her research work in Paris. Her obituary in the *Yorkshire Post* noted that Leslie was "one of the University's most distinguished women graduates."[64] Her former supervisor, Dawson, commented that her research reputation was "deservedly high" and that as a teacher she was "exceptionally gifted."[65]

---

[62] Anon. (1934). Obituary: Arthur Hamilton Burr. *Journal of the Royal Institute of Chemistry Proceedings* 68.

[63] (a) Burr, M. S. (née Leslie) (1925). *The Alkaline Earth Metals,* Vol. 3, Friend, J. N. (ed.), *A Textbook of Inorganic Chemistry*, Part 1, Charles Griffin, London; and (b) Gregory, J. C. and Burr, M. S. (née Leslie) (1926). *Beryllium and its Congeners*, Vol. 3, Friend, J. N. (ed.), *A Textbook of Inorganic Chemistry; Part 2*, Charles Griffin, London.

[64] *Yorkshire Post*, (6 July 1937), 5.

[65] Dawson, H. M. (1938). May Sybil Burr. 1887–1937. *Journal of the Chemistry Society* 151–152.

University College, Sheffield (University of Sheffield)

University College, Sheffield, was formed in 1897 by the merger of Sheffield School of Medicine, Firth College, and Sheffield Technical School.[66] University College, Sheffield, had applied to join the confederated Victoria University. However, the request was rebuffed.[67] At the breakup of Victoria University, Sheffield had proposed a Federal University of Yorkshire, having campuses at Leeds and Sheffield, but Leeds rejected the concept, opting to become an independent university. Sheffield then appealed for its own university status, which was granted in 1905.

Women Chemistry Students

Women students had been admitted since 1886 by the antecedent institutions of University College, Sheffield.[68] Yet, as elsewhere, co-education did not necessarily imply that women students were accepted as equals. Attitudes towards the women chemistry students at Sheffield by their male colleagues seemed to have evolved from bemusement in the early years through to hostility as the 20th century progressed.

In the Sheffield Student Newspaper, *Floreamus!* of 1907, there is a semi-fictitious discussion between four male students about the effect on the University of the presence of women students. The science student described the women students as being some separate species:

> 'I have never found the ladies such a nuisance,' said the Science man, 'They never make all that noise in the labs., and they're always extremely grateful if you help them to set up their apparatus, or show them how to do an experiment.'[69]

By the 1920s, the women students were being portrayed in a very negative light. A letter to the student paper commented that, despite decades of co-education at Sheffield, the activities of the women students seemed to be

---

[66] Anon. (1897). The foundation of University College, Sheffield. *Floreamus!* **1**: 3–5.

[67] Mathers, H. (2005). *Steel City Scholars: The Centenary History of the University of Sheffield.* James & James, London, pp. 33–34.

[68] Note 67, Mathers, p. 9.

[69] "Dizzy-Dotty" (March 1907). A dinner-table discussion. *Floreamus!* **3**: 129–132.

192 *Pioneering British Women Chemists: Their Lives and Contributions*

"chiefly dancing and scandal-mongering," leading to the conclusion that: "Without formulating it as a rule, and conceding many exceptions, it is undeniable that the influence of the woman student is on the whole deterrent to serious thought."[70]

Sheffield University Chemical Society

Up until the First World War, the University of Sheffield Chemical Society had a significant proportion of women members, some of whom gave presentations to the Society. Nevertheless, according to the rules of the Society, women still had to perform their traditional roles: "Women Students Only: All are expected to help to prepare tea, but beware lest thou are late in commencing operations, and so the hungry ones are kept waiting."[71]

By 1949, the student Chemistry Society had become exclusively male. According to a male writer in *By-Product*, the Journal of the Sheffield University Chemical Society, the woman chemistry student is regarded as simply a husband-hunter:

> In these days of man-shortage a girl is often forced into a career other than that of hunting and training the male. Consequently advice seems desirable on occupations for that period between algebra and the altar. This week research chemistry is my subject — a pleasant career involving little work, few restrictions and close contact with eligible men. ... If these points are followed carefully, the young [female] research chemist should perfect the arts of entertainment and cultivate the graces of a hostess in sufficient time to make an excellent wife for her Professor![72]

William Palmer Wynne

In Chapter 15, we will see that the First World War resulted in a demand for women chemists — in particular, for the small-scale production of fine

---

[70]"M/M" (1912–1914). Rules for the guidance of first-year science students. *Floreamus!* **6**: 133.

[71]"A.P.Q." (1920–1922). Correspondence. *Floreamus!* **10**: 56.

[72]"Auntie Ethyl" (May 1949). Careers for women-I: Research chemistry. *By Product: Journal of the University of Sheffield Chemical Society* **2**(2): 5–6.

English Provincial Universities   193

chemicals. Nowhere was this more evident than the Chemistry Department at the University of Sheffield under William Palmer Wynne.[73]

By 1915, Wynne was the sole remaining faculty member of the Chemistry Department, the others having departed for war work.[74] As a result, Wynne hired Emily Turner and Dorothy Bennett as Assistant Lecturers and Demonstrators; together with Annie M. Mathews, as Demonstrator and Lecture Assistant.[75] In addition to teaching duties, they synthesized large quantities of local anaesthetics for the war effort (see Chapter 15). Turner and Bennett were to remain key figures in the chemistry department for ensuing decades. As R. L. Wain remarked: "The two of them were, to my mind, a well-respected team, imposing strong discipline in classes. They were often referred to by students as 'the tartrate twins'."[76]

Emily Turner

Emily Gertrude Turner[77] was born in Manchester on 16 April 1888, to John Wesley Turner, gas meter fitter, and Annie Hague. She was educated at Wellgate Council School and then completed a B.Sc. in Chemistry at Sheffield in 1909, followed by an M.Sc. in 1911. From 1911 to 1912, Turner studied for an Educational Diploma, and then taught at a school in Newcastle-upon-Tyne.[78] It was while at Newcastle, that Wynne invited her back as an Assistant Lecturer and Demonstrator at Sheffield.

---

[73] Rodd, E. H. (1951). William Palmer Wynne, 1861–1950. *Obituary Notices of Fellows of the Royal Society* **7**: 519–531.

[74] Haworth, R. D. and Stevens, T. S. (1956). Schools of chemistry in Great Britain and Ireland–XXIV The University of Sheffield. *Journal of the Royal Institute of Chemistry* **80**: 269–274.

[75] Chapman, A. W. (1957–1958). The early days of the chemistry department. *By Product: Journal of the University of Sheffield Chemical Society* **12**: 2–5.

[76] Personal communication, letter, R. L. Wain, 27 February 1998. Turner and Bennett are referred to as the "tartrate twins" in a variety of sources.

[77] Chapman, A. W. (1957). Emily Gertrude Turner, 1888–1956. *Proceedings of the Chemical Society* 296.

[78] E. Turner to Dr. Rodd. Letter of biographical notes on W. P. Wynne which included asides on her own life. Turner Manuscripts, MS 280, Archives, University of Sheffield.

194   *Pioneering British Women Chemists: Their Lives and Contributions*

Turner stayed at Sheffield from 1915 to 1953. It was estimated that she had given "... some 4,000 lectures and spent at least 20,000 hours in laboratory supervision" during her 38 years of service.[79] Though her major responsibility was teaching, she was able to co-author six research papers between 1911 and 1941: three with James Kenner (see below), one with Wynne, one with G. M. Bennett, and one with Wynne and G. M. Bennett. Turner also acted as secretary to Wynne from about 1920 until his departure in 1931.

One of her colleagues, Peter A. H. Wyatt, described working with her:

> She fitted into the old order in the chemistry departments of those days and I don't think you would find anybody with quite her work description in any British university these days. Thus she had no ambitions whatsoever to be a research scientist and seemed outwardly content with her role as a teacher of practical classes only at the elementary level, never giving any lectures apart from the talks at the beginning of the practical classes to describe the day's work (some which I also delivered). In consequence she was never promoted beyond the grade of Lecturer (though most people towards the end of her time rose to be at least Senior Lecturers) and it could be that she may have felt some pangs of disappointment about that. If so, she never said.[80]

A former student, H. J. V. Tyrrell, recalled that she gave chemistry lectures to the health science students:

> Until her retirement she was an institution in the Department, presiding at tea in the little staff room, and giving lectures, lavishly illustrated with experimental demonstrations, to the first-year medical and dental students, supplemented by practical classes.[81]

Turner's nephew, Jeffrey Turner periodically visited her:

> ... as often as not [I] would meet my Aunt in her room adjoining the University Chem Labs, which she shared with Miss Dorothy Bennett, a life-long friend of E.G.T. and all the Turner families. EGT lived in a flat

---

[79] Anon. (1956–1957). Obituary. *By Product: Journal of the University of Sheffield Chemical Society* **10**: 4.

[80] Personal communication, letter, P. A. H. Wyatt, 24 March 1998.

[81] Personal communication, letter, H. J. V. Tyrrell, 9 April 1998.

in Sheffield throughout her university career, returning to her parents and sister at weekends. Holidays would be taken, usually with her sister, to English countryside resorts. Equipped with a 1 inch Ordinance Survey map containing the resort they picked their way along many of the footpaths in their walks in the surrounding area.[82]

Turner retired in 1952. She then went to live with her sister Beatrice at Bawtry Road, Rotherham, where she died 15 June 1956.

Dorothy Bennett (Mrs. Leighton)

Unfortunately, there is little information on Dorothy Marguerite Bennett.[83] Bennett was born in 1884 at Sheffield to Cornelius Bennett, commercial traveller, and Ada Bennett. She was educated at Sheffield High School for Girls then completed her B.Sc. in Chemistry in 1909 (the same year as Turner) and her M.Sc. in 1910. Again, like Turner, she initially proceeded into school teaching until requested to join the Chemistry Department of Sheffield in 1915. Bennett, too, kept her position of Assistant Lecturer after the end of the war, rising to Senior Lecturer, and in 1926 she was also appointed Tutor for women students. At some date, she was awarded a D.Sc. (Sheffield).

Bennett retired from her academic position in 1934, subsequently marrying Henry Birkett Leighton. However, she kept her post as Tutor for women students and added the role as Warden of University (Residence) Hall for Women (later called Halifax Hall) in 1936. She resigned from both these duties in 1947 and moved to Dartford. Bennett died on 11 May 1984 in Tunbridge Wells, 12 days after her 100th birthday.[84]

Annie Mathews

Annie Moore Mathews was born at Sheffield on 9 March 1891 to Samson Mathews, medical doctor, and Jane Wallace. Before the First World War,

---

[82]Family tree. Turner Manuscripts, MS 280, Series 2, 1/1, Archives, University of Sheffield, also containing biographical notes from Jeffrey Turner, nephew.

[83]Anon. (1984). Obituary: Mrs. D. M. Leighton. *University of Sheffield Newsletter* **8**(13): 2.

[84](a) (4 May 1984). Our wonderful centenarians. *Kent and Sussex Courier;* and (b) (25 May 1984). Obituary: Mrs. D. M. Leighton. *Kent and Sussex Courier.*

Mathews undertook research on polycarboxylic acids with James Kenner, Professor of Organic Chemistry at Sheffield, resulting in two publications in 1914.[85] In 1918, Mathews and Kenner were married. Lord Todd knew the family well and remarked about the son, George Kenner: "His mother, herself a chemist, I can recall only as a rather ebullient, talkative woman."[86]

The friction in the University of Sheffield Chemistry Department between Wynne and James Kenner became so extreme that Kenner felt he had to leave. In late 1924, Kenner was offered and accepted the position of Professor of Organic Chemistry (Pure and Applied) at the University of Sydney, and the family moved to Australia. They returned to England in January 1928 when Kenner was appointed Professor of Technological Chemistry at the Manchester College of Technology. During the Second World War, the family went on cycling holidays and it was during one of those in September 1943 that Mathews was killed in a road accident in Cornwall.

Mason College (University of Birmingham)

Mason College of Science opened in 1880.[87] At the inauguration of the new building in 1883, specific mention was made about the admission of women:

> Special interest was felt in the inauguration of the building on account of a feature which distinguishes it from any other Science College — namely, that it is open to women on exactly the same terms as men. All who have watched this experiment, if such it may be called, must be satisfied with the entirely successful way it has worked.[88]

In fact, according to the 1884–1885 *Mason Science College Calendar*, in the 1882–1883 session, the enrolment was 229 males and 137 females.

---

[85]Todd, Lord (1975). James Kenner, 13 April 1885–30 June 1974. *Biographical Memoirs of Fellows of the Royal Society* **21**: 389–405.

[86]Todd, Lord (1979. George Wallace Kenner, 16 November 1922–26 June 1978. *Biographical Memoirs of Fellows of the Royal Society* **25**: 391–420.

[87]Ives, E., Drummond, D. and Schwarz, L. (2000). *The First Civic University: Birmingham 1880–1980 an Introductory History*. University of Birmingham.

[88]Jordan, A. (1883). The Mason Science College: A sketch. *Mason College Magazine* **1**(2): 33–35.

There was a sense of community among the women as they could meet in the Ladies' Common Room as was described in 1899:

> The liveliest time in the Common Room is between the hours of 1 and 2 p.m., and 4 and 5.30 p.m. At those hours students may be seen making tea, coffee, or cocoa, according to their tastes, and lounging in the armchairs ....[89]

The name was changed to Mason University College in 1898. Then, following the granting of a Royal Charter by Queen Victoria in 1900, it was again renamed, this time to the University of Birmingham, making it the first civic or "red-brick" University to receive its own charter.

Women Chemistry Students

The presence of women in the chemistry laboratories is first mentioned in 1883. A commentary describes how, if one was short of chemicals or equipment, one appropriated them from a neighbour's bench. The writer added: "The lady students, too, though they are a little longer in acquiring this beautiful virtue, soon learn how dependent all members of society are on one another,"[90]

In another of the student magazines, *The Mermaid*, an account is given in 1909–1910 of the new woman chemistry student:

> The modern Chemistry Girl in the smart, brand-new Lab is, you may be sure, quite up-to-date, and, if the truth be known, a source of perpetual worry to most of the rather dull men, who have uneasy visions of their names on the terminal lists preceded by those of a whole string of young girls.[91]

---

[89] Anon. (1899). The ladies' common room. *Mason University College Magazine* **1**(2): 35.

[90] "An Average Specimen." (1883). Another point of view. *Mason College Magazine* **1**(5): 112–116.

[91] Gordon, A. (1909–1910). College encounters III: The chemistry girl. *The Mermaid: University of Birmingham* **6**: 151–153.

198 *Pioneering British Women Chemists: Their Lives and Contributions*

Chemical Society

The Mason College Chemical Society was formed in 1884,[92] and among two of the early members were the first woman Associate of the Institute of Chemistry, Emily Lloyd (see Chapter 9) and the first woman Student Member, Rose Stern (see Chapter 9). However, according to the reports in the *Mason College Magazine*, two Birmingham women preceded them as the first two women members of the Mason College Chemical Society. These were Jessie Charles and Constance Naden, both of whom were very active in the Society.

Jessie Charles (Mrs. White)

Born on 19 March 1865 to Andrew Charles, a hardware merchant, Jessie Charles[93] was educated privately before entering Mason College in 1882 and passing the external London B.Sc. (Hons.) in Mental and Moral Science in 1887. She entered Newnham College in 1890 and after completing Part 1 of the Science Tripos in 1893, she continued her studies in Breslau and Leipzig, passing the D.Sc. (London) Examination in Mental and Moral Science in 1898. In her multifaceted life, Charles (later Mrs. White) became a promoter of the Montessori School system, authoring a book on the subject.[94]

Constance Naden

Constance Caroline Woodhill Naden[95] was born on 24 January 1858 to Thomas Naden, architect, and Caroline Ann Woodhill.[96] In 1879, she

---

[92] Anon. (1884). Mason College Chemical Society. *Mason College Magazine* **2**(1): 21–22.

[93] White, A. B. (ed.), (1981). *Newnham College Register, 1871–1971*, Vol. II, *1924–1950*, 2nd edn., Newnham College, Cambridge, p. 104.

[94] White, J. (1913). *Montessori Schools as seen in the Summer of 1913*. Cornish Brothers Ltd., Birmingham.

[95] (a) Anon. (1890). Constance C. W. Naden, *Mason College Magazine* **8**(3): 49–55; and (b) Creese, M. R. S. (1998). *Ladies in the Laboratory? American and British Women in Science, 1800–1900*. Scarecrow Press, Lanham, Maryland, p. 362.

[96] Wikipedia. (29 January 2019). Constance Naden (accessed 29 February 2019).

attended the Birmingham and Midland Institute to study botany and French, then entered Mason Science College in 1882 to study physics, geology, chemistry, physiology, and zoology. Naden saw the world through poetry and was able to link science and poetry.[97] She had a long poem published in the *Mason College Magazine*, titled: "Free Thought in the Laboratory (Dedicated to the Demonstrator of Chemistry)" The rhyme began:

My mind was calm, my heart was light,
My doubts were few and fleeting,
Till I attended yesternight
An M. C. Chem. Soc. Meeting[98]

Naden continued to describe the presentation in verse. She reported that a member of the audience (her Chemistry Demonstrator, Thomas Turner) had disputed "in sad sepulchral tone"[98] the speaker's view of molecules. In particular, according to Naden, Turner contested the cyclic structure of benzene first proposed by Kekulé only 20 years earlier. Also in verse, Thomas Turner replied to Naden:

A solemn man with tomb-like voice
Would send Miss Naden greeting,
And thank her for her pleasant rhyme
On M.C. Chem. Soc. meeting
He's no regrets of causing doubts
of truth of benzene rings,
for doubts should only lead to faith,
in nobler, truer things.[99]

---

[97]Thain, M. (2003). "Scientific Wooing": Constance Naden's Marriage of Science and Poetry. *Victorian Poetry* **41**: 151–169.

[98]"C.C.W.N." [Naden, C.]. (1885). Free thought in the laboratory [Dedicated to the demonstrator of chemistry]. *Mason College Magazine* **3**(4): 83–84.

[99]"T.T" [Turner, T.]. (1885). Reply. *Mason College Magazine* **3**(4): 84.

200 *Pioneering British Women Chemists: Their Lives and Contributions*

Naden left Mason College in 1887 and focussed her life on philosophy and poetry. She had inherited a considerable fortune which enabled her to travel with her friend Madeline Daniell to the Middle East and to India. Returning to England in 1888, she became active in the women's suffrage movement. In 1889, Naden was diagnosed with infected ovarian cysts and though she survived the operation, she died 19 days later on 23 December 1889 of a related infection. A bust of Naden overlooks the archives room of the University of Birmingham.

Evelyn Hickmans

Another pioneering woman chemistry student at Birmingham was Evelyn Marion Hickmans.[100] Hickmans was born on 9 April 1882 at Coseley, Staffordshire, to David Hickmans, dairyman, and Mary E. Hickmans. She was educated at Leeds High School for Girls and then obtained a B.Sc. at the University of Birmingham in 1905 and an M.Sc. in 1906. Hickmans unsuccessfully applied for a Demonstratorship at Bedford College in 1906 and instead undertook research with Alexander Findlay at Birmingham. Hickmans completed an M.B. degree in 1918, then from 1919 until 1922, she was a Lecturer in the Household Science Department of the University of Toronto, Canada, returning to England in 1923.

Upon her return, Hickman's cousin, Leonard Parsons, a Professor specialising in children's diseases at the Children's Hospital, Birmingham, asked her if she would help him with his investigations.[101] She started the following day, a week later being requested to undertake blood chemical analyses for the whole Hospital. Hickmans was subsequently appointed Head of Department. Between 1924 and 1956, she authored and co-authored 13 publications in a wide variety of studies mostly on the relationship of abnormal blood chemistry to childhood diseases.

It was work in 1951 for which she became internationally renowned. At the time, Horst Bickel, a Research Fellow at the Hospital, was visited

---

[100] *Student Records*, Archives, University of Birmingham, and *Student Records,* Archives, Bedford College, files of the Chemistry Department.

[101] *History of the Clinical Chemistry Department.* Birmingham Children's Hospital, unpublished manuscript. Anne Green, Birmingham Children's Hospital is thanked for supplying a copy of the document.

by the mother of a 17-month-old girl, Sheila, who was diagnosed as suffering from phenylketonuria (PKU). The mother refused to accept that nothing could be done for her daughter. As a result of her persistence, Bickel, together with Hickmans, and John Gerrard devised a diet low in phenylalanine, having concluded it was an excess of phenylalanine that was the culprit in the metabolic process of PKU sufferers. The girl recovered.

Their discovery was published in 1954. The ground-breaking article has been considered so important that the journal *Acta Paediatrica* reproduced the first page of the original paper in an issue of 2001 on the 47th anniversary of the original publication.[102] The three researchers received the John Scott Medal for contributions to "the comfort, welfare and happiness of mankind" from the City Trust, Philadelphia, in 1962 for their work on methods of controlling phenylketonuria.[103] Hickmans retired in 1953, dying at Wolverhampton in 1972, aged 89.

University College, Bristol (University of Bristol)

To offer matriculants in the south-west of England the possibility of advanced academic courses, the Clifton Association for the Higher Education of Women was established about 1868.[104] When University College, Bristol, was founded in 1876, the Clifton Association transferred their students to the college, ensuring a significant female student population from the very beginning. In those early years, there was a competing institution of higher education: Merchant Venturers' Technical College. However, in 1906, the Technical College buildings burned down, leaving University College as the sole institution of higher education. Like the university colleges founded in other parts of England, University College, Bristol, offered degrees through the University of London until it obtained its own University Charter in 1909.

---

[102] Anon. (2001). Acta Paediatrica 47 Years Ago. *Acta Paediatrica* **90**: 13–56.

[103] Fox, R. (1968). The John Scott Medal. *Proceedings of the American Philosophical Society* **112**: 416–430.

[104] "E.S." (1876). University College, Bristol. *Journal of the Women's Educational Union* **4**: 124–126.

Women Students

Both the male and female founders and supporters of University College, Bristol, had been adamant that women were to be admitted from the first as equals to men. However, equal did not mean co-educational:

> The Lectures of the College will be open to the students of both sexes, a part of the lecture room being appropriated to women. Separate classes for women alone will also be held, in which the instruction will be of a more detailed and catechetical kind (see Footnote 104).

Marian F. Pease described in 1876 how segregation also occurred outside of the lecture room: "Between lectures we sat in the small women's cloakroom. ... It was furnished with three or four wooden chairs and a small deal table and with pegs for our heavy cloth waterproofs ... ."[105]

In reality, women students were accepted at Bristol only as long as they knew their place. One of the most shameful misogynist incidents in British university history occurred at University College, Bristol: the attack on the "Votes for Women" Office on Queen's Road. After throwing the furniture through the broken windows, the male students then pulled bales of hay on sledges to the Office, ignited the hay, and watched the shop burn. The Editor of the Bristol student newspaper, *The Bristol Nonesuch*, justified the attack on the grounds that: "The arm of the law is just too short to reach these female fanatics, but we were able to supply the missing inches... ."[106]

Women Chemistry Students

Marian Pease was one of the earliest women students to take chemistry courses. In her reminiscences, she commented on the chemistry laboratory of the time: "The Chemical laboratory under Dr. Letts was housed in the attic. They seemed to be always making sulfuric acid gas and the house was perpetually full of the smell of rotten eggs... ."[105] In 1898–1899, an

---

[105] Pease, M. F. (23 February 1942). *Some Reminiscences of University College, Bristol.* Bristol University Archives. n. pag.

[106] The Editor. (1913). Editorial. *The Bristol Nonesuch* **3**(8): 3.

anonymous contributor to the student magazine, *The Magnet*, commented:

> A research on the action of sodium in promoting the growth of plants is being carried on beneath one of the Laboratory windows under the direct supervision of the botanical gardener (i.e. Miss X. throws all her refuse metallic sodium out the window).[107]

The Chemical Society admitted women, but there seemed to be a divide between the sexes as was noted in 1911:

> An ordinary meeting of the Chemical or Physical Society is a somewhat dull affair. First of all, tea: two hungry parties, male and female, one on each side of a long table make the most of what is to be had, but ne'er exchange a word.[108]

Katherine Williams

As far as we could discover, the first woman chemist at University College, Bristol, was Katherine Isabella Williams.[109] Williams was born in 1848 at Llanwelly, Monmouthshire, daughter of Thomas Williams, Dean of Llandaff, and Elizabeth Davies. She was educated at King Edward VI High School for Girls, Birmingham and much later became a student at University College, Bristol in 1877 at the age of 29, passing the Cambridge Higher Local Examinations in 1885.

In the 1880s, she collaborated with William Ramsay and then she embarked upon her own research programme in food analysis. In his biography of Ramsay, William Tilden (see Chapter 9) noted as follows:

> At Bristol in the early days of the College there were but few advanced students capable of taking part in research. Among these Miss K. I.

---

[107] Anon. (1898–1899). Chemistry. *The Magnet, University College, Bristol* **1**(1): 27–28.

[108] Anon. (1911). The chemical and physical societies' social evening. *The Bristol Nonesuch* **1**(2): 33.

[109] Note 95b, Creese, pp. 266–267.

204   *Pioneering British Women Chemists: Their Lives and Contributions*

Williams, whose death took place in January 1917, deserves to be mentioned. Ramsay suggested to her an investigation into the composition of various food stuffs, cooked and uncooked, and this enquiry occupied her continued attention till the close of her life thirty-five years later.[110]

Morris Travers account of Ramsay's life and work, mentioned twice that Ramsay suggested Williams repeat the Cavendish experiment on air. It was the later repetition of Cavendish's experiment by others that resulted in the discovery of the noble gases:

> Miss Katherine Williams, who worked for many years in the department on the chemistry of cooked fish, came first under Ramsay. It is said that he suggested that she should repeat the Cavendish experiment on air, but she chose something easier, the determination of the oxygen dissolved in water (p. 69). Later, when he and Miss K. Williams, at Bristol, were investigating an alleged allotropic form of nitrogen (*Proc. Chem. Soc.*, 1886), he says that he suggested that she should repeat the Cavendish experiment; but the matter went no further.[111]

During Williams' career, she authored 10 papers on the chemistry of food over 14 years, obtaining a B.Sc. (Bristol) by research in 1910 in her early 60s. She was in Switzerland when the First World War commenced, and according to her obituarist, the "anxiety and actual hardships she suffered before being able to return to England told severely on her constitution."[112] Williams died in March 1917 at Bristol. Unfortunately the full account of her research on food chemistry was never published, as the following comment in the obituary described:

> Sir William Ramsay and others had induced her to write a popular account of her work, and Sir William had promised to write a preface. This was completed about a year ago, but he had passed away before the

---

[110]Tilden, W. A. (1918). *Sir William Ramsay: Memorials of His Life and Work.* Macmillan & Co., London, pp. 83–84.

[111]Travers, M. W. (1956). *A Life of Sir William Ramsay.* Edward Arnold, London, p. 100.

[112]Anon. (March 1917). In Memoriam. *The Bristol Nonesuch* **4**(18): 196–198.

English Provincial Universities   205

promise could be carried out, and Miss Williams died before the work had gone to the publishers (see Footnote 112).

Emily Fortey

Following after Williams at Bristol, Emily Comber Fortey,[113] was born in Chelsea, London, in 1866, daughter of Henry Fortey, Inspector of Schools (India). Fortey attended Clifton High School for Girls, then entered University College, Bristol, in 1892, receiving an external London B.Sc. (Hons.) in 1895. She was awarded a prestigious Science Research Scholarship of the Royal Commission for the Exhibition of 1851, which she used for research at Owens College, Manchester, over the period 1896–1898. Fortey's research, in conjunction with that of the Russian chemist Vladimir Markovnikov, showed that the cyclohexane fractions from American, Galician, and Caucasian crude oil deposits were identical.[114]

Upon her return to Bristol, Fortey commenced a 5-year collaboration with Sydney Young,[115] and between 1899 and 1903, she co-authored seven papers with Young on fractional distillation. Much of the research formed the basis of Young's book *Fractional Distillation*,[116] published in 1903. Despite Fortey's name and work being mentioned extensively in the book, her name is not one of those whose contributions were acknowledged in the Preface.

During that period, she must have also undertaken research with Tilden, as he noted:

> Finally, in 1902, a careful study of mixtures of the lower alcohols with water was carried out by Miss E. C. Fortey and myself [Tilden], and our

---

[113](a) (1961). *Record of the Science Research Scholars of the Royal Commission for the Exhibition of 1851, 1891–1960.* Published by Commissioners, London; (b) *Student Records,* Archives, University of Bristol; and (c) note 95b, Creese, p. 271.

[114]Fortey, E. C. (1898). Hexamethylene from American and Galician petroleum. *Journal of the Chemical Society, Transactions* **73**: 932–949.

[115]Atkins, W. R. G. (1938). Sydney Young, 1856–1937. *Obituary Notices of Fellows of the Royal Society* **2**: 370–379.

[116]Young, S. (1903). *Fractional Distillation.* Macmillan and Co., London.

206 *Pioneering British Women Chemists: Their Lives and Contributions*

results, taken in conjunction with those of Konowalow, afford strong evidence that no hydrate of any alcohol is formed, at any rate at temperatures above 0°C.[117]

In her 8-year research career from 1896 to 1904, she authored 14 papers and several shorter communications.

Fortey then left science completely. Her later-life was recounted in an Obituary in the *Leicester Evening Mail*:

> Later, while living in a workman's flat in St. Pancras she qualified in sanitary inspection, midwifery and also studied Sociology at the School of Economics. ... During World War One she was in charge of a small rescue home for girls in Le Havre under the auspices of the Y.M.C.A. She was elected unopposed for St Margaret's' ward [Leicester Council] in 1923 and served [on the Council] for 23 years. Her motion to the Council in 1929, that no married woman should be debarred from permanent employment with the City Council because she was married was lost.[118]

Fortey was a stern critic of the inadequacies of the [Leicester] Council's slum clearance policies during the 1930s. Although there was some justification for this, it was no doubt fuelled by her antipathy towards Dr. Charles Killick Millard, the Chief Medical Officer of Health who was an advocate of birth control and voluntary euthanasia. In the early 1930s she was reprimanded by the Labour Group for opposing the Council's promotion of birth control. She was Vice-President of the Leicester Temperance Society.[118] A portrait of Fortey hangs in the Leicester City Rooms and a special needs school in Leicester was named after her.[119] She died on 10 September 1947.

---

[117] Note 110, Tilden, pp. 80–81.

[118] Newitt, N. Emily Comber Fortey, in *The Who's Who of Radical Leicester* (accessed 29 July 2018).

[119] Arnott, J. and Price, J. (1981). *The Emily Fortey School, Glenfield Road, Leicester.* n. Pub.

Millicent Taylor

As we have seen, many of the provincial universities had a woman chemist who devoted her life to the chemistry department of that institution, and for Bristol, this was Millicent Taylor.[120] Taylor was born on 17 October 1871 at Kingswood, Surrey, to George Taylor, farmer, and Alice Drury. She attended Cheltenham Ladies' College (CLC) as a student between 1888 and 1893. It was from Cheltenham that she obtained an external London B.Sc. in 1893, the same year that she was appointed to the staff of CLC. The following year, Taylor was made Head of the CLC Chemistry Department, and then Head of the Science Department in 1911.

Between 1898 and 1910, Taylor devoted most of her spare time to research work in organic and physical chemistry at University College, Bristol producing a range of papers in those fields. On weekends, she would often cycle to and from the Bristol chemistry laboratories, an eighty-mile round trip.[121]

Taylor received an M.Sc. (Bristol) in 1910 and a D.Sc. (Bristol) in 1911. She then travelled to northern Ontario, Canada, a region where transportation was primitive and unaccompanied single-women visitors were unknown. Ever the academic, she wrote a monograph describing her exploits.[122] In 1914, Taylor unsuccessfully applied for the position of Reader in Chemistry at the Home Science Department of King's College for Women.

---

[120]Files in the Chemistry Library Archives, University of Bristol: (a) Millicent Taylor to the Senate of the University of London, (December 1914). Application of Millicent Taylor, Head Lecturer in Chemistry at the Ladies' College, Cheltenham, Candidate for the Readership in Chemistry in the Home Science Department of the King's College for Women, University of London; (b) "D.T." (January 1961). Millicent Taylor, D.Sc. *Incorporated Guild of the Cheltenham Ladies' College.* leaflet 182; (c) Wallace, E. S. (05 May 1961). An appreciation of Dr. Millicent Taylor. unpublished; and (d) Letter, Dykes, D. W. (Administrative Assistant, University of Bristol) to Mrs Reginald Temperley (06 January 1961). Somerset.

[121]Baker, W. (1962). Millicent Taylor, 1871–1960. *Proceedings of the Chemical Society* 94.

[122]Taylor, M. (1912). *The Mining Camps of Cobalt and Porcupine.* Thomas Hailing Ltd., Cheltenham.

208 *Pioneering British Women Chemists: Their Lives and Contributions*

During the First World War, Taylor was involved in production of β-eucaine and then in 1917, she was appointed a Research Chemist at H. M. Factory, Oldbury, returning to her post at Cheltenham at the end of the war in 1919. In 1921, she left CLC to accept an appointment as Demonstrator in Chemistry at the University of Bristol, being promoted to Lecturer in 1923. Taylor continued research in physical chemistry, being the sole author of most publications, while one was co-authored with Mary Laing (see Chapter 14). During the Spring and Summer terms of 1934–1935, Taylor was acting Warden of Clifton Hill House, the Women's Hall of Residence.

Retirement in 1937 was not the end for her, as she was given the use of a small laboratory in an army hut on the grounds of the Bristol Chemical buildings. Taylor continued research, her last publication being in 1951 at age 80. Sadly, an accident in November 1960 ended her active life, with her death occurring on 23 December 1960, aged 90.

Millicent King

The fourth Bristol woman chemist of note was Annie Millicent King.[123] King, born on 12 September 1900 in Dursley, Gloucestershire, was the daughter of Joseph William King, law clerk, and Elizabeth Maude Maria Preator. Educated at Redland High School, she entered the University of Bristol intending to complete an arts degree. This was towards the end of the First World War, and King changed to a chemistry degree to help meet the need for scientists.

King graduated with a B.Sc. in chemistry in 1922 and then completed a Ph.D. degree in 1927 under James William McBain.[124] Her work with McBain consisted of investigations into the detergent action of soaps and a study of the conductivity of glass surfaces in solutions of potassium chloride. McBain left the University of Bristol in 1926 (see Chapter 14).

---

[123] (a) Garner, W. E. (1954). Millicent King 1900–1952. *Journal of the Chemical Society* 2160; (b) Anon. (Summer 1953). Obituary: Dr. A. M. King. *Redland High School Magazine* 45.

[124] Rideal, E. K. (1952–1953). James William McBain, 1883–1953. *Obituary Notices of Fellows of the Royal Society* 8: 529–547.

King continued her research but now under William Edward Garner,[125] co-authoring six papers on the specific heats and heats of crystallisation of a number of homologous series, including hydrocarbons, fatty acids, methyl and ethyl esters, and amides. The work was of importance in relation to the cause of the alternation in melting points of the homologous series. She also became Secretary and Librarian to the Chemistry Department at Bristol with the title of Research Assistant.

King ceased experimental work at the beginning of the Second World War, when her duties as Librarian and Secretary took up all her time. In addition, she was an ambulance driver during the air raids on Bristol. King died on 17 December 1952.

Armstrong College (University of Newcastle upon Tyne)

There was another English provincial college, Armstrong College. Unlike the other provincial colleges which soon gained university status, Armstrong was not granted university status as the University of Newcastle upon Tyne until 1963. The delay was caused by Armstrong College being an appendage of the University of Durham.

The University of Durham had been founded in 1832, with a philosophy resembling that of an arts college on the Oxbridge model.[126] A School of Physical Science had been opened in Durham in 1865, but it closed in 1871.[127] With the founding of the Newcastle Chemical Society (later the Society of Chemical Industry, Newcastle section) in 1869, pressure mounted for a science college in that city; as a result, the Durham College of Physical Science was opened in Newcastle in 1871. Three years later, the school was affiliated with the University of Durham as Armstrong College.

---

[125]Bawn, C. E. H. (1961). William Edward Garner, 1889–1960. *Obituary Notices of Fellows of the Royal Society* **7**: 85–94.

[126]Hird, M. (ed.), (1982). *Doves & Dons: A History of St. Mary's College Durham.* University of Durham.

[127]Clemo, G. R. and Brown, N. S. (1956). Schools of Chemistry in Great Britain and Ireland–XXII The University of Durham. *Journal of the Royal Institute of Chemistry* **80**: 14–21.

Women Science Students

The College of Physical Science, Newcastle, admitted women to lectures and laboratories from its opening in 1871. Women were able to register for the qualification of Associate of Science (A.Sc.), but Durham barred them from admission to a bachelor's degree. As Marilyn Hird noted: "When one enterprising young lady completed the qualifying examinations and presented herself, without success, for admission to the degree of B.Sc., the whole question was raised."[128] Finally, in 1895, a Supplemental Charter was approved which gave the University of Durham the power to confer degrees on women.

The Suffrage movement was an active issue at Armstrong College. In 1906, a very heated debate among the women students resulted in a vote of 29 in favour of women's suffrage and 55 against (see Footnote 136). As the author of the report noted: "Some of the suffragettes did not look very happy when the result was announced, however, several converts to the cause were made."[129]

Friction between the male and female students arose in 1912, which prompted a retort in the student magazine, *The Northerner*:

> Certainly if we were to take the College as an example, there can be little doubt that man is a less industrious creature than woman. Who do the most work in college, men or women? Ask the lecturers, their opinion would be unanimous — the women.[130]

Charlotte Schofield (Mrs. Cole)

The first woman on the chemistry staff at Armstrong College was Charlotte Bean Schofield.[131] She was born on 14 September 1894 at Morpeth,

---

[128] Note 133, Hird, n. pag. We have been unable to ascertain the identity of the "enterprising young lady."

[129] "A Female Girl." (December 1906). Debating society. *The Northerner: The Magazine of Armstrong College* 7(2): 42–43.

[130] "Umph." (March 1908). Woman and man. *The Northerner: The Magazine of Armstrong College* 8(4): 117–118.

[131] (a) *Student Records*, Archives, University of Newcastle-upon-Tyne; and (b) Anon. (1974). Personal News: Deaths. *Chemistry in Britain* 10: 271.

Northumberland, to Frederick E. Schofield, chemist & druggist, and Alice Grey. Schofield was a chemistry student at Armstrong College, receiving a B.Sc. (Durham) in 1915, then an M.Sc. (Durham) in 1917. In 1919, Schofield was appointed Assistant Lecturer and Demonstrator in Chemistry at Armstrong College. After 1 year, she entered the College of Medicine, Newcastle-upon-Tyne, receiving an M.B. and B.S. (Durham) in 1925. The following year, Schofield married Alfred H. Fletcher Cole, a South African medical doctor. They moved to South Africa, with Schofield retaining her medical practice. She died on 1 August 1972 at Cape Town, South Africa.

Grace Leitch

Replacing Schofield at Armstrong College was Grace Cumming Leitch.[132] Leitch, who provided the backbone of much of the teaching in the first part of the 20th century, was born on 14 July 1889 at Cupar, Fife, to David Leitch and Agnes K. Leitch. She was educated at Bell Baxter School in Cupar and then obtained a B.Sc. from St. Andrews in chemistry in 1913, followed by a Ph.D. in 1919. Leitch's research work on the structure of sugars was undertaken with Norman Haworth.[133] During the First World War, she was appointed Junior Lecturer, and over the same period, she also undertook work on mustard gas. However, the war had a more direct impact on Leitch in that her fiancé was killed in France just before the Armistice.

When Haworth left St. Andrews and took up a position at Armstrong College in 1920, Leitch accompanied him, herself being appointed Lecturer in Organic Chemistry. She continued her research on sugars, co-authoring three papers with Haworth and being the sole author of another. When Haworth subsequently moved to Birmingham, Leitch joined the research group of George Clemo[134] co-authoring three papers

---

[132]"C.E.M." [Mallen, Catherine]. (1942). Dr. Grace Cumming Leitch. *The Northerner: The Magazine of Armstrong College* **42**(3): 24.

[133]Hirst, E. L. (1951). Walter Norman Haworth, 1883–1950. *Obituary Notices of Fellows of the Royal Society* **7**: 372–404.

[134]Lythgoe, B. and Swan, G. A. (1985). George Roger Clemo, 2 August 1889–2 March 1983. *Biographical Memoirs of Fellows of the Royal Society* **31**: 60–86.

212 *Pioneering British Women Chemists: Their Lives and Contributions*

on alkaloids. However, as she was responsible for a substantial part of the teaching load, there was little time for research.

In 1926, Leitch became the sub-Warden of the women's residence, Easton Hall. She became seriously ill in the mid-1930s, and though she never fully recovered, she continued to hold both Lectureship and sub-Warden appointments until her health deteriorated again in July 1941. She died on 12 March 1942 in her home town of Cupar. As was noted by her friend and mentor, Haworth:

> Among scientific workers she had a wide acquaintance, fostered by her regular attendance at the annual British Association meetings. Few could have failed to appreciate her gaiety of spirit, warm enthusiasm and loyalty, her capacity for friendship, and her vigorous personality; for it was these qualities which commended her to her students and colleagues, by whom she will be greatly missed.[135]

In 1944, a Memorial Fund in Leitch's name raised enough funds for a travelling bursary for a science student in alternate years.[136]

Catherine Mallen (Mrs. Elmes)

When Grace Leitch arrived at Newcastle, Catherine Eleanor Mallen[137] had already taken up her appointment as Assistant Lecturer and Demonstrator. Mallen and Leitch became close friends, and it was Mallen who wrote one of the obituaries of Leitch.[136] Mallen was born on 22 April 1896 at Sunderland to Robert Coxon Mallen, hosier dealer, and Margaret Jane Robson. She was educated at the Bede School for Girls, Sunderland before entering Newnham College in 1915. Mallen completed the chemistry degree requirements in 1919 and obtained the appointment at Armstrong College in the same year. Mallen was promoted to Lecturer in 1939, a rank which she held until her retirement in 1960. In addition to

---

[135] Haworth, W. N. (1941). Grace Cumming Leitch. *Journal of the Chemical Society* 341.

[136] Anon. [Mallen, C.] (March 1944) Dr. G. C. Leitch. *Durham University Journal* **2**: 70

[137] (a) White, A. B. (ed.), (1979). *Newnham College Register, 1871–1971*, Vol. I, 1871–1923, 2nd edn., Newnham College, Cambridge, p. 265; and (b) *Calendars*. University of Newcastle-upon-Tyne.

teaching, Mallen authored two research publications — one alone, and one with Australian chemist, Thomas Iredale, who was a Lecturer in Chemistry at Newcastle from 1925 until 1927.

Mallen married Ralph S. Elms, also a Cambridge graduate and a Lecturer in English at Newcastle, in August 1943. After retirement, they moved to Wensleydale where Mallen became active with the Wensleydale Society until her death on 27 April 1984, aged 88. As her obituarist, Stella Buckley, explained:

> This [the Wensleydale Society] had evolved from a series of lectures on local history and had become a most successful institution providing lectures by well-known specialists on the archaeology, geology, botany and history of the district. Catherine was no mean botanist and greatly enjoyed our walks and expeditions until well over eighty.[138]

Southampton University College

The University College in Southampton actually began as the Hartley Institution.[139] Henry Robinson Hartley left instructions in his Will for the founding of an establishment for the study and advancement of the sciences in his property on Southampton's High Street. The Institution opened its doors in 1862. It changed its name to Hartley University College in 1902, then to Southampton University College in 1914 (not receiving full university status until 1952).

Ishbel Campbell

Southampton University College, too, had an active chemistry department,[140] a central role being played by Ishbel Grace MacNaughton

---

[138]Buckley, S. (1985). Catherine Eleanor Elmes (née Mallen), 1896–1994. *Newnham College Roll Letter* 59.

[139]Paterson, A. T. (1962). *The University of Southampton. A Centenary History of the Evolution and Development of the University of Southampton, 1862–1962.* University of Southampton.

[140]Adam, N. K. and Webb, K. R. (1956). Schools of chemistry in Great Britain and Ireland–XXIII The University of Southampton. *Journal of the Royal Institute of Chemistry* **80**: 133–140.

214 *Pioneering British Women Chemists: Their Lives and Contributions*

Campbell.[141] Campbell was born on 13 October 1905 in Kirkcaldy and graduated from St. Andrews with a B.Sc. in 1927 and a Ph.D. in 1931. She then spent a year at Cornell University, New York, where she held one of the first Commonwealth Fellowships awarded to a woman.

On her return, Campbell accepted a Lectureship in Chemistry at Swanley Horticultural College. In 1936, she joined Bedford College as a Demonstrator and Teacher before making her final move in 1938 to University College, Southampton, first as Lecturer, and later as Reader of the University of Southampton.

One of her former students, Martin Hocking, recalled:

> 'Ish' was experimentally well known for her ability to coax more-or-less pure crystals of a new substance from tiny amounts of solution of an unlikely looking, gluey reaction product. It was rumoured that her success was the beneficiary of traces of her cigarette ash that provided nuclei in the crystallization test tube to help initiate the crystallization process aided by temperature changes and by scratching the side of the tube with a glass rod.[142]

Hocking noted also that Campbell provided students with a sense of community:

> "Ish" as she was colloquially known, was a slight, but physically tough, proud Scottish spinster who regularly played tennis until well into her 80s. She also loved long walks in the hills around Southampton, once a year or so inviting a small group of students to join her. These walks were followed by a most enjoyable picnic at some high point in the walk, where the view could be enjoyed along with the conversation.[142]

During the Second World War, despite heavy air attacks on the town, teaching continued, with evening classes for part-time students being

---

[141] (a) Cookson, R. C. (April 1998). Ishbel Campbell, 1906–1997. *Chemistry in Britain* **34**: 72; and (b) Harris, M. M. (2001). Chemistry. In Crook, J. M. (ed.), *Bedford College: Memories of 150 Years.* Royal Holloway and Bedford College, p. 87.

[142] Personal communication. e-mail, M. B. Hocking, 29 August 2007.

transferred to weekends. While at Southampton, Campbell undertook a tremendous quantity of research, primarily on organometallic compounds of Group V and VI. A total of 19 publications resulted from her work between 1945 and 1966: 15 with her as senior author and 4 with her as sole author. According to Hocking, Campbell never really retired:

> Long after her official retirement Ish enthusiastically gave us a tour of the new medical faculty at Southampton and where she had volunteered to teach courses in chemistry to new medical students. "It keeps me young" she said, and it certainly worked! (see Footnote 142)

Campbell died on 10 October 1997 at age 91.

Commentary

We have shown that at each of the provincial universities, there were significant numbers of women students in chemistry. At most of the universities in the early years, there were one or more chemistry lecturers willing to be supportive of the women, as examples, Haworth at Newcastle and Birmingham; Lapworth at Manchester; Smithells at Leeds; and Wynne at Sheffield.

At the same time, many of the male students perceived chemistry as an exclusively male domain with its own social society, being hostile to the new arrivals (the riot at Bristol being the most extreme manifestation). Indeed, the male-only chemical-musical-and-social events resembled a men's club or fraternity rather than a chemical society at each of the northern universities. This point was actually stated at the Manchester University Chemical Society: "It was decided that women students be not admitted as members of the Society as it would spoil the social nature of the meetings."[143] Of note, at Manchester and Liverpool, it was the 1910 period when both Chemical Societies dramatically reversed themselves and admitted women chemists.

---

[143] Anon. (1907). Chemical Society notes. *Manchester University Magazine* 23.

Some men obviously did not adapt to the new environment, as can be read from the quotes, earlier in this chapter, in the various student magazines. There were also the male-only gatherings which each university chemistry department continued to possess: the Organic Lab "club" at Leeds (which had been the male sanctuary ever since the first arrival of women chemistry students) and the Research Men's Club at Liverpool, while at Manchester, the male chemists retreated to the safety of the Men's Student Union, which survived as a male-only environment until 1957.[144]

---

[144]Pullan, B. and Abendstern, M. (2001). *A History of the University of Manchester, 1951–1973*. Manchester University Press, Manchester, p. 71.

Chapter 7

Universities in Scotland and Wales

Christina Cruikshank Miller, FRSE (29 August 1899–16 July 2001)

Up to this chapter, we have reviewed the lives and work of women chemists in England. We showed how certain girls' schools played a major role in providing the teaching environment that encouraged girls to choose chemistry degree programmes at university. In this chapter, we shall see that the cause of higher education of women in Scotland and Wales followed different paths.

Within Scotland, there was a diversity of approaches from the autonomous women's Queen Margaret College of the University of Glasgow to the tougher fight for acceptance at the University of Edinburgh. In Wales, it was predominantly English women students who entered the University Colleges of Wales.

Entry of Women to Scottish Universities

In Scotland, by the late 1850s and early 1860s, there was considerable discussion in the pages of newspapers about the social advantages and the personal justice of admitting middle-class women into paid employment. There was particularly wide support for the training of women in medicine as it was contended that it was more suitable for women and adolescent girls to be intimately examined by women.[1] In addition, it was believed that middle-class women medical practitioners could bring hygienic and moral guidance to working-class women. The issue of the admission of women students into Scottish medical schools first arose in 1861. It was Elizabeth Garrett[2] who applied to enter the medical school at St. Andrews University. The university initially approved her admission, and then withdrew its consent when some faculty voiced their antagonism to the presence of women on campus.

The Edinburgh Seven

It was in March 1869 that Sophia Jex-Blake[3] applied for admission to the University of Edinburgh to study medicine. Her request was rejected upon the grounds that the university could not make special arrangements for one lady. She advertised to ask if any others would join her for a group submission. Seven names were approved by the University Court: Sophia Jex-Blake, Isabel Thorne, Edith Pechey, Matilda Chaplin, Helen Evans, Mary Anderson, and Emily Bovell. In November 1869, the "Edinburgh Seven,"[4] as they came to be called, were the first women admitted to a British university.

Many of the male students expressed extreme hostility to the women, and this antagonism reached its climax with the Riot of 18 November

---

[1] Moore, L. (1992). The Scottish universities and women students, 1862–1892. In Carter, J. J. and Withrington, D. J. (eds.), *Scottish Universities: Distinctiveness and Diversity.* John Donald Publishers, Edinburgh, pp. 138–146.

[2] Manton, J. (1965). *Elizabeth Garrett Anderson.* Methuen, London, 1965.

[3] Roberts, S. (1993). *Sophia Jex-Blake: A Woman Pioneer in Nineteenth Century Medical Reform.* (The Wellcome Institute Series in the History of Medicine), Routledge, London.

[4] Wikipedia (20 January 2019). Edinburgh Seven (accessed 26 February 2019).

Universities in Scotland and Wales 219

1870. This event occurred when the women took their anatomy examination and, like Newnham (see Chapter 3) and Bristol (see Chapter 6) riots, showed the violent misogynist attitudes of many of the male students. The rioters had been called together by a rabble-rousing anti-women pamphlet from the "Chemistry Class of the University" with the encouragement of one of their professors.[5]

As the women students approached the Surgeons' Hall, where the examinations were to be written, they were mobbed by drunken students. However, it was the aftermath of the exam which was truly frightening for the women. Isabel Thorne recounted the event:

> By the end of the examination it was dark and a crowd had again gathered around the gates. We were asked if we would leave by a private door; but we felt it would not do to be intimidated, and relying on the support of our class mates, who formed a sort of bodyguard around us, arming themselves, in default of other weapons, with osteological specimens, we passed quietly through the mob, only our clothes being bespattered with the mud and rotten eggs thrown at us.[6]

The most vociferous supporter of the women was Robert Wilson of the Royal Medical Society of Edinburgh. He sent a letter to Pechey following the riot:

> I wish to warn you that you are to be mobbed again on Monday. A regular conspiracy has been, I fear, set on foot for that purpose.... I have made what I hope to be efficient arrangements for your protection.... I had a meeting with Micky O'Halloran who is leader of a formidable band, known in college as 'The Irish Brigade' and he has consented to tell off a detachment of his set for duty on Monday.... May I venture to hint my belief that the real cause of the riots is the way some of the professors run you down in their lectures. However, as I tell you, you

---

[5]Ross, M. (1996). The Royal Medical Society and medical women. *Proceedings of the Royal College of Physicians, Edinburgh* **26**: 629–644.

[6]Thorne, I. (January–March 1951). The time in Edinburgh: 1869–1874. *Royal Free Hospital Magazine* **13**: 102.

and your friends need not fear, as far as Monday is concerned. You will be taken good care of.[7]

In fact, the "Irish Brigade" continued their escort duties of the women between accommodations and lectures for some time afterwards.

Though the women passed all the examinations, the University of Edinburgh refused to grant them degrees. The group then took legal action against the University and initially won their case, the judges being scathing in their condemnation of the University.[8] Unfortunately, on appeal, it was ruled that women should never have been admitted to the University and therefore could not graduate. As a result, the British Medical Association refused to register the women as doctors. Jex-Blake moved to London and was active in founding the London School of Medicine for Women (see Chapter 5).

A Private Member's Bill was then introduced at Westminster that would allow the universities in Scotland to admit women as students and permit them to grant degrees to women. Much of the opposition to the Bill came from the Universities of Edinburgh and Glasgow, though many faculty at both institutions supported the women's cause. The Bill was defeated in March 1875.

The Battle Finally Won

Campaigns followed by ladies' educational associations in each of the Scottish cities for a higher qualification for women. It was St. Andrews who led the way by introducing the Lady Licentiate in Arts (LLA).[9] This qualification proved hugely popular, attracting women students from across Scotland, much to the annoyance of the other Scottish universities.

---

[7] Todd, M. (1918). *The Life of Sophia Jex-Blake*. Macmillan and Co., London, pp. 293–294.

[8] Case 138. (28 June 1873). Sophia Louisa Jex-Blake and Others, Pursuers; The Senatus Academicus of the University of Edinburgh, Defenders. *Cases decided in the Court of Session, Teind Court, &c. Third Series*, Vol. 11, T. & T. Clark, Law Booksellers, London, pp. 784–802.

[9] Smart, R. N. (June 1968). Literate ladies — A fifty-year experiment. *St. Andrews Alumnus Chronicle* **59**: 21–31.

In 1889, the Universities (Scotland) Bill was passed in Westminster which finally permitted the universities to admit and graduate women. The first eight women graduated in 1890.[10] The struggle was finally over, and by the beginning of the 20th century, a substantial proportion of students at Scottish universities were women.

Life of Scottish Women Students

On the basis of their father's occupation, women students came predominantly from the middle class like their male colleagues.[11] Some of the upper-class women used university as more of a finishing school, and this was particularly true of some of those attending Queen Margaret College, Glasgow.[12]

Through oral interviews, Sheila Hamilton has shown how life as a middle-class Scottish woman student (Mrs. M) in the 1920s was totally different to that of a working-class woman student (Mrs. G).[13] The middle-class student, led an active social life: "Dancing was very much to the fore in the 1920s and this was reflected within the social life of Mrs. M. She attended sixty-three dances in one winter, although these were not all university functions."[14]

Whereas Mrs. M's father was a graduate of Edinburgh, and M's three siblings (all girls) went to university, the working-class interviewee came from a very different background. Mrs. G., one of six children, was the first and only member of her family to go to university.

For Mrs. M., staff took care of domestic work, such as cooking and cleaning. However, in addition to her studies, Mrs. G. had to shoulder part of the family duties of women's work: "One weekend in particular, she

---

[10] Watson, W. N. B. (1967–1968). The first eight ladies. *University of Edinburgh Journal* **23**: 227–234.

[11] McDonald, I. J. (1967). Untapped reservoirs of talent? Social class and opportunities in Scottish higher education 1910–1960. *Scottish Educational Studies* **1**: 53.

[12] Crichton, A. C. (1967). Finishing school for young ladies. *The College Courant* (*Journal of the Glasgow University Graduates Association*) **19**(38): 19–21.

[13] Hamilton, S. (1982). Interviewing the middle class: Women graduates of the Scottish Universities c.1910–1935. *Oral History* **10**(2): 58–67.

[14] Note 13, Hamilton, p. 62.

recalled that she was nearly in tears because she had all the men's shirts to iron for four brothers and father, and she should have been studying for an examination."[15]

University of St. Andrews

The University of St. Andrew's was founded in the Kingdom of Fife in 1411, making it the oldest university in Scotland and the third oldest in the United Kingdom.[16] It was the election of Thomas Purdie[17] as Chair of Chemistry in 1884 which marked the real founding of the modern Chemistry Department. Purdie saw the future as being organic chemistry and, in particular, stereochemistry. This proved to be a brilliant move, producing research on carbohydrates that became renowned throughout Europe.[18] With Purdie's resignation in 1909, his successor was his former student, James Colquhoun Irvine,[19] who continued with Purdie's research direction on sugar structures.

Life of Women Students

St. Andrews was the only Scottish University to strongly champion the admission of women. An article in the pages of *Girl's Realm* commented:

> Not only did St. Andrews University early open its doors to women, it has treated them royally ever since; and today [1911] it holds the unique position of being the one college in which men and women students

---

[15]Note 13, Hamilton, p. 65.

[16]Reid, N. (2011). *Ever to Excel: An Illustrated History of the University of St Andrews*. Edinburgh University Press.

[17]"P.F.F." (1922). Obituary Notices of Fellows Deceased: Thomas Purdie, 1843–1916. *Proceedings of the Royal Society of London, Series A* **101**: iv–x.

[18]Read, J. (1953). Schools of Chemistry in Great Britain and Ireland–I: The United College of St. Salvator and St. Leonard, in the University of St. Andrews. *Journal of the Royal Institute of Chemistry* **77**: 8–18.

[19]Read, J. (1953). Sir James Colquhoun Irvine, 1877–1952. *Obituary Notices of Fellows of the Royal Society* **8**: 459–489.

stand upon an absolute equality.... There are those who have shaken their heads over a Senate which thus decrees that Jill shall enter into the full college life enjoyed by Jack; but time has proved it wise in its generation, and the University of St. Andrews is justified of all her children. Sharing alike the work and play of the men gives the women a broader, fuller outlook upon life; while the gentler, more refining influence of the girl students is by no means lost upon even the wilder spirits among the male undergraduates.[20]

Nevertheless, it did not mean women students were necessarily welcomed by the male contingent, according to a complaint in the *St. Andrews University Magazine* of 1890:

> It seems to me that I scarcely know a young lady now-a-days who has not, to some undefined extent, a craving in the direction of higher education and deep thought, which renders the pleasure of her companionship somewhat dubious. One can pardon a lady who desires to read Plantus in the original so long as one has not the privilege of her acquaintance... It is one of the most trying ordeals I can think of, to be in the company of a <u>femme savant</u> for more than a few minutes at a time.[21]

Women Chemistry Students

Among Irvine's research students were several women chemists: Agnes M. Moodie, M. E. Dobson, B. M. Paterson, Ettie S. Steele, Helen S. Gilchrist, and J. K. Rutherford. It is not surprising that Irvine had so many women researchers, as by the 1910–1911 academic year, the first-year cohort in chemistry consisted of 36% women (27 M, 15 F).[22] Irvine was appointed Principal of the University in 1921; nevertheless, researchers worked with him until 1935.

---

[20] Sloan, S. (1910–1911). University Hall, St. Andrews. *Girl's Realm* **13**: 893.

[21] "Selah" (27 February 1890). My erudite lady-friends. *College Echoes: St Andrews University Magazine* **15**: 125. Courtesy of the University of St. Andrews Library.

[22] Cant, R. G. (1970) . *The University of St Andrews: A Short History*. Chatto & Windus.

Agnes Moodie

One of the early women chemistry graduates was Agnes Marion Moodie,[23] born on 6 October 1881 at Arbroath, Scotland, to Robert Moodie, mathematics and physical science teacher, and Mary Lithgow Mackintosh. She completed her M.A. at St. Andrews in 1902 and a B.Sc. in 1903. Moodie stayed on at St. Andrews to undertake research with Irvine, co-authoring five publications with him on alkylated sugars. She was awarded a Carnegie Scholarship in 1907, but we know nothing of the intervening years until she retired from the Ministry of Education in 1946, moving to Hove, Sussex where she died in 1969, aged 87.

Jannette (Ettie) Steele

In the history of the Chemistry Department of St. Andrews, there was one woman who made an indelible imprint — Ettie Stewart Steele.[24] Steele was born on 5 October 1890, at Dunfirmline, Scotland to William and Jessie Steele. She was educated at Dunfirmline High School and then entered St. Andrew's University, obtaining an M.A. followed by a B.Sc. in 1914. After graduation, she commenced research with Irvine on the properties of mannitol derivatives.[25] Appointed as University Assistant in 1919, Steele received a Ph.D. in 1920 for her research work. She was promoted to Lecturer and Assistant in 1921, and to Lecturer in 1924. Steele contributed to five research papers on carbohydrates including one under her name alone.

When Irvine became Principal of the University in 1921 and Robert Robinson arrived to assume the Professorship in Chemistry, Robinson noted:

> I worked in the private laboratory also occupied partly by Dr. Catherine [sic] Steele, a colleague of Irvine's who kept alive some of his work on sugars, but this petered out on account of the onerous duties [of Irvine].[26]

---

[23] Anon. (June 1953). (a) *St. Andrews Alumnus Chronicle* 40; and (b) *Student Records*, Archives, University of St. Andrews.

[24] K.M.M. (June 1984). Obituary. *St. Andrews Alumnus Chronicle* **75**: 24.

[25] Anon. (January 1953). *St. Andrews Alumnus Chronicle* **39**: 12; and Anon. (June 1959). *St. Andrews Alumnus Chronicle* **50**: 17.

[26] Robinson, R. (1976). *Memoirs of a Minor Prophet: 70 Years of Organic Chemistry*. Vol. 1, Elsevier, Amsterdam, p. 127.

In 1930, Steele was appointed warden of the women's students' residence, then called Chatten House (later McIntosh Hall). Her duties filled much of her time, as her obituarist noted:

> Each alumnus who knew Ettie Steele will have a particular and personal recollection — of her penetrating blue eyes, mirrored but not eclipsed by the blue of her Ph.D. gown, which she wore with such pride and distinction; of her firmness in dealing with young male intruders in McIntosh Hall; of her infinite kindness and concern for students, a caring interest maintained to the end of her life as her Christmas correspondence bore witness.[27]

Steele also filled a third role, that of secretary to Irvine during all his years as principal. The University of St. Andrews was her life:

> No day passed, even in retirement, when the institution was not central to her thoughts, and from her flat in St. Andrews she jealously watched over the affairs of the University and was firm in her comments upon current issues — comments which were trenchant and apposite.... To Ettie Steele life was full of fortunate and happy events. In her own eyes her association with the University was one of supreme privilege and she considered herself unworthy of the responsibilities she carried with such competence and humour.[27]

Steele continued as Lecturer and Researcher until her retirement in 1956 though she continued as Warden for an additional 3 years. She died in July 1983.

Helen Gilchrist (Mrs. Childs)

A later graduate was Helen Simpson Gilchrist, born in 1896 at Largo, Fifeshire, to James Gilchrist, stone and brick Builder, and Elizabeth Gillis.[28] Gilchrist completed a B.Sc. degree in Botany and Chemistry in 1917 and then became a Berry Scholar in Chemistry to pursue chemical research. The following year, Gilchrist was awarded a Carnegie

---

[27] Note 24, Anon., p. 24.

[28] *Student Records*, Archives, University of St. Andrews.

226 *Pioneering British Women Chemists: Their Lives and Contributions*

Scholarship but soon after moved to the Ministry of Munitions, where she undertook research connected with the war, which included the preparation of novocaine and "mustard gas." In 1919, she resumed her Carnegie Scholarship. Later the same year, Gilchrist was appointed as a Research Chemist under the Food and Investigation Board of the Department of Scientific and Industrial Research (DSIR), a government research agency which had been instituted in 1917. It was for her research with DSIR that Gilchrist was awarded a Ph.D.

In 1927, Gilchrist applied for a position at the Royal Institution. W. H. Bragg, Professor at the Royal Institution, wrote to Irvine:

> I think I could promise that she would not be merely a hewer of wood and drawer of water. The work she would be doing would give her a chance of carrying out investigations of her own though, of course, she would be here for a definite purpose. I always like the people here to publish under their own names as much as possible.[29]

Gilchrist accepted the position, working on the separation of long-chain hydrocarbons with the Austrian scientist Berta Karlik[30] and then on the synthesis of fats containing a carbohydrate unit. She married William H. J. Childs in 1934 at Chelsea. Gilchrist died on 31 March 1984 at her hometown of Largo, a year after her spouse.

University of Glasgow and the Royal Technical College, Glasgow

The University of Glasgow (UG), the second oldest university in Scotland, was founded in 1451.[31] In 1796, UG was joined in Glasgow by a new institution: the Royal Technical College (RTC), now the University of Strathclyde. The RTC was the oldest technical college in the world and the

---

[29]Letter, W. H. Bragg to J. C. Irvine, 17 January 1927. Bragg Collection, Royal Institution Archives.

[30]For information on Karlik, see: Rentetzi, M. (2004). Gender, politics, and radioactivity research in interwar Vienna: The case of the Institute for Radium Research. *Isis* **95**: 359–393.

[31]Brown, A. L. and Moss, M. S. (1996). *The University of Glasgow, 1451–2001*. Edinburgh University Press.

first to open its doors to women.[32] Students at the RTC were granted degrees from the University of Glasgow.

Anderson's Institution

Chemistry for women in Glasgow actually goes back to the beginning of the 1800s.[33] In the Will of John Anderson, a political radical and professor at the University of Glasgow, he left money for the founding of an institution, focussing on the sciences, which would be open to ladies. Thus was founded Anderson's Institution[34] in 1796 open to both women and men, women making up a significant proportion of the enrolment.

Andrew Ure, Professor of Chemistry and Natural Philosophy from 1804 to 1830, reported that his course of lectures of popular illustrations of Physical Philosophy [Chemistry]: "... is attended by ladies and gentlemen, the average number of auditors being about 400."[35] The Institution went through a series of name changes — Glasgow Mechanics' Institution (1823), Anderson College (1877), College of Science and Arts (1881) — before becoming the Royal Technical College, Glasgow, in 1912.

In 1842, a "Ladies' Academy" also opened in Glasgow. Named the Queen's College for the Education of Ladies,[36] it had Frederick Penny as President and Professor of Chemistry. Strangely, the College disappeared without trace in the late 1840s or early 1850s.

Queen Margaret College

In 1877, the Glasgow Association for the Higher Education of Women[37] was founded and they organized courses for women, some of the lectures being

---

[32] Cranston, J. A. (1954). Schools of chemistry in Great Britain and Ireland–X The Royal Technical College, Glasgow. *Journal of the Royal Institute of Chemistry* **78**: 116–124.

[33] Smith, S. J. (2000). Retaking the register: Women's higher education in Glasgow and beyond, c. 1796–1845. *Gender and History* **12**: 310–335.

[34] Butt, J. (2001). *John Anderson's Legacy: University of Strathclyde and its Antecedents*. Tuckwell Press, Strathclyde.

[35] Note 32, Cranston, p. 117.

[36] Note 33, Smith, pp. 322–325.

[37] Myers, C. D. (2001). The Glasgow Association for the Higher Education of Women, 1878–1883. *The Historian* **63**: 357–371.

228 *Pioneering British Women Chemists: Their Lives and Contributions*

given at the university while others were given in rented rooms. Lectures in chemistry were only given in the 1879–1880 academic year.[38] With the success of this endeavour, the association was incorporated in 1883 as Queen Margaret College (QMC-UG) for the education exclusively of women. The name was chosen to commemorate the 11th century Queen Margaret who was reputed to have brought education and culture to Scotland.

The College steadily expanded, adding science laboratories in 1888 and a medical school in 1890. In 1892, the College was merged into the University of Glasgow. A student, who had come from King's College, Lady's Department, London, reported in 1902: "As a general rule the women attend the same classes as the men, occupying the front benches."[39]

By the 1920s and 1930s one would have expected that the presence of women at Glasgow was an established "fact," but this was not the case. In 1924, Miss Sheavyn, Director of Women Students at the Victoria University of Manchester, sent a letter to Frances Helen Melville, Mistress of QMC-UG, enquiring whether the College would like to participate in a survey on women students at university and particularly any health issues. Melville's reply was very negative:

> I fear that the women in this University are not yet sufficiently regarded as a matter of course to make it safe to begin an investigation upon them. In any case, I should have doubts about such an investigation being closely associated with medical men, whom I have, so far, seldom found unbiased in their judgements as to the causes of any ill-health in women students. They invariably put it down to the "effects of study", whereas a little enquiry would show a plurality of causes at work, including the effects of much dancing with late hours, in conjunction with study.[40]

The success of QMC-UG was its own downfall, with the number of women students increasing from 857 in 1917–1918 to 1708 in 1929–1930. To cope with the larger and larger class sizes, more and more of the courses had to be offered in the large lecture halls on the UG campus.

---

[38] Note 37, Myers, p. 370.

[39] Anon. (1902). Women students at the Scottish university. *King's College Magazine, Ladies' Department* **16**: 11–16.

[40] Letter, Miss Melville to Miss Sheavyn, 15 March 1924. University of Glasgow Archives, Special Collections GB248 DC 233/2/6/2/17.

In fact, by the 1920s and 1930s women were taking more courses on the UG campus than at QMC-UG. Considerable amounts of time were being spent by the women students trudging the 15 minutes' walk back-and-forth between the campuses. To solve the problem, the university authorities decided in 1934 to absorb the women students (and Queen Margaret College) into the university. The only Scottish experiment of a separate women's college had come to an end.[41]

Women Chemistry Students

The University of Glasgow had a student chemistry society, the Alchemists' Club, founded in 1916 by Catherine F. Davidson (later Mrs. Jones).[42] The Club was formalized in 1918, at which time one of the two Vice-Presidents was a woman (Davidson) as were four of the six Ordinary Members of Council.[43]

The Club had regular social events, particularly dances. For example, in 1925, it was reported that:

> A new reaction was tried by several women students in the vicinity of Lab. III on 19th. November. It is called the French Tango Reaction. Several complicated movements of the Ions took place, usually they moved in pairs. The Results of the experiment were demonstrated at the Dance that evening.[44]

The dances were not always well received by parents. A Mrs. V. A. Grant complained to *The Alchemist* that the goal of "social intercourse among its members" was not being met:

> The experience of my dear daughter, Audrey, simply testifies to that. ...
> Are the Dance Committee promoting social intercourse when they allow

---

[41] Melville, F. (1949). Queen Margaret College. *The College Courant (Journal of the Glasgow University Graduates Association)* **1**(2): 99–107.

[42] University of Glasgow Archives, DC306/4/1.

[43] *Minute Book 1918–1925* of the Glasgow University Alchemists' Club. University of Glasgow Special Collections, DC306/1/1.

[44] Anon. (December 1925). Laboratory gossip. *The Alchemist* (Glasgow University) **1**(2): 12.

230 *Pioneering British Women Chemists: Their Lives and Contributions*

my daughter to dance with the same partner for the whole evening? Whilst discussing this dance, I may say that a dance is rather prolonged if carried on till 4 a.m. My daughter did not arrive home until 4.15 a.m.[45]

By 1929, the welcome for women seemed to be fading when *The Alchemist* carried an "imaginary" conversation between two Glasgow University male chemistry students:

> Coeducation would not be so bad if the women stayed at Q.M. and only emerged perhaps for tennis, golf or dances ... What woman, think you, is intelligent enough to grasp the principles of chemistry — you've only got to look at the exam lists. No, woman's brain is too feeble, her temperament too volatile, her nature too erratic, for her to be of any use to Chemistry or Industry, even if, by some lucky fluke, she does get her degree. I repeat, women ought to be abolished from the labs., yea, even from the University.[46]

No rebuttal could be found in later issues from women chemists, and it is curious that the author claims women students were at the bottom of the exam lists when the usual complaint was that women chemists dominated the upper ranks.

Ruth Pirret

Ruth Pirret[47] was the first woman graduate in science from the University of Glasgow, obtaining a B.Sc. Honours degree in Pure Science (mainly chemistry) in 1898. She was born on 24 July 1874 in Glasgow to David

---

[45] Grant, (Mrs.) V. A. (November 1927). Letters to the Editor: Protest from an Alchemist's mother. *The Alchemist* (Glasgow University) **3**(1): 16.

[46] "Nitrogen Iodide." (February 1929). Chemistry women. *The Alchemist* (Glasgow University) **4**(4): 63–64.

[47] (a) Ewan, E., Innes, S., and Reynolds, S. (2006). *The Biographical Dictionary of Scottish Women from the Earliest Times to 2004*. Edinburgh University Press, p. 292; (b) Rayner-Canham, M. and Rayner-Canham, G. (1997).... And some other women of the British group. In Rayner-Canham, M. and Rayner-Canham, G. (eds.), *A Devotion to Their Science: Pioneer Women of Radioactivity*. Chemical Heritage Foundation, Philadelphia, and McGill-Queen's University Press, Montreal, pp. 159–160.

Pirret, Minister, and Violet Brown. Pirret was an outstanding student throughout her undergraduate work, and after graduation, she became a school teacher at Kilmacolm High School in Newcastle-upon-Tyne and at a school in Arbroath. Starting in October 1909, Pirret worked for at least 6 months with Frederick Soddy.[48] Lord Alexander Fleck, another researcher with Soddy at the time, commented: "T. D. MacKenzie was the research student who was the most important worker in those early years. Miss Ruth Pirret was also a contributor, but on a less prominent scale."[49]

Pirret's studies were on the ratio of uranium to radium in minerals, an extension of the work of Ellen Gleditsch,[50] the famous Swedish researcher in radiochemistry. It had been argued at the time that the older the mineral, the higher the proportion of radium, and Pirret and Soddy confirmed some of Gleditsch's findings. However, during their studies, they made a much more important discovery: that uranium existed in nature as a mixture of two isotopes. As they stated in the paper:

> We are therefore faced with the possibility that uranium may be a mixture of two elements of atomic weights 238.5 and 234.5, which, like ionium, thorium, and radio-thorium [thorium isotopes 230, 232, and 228], are chemically so alike that they cannot be separated.[51]

In July 1913, Pirret moved to Manchester, after accepting an appointment as Vice-Warden of Ashburne Hall,[52] the women's residence of Manchester

---

[48]Fleck, A. (1957). Frederick Soddy. *Biographical Memoirs of Fellows of the Royal Society* **3**: 203–216.

[49]Fleck, Lord. (1963). Early work in the radioactive elements. *Proceedings of the Chemical Society* 330.

[50](a) Kubanek, A.-M. W. (2010). *Nothing Less Than an Adventure: Ellen Gleditsch and Her Life in Science*. CreateSpace Independent Publishing Platform, Montreal, Canada; and (b) Lykknes, A., Kragh, H. and Kvittingen, L. (2002). Ellen Gleditsch: Pioneer woman in radioactivity. *Physics in Perspective* **6**: 126–155.

[51]Soddy, F. and Pirret, R. (1910). The ratio between uranium and radium in minerals *Philosophical Magazine* **20**: 345–349; a subsequent paper was Pirret, R. and Soddy, F. (1911). The ratio between uranium and radium in minerals II. *Philosophical Magazine* **21**: 652–658. It is noteworthy that in this second paper, Soddy gave Pirret's name priority.

[52]Tout, M. (1949). *Ashburne Hall: The First Fifty Years, 1899–1949*. n. pub.

232   *Pioneering British Women Chemists: Their Lives and Contributions*

University.[53] During the First World War, she helped the Warden, Dr. Phoebe Shaven, organise voluntary war work amongst students, such as sewing parties and fruit picking. Of particular note, Pirret worked with the Voluntary Aid Detachment (V.A.D.) to train student nurses and others to meet troop trains during the night to assist the wounded arriving back from France. During 1916, the Warden, Dr. Shaven, became ill and Pirret became acting Warden for most of the year.

In the 1920s, Pirret moved to London, sharing a house with her sister, Mary Janet Pirret, an early medical school graduate. Ruth Pirret joined the research group of Guy Dunstan Bengough[54] at Imperial College. Bengough was one of the leading researchers on corrosion in marine boilers, a vital topic in the days of steam-driven ships. She co-authored at least three of the group's publications between 1920 and 1924. Pirret died in South Kensington on 19 June 1939.

Maggie Sutherland

Maggie Millen Jeffs Sutherland[55] was the first woman Lecturer in Chemistry at the RTC. Daughter of Andrew Sutherland, manufacturer, and Frances Sutherland, she was born on 18 October 1881 in Glasgow. Sutherland was educated at Westbourne Terrace School, Glasgow, and the Lenzie Academy, Lenzie. She obtained a B.Sc. (Glasgow) in 1908 and was then awarded a Carnegie Research Fellowship. Sutherland undertook research with George Gerald Henderson[56] on the chemistry of terpenes. Henderson had been appointed in 1889 as the Lecturer in Chemistry at

---

[53] We thank Sheila Griffiths, Ashburne Association Archivist, Ashburne Hall, University of Manchester for providing information on Pirret's time at Ashburne Hall. The original sources were issues of the early Ashburne Hall student magazine: *Yggdrasill, the Tree of Knowledge.*

[54] Desch, C. H. (1945). Guy Dunstan Bengough, 1876–1945. *Obituary Notices of Fellows of the Royal Society* **5**(14): 168–178.

[55] (a) Anon. (1921). Certificates of candidates for election at the ballot to be held at the ordinary scientific meeting on Thursday, February 17th. *Proceedings of the Chemical Society* 14–15; and (b) *Student Records*, Archives, University of Glasgow.

[56] Irvine, J. C. and Simonsen, J. L. (1944). George Gerald Henderson, 1862–1942. *Obituary Notices of Fellows of the Royal Society* **4**: 491–502.

Queen Margaret College and had held the position for 3 years before being promoted to Chair in Chemistry at the RTC.[57]

In 1913, Sutherland was appointed as Assistant to the Professor of Chemistry (Henderson) at the RTC and was awarded a D.Sc. (Glasgow) the following year. In 1920, she was promoted to Lecturer, and then to Senior Lecturer in 1942. During her career, she co-authored five publications with Henderson between 1910 and 1914. Sutherland then had six publications with Forsyth James Wilson,[58] Henderson's successor, on semicarbazones and acridines between 1924 and 1943. Sutherland authored Wilson's obituary, concluding with the statement:

> It has been a privilege of the writer to collaborate with Professor Wilson in much of his later research and she therefore takes this opportunity to pay tribute to a kind and inspiring teacher and friend.[59]

Sutherland retired in 1947 and died in 1972 in her hometown of Lenzie in nearby Stirlingshire.

Mary Andross

Among the early graduates from UG was Mary Ann Andross.[60] Born on 15 March 1893 in Ayrshire, she was the daughter of Henry Andross, cashier. Andross graduated from UG in 1916, before briefly joining Henderson's research group in the Chemistry Department at the RTC. She worked as a day teacher at Irvine Royal Academy for the 1916–1917 academic year and then transferred to war work at the Ministry of Munitions Inspections Department on poison gases from 1917 until 1919, before becoming a Chemistry Assistant at Glasgow University from 1919 to 1923.

In 1924, Andross was appointed Lecturer in the Science Department of the Glasgow and West of Scotland College of Domestic Science

---

[57] Note 56, Irvine and Simonsen, pp. 492–493.

[58] Sutherland, M. M. J. (1945). Obituary Notices: Forsyth James Wilson, 1880–1944. *Journal of the Chemical Society* 723–724.

[59] Note 58, Sutherland, p. 724.

[60] (a) Cuthbertson, D. P. (31 August 1968). Mary Andross. *Chemistry and Industry* 1190; and (b) *Student Records*, Archives, University of Glasgow.

234 *Pioneering British Women Chemists: Their Lives and Contributions*

(the Scottish Education Department having stated that "they would have no objection to the College's request to appoint a lady lecturer").[61] Later renamed the Nutrition Department, she was promoted to Head in 1940. Andross held that position until retirement in 1965. In addition to pioneering courses for the training of dieticians, she was an active researcher on the chemical composition and nutritional value of foodstuffs, publishing five papers solely under her name. Andross was elected Fellow of the Institute of Food Science Technology. During the Second World War, she promoted the use of rose hips as an excellent source of Vitamin C, as was described in a newspaper article of the time:

> Miss Mary Andross, nutrition expert at the Glasgow College of Domestic Science, has made some remarkable discoveries about the properties of the dog-rose hips, and hospitals all over the country are planning to take up this 'fruit' in a big way.[62]

Andross was a colourful character, and in addition, she was very active in social events as her obituarist, David Cuthbertson noted:

> The revival of the Ramsay Dinner in Glasgow was largely due to her enthusiasm and energy and she acted as convenor on many an occasion. ... Many would have liked her to be Chairman of the Glasgow Section of the Society for Chemical Industry, but she emphatically refused the honour.[63]

In her youth, Andross had been extremely athletic. While still quite young, she was seriously injured by a car and lost much of the use of her legs. This did not diminish her interest in sports, in particular, fishing on the island of Harris. Her independence is illustrated by the story how a tyre on her car had a puncture one dark night in the country. She changed the tyre without any light source, storing the grimy wheel nuts in her mouth for fear of losing them. Andross died on 22 February 1968.

---

[61] McCallum, C. and Thompson, W. (1998). *Glasgow Caledonian University: Its Origins and Evolution.* Tuckwell Press, East Linton, Scotland, p. 49.

[62] Note 61, McCallum and Thompson, pp. 60–61.

[63] Note 60(a), Cuthbertson, p. 1190.

Aberdeen University

The University and King's College of Aberdeen was founded in 1495, but it was a merger with Marishal College (founded in 1593) in 1860 which resulted in the formation of Aberdeen University, itself.[64] In the early years, individual professors had allowed women to attend courses at Aberdeen; in fact, it was said that one young women had attended her father's regular chemistry classes in the 1820s at one of the constituent colleges.[65] However, in 1873, Aberdeen University students voted against university degrees being open to women, and the (male) students periodically voted the same way for the next two decades.

Despite the efforts of the Aberdeen Ladies' Educational Association[66] and others, women's access to university in the north-east lagged behind that in other parts of Scotland, as Lindy Moore had concluded: "… Aberdeen's efforts to promote their [women's] higher education were too late, too tentative, and intrinsically unsuited to the requirements of most girls in north-east Scotland."[67]

Life of Women Students

Mary Paton Ramsay commented on the life of the first-year woman student, and in particular the segregated seating:

> The Session opens about the third week of October, and the eager Bajanella (the feminised form of Bajan, or first year's student) begins her College life. The gown and trencher are not compulsory, but most girls prefer to wear them. The scarlet gown, and the trencher with scarlet tassel, (the men wear the tassel black) is becoming to most girls …

---

[64]Strathdee, R. B. (1953). Schools of chemistry in Great Britain and Ireland–V The University of Aberdeen. *Journal of the Royal Institute of Chemistry* **77**: 220–231.

[65]Moore, L. (1991). *Bajanellas and Semilinas: Aberdeen University and the Education of Women 1860–1920*. Aberdeen University Press, Aberdeen, p. 43.

[66]Moore, L. (1977). The Aberdeen Ladies' Educational Association, 1877–1883. *Northern Scotland* **3**: 123–157.

[67]Moore, L. (1980). Aberdeen and the higher education of women. *Aberdeen University Review* **48**(3): 280–303.

As one's curriculum draws to a close, one looks back with mingled amusement and melancholy to the excitement of these early days of College life; the shyness of the first encounters in the crowded women students' room, and the dread of being late for a lecture and of having to pass the crowded benches of men students in order to reach the ladies' seats beyond at the back.[68]

Ramsey noted that the greater proportion came from "the country" and resided in lodgings. As to gender relations at Aberdeen, she proclaimed them to be far superior than elsewhere:

A strong spirit of unity, a keen "esprit de corps", exists among the women students, ... All the great Scholarships, Fellowships, Bursaries, and Prizes are open equally to men and women, and though competition is keen, the courtesy shown by the men students to the women students has always been perfect. The same courtesy has not always been shown to women students at other Scottish Universities, ...[68]

Women Chemistry Students

As we mentioned earlier, a significant number of women studied chemistry at Aberdeen in the early part of the 20th century and specifically between 1912 and 1920.[69] The First World War period also opened up positions for the first women staff — but on a temporary basis as Kenneth Page noted:

In order to meet staff shortages in 1917, history was made by the appointment of the first women ever to hold official posts in the department. Miss Jeannie Ross and Miss Beatrice Simpson both joined the department as assistants during the winter term of 1917.... In 1918 Dr. Marion Richards took up the post of senior assistant in chemistry.[70]

---

[68] Ramsay, M. P. (November 1906). Women students in Aberdeen University. *The World of Dress & Women's Journal* 11.

[69] Page, K. R. (Autumn 1979). Frederick Soddy: The Aberdeen interlude. *Aberdeen University Review* **48**(2): 127–148.

[70] Note 69, pp. 136–137.

Alice Bain

The first woman to complete a chemistry degree at Aberdeen was Alice Mary Bain.[71] Bain, daughter of John Bain and Jessie Bain, was born on 31 August 1875 at Greenock. Entering Aberdeen University in 1898, she completed an M.A. in 1899 and a B.Sc. in 1901. Upon graduation, Bain was appointed Junior Science Mistress at Croydon High School. In 1903, Bain went to the Northern Polytechnic, London, as Assistant Lecturer and Demonstrator in Chemistry, her research on optically active compounds being with William Mills,[72] which resulted in four publications.

In September 1916 at age 40, Bain travelled to India to take up an appointment of Professor of Chemistry at Lady Hardinge Medical College, Delhi.[73] She left that position in 1923, her replacement being Sosheila Ram (see Chapter 3). The same year, Bain accepted an appointment as Reader in Chemistry at the University of Delhi.[74] Bain was still teaching there in 1928, but no later records of her could be traced.

Ada Hitchins (Mrs. Stephens)

Ada Florence Remfry Hitchins[75] was the principal research student of Frederick Soddy.[76] It was her toil, isolating lead samples from uranium

---

[71] (a) Masson, M. (1966). Early women chemistry students at Aberdeen. In Masson, M. R. and Simonton, D. (eds.), *Women and Higher Education: Past, Present and Future.* Aberdeen University Press, Aberdeen, pp. 316–318; and (b) *Student Records*, Archives, University of Aberdeen.

[72] Mann, F. G. (1960). William Hobson Mills, 1873–1959. *Biographical Memoirs of Fellows of the Royal Society* **6**: 200–225.

[73] Mathur, N.N. (1989). Indian Medical College: Lady Hardinge Medical College, New Delhi. *The National Medical Journal of India* **11**(2): 97–100.

[74] Watt, T. (compiler). (1935). *Roll of the Graduates of the University of Aberdeen 1901–1925.* Aberdeen University Press.

[75] Rayner-Canham, M. and Rayner-Canham, G. (1997). Ada Hitchins: Research assistant to Frederick Soddy. In Rayner-Canham, M. and Rayner-Canham, G. (eds.), *A Devotion to Their Science: Pioneer Women of Radioactivity.* Chemical Heritage Foundation, Philadelphia, and McGill-Queen's University Press, Montreal & Kingston, pp. 152–155.

[76] Note 48, Fleck, p. 206.

238 *Pioneering British Women Chemists: Their Lives and Contributions*

ores, which contributed to the discovery of isotopes and Soddy's subsequent Nobel Prize. Hitchins was born on 26 June 1891 at Tavistock, Devon, daughter of William Hedley Hitchins, Supervisor of Customs and Excise. The family moved to Campbeltown, Scotland, and it was there that she attended high school, matriculating in 1909. Hitchins obtained a B.Sc. (with special honours in chemistry) from the University of Glasgow in 1913.

Hitchins commenced working with Soddy in 1913 and moved with him to Aberdeen in 1914, where she held an appointment as Carnegie Research Scholar. It was anticipated that Soddy, when appointed to the Chair of Chemistry at Aberdeen, would build up a strong school of radiochemistry. This did not happen, as R. B. Strathdee quoted Soddy's other student, John A. Cranston,[77] as saying:

> During his sojourn in Aberdeen, 1914–19, I am not aware of any other work done by Soddy in the field of radioactivity-apart from supervising the completion of some work by his research student, Miss Hitchens [sic], on the growth of radium from uranium.[78]

At the time, Soddy was seeking evidence that lead from thorium ores had different atomic weights from "normal" lead. When Soddy announced the discovery of a sample of lead of atomic mass 207.74, he acknowledged the contribution of Hitchins for the separation and analysis work. Thus, Hitchins precise and accurate measurements on the atomic masses of lead from different sources was among the first evidence for the existence of isotopes.[79] In addition, Hitchins took over the research on protactinium from Cranston, when the latter was drafted for the First World War.

---

[77]Cumming, W. M. (1972). John Arnold Cranston 1891–1971. *Chemistry in Britain* **8**: 388.

[78]Cranston, J. A. cited by Strathdee, R. B. (1963). Chemistry: From retort to grid — 1860–1960. In Simpson, W. D. (ed.), *The Fusion of 1860: A record of the Centenary Celebrations and a History of the United University of Aberdeen 1860–1896*. Oliver and Boyd, Edinburgh, pp. 300–307.

[79]Rayner-Canham, M. and Rayner-Canham, G. (2000). Stefanie Horovitz, Ellen Gleditsch, Ada Hitchins, and the discovery of isotopes. *Bulletin for the History of Chemistry* **25**(2): 103–108.

In 1916, Hitchins herself was drafted to work in the Admiralty Steel Analysis laboratories.[80] When the former male occupants of the analytical laboratories returned upon the end of hostilities, Hitchins lost her position. However, the wartime analytical experience enabled her to find employment as a chemist with a Sheffield steel works until Soddy, then at Oxford, obtained funding to rehire her.[81]

Despite a Nobel Prize in Chemistry, Soddy had great difficulty in attracting graduate students to work with him at Oxford[82]; thus, Hitchins played a crucial role in Soddy's research program. Initially appointed as Technical Assistant, she was promoted to private Research Assistant in 1922. Soddy noted: "... she has also charge of my radioactive materials ... and has worked up considerable quantities of radioactive residues and other materials for general use."[83]

Hitchins finally left Soddy's employ in 1927, emigrating to Kenya with her family, and becoming Government Assayer and Chemist for the Mining and Geological Department of the Colonial Government in Nairobi. After her retirement in 1946, she married John Ross Stephens of Amboni Bend Farm, Nyeri Station, Kenya. Subsequently returning to England, she died on 4 January 1972 at Bristol.[84]

Rowett Research Institute

Horticultural and agricultural research institutes were avenues for employment of women chemists.[85] Muriel Wheldale (see Chapter 11) was encouraged to join William Bateson's research group at the John Innes

---

[80]Letter, J. C. W. Humfrey, Admiralty Inspection Officer, Sheffield to Dr. Desch, 6 January 1916. Glasgow University, "We have written to Miss Hitchins today offering her an appointment," University of Sheffield archives, CHD14/WW/180.

[81]Papers of Professor Frederick Soddy, Oxford University Archives, Bodleian Library, File 201.

[82]Cruickshank, A. D. (1979). Soddy at Oxford. *British Journal for the History of Science* **12**: 277–288.

[83]Soddy, F. undated reference for A. F. R. Hitchins, Oxford University Archives.

[84]Anon. (1972). Personal news: Deaths. *Chemistry in Britain* **8**: 376.

[85]Anon. (1938). Prospects of employment for women science graduates: Part II, Opportunities in Agricultural Research Stations. *Journal of Careers* **17**: 154–158.

240   *Pioneering British Women Chemists: Their Lives and Contributions*

Horticultural Institute,[86] while Frances Micklethwait (see Chapter 4) attended, then taught at, Swanley Horticultural College,[87] and Elfreida Cornish (see Chapter 16) joined the research group at the Reading Agricultural Institute.

Several Scottish women chemistry graduates found employment with the Rowett Research Institute, near Aberdeen.[88] It was in 1875 that the Aberdeen Agricultural Association (subsequently renamed the Agricultural Research Association) was established. Relying on donations from landowners, the Association ran an experimental station and laboratory, and, for a few years, had an experimental farm. The Association chemist, Thomas Jamieson, argued that insoluble phosphate was a more useful fertilizer than was the current view. Such heresy brought him into conflict with the Director of the Rothamsted Experimental Station. Lacking any continuing funding, the Agricultural Research Association ceased operation in the 1910s.

However, in 1913, a Joint Committee of the University of Aberdeen and the North Scotland College of Agriculture proposed an agricultural research institution focussing on animal nutrition. John Boyd Orr was appointed to develop the proposal and obtain funding for such an endeavour. In 1919, Orr persuaded the British Government to provide half of the funding necessary to construct the Research Institute if he could find a private donor for the other half. The Government agreed, and John Quiller Rowett provided the remaining funds. The now-named Rowett Institute was formally opened in 1922.

Marion Richards

The Rowett Institute attracted a significant number of women researchers in the field of biochemistry, one of the early notables being Marion Brock

---

[86] Olby, R. (1989). Scientists and bureaucrats in the establishment of the John Innes Horticultural Institution under William Bateson. *Annals of Science* **46**: 497–510.

[87] Opitz, D. L. (2013). "A triumph of brains over brute": Women and science at the Horticultural Institute, Swanley, 1890–1910. *Isis* **104**: 30–62.

[88] Smith, D. (1998). The Agricultural Research Association, the development fund, and the origins of the Rowett Institute. *Agricultural History Review* **46**(1): 47–63.

Richards.[89] Richards, born on 17 July 1885, was the daughter of Robert Richards, a commercial traveller, and Mary Carmichael. She obtained her M.A. from the University of Aberdeen in 1907 and a B.Sc. in 1909. Richards was the first woman chemistry graduate of Aberdeen to undertake research, completing a D.Sc. with Professor Freundlich at the University of Leipzig in 1916.

From 1916 to 1918, Richards taught and lectured at Leeds[89]; in 1918, she was appointed Senior Assistant in Chemistry to Soddy at Aberdeen, taking over the responsibility for the organic chemistry course. She joined the staff of the Rowett Institute as a biochemist in 1920, a move forced upon her when Soddy left Aberdeen for Oxford. Richards remained at the Rowett until her retirement in 1951.[90]

Beatrice Simpson

Beatrice Weir Simpson[91] was born on 15 February 1893 at Inveravon, Banffshire, the daughter of James S. Simpson, excise officer, and Elizabeth Hannah. Admitted to Aberdeen University, Simpson completed her M.A. in 1913 and her B.Sc. in 1917. Following graduation, she was appointed temporary Assistant in the Chemistry Department upon the recommendation of Soddy.[92]

From 1923 until her retirement in 1956, Simpson was on the staff of the Rowett Research Institute in Aberdeen. In the 1930s, the Biochemistry Department of the Rowett consisted solely of William Godden (the Head), Richards, and Simpson.[93] Her published research covered a variety of topics and included a joint paper with Richards.[94] Simpson died on 14 February 1981, aged 87.

---

[89] Note 71, Masson, p. 317.

[90] Anon. (Autumn 1964). Obituary: Marion Richards. *Aberdeen University Review* **40**: 395.

[91] *Student Records*, Archives, Aberdeen University.

[92] *Minutes of Meeting.* (1920). *The Senatus Academicus of the University of Aberdeen* 18 December 1917, item 2, The Aberdeen University Press Ltd.

[93] Orr, J. B. (ed.), (1930). Members of staff. *The Rowett Research Institute Collected Papers.* Vol. II, Edmond & Spark, Aberdeen, p. vi.

[94] Richards, M. B. and Simpson, B. W. (1934). The curative method of vitamin A assay. *Biochemical Journal* **28**: 1274–1292.

University of Edinburgh

The University of Edinburgh is actually the youngest of the "old" Scottish Universities, founded in 1583.[95] Thomas Charles Hope was appointed as the sole Lecturer in Chemistry in 1797.[96] In addition to teaching the regular students, in 1826 he introduced "a Short Course of Lectures for Ladies and Gentlemen."[97] The presence of women on campus was not appreciated by many academics, for example, Lord Cockburn wrote to a T. F. Kennedy:

> The fashionable place here now is the College; where Dr Thomas Charles Hope lectures to ladies on Chemistry. He receives 300 of them by a back window, which he has converted into a door. Each of them brings a beau, and the ladies declare that there was never anything so delightful as these chemical flirtations.[98]

Life of Women Students

At the beginning of this chapter, we described the "Surgeon's Riot" at the University of Edinburgh. It was not just in medicine that women students were treated badly. In 1896, a plea was made to treat women students in a more civilized fashion:

> It is unfortunately a matter of not infrequent observance that the treatment of lady students at the hands of their *confrères* has been utterly out of keeping with the cherished canons of gentlemanly conduct. It is a painful fact that ever since the portals of our University were opened to the lady students, as the result between sweet reason and dogmatic

---

[95] Anderson, R. D., Lynch, M. and Phillipson, N. (2003). *The University of Edinburgh: An Illustrated History, 1582–present*. Edinburgh University Press.

[96] Hirst, E. L. and Ritchie, M. (1953). Schools of chemistry in Great Britain and Ireland– VII The University of Edinburgh. *Journal of the Royal Institute of Chemistry* **77**: 505–511.

[97] Morrell, J. (23 September 2004). Thomas Charles Hope (1766–1844), *Oxford Dictionary of National Biography* (accessed 18 January 2019).

[98] Cockburn, H. (1874). *Letters Chiefly Connected with the Affairs of Scotland*. Ridgway, London, pp. 137–138.

rigidity, they have been the victims of gratuitous annoyance. Their entry into the class-rooms opens the floodgates of British chivalry. The tapers and tadpoles of the back benches begin to howl and screech with all the lustiness of rural louts, and seem to be as much amused as if they saw a picked company of the far-famed Dahomeyan Amazons march in all the glory of their military attire.[99]

Edith Pechey (Mrs. Pechey-Phipson)

Mary Edith Pechey,[100] one of the "Edinburgh Seven," had been an outstanding chemistry student in the 1869–1870 session at the University of Edinburgh.[101] She was born on 7 October 1845, the daughter of William Pechey, a Baptist minister at Langham, near Colchester, and Sarah Rotton, a well-educated nonconformist lawyer's daughter. Pechey had worked as a governess and a teacher before entering medical school at Edinburgh.

Chemistry was required for admission to medical school, and the Professor of Chemistry, Alexander Crum Brown,[102] gave separate lectures to the women students, insisting they were identical to those that he was concurrently giving to the men. Pechey attained the third place in the chemistry examinations. Forty years earlier, Professor Hope had instituted awards known as the Hope Scholarships, and these were presented to the top four students in the first-year chemistry examinations. The recipients were entitled to free use of the facilities of the university chemistry laboratory for the next term. The two students above Pechey in the list were repeating the course and were therefore ineligible. Crum Brown, probably

---

[99]"Sarat Mullick." (13 February 1896). A plea for the rational treatment of women students, *The Student: Edinburgh University* **10** (new ser.): 60–61. Centre for Research Collections, Main Library, University of Edinburgh.

[100]Lutzker, E. (1973). *Edith Pechey-Phipson, M.D.: The Story of England's Foremost Pioneering Woman Doctor*. Exposition Press, New York.

[101]Cited from Note 3 in Hamilton, S. (1983). The first generations of university women 1869–1930. In Donaldson, E. (ed.), *Four Centuries: Edinburgh University Life 1869–1983*. University of Edinburgh, Edinburgh, p. 114.

[102]"J.W." (1924). Obituary Notices of Fellows Deceased: Alexander Crum Brown, 1838–1922. *Proceedings of the Royal Society of London, Series A* **105**: i–v.

surprised by her outstanding marks, then proclaimed that Pechey was ineligible as she had been taught in a separate class, contradicting his earlier statements.[103]

Following from his decision, Crum Brown then refused to issue the group of women students the usual certificates of attendance required for admission to medical school, instead giving them his own certificates, which Jex-Blake referred to as the Professor's "strawberry jam labels."[104] Finally, in 1870, by a one-vote margin the University Senate approved the issuing of attendance certificates to the women, while also by a margin of one vote, the Senate denied Pechey the Hope Scholarship.

The issue of Pechey's disqualification rapidly escalated, gaining national attention and even being a front-page article of the American newspaper, *New Era*.[105] It was the outcry against this injustice that resulted in the following poem being published in the women-supportive London review magazine, *The Period*. Verses 1 and 8 are provided here:

Shame upon thee, great Edina! shame upon thee, thou hast done
Deed unjust, that makes our blushes flame as flames the setting sun.
You have wrong'd an earnest maiden, though you gave her honours crown,
And eternal shame must linger round your name, Professor Brown.

And I blush to-day on hearing how they've treated you, Miss P.,
How that wretched old Senatus has back'd up Professor B.
Ah! the 'Modern Athens' surely must have grown a scurvy place,
And the 'Varsity degraded to incur such dire disgrace.[106]

In 1873, Pechey asked the College of Physicians in Ireland to allow her to take exams leading to a licence in midwifery, which they did.

---

[103] Jex-Blake, S. (1886). *Medical Women: A Thesis and a History*. Oliphant, Anderson, & Ferrier, Edinburgh, p. 60.

[104] Note 3, Roberts, p. 84.

[105] Anon. (16 June 1870). Woman's Rights in Scotland. *New Era*. Washington, D.C., 1.

[106] Anon. (14 May 1870). A Cheer for Miss Pechey. *The Period*. pp. 12–13.

Despite the lack of an official qualification, she then worked at the Birmingham and Midland Hospital for Women. In 1877, Pechey obtained a medical diploma from the Irish College of Physicians and, in the same year, an M.D. from the University of Bern.[107] For the next 6 years, she practiced medicine in Leeds, specializing in abdominal surgery.

It was in 1883 that George Kittredge was visiting England from India, tasked with finding a British woman doctor for the post of Senior Medical Officer at the new Pestonjee Hormusjee Cama Hospital for Women and Children in Bombay (now Mumbai).[108] He was having no success until Elizabeth Garrett Anderson suggested the name of Pechey. At the time, Pechey was in Vienna undertaking surgical practice, and after Kittredge contacted Pechey, they arranged to meet in Paris. After some deliberation, Pechey accepted the position.

Pechey arrived in Mumbai in December 1883, quickly becoming fluent in Hindi. She married Herbert Musgrave Phipson, a reformer as well as wine merchant and naturalist, in 1889. Her obituarist, Margaret Todd commented:

> From the first, Dr. Pechey found the climate of Bombay very trying, aggravated as were its effects by hard work and heavy responsibility. After some ten years of arduous service, she retired from her post with broken health, but on the outbreak of the plague in 1896, she was among the first to organise and take part in the house-to-house visitation.[109]

Accompanied by her husband, Pechey returned to Britain in 1905, becoming active in the Leeds suffrage movement. In 1907, she needed surgery, the surgeon being May Thorne, a graduate of the London School of Medicine for Women and the daughter of Pechey's former classmate of the "Edinburgh Seven," Isabel Thorne. Though the operation was

---

[107]Todd, M. (1908–1909). Obituary: Dr. Edith Pechey Phipson. *Magazine of the Royal Free Hospital and London School of Medicine for Women* 882–884.

[108]Kittredge, G. A. (1889). *A Short History of the "Medical Women for India" Fund of Bombay*. Education Society's Press, Bombay, pp. 14–15.

[109]Note 107, Todd, pp. 883–884.

246 *Pioneering British Women Chemists: Their Lives and Contributions*

successful, Pechey never fully recovered and she died in 1908 at Folkestone, Kent. Her husband set up a scholarship at the London School of Medicine for Women in Pechey's name which was granted regularly up to 1948.

Christina Miller

There was one star student who became a key member of the staff of the Chemistry Department — Christina Cruickshank Miller.[110] Christina Miller was born on 29 August 1899 in Coatbridge. In grade school, Miller developed an interest and aptitude towards mathematics. At the time, she was told that, as a woman, she could only use her mathematical aptitude towards a career as a school teacher. Unfortunately, she had become deaf as a result of childhood measles and rubella, and this was considered a major impediment to a school teaching career. She read an article in a magazine that mentioned the employment potential for women as analytical chemists, and this possibility determined her future.

Miller won an entrance scholarship to the University of Edinburgh. Despite her hearing difficulty (and her workload), she won the class medal in the university advanced chemistry course, graduated with a B.Sc. with special distinction in Chemistry. Miller was awarded a Vans Dunlop Research Scholarship which allowed her to pursue a higher degree. James Walker,[111] Professor of Chemistry at Edinburgh, accepted Miller as a graduate student, and from 1921 until the completion of her Ph.D. in 1924, she worked with him on the process of diffusion in solution.

To give her better training for an industrial position, Miller had also been taking courses towards a 4-year industrial chemistry diploma program at Heriot-Watt College. Heriot-Watt College had been founded in

---

[110](a) Chalmers, R. A. (1993). A mastery of microanalysis. *Chemistry in Britain* **29**: 492–494; (b) Mango, K. (Spring 2003). Dr. Christina Miller: A beacon of knowledge and strength. *Hearing Health*, 16–21; and (c) Anon. (7 August 2002). University of Edinburgh, news release.

[111]Kendal, J. (1932–1935). Sir James Walker. *Obituary Notices of Fellows of the Royal Society* **1**: 537–549.

Universities in Scotland and Wales 247

1821, the first institute in Britain for the education of manual workers, and one which welcomed women as early as 1869.

Miller's research with Walker was actually undertaken in the Heriot-Watt chemistry laboratory and for the 1920–1921 Session, she was appointed as Research Scholar and Demonstrator at Heriot-Watt. In 1921, Miller received a Diploma of Applied Chemistry and the following year, an Associateship of Heriot-Watt College in Applied Chemistry.

In 1922, Miller was awarded a 2-year Carnegie Research Fellowship to undertake independent research. The position enabled her to study a problem that had long fascinated chemists, the glow produced when tetraphosphorus hexaoxide oxidized. During this time, she applied for a lectureship at Bedford, but was rejected on the grounds of her deafness. Walker suggested Miller accept a post as an Assistant in the Chemistry Department at the University of Edinburgh. The position involved the supervision of undergraduate students but she was allowed to continue with research in her spare time.

By 1929, Miller's thorough research on the oxides of phosphorus showed that traces of elemental phosphorus caused the glow, not the oxides themselves. As a result of her five publications, Miller was awarded a D.Sc. degree, the prestigious Keith Prize by the Royal Society of Edinburgh, and a Lectureship with tenure. Shortly afterwards, during her research, a glass bulb exploded, the fragments blinding her in one eye.

Deciding that avenue of research was too dangerous, Miller developed an interest in micro- and semimicro-analysis, and in 1933, she was appointed Director of the inorganic laboratory. During the next 28 years, she became renowned for her innovative undergraduate microscale analytical techniques. She was always looking to refining and improving analytical methods, and her many research papers were devoted to this topic, such as the definitive work on 8-hydroxyquinoline as a reagent for magnesium.

Sadly, ill-health in the form of otosclerosis, together with family commitments (care of a relative), caused Miller to take early retirement in 1961. Never one to remain idle, she pursued interests in genealogy and in the history of Edinburgh. Her biographer, Robert Chalmers, commented:

> It is arguable that had she been a man ... she might have become one of the UK's first professors of analytical chemistry. At a time when

248   *Pioneering British Women Chemists: Their Lives and Contributions*

analytical chemistry was practically non-existent in UK universities she provided courses that would stand comparison with the best available today. ... Her work was highly esteemed by many internationally renowned analysts.[112]

Miller died on 16 July 2001 at age 101.

Elizabeth Kempson (Mrs. Percival, Mrs. McDowell)

The only woman to be mentioned in the history of the Chemistry Department at the University of Edinburgh was Ethel Elizabeth (Betty) Kempson.[113] Born on 3 January 1906 in Coventry to Frank George Kempson, Foreman, and Emily Marsalina Holloway, she was educated at Wolverhampton High School. Kempson then entered the University of Birmingham, obtaining an Honours degree in Chemistry in 1928. She stayed at Birmingham, undertaking research with Norman Haworth.[114] Haworth's Senior Research Assistant was Edmund George Vincent Percival, and in 1934, Kempson and Percival married.

That same year, Percival was appointed to a lectureship at the University of Edinburgh where Kempson started research with him on the synthesis and reactions of carbohydrates for which she received a Ph.D. in 1941. With the founding of the Scottish Seaweed Research Association, their interest turned to marine polysaccharides. Percival died in 1951 and Kempson took over the research, as well as being appointed Lecturer and raising two children. For her contributions to polysaccharide chemistry, she was elected a Fellow of the Royal Society of Edinburgh.

In 1962, Kempson married Richard McDowell of Alginate Industries Ltd., and together, they wrote the monograph *Chemistry and Enzymology*

---

[112] Note 110a, Chambers, p. 494.

[113] (a) Weigel, H. (1978). Elizabeth E. Percival, F.R.S.E. *Carbohydrate Research* **66**: 7–8; and (b) Manners, D. J. (May 1998). Betty Percival 1906–1997. *Chemistry in Britain* **34**: 61.

[114] Hirst, E. L. (1951). Walter Norman Haworth, 1883–1950. *Obituary Notices of Fellows of the Royal Society* **7**: 372–404.

of Seaweed Polysaccharides.[115] Kempson then moved to Royal Holloway College as an Honorary Lecturer where she built up a new research group. Her collaborators included 26 Ph.D. students and resulted in more than 100 publications.

During her lifetime, an issue of the journal *Carbohydrate Research* was dedicated to her life and work. In the introduction, one of her former students, Helmut Weigel, commented:

> Dr. Percival is an ardent traveller. In her journeyings, which have taken her to all continents, she always combines a taxing lecture programme with the collection of seaweeds-sometimes a very hazardous occupation-and with visits to former students and friends. Her zest for travel has made her an enthusiastic camper, and not many of us can claim, as she can, to have cooked a three-course meal whilst stranded by a blizzard on top of the Atlas mountains. … She plays tennis, enjoys her piano and, in the Royal Holloway College, Dr. Percival has been President of the Women's Club. However, her chief interest remains the pursuit of chemistry, and she takes pleasure in communicating her own enthusiasm to younger students of chemistry.[116]

Kempson died on 16 April 1997 in Surrey, aged 91.

University College, Dundee

University College, Dundee (the University of Dundee since 1967), was founded in 1881 as a satellite campus of the University of St. Andrews with the mission of "promoting the education of persons of both sexes."[117] It was the organic chemist, Alexander McKenzie,[118] Professor of Chemistry, who attracted a significant number of women chemistry

---

[115]Percival, E. and McDowell, R. H. (1967). *Chemistry and Enzymology of Marine Algal Polysaccharides*. Academic Press, London.

[116]Note 113a, Weigle, p. 8.

[117]Shafe, M. (1982). *University Education in Dundee 1881–1981*. University of Dundee, Dundee.

[118]Reed, J. (1952–1953). Alexander McKenzie, 1869–1951. *Obituary Notices of Fellows of the Royal Society* **8**: 207–228.

250 *Pioneering British Women Chemists: Their Lives and Contributions*

researchers. During his "reign" from 1914 to 1938, they included Nellie Walker, Isobel Smith, Mary Lesslie (see Chapter 5), Agnes Grant Mitchell, Ethel Luis, and E. R. L. Gow.

Life of Women Students

Though it attracted a high proportion of women students, gender relations were not always smooth, particularly in the 1920s as "An Undergraduette" commented:

> Before I came to College I had no idea any community of men could be so collectively dull.... The rock-bottom trouble is that you University men continue to treat us University women as survivals of the Victorian age. You continue your pre-historic drawing-room attitudes in class-room, dance and corridor, simpering like a bunch of be-whiskered Lord Dundrearys before us, and forgetting that you are addressing the new, jazz-infected, bobbed and madcap daughters of this generation. Oh, I know there are some Victorian survivals, sitting primly acidulate in the Women's Union, but the rest of us don't pay attention to them, and when they do cut in with their reproofs and reminders, you can take it from me, we moderns just ignore them. They always have the Christian Union to occupy most of their spare time.[119]

It is interesting to note this division among the women students of the "moderns" and the "traditionalists." Another correspondent, "Alphabeta" sounded even more bitter about her male colleagues:

> Time was when I thought I should get me a husband and be happy. Husbands are not nearly so difficult to get as most men fondly imagine, and I feel sure that, had I been so inclined, I could have been married once legally and twice bigamously within the past two years. But I have changed, and no longer do I care for men. Rather would I be an old maid and live my life alone, uncared for and uncuddled, but free from the baneful influence of men and their stupid brutish ways. That is what

---

[119]"An Undergraduette." (December 1924). Sheiks and shrieks. *The College: The Magazine of University College, Dundee* **22**(1): 15.

a University education has done for me. ... As for 'Varsity men, I should no more think of accepting one of their illustrious number in marriage than I would of knitting a jumper of jute.[120]

Women Chemistry Students

During the early years of the college, *The College: The Magazine of University College Dundee* contained periodic remarks on women chemists. In the 1890s, *The College* describes a ladies-only room:

> Coming, as the greater number of us do, from the unsavoury and often poisonous atmosphere of heated laboratories, to enjoy an hour of chat and gossip (with its accompaniment of afternoon tea), we long for a complete change of environment.[121]

The novelty of women in chemistry prompted the comment: "The junior chemistry class is a study of blouses,"[122] and also: "The fame of Marie Curie seems to have stimulated the female sex to pursue the study of chemistry. The advanced class is composed of four women students, and one solitary male. Changed days indeed."[123]

The pages of the magazine were also used for cutting comments about the women chemistry students: "Rumour has it that the female section is experiencing difficulty in sucking up liquids into pipettes. They say they don't know when to stop. That's the worst of practicing with straws in the summer."[124] Objections to the behaviour of women students during glassblowing was reported: "We would remind the female elementary chemists

---

[120] "Alphabeta." (February 1922). A maiden speaks. *The College: The Magazine of University College, Dundee* **19**(2): 4.

[121] "Lady Students." (8 February 1889). *The College, Magazine of University College, Dundee* **1**(2): 64.

[122] Anon. (January 1904). (Untitled). *The College, Magazine of University College, Dundee* **1**: 1.

[123] Anon. (November 1904). Department notes: Science. *The College, Magazine of University College, Dundee* **2**(1): 19.

[124] Anon. (December 1904). Department notes: Science. *The College, Magazine of University College, Dundee* **2**(2): 42.

252 *Pioneering British Women Chemists: Their Lives and Contributions*

that bulb-blowing can be done quite well without screeching and howling, or any sort of accompaniment."[125]

The absence of women from an alcoholic social was also noted:

> This [chemistry] society, ever mindful of the comforts of its members, prefaced the month of December with an instruction by Mr Young on the merits and demerits of alcohol in its many and varied forms. It was notable that none of the women were present. Perhaps, after the manner of their sex, they wished to convey the suggestion that they needed no instruction in these matters.[126]

Nellie Walker (Mrs. Wishart)

The first woman staff member was Nellie Walker.[127] The daughter of William Walker, she was educated at Dundee High School and entered University College, Dundee in 1908 at age 17. She obtained an M.A. (St. Andrews) in 1911 and a B.Sc. (St. Andrews) in 1913. For the next 2 years, she was a research scholar at St. Andrew's, and then briefly in 1915 undertook war work with the Royal Society War Committee. Later that year, she obtained a position as Demonstrator in Chemistry at Bedford College. At the end of the First World War, Walker returned to Dundee where she completed a Ph.D. (St. Andrews) in 1920.

Walker was then appointed as University Assistant. The Editor of *The College, Magazine of University College, Dundee* commented:

> For two years a Carnegie research scholar, she has now been appointed assistant in our own Chemistry Department, and we rejoice at her promotion. Successful as a student, she will be as successful in the capacity of teacher. In the social life of the College, too, Miss Walker has always taken an active part. No society for which she is eligible has lacked her

---

[125] Anon. (February 1906). Department notes: Science. *The College, Magazine of University College, Dundee* 3(4): 77.

[126] Anon. (December 1906). Society notes: Scientific society. *The College, Magazine of University College, Dundee* 4(2): 37.

[127] *Student Records*, Archives, University College, Dundee, (University of St. Andrews); and *Staff Records*, Archives, Bedford College.

support... Quiet, unassuming, deliberate in all her words and actions, she has a very real influence on all around her. If she be no brilliant conversationalist, to her at least belongs that gift so priceless, yet so rare, the genius for listening.[128]

In 1921, Walker was promoted to Lecturer and Assistant, a post that she held until 1927 when she resigned and married W. G. Wishart.

Isobel Agnes Smith

Isobel Agnes Smith[129] was the stalwart of Dundee just as Ettie Steele was the "backbone" of St. Andrews Chemistry Department. Born in 1897, daughter of Robert Graham Smith, she entered Dundee in 1914 at the age of 17. Completing an M.A. (St. Andrews) in 1918, a B.Sc. in 1920, and a Ph.D. in 1922, she was awarded Carnegie Fellowships for 1922–1924.

Smith was appointed Assistant Lecturer at Dundee in 1928 and promoted to Lecturer in 1937. She co-authored 13 research papers over the period 1922–1943, all on optical activity of organic compounds. She held the rank of Senior Lecturer at the time of her retirement in 1961. Smith died on 18 January 1981, aged 84.[129c]

Ethel Luis

For the duration of the Second World War, a second woman was appointed to the chemistry teaching staff at Dundee: Ethel Margaret Luis.[130] Luis was born on 28 August 1899, daughter of Theo. G. Luis, spinner and manufacturer of Bloomfield, Dundee. She entered the Royal Holloway College in 1918, obtaining an B.Sc. (London) in 1921, while for the 1923–1925 period, Luis is listed as attending the Royal College of

---

[128] Anon. (December 1915). The Editor. *The College, Magazine of University College, Dundee* **13**:

[129] (a) *Student Records*, Archives, University of St. Andrews; (b) *Calendars*, University of St. Andrews; and (c) Anon. (1981). Personal news: Deaths. *Chemistry in Britain* **17**: 270.

[130] (a) *Student Records*, Archives, University of St. Andrews; (b) (1951). *Old Students and Staff of the Royal College of Science,* 6th edn., Royal College of Science, London; and (c) Anon. (October 1998). Personal news: Deaths. *Chemistry in Britain* **34**: 91.

254   *Pioneering British Women Chemists: Their Lives and Contributions*

Science, Imperial College from which she received an Honours B.Sc. in 1925.

Luis then returned to Scotland to begin research with McKenzie at Dundee, being awarded a Ph.D. (St. Andrews) in 1931. She continued research at Dundee, authoring and co-authoring nine papers between 1929 and 1941 and was appointed Demonstrator in 1938. In 1939, at the start of the Second World War, Luis was promoted to Assistant Lecturer to replace male faculty who had departed on war duties, her employment being terminated with the end of the War in 1945. Luis died on 30 May 1998 at Broughty Ferry, aged 98.

The University Colleges of Wales

Four university colleges were founded in Wales during the late 19th century: the University College of Wales, Aberystwyth; the University College of North Wales, Bangor; the University College of South Wales and Monmouthshire, Cardiff; and University College, Swansea. These colleges were combined in 1894 under the name of the University of Wales. The Charter was clearly progressive on equal opportunities for women:

> Women shall be eligible equally with men for admittance to any degree which the University is by this our charter authorised to confer; every office hereby created in the University and membership of every author-ity hereby constituted shall be open to women equally with men.[131]

Nevertheless, in the early years, perhaps due to the lack of academic schools for Welsh girls, the women who populated the Welsh University Colleges came largely from such English schools as: Cheltenham Ladies' College; North London Collegiate School; Wyggeston School, Leicester; Orme's School, Newcastle-under-Lyme; Haberdashers' Aske's Hatcham School; and the Girls' High Schools in north-west England such as those at Blackburn, Bury, Dulwich, and Southport.[132]

---

[131] Zimmern, A. (August 1898). Women at the universities. *Leisure Hour* p. 440.

[132] Evans, W. G. (1990). *Education and Female Emancipation: The Welsh Experience, 1847–1914*. University of Wales Press, Cardiff.

University College of North Wales, Bangor

The location for a North Wales College was settled in 1884, with Bangor being chosen.[133] Though co-educational, life for the women students was strictly controlled. J. Gwynne Williams elaborated:

> Bangor was by no means unusual in drawing up careful rules to govern relations between men and women students. If they talked to one another for more than a few moments even between lectures, they were in danger of being reported for 'skylarking' and in lectures they sat quite separately.... There was a rigid system of chaperonage, the women's warden attending all mixed gatherings.[134]

From 1903 until his death in 1930, Kennedy Joseph Previté Orton[135] held the Chair of Chemistry and the women chemists who came to Bangor all undertook research with him (including Phyllis McKie, see Chapter 15). Orton was a strong believer in the historical context of chemistry, and a quote by one of his obituarists is of note, as it mentions the book written by Ida Freund (see Chapter 3):

> Historical chemical characters were made to live. Who will ever forget the lecture on Mendeléeffs predication of the properties of the missing elements and their fulfilment by later discoveries, a discourse largely based on the account in Ida Freund's 'The Study of Chemical Composition' — a unique treatise for which Orton had the highest regard.[136]

---

[133] Angus, W. R. (1954). Schools of chemistry in Great Britain and Ireland–XI University College of North Wales, Bangor. *Journal of the Royal Institute of Chemistry* **78**: 291–298.

[134] Williams, J. G. (1985). *The University College of North Wales: Foundations 1884–1927*. University of Wales Press, Cardiff, p. 308.

[135] (a) "H.K." [probably Harold King, a former student of Orton's] (1931). Obituary Notices: Kennedy Joseph Previté Orton. *Journal of the Chemical Society* 1042–1048; and (b) "F.D.C." (1930). Obituary Notices: Kennedy Joseph Previté Orton, 1872–1930. *Proceedings of the Royal Society of London, Series A* **129**: xi–xiv.

[136] Note 135a, "H.K." p. 1043.

256   *Pioneering British Women Chemists: Their Lives and Contributions*

Alice Smith

One of Orton's students was Alice Emily Smith. Smith was born on 18 June 1871 in Louth, Ireland, daughter of Thomas Smith, Commission Agent of Warrenpoint, Co., Down, Ireland, and she was educated at Crescent House School, Bedford, Ireland.[137] She entered University College of North Wales, Bangor, in 1897 and completed a B.Sc. (London) in Chemistry in 1901. Smith was awarded an 1851 scholarship which she chose to use from 1901 to 1903 at Owens College, Manchester, where she worked with William Perkin Jr,[138] her research resulting in four substantial papers.

Smith returned to Bangor as Assistant Lecturer in Education and Assistant Lecturer in organic chemistry where she collaborated with Orton on reaction mechanisms, five papers resulting from her research. In 1914, she relinquished her position at Bangor, being succeeded by May Leslie (see Chapter 6). For the next 3 years she held an appointment as Lecturer in Science at the Maria Grey Training College, London (see Chapter 10).

Perhaps answering the wartime need for women chemists in industry (see Chapter 15), Smith left the Training College in 1917 to become a research chemist at Messrs Cooper's Laboratory. With the War ended and the male chemists reclaiming their pre-war posts, she lost her position. Smith returned to teaching in 1920, becoming Principal at a private school in Ilkley, Yorkshire. She held this position until her death in 1924.

Commentary

As in England, we find that in Scotland and Wales mentors played an important role for women chemistry students: James Irvine at St. Andrews;

---

[137] Anon. (1961). *Record of the Science Research Scholars of the Royal Commission for the Exhibition of 1851, 1891–1960*. The Commissioners, London; (December 1900). University College of North Wales, Bangor, *Magazine* p 38.; and *Student Records*, Archives, University of Wales, Bangor. E. W. Thomas, Archivist, is thanked for supplying a copy of this document.

[138] "J.F.T." (1931). Obituary Notices: William Henry Perkin, 1860–1929. *Proceedings of the Royal Society of London, Series A* **130**: i–xii.

George Henderson at the Royal Technical College, Glasgow; Frederick Soddy at Aberdeen and Glasgow; James Walker at Dundee and Edinburgh; Alexander McKenzie at Dundee; and Kennedy Orton at Bangor.

It seems that the smaller institutions were more receptive to the hiring of women faculty. Women chemistry graduates from St. Andrews, Dundee, and Bangor flourished and in some cases became the stalwarts of the department, such as Ettie Steele at St. Andrews, Isobel Smith at Dundee, and Alice Smith at Bangor. The exception was Christina Miller, under the mentorship of Walker, being appointed to an academic position at "big" Edinburgh.

Some of those who left St. Andrews, Dundee, and Bangor, were successful in small academic institutions in England: for example, Ishbel Campbell of St. Andrews became a Lecturer at University College, Southampton (Chapter 6); Mary Lesslie of Dundee, a Lecturer at Bedford College (see Chapter 5); Nellie Walker of Dundee, a Lecturer at Bedford College, before returning to Dundee (see Chapter 6); and Phyllis McKie of Bangor, being appointed a Lecturer at Bedford College, then at Westfield College, London (see Chapter 15).

Chapter 8

Domestic or Household Chemistry

Margaret Seward (22 January 1864–29 May 1929)

One avenue for employment for women chemistry graduates was the teaching of domestic chemistry at domestic science institutions. In Chapter 4, we described how the women's campus of King's College was transformed into a Home Science and Economics Department. In fact, there were many such departments and institutions, some strong in "real science" and others not. In doing so, we will document some of the

260 *Pioneering British Women Chemists: Their Lives and Contributions*

prominent women chemistry teachers at these institutions. But first, we will provide some context.

The Domestic Science Controversy

In Britain, during the 1880s and 1890s, there had been an increasing interest in the teaching of domestic subjects to girls.[1] Initially, it was domestic economy — cooking, laundry work, and so on. Educators thought such mandatory instruction was important for two reasons: first, it was believed that the squalor and drunkenness that prevailed among the lower classes could be prevented by education in "home-thrift" and economic cookery; and second, there was a fear of a shortage of domestic servants for upper-middle-class homes. There were two subsequent reports: the Interim Report on Housecraft in Girls' Secondary Schools in 1911 and the Consultative Committee on Practical Work in Secondary Schools in 1913. Both reports stated that, in the new "scientific age," the teaching of domestic subjects should have a strong foundation in science and become domestic or household science.[2] At the core of domestic science was chemistry: especially the chemistry of foodstuffs and household cleaners.

Women had only a few years earlier gained admission to university to take academic chemistry. As a result, a fierce debate arose in England among the first generation of women chemists and their supporters as to the type of chemical education most appropriate for young women. That is, should the next generation of girls learn "real" chemistry, which would continue to give them access to the same opportunities as men? Or should they learn domestic chemistry as a component of domestic science, which would enable them to undertake their role as wives and mothers in a scientific manner? The essentiality of the latter was pointed out in an article of 1903 in *Child Life*:

> A short time ago a very young bride was left, through illness, without her maids. She tried to manage the evening dinner, but it ended in her

---

[1] Dyhouse, C. (1977). Good wives and little mothers: Social anxieties and the schoolgirl curriculum, 1880–1920. *Oxford Review of Education* **3**: 21.

[2] Dyhouse, C. (2013, reprint). *Girls Growing Up in Late Victorian and Edwardian England*. Routledge, New York, pp. 166–167.

suggesting to her husband that they should go to a hotel, as she found that she could not roast a fowl or make gravy, and that potatoes were utterly beyond her.[3]

A leading proponent of domestic science for girls was Arthur Smithells[4] (see also Chapter 6) Professor of Chemistry at the University of Leeds. Smithells, who had given lectures at Manchester High School for Girls, was a strong champion of education for girls.[5] He saw domestic science as a means of bringing an applied aspect that would, in particular, be beneficial for women's roles in society.[6] Henry Armstrong (see Chapter 10) was another believer in the importance of teaching domestic science in schools, but to both boys and girls.[7] Armstrong wrote a series of articles in the *Journal of Education*, the last in 1934, expounding the importance of teaching household science including topics such as the chemistry of milk, vegetables, and even "silken stockings."[8]

Having fought so hard for getting girls an academic education equal to that of boys, many women scientists saw domestic science as a reversal of those gains, limiting girls' aspirations and opportunities to that of domesticity. Ida Freund, Lecturer in Chemistry at Newnham College (see Chapter 3), was one of the most vociferous opponents of the teaching of science to girls through the context of domestic science. In particular, she authored a lengthy denunciation in the feminist publication, *The Englishwoman*:

> It was erroneous to think that through the study of the scientific processes underlying housecraft and especially cookery, you can teach science, that is, give a valuable mental training which should enable the

---

[3]"E. E. L. B." (1903). Is the teaching of cooking desirable in girls' schools? *Child Life* **5**(17): 37.

[4]Raper, H. S. (1940). Arthur Smithells. 1860–1939. *Obituary Notices of Fellows of the Royal Society* **3**: 96–107.

[5]Flintham, A. J. (1977). The contributions of Arthur Smithells, FRS, to science education. *History of Education* **6**: 195.

[6]Bailiss, R. A. (1976). Flexner, Smithalls and home economics. *Journal of Educational Administration and History* **8**(2): 9–18.

[7]Bailiss, R. A. (1983). Henry Armstrong and domestic science. *Journal of Consumer Studies and Home Economics* **7**: 299–305.

[8]Armstrong, H. E. (1934). Silken stockings. *Journal of Education* **66**: 386–488.

pupils in after life to judge whether an alleged connection between effect and cause has been established or not.[9]

Most of the influential Headmistresses of girls' schools similarly opposed the introduction of domestic science. For example, Lilian Faithful, then-Principal of Cheltenham Ladies College (CLC) concurred:

> The foundations of a knowledge of chemistry and physics should be built up on a well-ordered system which must not be subordinated from the outset to the requirements of home science. The teaching of science during the school years should be such as to prove equally useful to the pupil who elects to take at a later stage a university course in science and to the pupil who enters upon the home science course.[10]

In terms of the chemistry component, there were two parallel threads to the debate: the type of chemistry taught at girls' secondary schools and the offering of courses in domestic chemistry at colleges, polytechnics, and universities. Catherine Manthorpe has provided a detailed discussion of the former,[11] but the latter, in particular, the chemistry content of domestic science programs in higher education, has not previously been researched.

The debate about the college teaching of domestic chemistry is illustrated by an exchange in 1911 initiated by Lucy Hall and Ida Grünbaum, Science Lecturers at Avery Hill [Teachers] Training College, Eltham. They contended that incoming women students in domestic science programs required only very basic chemistry before being taught household chemistry:

> Before "domestic" chemistry can be introduced with profit, they [college students] must understand the composition of air and water and the nature and reactions of acids, bases, and salts. In the short time at our

---

[9]Freund, I. (1911). Domestic science — A protest. *The Englishwoman* **10**: 147.

[10]Faithfull, L. M. (1987). Home science. In Spender, D. (ed.), *The Education Papers: Women's Quest for Equality in Britain, 1850–1912*. Routledge & Kegan Paul, New York, p. 326.

[11]Manthorpe, C. (1986). Science or domestic science? The struggle to define an appropriate science education for girls in early 20th century England. *History of Education* **15**(3): 195–213.

disposal we do not think that chemical formulæ and equations can be explained with any advantage, nor do we consider such explanation absolutely necessary. When the effects of air and of water on ordinary substances have been grasped, the methods of cleaning such substances can be deduced and practiced on all the available household appliances. The lessons on natural waters teach the methods of softening and make an introduction to the chemistry of laundry work.[12]

Among the respondents was Hilda J. Hartle (see Chapter 10) of Homerton College, Cambridge, another teachers' training college. Hartle was opposed to the whole concept of domestic science, arguing that it did not have a basis in science. She pointed out:

> The science of cookery and of laundry work is yet in its infancy. No literature of the subject exists. Not even the most brilliant organic chemist can be said to "know" the chemistry of foods, still less can such a subject be within the grasp of students in training.[13]

Teaching of Domestic Chemistry

Despite the opposition, the teaching of domestic science thrived in some English institutions of higher education for many decades. There was a diversity of programs from practical to academic. At the academic end was the B.Sc. in Domestic Science offered jointly between Bristol University and the Gloucestershire Training College of Domestic Science.[14] First offered in 1926, a regular science curriculum was followed for the first 3 of the 4 years, the only specific additions being the chemistry of foods in the third year and special reference to bacteriology in second-year biology. Employment for the graduates of the program was seen as for: "... educated women capable of the efficient and

---

[12]Hall, L. and Grünbaum, I. (1911). Correspondence: Housecraft in training colleges. *Journal of Education* **33**: 678.

[13]Hartle, H. J. (1911). Correspondence: Housecraft in training colleges. *Journal of Education* **33**: 849.

[14]Bird, E. (1998). High class cookery: Gender, status and domestic subjects, 1890–1930. *Gender and Education* **10**(2): 117–131.

264 *Pioneering British Women Chemists: Their Lives and Contributions*

scientific direction of college, school and hospital kitchens, industrial canteens etc."[15]

Here, we will contrast the rise and fall of the chemistry content of domestic science programs at four well-respected institutions of higher education in the London area. These programs were at Berridge House, a college for working-class girls; at two polytechnics with very different programs, both aimed at middle-class young women; and at King's College for Women, designed for upper-middle-class women students.

Domestic Chemistry at Berridge House, Hampstead

In the 1890s and 1900s, some colleges were established specifically to teach domestic subjects to girls.[16] The women students were primarily recruited from the lower classes of society and many, upon graduation, obtained employment as maids with "fine families." The emphasis at these institutions was less on science than on domestic training in a "scientific manner." For example, Elizabeth Atkinson, teacher at the Manchester Municipal Training College of Domestic Economy, described in her book, *The Teaching of Domestic Science*, that a course of laundry-work should contain theoretical and practical studies on the laundry roles of starch, bran, water, soap, soda, salt, bleaching, patent cleaners, stain-removing, and paraffin wash.[17]

The most renowned institution of this type was the Training College of Domestic Subjects, Berridge House, Hampstead. Berridge House was opened in 1909 by the National Society for Promoting Religious Education. In 1911, the magazine, *Girls' Realm*, devoted a whole article to the Domestic Science program at Berridge House. Besides the more traditional domestic science topics of cookery, needlework, and housewifery, the magazine lauded the chemistry component of the program:

---

[15]Gloucestershire Training Prospectus (1902). Cited in Note 14, Bird, p. 124.

[16]Turnbull, A. (1994). An isolated missionary: The domestic subjects teacher in England, 1870–1914. *Women's History Review* **3**: 81.

[17]Atkinson, E. (1938). *The Teaching of Domestic Science*. Methuen & Co., London, 2nd edn., p. 60.

Specially interesting is the laboratory, where the students actually make their own tests, classify foodstuffs, ascertain the chemistry of bread-making, the composition of soap, the properties of starch, borax, soda, etc., as applied in washing and naturally manufacture for themselves such household commodities as baking powder and furniture polish according to their own tried formulæ.[18]

Berridge House was proud of its well-equipped Science Laboratory, and it was the first Domestic Science Training College in Britain to appoint a Lecturer with a science degree, Miss Marshall. From 1911 onwards, she took the students annually to a soap manufacturing company. The students watched each of the steps involved in producing the different types of soap. One of the students added: "In the Chemistry Lab we saw the experiments for testing the purity of soaps, and also saw growth of disease germs and action of disinfectants..."[19]

In 1964, Berridge House was merged with St. Katherine's, Tottenham, to form The College of All-Saints, Tottenham. The Berridge House site was closed. The combined institution became a teachers' training college offering home economics and general science, but the domestic science program never survived the merger.[20]

Domestic Chemistry at South-Western Polytechnic

The undefined nature of "domestic science" meant that the chemistry component at each polytechnic differed considerably and also varied over time at any particular institution. We have chosen to contrast the domestic chemistry content at South-Western Polytechnic (Chelsea College, as of 1922) and at Battersea Polytechnic Institute, the former being chemistry-deficient and the latter being chemistry-rich. We will leave the discussion of Battersea until after King's as many of the King's chemistry staff later moved to Battersea.

---

[18] Bingen, L. (1911). Careers for girls: XIII-Domestic science. *Girls' Realm* **13**: 887.

[19] Anon. (1911). College notes. *The Berridge Magazine* (3): 10 (accessed at the London Metropolitan Archives, file: GB 0074 ACC/0900).

[20] Handley, G. (1978). *The College of All Saints: An Informal History of One Hundred Years, 1878–1978*. John Roberts Press, London.

266 *Pioneering British Women Chemists: Their Lives and Contributions*

In 1899, the offering of domestic science at South-Western Polytechnic was noted in the journal, *Nature*. This discussion was in the context of new diploma offerings aimed at middle-class women. It reflected the growing scarcity of domestic workers who were finding less onerous employment elsewhere:

> In this connection may be cited the work now being done on the women's side of the institute in the direction of offering ladies of the middle classes such instruction in domestic science as will make them independent of servants.[21]

Whatever the views of the author, such training also opened up new opportunities for the employment of middle-class women as supervisors in domestic and catering organizations.

The diploma program at South-Western Polytechnic became the autonomous School of Home Training in Domestic Science in the 1903–1904 academic year.[22] In its second year of existence, the program included a course titled "Household Chemistry" consisting of 25 lectures. By 1909–1910, the chemistry content had decreased and the course had been renamed "Household Science." During the 1920s, that course also disappeared to be replaced by one titled "Applied Electricity" later renamed "Domestic Electricity."

In the 1913–1914 year, the school had also changed its name to the School of Training in Housecraft and Household Management. Nevertheless, the near-science-less Domestic Science Department continued on until the 1940s when there was increasing pressure for the college to discontinue non-degree programs. As the anonymous biographer of Chelsea College noted: "The domestic science department was the first to go, in 1949, to provide space for pharmacy; vocational work was transferred to Battersea [Polytechnic], and non-vocational work to a women's institute."[23]

---

[21] Simmons, A. T. (1899). The South-Western Polytechnic. *Nature* **61**: 208.

[22] Information obtained from the collection of South-Western Polytechnic and Chelsea College Calendars held at the King's College, London, Archives, file: GB 0100 KCLCA.

[23] Anon. (1977). *Chelsea College — A History*. Chelsea College, London, p. 49.

Domestic Chemistry at the Women's Department of King's College, Kensington

In Chapter 4, we described how the King's College for Women, University of London, was located not on the Strand with King's University itself, but in Kensington. It was in 1907 that a proposal was made for a Department of Household Science. The aim was to attract upper-middle-class women who would, upon graduation, become high school teachers of domestic science.[24]

A report in the *King's College Magazine, Women's Department* noted the major role of Arthur Smithells:

> At the second meeting [on the Scheme] Prof. Smithells, of Leeds University, was the chief speaker. He may be called the pioneer of this movement, as he was the first to start special classes in practical science for teachers of Domestic Economy and others in Leeds. He has been appointed Honorary Advisor to the Committee in charge of the King's College Scheme.... He was led to see the importance of this form of training by being asked to assist in the future teachers of Cookery and other Domestic Arts in the ordinary branches of physical science.[25]

The newly formed department offered a 3-year program, initially as a College Certificate. The chemistry instructor of the time, Margaret McKillop (see Chapter 4), wrote an enthusiastic account of the program and of its possible conversion to full degree status (which occurred in 1921):

> There is no doubt that the idea of the possible new degree, with as good a standing as that to which engineering and agriculture have now established their claim, is gaining ground with most people. Meanwhile headmistresses have begun to ask, much too early for our present

---

[24] Anon. (1908). The higher education of women in home science and household economics. *Journal of Education* **30** (new series): 380; and Anon. (1908). College notes. *King's College Magazine, Women's Department* (33): 3, K/SER1/170.

[25] Anon. (1908). The home science and economics scheme. *King's College Magazine, Women's Department* (33): 8–9, K/SER1/170. See also: Flintham, A. J. (1975). Chemistry or cookery? *Housecraft* **48**: 105.

268 *Pioneering British Women Chemists: Their Lives and Contributions*

achievements, for the 'new sort of domestic science teacher.' They mean, or ought to mean, someone who teaches science with constant reference to home life, a practical-minded woman who can also be a good form mistress and bring a little college atmosphere; but at present, it is true, they are a little inclined to expect a first class chemist combined with a first class cook, who can also take odd sciences and other subjects throughout the school! There is no doubt that many girls' schools are going to have Domestic Science now put right into the ordinary curriculum instead of being left as a top-dressing for a possible (but unusual) last year.[26]

The chemistry content of the program was very strong as exemplified by the requirements in the *1912–1913 King's College, Women's Department Calendar*: First Year General Chemistry (60 lectures and 120 hours of practical work); Second Year Organic Chemistry (60 lectures and 150 hours of practical work); and Third Year Applied Chemistry (60 lectures and 180 hours of practical work). The Applied Chemistry course consisted of the following:

The constituents of the atmosphere and methods of estimation — water analysis with special reference to its use for drinking purposes, cooking, and in the laundry — the constituents of foods, adulterants, and preservatives, with a value to determining their wholesomeness — the chemistry of cooking and of the materials used in cooking — the chemical changes caused by organized and unorganized ferments, applied to the preparation, preservation, and deterioration of foods and to digestion — the chemistry of laundry work and other cleansing processes — the nature and quality of textile fabrics in common use; the physical and chemical properties of their constituent fibres — disinfectants and antiseptics — scientific principles underlying the care and preservation of the chief materials used in the structure and equipment of a house.[27]

---

[26] McKillop, M. (1909). Notes and news. *Somerville Students' Association*, 22nd Annual Report and Oxford Letter 47.

[27] Information obtained from the collection of King's College for Women Calendars held at the King's College, London, Archives, file: GB 0100 KCLCA.

The Household Chemistry Controversy at King's

As mentioned earlier in this chapter, the teaching of domestic science to girls was a very controversial topic. There was a very specific debate on the teaching of "household chemistry." Christina Bremner, famed advocate of female education,[28] believed that graduates from this program would have excellent employment possibilities in hospitals, schools, and other public organizations, and, of course, such graduates would excel at scientific homemaking. She described the chemistry component:

> Students of chemistry must learn to perform simple analyses, to study hydrocarbons, alcohols, acids, and so forth, so that in the final year they may deal effectively with water analysis, constituents and relative values of different foods, the chemical changes of ferments, preservation and deterioration of food, purity of milk, and so forth.[29]

Bremner assailed those who argued that only "pure" or "men's" chemistry should be taught to women students:

> It would be interesting to know precisely how far feminism and opposition to a Domestic Science course in a University coincide. I cannot think the lines of demarcation correspond perfectly, for I have known advanced feminists, and count myself amongst them, who for years have bitterly complained that so little of the money devoted to technical training has been spent on women, and also how very lacking in thoroughness have been many domestic science courses carried on all over the country.[29]

The pro-and-con divide did not, indeed, correspond perfectly to the division among women in society. Some feminist chemists, such as Ida Smedley Maclean (see Chapter 9), supported the teaching of domestic studies on a scientific basis.[30]

---

[28] Bremner, C. S. (1897). *Education of Girls and Women in Great Britain*. Thompson Press, London (reprinted 2009).

[29] Bremner, C. S. (1913). King's College for Women: The department of home science and economics. *Journal of Education* **35** (new series): 72.

[30] Maclean, I. S. (1913). Household science in the Universities. *Girls' Own Annual* **35**: 424.

270   *Pioneering British Women Chemists: Their Lives and Contributions*

The optimistic and rosy view of the King's program expressed by Bremner was challenged by Rona Robinson. Robinson, a chemistry graduate, and at the time a Gilchrist Postgraduate Scholarship holder at King's College for Women, wrote a fierce rebuttal, first of all noting that Bremner's rosy description of the program was based on a one-day visit. In particular, Robinson also took the college to task for claiming that the first- and second-year chemistry was strong enough to provide sufficient theory for the third-year Applied Chemistry:

> Such 'applied Chemistry' is far beyond the reach of beginners in science, and it is nothing short of charlatanry and deception on the part of the authorities to state that they teach anything of this nature. To talk of the students *applying* the knowledge of such matters in the third year is to apply knowledge which they do not possess. The student who is going to work on the *chemistry of foodstuffs* would have first to do an amount of pure chemistry that would shatter the whole curriculum of this course.[31]

Smithells responded to Robinson's attack on the program:

> I think it is hardly necessary to assure your readers that the somewhat elementary educational questions raised by Miss Robinson have not escaped the notice of those who are responsible for the course. We have had many difficulties to face and still have problems to solve; we shall, no doubt, continually mend our ways. But the suggestion that the courses at King's College for Women are superficial or unsound scientifically is one that I am sure would not have been made had Miss Robinson continued her studies.[32]

Despite the criticisms by Robinson, the King's program prospered. We have an account of the experiences of a domestic science student, Lucy Smart, taking the first-year chemistry course:

---

[31] Robinson, R. (1913). The department of home science and economics, King's College for Women. *Journal of Education* **35** (new series): 284.

[32] Smithells, A. (1913). The department of home science, King's College for Women. To the Editor. *Journal of Education* **35** (new series): 313.

On other days we are startled by flames of burning ether and explosions in treacle tins — during the so-called Chemistry lecture. After spending several hours staining our hands in trying to detect arsenic, we are allowed to go to the 'Workhouse', where we learn how to remove the same stains and how to wash woollens.[33]

Another student, Susan Lovell, commented: "There is far more Science attached to the Household than one would think. Physics, Chemistry and Biology take a far more prominent place during the first year than Household Arts."[34]

As mentioned in Chapter 4, the Arts and Science Departments of King's College for Women were transferred to the Strand in 1915. The surviving portion, the Household and Social Science Department of King's College for Women became completely independent on a new site at Camden Hill Road, Kensington. This orphan unit became the King's College of Household and Social Science (KCHSS). During the inter-War period, there was a steady demand for graduates of the program to take positions in hospital dietetics.[35]

In 1953, KCHSS was renamed Queen Elizabeth College. Along with a change in name, came a broadening of mandate, including the formation of departments in each of the pure sciences. That same year, the B.Sc. (Household and Social Science) was replaced by two separate degrees: B.Sc. (Household Science) and B.Sc. (Nutrition).[36] The Nutrition degree was far more popular among incoming students. As a result, in 1966, the Household Science Department was renamed the Department of Food and

---

[33] Smart, L. (1931). News from the Universities: King's College of Household and Social Science. *Croydon High School Magazine* (42): 58.

[34] Lovell, S. (1943). News from the Universities: King's College of Household and Social Science. *Croydon High School Magazine* (45): 68.

[35] Blakestad, N. (1996). King's College of Household Science and the origins of dietetics education. In Smith, D. F. (ed.), *Nutrition in Britain: Science, Scientists, and Politics in the Twentieth Century.* (Studies in the Social History of Medicine), Routledge, 1997, pp. 75–98.

[36] Marsh, N. (1986). *The History of Queen Elizabeth College: One Hundred Years of Education in Kensington*, King's College, London, pp. 231–232.

272 *Pioneering British Women Chemists: Their Lives and Contributions*

Management Science. Domestic science, and domestic chemistry, in particular, was no more.

The Chemistry Department at King's College for Women

As we mentioned in Chapter 4, it was Margaret Seward (Mrs. McKillop) who was the first chemistry Lecturer at King's College for Women. By 1912, the number of Home Science students exceeded the number of B.Sc. women Chemistry students. At this time, Henry L. Smith was appointed full-time Lecturer in Chemistry and Helen Masters was appointed Demonstrator.[37]

It was with the move in 1921 to the new site on Campden Hill that Charles Kenneth Tinkler[38] was appointed to take over the department. It was Tinkler, together with Masters, who oversaw the real development of the program, as his obituarist, Agnes Browne explained: "In collaboration with Miss Helen Masters he [Tinkler] created a course in Applied Chemistry unique in its scope and to the present day regarded as a model in Home Science departments throughout the world."[39] The content of the course was published in 1926 as a two-volume set, *Applied Chemistry*.[40] This text became the standard reference work on analytical procedures for chemistry related to the home, and it was still being reprinted in 1948, while the first American edition was published in 1950.

In 1917, Tinkler appointed Marion Soar as a second Assistant Lecturer and Demonstrator. About the same date, Phyllis Garbutt was added to the Department as a Research Worker and Assistant. Garbutt resigned in 1920 to accept a position at Battersea Polytechnic (see the respective section) as did Soar in 1921 (see the respective section). Soar's successor was Agnes Browne. Masters followed Garbutt and Soar to Battersea in 1924.

---

[37] Anon. (1912). College notes. *King's College Magazine, Women's Department* (46): 4, K/SER1/170.

[38] Jackman, A. (1952). Charles Kenneth Tinkler 1881–1951. *Journal of the Chemical Society* 1191–1192.

[39] Note 38, Jackman, p. 1192.

[40] Tinkler, C. K. and Masters, H. (1926). *Applied Chemistry: A Practical Handbook for Students of Household Science and Public Health,* Vol. I, *Water, Detergents, Textiles, Fuels, etc.,* Vol. II, *Foods.* The Technical Press, London.

With Tinkler becoming ill and Masters having departed, Browne was temporarily appointed as Lecturer in Chemistry, the position only being made permanent in 1944. In 1938, Mary Thompson was hired as Demonstrator and Assistant Lecturer.

Agnes Browne (Mrs. Jackman)

Agnes (Lillie) Browne,[41] daughter of John Browne, a farmer in Londonderry, Ireland, and Jane Eakin, was born on 31 August 1896 at Moneymore, Ireland. Educated at Banbridge Academy, Banbridge, Ireland, she entered the Royal College of Science, Dublin in 1915 and graduated with a degree from the University of London in 1919. After graduation, Browne remained there for a year doing research in organic chemistry, supported by a grant from the Department of Scientific and Industrial Research (DSIR), under the supervision of A. G. G. Leonard[42] and W. E. Adeney.[43] In 1920, they arranged for Browne to do research with Frederick G. Donnan at University College, London. According to her daughter, the crystallographer, Mary Truter (née Jackman): "She [Browne] was unhappy there partially because London was a lonely unfriendly place compared with Ireland and partially because she had no idea what she was supposed to do in the lab."[44]

Browne was appointed to King's College of Household and Social Science as Assistant Lecturer in Chemistry in December of 1921. In April 1924, Browne married Douglas Jackman, who she had met in the Chemistry Department at University College. Truter commented: "[She] did not give up her job to the surprise and annoyance of several people.

---

[41] Truter, M. R. (1987–1988). One of the First Women Fellows of the Chemical Society Agnes Jackman (née Browne). *Journal of the Chemical and Physical Society, University College* 12–13.

[42] Thornton, H. D. (1967). Obituary: Alfred Godfrey Gordon Leonard, 1885–1966. *Chemistry in Britain* **3**: 224.

[43] Anon. (1935). Obituary. Walter Ernest Adeney, *Journal and Proceedings of the Institute of Chemistry of Great Britain and Ireland* **59**: 326.

[44] Personal communication, letter, M. R. Truter (Lady Cox, daughter of Jackman), 31 December 1999.

274 *Pioneering British Women Chemists: Their Lives and Contributions*

Pregnancy forced her to resign in September 1925. ... There followed nine months of boredom and loneliness ..." (see Footnote 44).

However, Browne was asked to return in October 1926 on a temporary basis because Tinkler was ill and Masters had resigned to take up the position at Battersea, so there was no one left with the qualifications to teach the specialised domestic chemistry courses.

Browne remained on annual appointments until February 1944 when, at last, the position was made full time. Truter described how hard the War period was for Browne:

> In 1939 the College was moved was evacuated to Cardiff. ... After two terms the College moved to Leicester. Mother spent the week in digs [rented accommodations] and the weekends in London with her husband and daughter [Truter]. There she worked hard cooking to keep them going for the week; as food was rationed this required thought as well as time (see Footnote 44).

While at the college, she co-authored a text on *The Principles of Domestic and Institutionalised Laundry Work*.[45] Browne retired in 1956 at the rank of Senior Lecturer in Chemistry at what had now become Queen Elizabeth College, and she died on 6 November 1978.

Mary Thompson (Mrs. Clayton)

Born on 20 December 1910 at Herne Hill, Surrey, Mary Christina Thompson[46] was the daughter of William George Thompson, an accounts clerk, and Elsie Rudd. She was educated at James Allen's Girls' School, Dulwich and entered Bedford College in 1930. Thompson completed her B.Sc. in 1933, then undertook research with Eustace Turner.[47]

---

[45] Jackman A. and Rogers, B. (1934). *The Principles of Domestic and Institutionalised Laundry Work*. E. Arnold and Co., London.

[46] (a) (1933). Certificates of Candidates for Election at the Ballot to be Held at the Ordinary Scientific Meeting on Thursday, December 7th, *Proceedings of the Chemical Society* 72; (b) *Student Records*, Archives, Bedford College.

[47] Ingold, C. (1968). Eustace Ebenezer Turner, 1893–1966. *Biographical Memoirs of Fellows of the Royal Society* **14**: 449–467.

Domestic or Household Chemistry 275

Just before Christmas 1933, Thompson had an accident in one of the Bedford chemistry labs. She was distilling benzene using a Bunsen burner when the benzene caught fire. She extinguished the flames and then, as she told Turner, with her clothes on fire:

> She threw herself on the floor 'as this was the proper thing to do', and, owing to her prompt action and that of Miss Cook and Miss Lockhart in wrapping her up in the fire blanket which was only a few feet away, Miss Thompson was saved from a much more serious accident.[48]

Following the accident, Thompson returned to her studies in May 1934, completing her Ph.D. in 1937. After which she took a position at Roedean School, but left after 1 year as she was expected to teach biology and did not know enough of that subject.[49] It was then that Thompson obtained her position at King's where she remained for many years. She married David Clayton in 1956, while in 1966, she was noted as living in Middlesbrough where she was employed as a school teacher.

Domestic Chemistry at Battersea Polytechnic Institute

By contrast with South-Western Polytechnic, the chemistry component of the domestic science program at Battersea Polytechnic Institute was much stronger. Battersea Polytechnic (see Chapter 4) introduced a School of Domestic Economy for Girls, and a Training School for Domestic Economy in the early 1890s.[50] The school subsequently became the Department of Domestic Science.

From its very inception, the chemistry of food and cookery was a significant part of the syllabus. The *Battersea Polytechnic Magazine* reprinted an article from the British women's weekly, the *Gentlewoman*, lauding their domestic science program:

---

[48] (19 December 1933). Letter, E. E. Turner to Miss Monkhouse, Bedford College Archives (held at Royal Holloway College).

[49] *Student Records*, Archives, Bedford College.

[50] Watson, L. (2 July 1904). The Battersea Polytechnic. *Girl's Own Paper* **25**: 628.

276 *Pioneering British Women Chemists: Their Lives and Contributions*

> One of the most thorough and up-to-date establishments for training in the science of domesticity is the Women's Department of the Battersea Polytechnic, Battersea Park Road, which is staffed by highly trained teachers under the control of Miss M. E. Marsden. Thither flock girls from all parts of the world, even from South Africa and Japan, and many of them, especially those who intend to follow domestic science as a profession, take the three-year course.... Special stress is laid on the scientific principles underlying household processes, and the work of the kitchen and laundry is co-ordinated with that of the scientific laboratory and the lecture-rooms.[51]

In previous chapters, we have described the rigid rules required of women students in residence and such rigid requirements also existed at Battersea. A hostel was established in 1910 mainly for women students of the Domestic Science Training College. The *Battersea Polytechnic Magazine* reported in 1911 on the strict daily regime:

> The day in the Hostel begins with morning prayers at 7.45. Breakfast is served at 8 a.m. and by 9 o'clock the students have all left to begin their day's work at the Polytechnic. By 4.30 p.m. the hum of student life is again heard in the hostel where tea is served till 5.30 p.m. This meal is made as informal as possible. The study room is available for study till 6.30, when the dressing bell rings, and at 7 o'clock dinner is served. After dinner students who wish to resume their study may do so, while others engage in social intercourse, music, and other occupations. At 10 o'clock all public room lights are extinguished and by 10.30 bedroom and cubicle lights are out... On Sunday it is expected that each student shall attend a place of worship.[52]

Chemistry in the Department of Women's Subjects

In the early decades, there were six departments at Battersea Polytechnic: Mechanical Engineering and Building Trades, Electrical Engineering and

---

[51] Anon. (May 1910). Domesticity without dulness (sic). *Battersea Polytechnic Magazine* **2**(11): 43 (U. Surrey Archives, BA/CR/2/2/2).

[52] Anon. (October 1911). Hostel Life. *Battersea Polytechnic Magazine* **4**(19): 63 (U. Surrey Archives, BA/CR/2/2/4).

Physics, Chemistry, Women's Subjects, and Art and Music. Thus, it would appear that the chemistry staff of the School of Domestic Economy were autonomous or independent of the Department of Chemistry. In fact, in the obituary of the long-time Chair of the Chemistry Department, Joseph Kenyon,[53] no mention is made of the chemists of the Domestic Science program.

By 1919, in addition to traditional general and organic chemistry courses, a course "Chemistry as Applied to Household Processes" appears, containing the following topics:

Air. Water. Chemical theory. Acids, alkalis and salts. Carbon and its oxides; fuels. Soaps. Textile fabrics. Water softeners. Sugars, starch, alcohol, acetic acid. Proteins. Fats. Vitamines. Yeasts, moulds, and bacteria. Study of certain foods. Preservation and sterilisation of food stuffs. The practical work will be partly illustrative of the lectures and partly experimental craft work, i.e.:-

Experimental Housewifery — Study of metals, causes of tarnish, metal polishes and preservers, stainless cutlery. Study of woods, dry rot, furniture polishes, stains, paints and varnishes. French polish. Lacquers. Care of leather. Materials used in making floor coverings, and scientific reasons for methods of cleaning and preserving them. Household disinfection.

Experimental Laundrywork — Comparative value of methods of softening water for laundry purposes. Study of detergents and their action on textile fabrics. Methods of testing fabrics, and the reactions of laundry reagents on them. Experimental removal of stains; bleaching and dyeing. Laundry blues. Microscopic and chemical examination of starches. Disinfection of clothing.

Experimental Cookery — Examination of the chemical and physical natures of various foodstuffs, e.g., flour, fat, fish, meat, eggs, vegetables, pulses, milk. The effects of heat, and of different methods of cooking on these food stuffs. Study of yeast and its action on bread making.

---

[53]Turner, E. E. (1962). Joseph Kenyon, 1885–1961. *Biographical Memoirs of Fellows of the Royal Society* **8**: 49–66.

278 *Pioneering British Women Chemists: Their Lives and Contributions*

Examination of sugar substitutes. Experiments to attempt the solution of problems encountered in the kitchen.[54]

As elsewhere, the chemistry students resorted to poetic verse to express their thoughts. Parts of a lengthy and poignant poetic verse are given in the following:

Oh, Chemy Lab. of strange renown! compound of bliss and woe!
Where the soporific student spends her day
'Mid retorts that are not verbal (though we oft could wish them so),
And experiments that lead the mind astray
Yet often, 'mid the harassments of those who rashly try
To train the idea to shoot with skill,
We call to mind your labours, with a half-regretful sigh,
And wish we were experimenting still.

For we proved the utter hollowness of all that men aver,
And we faced malicious milkmen with their sin;
And we traced their interference with the sly lactometer,
And then we grubbed about for formalin.
We heated up the butter, and we noted with a smile
Its behaviour much resembled margerine's;
And we hunted for our protein in a scientific style,
And tracked it to its lair in our beans.

We analysed the powders that have made themselves a name
As equal to the treasure of the hen;
And we proved they all were liars, and we left them to their shame
As mere starches and bicarbonates; and then
We rushed on meaty extracts, and we stripped them of their pride,
And we laid the boast of beef-tea in the dust;
And the glory of the rice-grain found we chiefly in the hide,
While the staple of the loaf was in the crust.

---

[54]Information obtained from the collection of Battersea Polytechnic Institute Calendars held at the University of Surrey Archives, file: (U. Surrey Archives, UA/PN/5/20).

And we made our soap and measured it and analysed it out,
And we filled the Lab. with devastating whiffs,
And we choked us with formaldehyde, and placed beyond a doubt
The identity of chlorine by our sniffs.
And we followed up the starch through a wilderness of ways,
And puzzled out its destiny and change
To dextrin and to maltose and to glucose, through a maze
Of phenomena both strenuous and strange.
. . . . .

Well, our little heads grew swollen with the wisdom of the
spheres,
And we mused a secret grievance (known to none)
'Gainst the people who had flourished in the unhygienic years,
Ere the soaring flights of science had begun;
When they didn't work out diets, and their minds were fancy free
Of the prevalence of protein in their 'prog';
And they lived in unsuspiciousness of tannin in the tea,
And furfural in hiding in their grog;

When they thought the air an element, and didn't know a jot
Of the mechanism of their own inside —
And in spite of all this ignorance, they lived (as they ought not,
For if they'd done their duty they'd have died).
Well, a century from hence, no doubt, posterity will speak
Of "those grandmas who existed in the dark,
Whose experiments were faulty and whose theory was weak,
And who thought they knew some science — what a lark!"[55]

The continued strength of the chemistry content at Battersea from 1919 until 1948 seems to have been the exception among domestic science programs. It is of note that all the Domestic Science chemistry staff throughout the program's history were women. Claudia McPherson was the senior chemistry instructor from 1915 until 1948 and every year the

---

[55]"L.P." (July 1917). Oh Chemy Lab. *Battersea Polytechnic Magazine* **9**: 111–112 (U. Surrey Archives, BA/CR/2/2/9).

280 *Pioneering British Women Chemists: Their Lives and Contributions*

junior instructor(s) were also women; specifically Marion Soar and Phyllis Garbutt. In addition, from 1926 until 1948, the Head of the Department of Domestic Science was a woman chemist: Helen Masters. Both Masters and McPherson retired in 1948, and it seems quite probable that the survival of a strong component of domestic chemistry until that year was the result of their influence.

In 1948, the Department of Domestic Science became a separate entity: the Battersea College of Domestic Science. Thereafter, the syllabus no longer included any specific mention of chemistry, instead there was a course "Science, Physiology, and Nutrition." In 1963, the college was transformed into the Battersea Training College for Primary Teachers, offering courses leading to a Teachers' Certificate with special reference to domestic subjects.[56]

Claudia Cox (Mrs. McPherson)

Claudia Evelyn Cox was born on 21 January 1898, daughter of John Cox, house decorator, and Ada Glade. She entered The Ladies' Department, King's College, London in 1906, graduating in 1909 with a B.Sc. in Chemistry and Physics. Cox taught in a secondary school until 1918, when she was appointed to the Domestic Science Department of Battersea Polytechnic, the same year that she married Graeme McPherson.

Marion Soar

Marion Crossland Soar,[57] daughter of Henry James Soar and Harriett Cobden, was born on 2 March 1895 at Streatham, London. She was educated at the County School for Girls, Bromley, Kent. Soar entered

---

[56] Arrowsmith, H. (1966). *Pioneering in Education for the Technologies: The Story of Battersea College of Technology 1891–1962*. University of Surrey, Surrey.

[57] (1920). Certificates of Candidates for Election at the Ballot to be Held at the Ordinary Scientific Meeting on Thursday, December 2nd, *Proceedings of the Chemical Society*, 97; and *Student Records*, Archives, University of Wales, Bangor. We thank. E. W. Thomas, Archivist for this information.

University College of North Wales, Bangor, in 1913 and graduated with a B.Sc. in Chemistry in 1917. Upon graduation, she accepted the offer of a position as Assistant Lecturer in Chemistry at KCHSS.

In addition to teaching, Soar undertook research with Tinkler, her most famous publication being in the *Biochemical Journal* on the formation of ferrous sulphide in eggs during cooking.[58] As described in the previous section, she resigned in 1921, accepting a position as Lecturer in Chemistry at Battersea Polytechnic. Nothing could be found on her later life. Soar died in 1979 at Hastings, Sussex.

Helen Masters

Helen Masters[59] was born on 26 March 1887 at Hammersmith, London to John A. Masters, Surgeon Physician, and Catherine Simpson. She entered King's College, Women's Department, as a student in 1906. After completing a B.Sc. in Applied Chemistry in 1909, she stayed for a year of postgraduate studies in Applied Chemistry, Domestic Arts, and Practical Chemistry.

For the 1910–1911 year, Masters was Demonstrator in Physics at Cheltenham Ladies College; then, in 1911, she was appointed Demonstrator in Applied Chemistry back at King's in Kensington. As well as teaching, Masters undertook a range of research at King's related to food chemistry, resulting in at least eight publications. The first two papers were with Henry L. Smith; four on her own, two of which were reports on the solubilisation of lead from lead-glazed casseroles, and one each with Phyllis Garbutt (see the next section) and Marjory Maughan, two research students at the College.

---

[58]Tinkler, C. K. and Soar, M. C. (1920). Formation of ferrous sulfide in eggs during cooking. *Biochemical Journal* **14**: 114–119.

[59]Anon. (1910). College notes. *King's College Magazine, Women's Department* (38): 21 and (39): 17; Anon. (1910). Appointments. *King's College Magazine, Women's Department* (40): 31; and Anon. (1911). Appointments. *King's College Magazine, Women's Department* (43): 27, K/SER1/170.

282 *Pioneering British Women Chemists: Their Lives and Contributions*

Becoming more and more interested in the chemistry of cooking, Masters spent the summer of 1924 visiting Household Science Departments in Canada and the United States. Her growing fascination with the subject led her to resign her position and take up an appointment as Head of Domestic Science at Battersea Polytechnic.

Arrowsmith's history of Battersea Polytechnic specifically mentions the hiring of Masters in 1926 as a key figure in the success of the Department:

> The Governors might well have feared that they could not possibly find anyone who could fill this important position [Head of Domestic Science] in the Polytechnic organization, but in Miss Helen Masters, B.Sc. (Lond.) they did. ... She had been on the staff of King's College for Women for 15 years, as a Lecturer in Chemistry in the Household and Social Science Department, and had published research on Cookery and Laundry work. She thus brought to the post at Battersea a training and experience, allied to a wide outlook on Domestic Science, which seemed to make her the right person to follow Miss Marsden. Miss Masters filled this post with great distinction and brought added lustre to the high reputation the Department already enjoyed.[60]

From 1933 until her resignation in September 1948, Masters was also Supervisor of the Battersea Polytechnic Hostel.

Phyllis Garbutt

Phyllis Louisa Garbutt[61] was born on 19 May 1892 at Hornsey, London to Henry John Garbutt, Bank Manager, and Bertha Maud Restarick. Little could be found on her education, just that she was listed as a Research Worker/Assistant at King's College for Women. As described earlier, Garbutt left King's in 1920, being appointed Assistant Science Lecturer at Battersea. She resigned her position in 1924 to become Principal of the

---

[60]Note 56, Arrowsmith, pp. 53–54.
[61](a) *Register*. Royal Institute of Chemistry, 1926 to 1948; and (b) Anon. (1971). Personal news: Obituaries. *Chemistry in Britain* **7**: 78.

Good Housekeeping Institute. In 1926, Garbutt co-authored a book, *Food Wisdom*.[62] She died in 1970 at Worthing, Sussex.

Commentary

In the late 19th and the first half of the 20th century, domestic chemistry was taught in some English institutions of higher education as part of domestic science programs. Women chemists were divided about the validity of "domestic chemistry." Nevertheless, for that period of time, the subject of domestic chemistry, usually taught by women chemists, existed.

Domestic science never did become defined as an independent science, nor did domestic chemistry. Over time, some of the domestic science departments tended to become orphaned from their parent institutions. Nancy Blakestad summed up the cause of the decline in household science:

> Household science had sought to work within modern scientific paradigms and to develop a new type of scientific expert, yet its interdisciplinary approach to social problems, based on a similar holistic notion of women's domestic roles, was equally subject to displacement by specialist experts as the twentieth century progressed.[63]

It was as if domestic science as a claimed science had become an embarrassment. Thus, by the end of the 1960s, this chapter in the history of chemistry for women had come to an end.

---

[62] Cottington-Taylor, D. D. and Garbutt, P. (1926). *Food Wisdom: A Book for the Housewife and Everyone Responsible for the Care and Preparation of Food.* Sir Isaac Pitman & Sons, London.
[63] Note. 35, Blakestad, p. 92.

Chapter 9

The Professional Societies

Ida Smedley (14 June 1877–2 March 1944)

These days we take for granted that scientific organizations are open to both men and women, but this was not always the case.[1] It is hard to realize that the admission of women chemists to chemical organizations was once a contentious issue. For example, in 1880, the American Chemical Society even held a formal Misogynist Dinner.[2]

---

[1] Noordenbos, G. (2002). Women in academies of sciences: From exclusion to exception. *Women's Studies International Forum* **25**: 127–137.
[2] Kauffman, G. B. (1983). The misogynist dinner of the American Chemical Society. *Journal of College Science Teaching* **12**: 381–383.

286 *Pioneering British Women Chemists: Their Lives and Contributions*

During the 19th and early 20th centuries in the United Kingdom, there were a number of organizations that catered to the professional and social needs of chemists, the two aspects overlapping in the male club culture of the time.[3] What is particularly interesting is the very wide range of responses by the organizations when the issue of the admission of women arose.

Each society treated the problem of the admission of women in a different way. In this chapter, we will focus particularly on the lives of the British women who led the fight for professional acceptance.[4] The women chemists could not have made any progress without the support and votes of male chemists. The key role of the men's campaigns for women's rights is often overlooked.[5] In this chapter, we will ensure that credit is given to these very supportive men who often faced outright hostility from their own sex and who were "traitors to the masculine chemistry cause."[6] The saga begins with the short-lived London Chemical Society.

London Chemical Society

Events started promisingly for women. The London Chemical Society seemed to take pleasure in noting that women had participated in its events. At a pre-inaugural lecture on 7 October 1824, it was reported that: "Several ladies were present, taking a warm interest in all that was said, encouraging the lecturer by their smiles, and ensuring order and decorum by their presence."[7]

---

[3] Gay, H. and Gay, J. W. (1997). Brothers in science: Science and fraternal culture in nineteenth-century Britain. *History of Science* **35**: 425–453.

[4] Part of this work has been published as: Rayner-Canham, M. and Rayner-Canham, G. (2003). Pounding on the doors: The fight for acceptance of British women chemists. *Bulletin for the History of Chemistry* **28**(2): 110–119.

[5] Strauss, S. (1982). *Traitors to the Masculine Cause: The Men's Campaigns for Women's Rights*. Greenwood Press, Westport, CT.

[6] Rayner-Canham, M. and Rayner-Canham, G. (2009). Fight for Rights. *Chemistry World* **6**(3): 56–59.

[7] Anon. (1824). The London Chemical Society. *The Chemist* **2**: 56.

At the subsequent inaugural lecture, it was mentioned that among the 300 persons attending, there were "a great many ladies."[8] The address was given by George Birkbeck,[9] who specifically welcomed the participation of women:

> It may not be out of place here to state, that chemistry is not only intended to be confined to *learned* men but not even to *men* exclusively. Hitherto, ladies have conferred the honour of their presence upon all our public proceedings; and we are extremely desirous, although it is not consistent with the present constitution of the Society, that they should hereafter become participators also, as members.[10]

Birkbeck continued his address by pointing out the contributions from the late 18th century of Elizabeth Fulhame (see Chapter 1) and Jane Marcet (see Chapter 1):

> That they are well qualified for pursuing this branch of science, I may adduce, as evidence, the very able Essay, by Mrs. Fulhame, 'On Combustion'.... In further evidence, I may adduce the 'Conversations on Chemistry', by Mrs. Marcet; which, as an interesting and instructive elementary work for the uninitiated, has never yet been equalled.[10]

Unfortunately, the London Chemical Society, with its women-supportive attitude ceased to exist shortly afterwards.[11]

Society for Analytical Chemistry

Women gained admittance to the Society for Public Analysts (later called the Society for Analytical Chemistry) without any problem. In 1879,

---

[8] Anon. (1824). The London Chemical Society. *The Chemist* **2**: 162.
[9] Lee, M. (23 September 2004). Birkbeck, George (1776–1841), physician and educationalist. *Oxford Dictionary of National Biography*. Oxford University Press (accessed 18 January 2019).
[10] Anon. (1824). The London Chemical Society. *The Chemist* **2**: 164.
[11] Brock, W. H. (1967). The London Chemical Society 1824. *Ambix* **14**: 133–139.

288 *Pioneering British Women Chemists: Their Lives and Contributions*

5 years after the founding of the organization, the comment was made in the society journal, *The Analyst*, that: "We are liberal enough to say that we would welcome to our ranks any lady who had the courage to brave several years' training in a laboratory ..."[12] However, we were unable to find any evidence that this invitation produced any results. It was not until the 1920s that significant numbers of women started to join the Society as a result of their entry into analytical positions in the industry and the government.[13]

Isabel Hadfield

In fact, the most prominent woman in analytical chemistry, Isabel Hodgson Hadfield,[14] did not join the Society until 1944. She was born 29 January 1893 in Welling, Kent to George William Hadfield, a schoolmaster, and Annie Hodgson. She graduated from East London College (later Queen Mary College) in 1914. The following year, Hadfield obtained a Diploma of Education from the London Day Training College and became a Chemistry Mistress with the Birmingham Education Council.

During the First World War, the need for scientists overrode the traditions of society and women were pressed into scientific employment (see Chapter 15). Thus, in 1917, Hadfield joined the staff of the National Physical Laboratory (NPL), Teddington. Studying chemical problems relating to the fabric surfaces of aircraft, she contributed many reports to the Fabrics Research Co-ordinating Committee of the Department of Scientific and Industrial Research (DSIR). Exceptionally, Hadfield retained her NPL position at the end of the War. In 1923, she was awarded an M.Sc. (London) on the basis of her research at the NPL.

---

[12] Chirnside, R. C. and Hamence, J. H. (1974). *The 'Practicing Chemists': A History of the Society for Analytical Chemistry 1874–1974*. The Society for Analytical Chemistry, London, p. 87.

[13] Horrocks, S. M. (2000). A promising pioneer profession? Women in industrial chemistry in inter-war Britain. *British Journal for the History of Science* **33**: 351–367.

[14] (a) Butterworth, D. E. (1965). Obituary: Isobel Hodgson Hadfield. *Proceedings of the Society for Analytical Chemistry* **2**: 101; and (b) *Student Records*, Archives, Queen Mary College.

The Professional Societies   289

Much of her analysis work involved small samples and, as a result, Hadfield became a pioneer in the development of microanalytical measurements.[15] In the 1930s, she was a founder member of the Microchemical Club. Hadfield retired from NPL with the rank of Principal Scientific Officer in 1953 in order to look after her elderly father. After his death, she went to live with a friend in Hampshire where she stayed until her own death on 6 February 1965.

Royal Institute of Chemistry

The entry of women into the Institute of Chemistry (later the Royal Institute of Chemistry) can best be regarded as "accidental." The Institute had been founded in 1877, and the successful sitting of an examination was a prerequisite for admission. In November 1888, the Council recorded a minute noting that they did not contemplate the admission of women candidates to the examinations.[16]

Emily Lloyd

It was only 4 years later that Emily Jane Lloyd[17] became the first woman Associate. Lloyd was born in 1860 in Birmingham, the daughter of Martin Lloyd, nail manufacturer, and Dinah Greenway. Although some of the information on her early life is contradictory, we do know that she attended a private school in Leamington. In 1881, Lloyd was listed in the decennial Census as being a teacher of English at a "Ladies School" in Swansea. She was admitted as a student to Mason Science College (later, the University of Birmingham) in 1883 at the age of 23, spending only 1 year there before

---

[15]For example, see Hadfield, I. H. (1942). Two simple micro burettes and an accurate wash-out pipette. *Journal of the Society for Chemical Industry* **61**: 45–50.

[16]Pilcher, R. B. (1914). *The Institute of Chemistry of Great Britain: History of the Institute 1877–1914*. Institute of Chemistry, London, pp. 113–114.

[17]Anon. (1913). Obituary: Miss Emily Jane Lloyd. *Proceedings of the Institute of Chemistry of Great Britain and Ireland* 32–33; The following are thanked for information on Lloyd: I. Salmon, (1998). Assistant Registrar, University of Wales, Aberystwyth; P. Bassett, (1998). Archivist, Special Collections, University of Birmingham; and N. Jeffs, (1998). Archivist, Special Collections, University of London Library.

290 *Pioneering British Women Chemists: Their Lives and Contributions*

transferring to University College, Aberystwyth (now the University of Wales, Aberystwyth). Lloyd remained at Aberystwyth until 1887, gaining a pass in the University of London Intermediate examination in the sciences. She is the only woman student mentioned in a history of the Chemistry Department of University College, Aberystwyth.[18]

Lloyd returned to Mason College where she was awarded a London B.Sc. in 1892. In that same year, she applied under the name of E. Lloyd to sit the associateship examination of the Institute of Chemistry. As the committee would have had no expectation that she was a woman, Lloyd was permitted to write the paper, which she duly passed. Once she had passed, the Institute had no means of denying her admission to the Society, and having admitted one woman, there was no feasible route of barring subsequent women applicants (see Footnote 16). In fact, it was only 2 years later in 1894 that Lucy Everest Boole (see Chapter 5) was elected as the first woman Fellow of the Institute.

Having gained her associateship in 1892, Lloyd applied to take the required Institute of Chemistry examination to qualify as a public analyst.[19] The admission of women candidates to this examination had not been contemplated by the Institute's Council, but they had no excuse to refuse her. She was admitted to, and passed, this hurdle as well. In 1893, she obtained an appointment as Science Mistress in a public school for girls at Uitenhage, Cape Colony (now South Africa). Lloyd remained there for 4 years and then returned to Wales, teaching at a school in Llanelly until about 1909, when she retired due to ill health. She died on 14 November 1912 at Kings Norton, aged 52.

Rose Stern

The first woman Student Member of the Institute of Chemistry, Regina Rosa (Rose) Stern, was elected in 1893.[20] Born in 1869 at Birmingham,

---

[18]James, T. C. and Davies, C. W. (1956). Schools of chemistry in Great Britain and Ireland–XXVII The University College of Wales, Aberystwyth. *Journal of the Royal Institute of Chemistry* **80**: 569.

[19]Creese, M. R. S. (1998). *Ladies in the Laboratory: American and British Women in Science 1800–1900*. Scarecrow Press, Lanham, Md., pp. 271–272.

[20]*Minutes.* (17 February 1893). Institute of Chemistry.

The Professional Societies 291

Stern was the daughter of Moritz Stern, general merchant, and Fanny Schwartz. Stern was educated at King Edward VI High School for Girls (KEVI), Birmingham. In 1889, she entered Mason College, Birmingham, to study chemistry and it was while there that she was elected as the first woman Student Member. Stern completed a London B.Sc. at Mason College in 1894.[21] From 1895 to 1896, she studied towards a teaching diploma at Cheltenham Ladies' College. Stern obtained a position as a Science Mistress at Bangor County School for Girls in 1897 where she stayed until 1902, when she was offered a Science Mistress position at the prestigious North London Collegiate School for Girls (NLCS)[22] (see Chapter 2).

Stern provided the chemical inspiration for many young women at the NLCS during the early decades of the 20th century. One of her students later recalled:

> "Stern by name and stern by nature" were the words in which she introduced herself to me. Luckily Miss Stern also had a sense of humour and a twinkle in her eye for many sinners. Although she was a merciless judge of slackness and inefficiency, she had a very real patience with the hard working but not so bright pupil. She taught her classes to be independent and nothing pleased her more than to find that they were really experimenting and carrying on without too much help.[23]

Following her retirement in 1920, Stern authored *A Short History of Chemistry*.[24] In 1939, she was noted as living in Greenwich. As a result of breaking a leg during the Second World War, she became less and less mobile, dying at Uckfield in October 1953, aged 83.

---

[21] The following are thanked for information on Stern: P. Bassett (1998). Archivist, Special Collections, The University of Birmingham; and N. Jeffs (1998). Archivist, Special Collections, University of London Library.

[22] Rayner-Canham, M. and Rayner-Canham, G. (2017). *A Chemical Passion: The Forgotten Story of Chemistry at British Independent Girls' Schools, 1820s–1930s*. Institute of Education Press, London, pp. 47–50; 62, 65–66; 93–95; 97; A66–A70.

[23] "K.N.H.H." (1954). Death. *N.L.C.S. Magazine* (3): 44.

[24] Stern, R. (1924). *A Short History of Chemistry*. J. M. Dent & Co., London.

The Biochemical Society

The Biochemical Club, as it was initially called, was founded in 1911. At the first meeting, the second item on the agenda concerned the admission of women.[25] A letter had been received from "a lady" (possibly Ida Smedley or Harriette Chick, see the respective section) requesting permission to become an original member. An amendment was therefore proposed to the rules that only men were eligible for membership. The amendment passed by a vote of 17 to 9. This vote was challenged, and at a committee meeting the following year, the Club reversed its position, voting by 24 to 7. In 1913, the Club held a meeting to elect new members, and of the seven elected, three were women: Ida Smedley, Harriette Chick, and Muriel Wheldale (see Chapter 11). Smedley was later (1927) to become the first woman Chairman of the Biochemical Society Committee.

Ida Smedley (Mrs. Smedley Maclean)

Ida Smedley[26] was a key individual in the early advancement of women in chemistry and biochemistry. As we mentioned earlier, she was one of the first women members of the Biochemical Society; but of more importance, as we will see shortly, she was one of the two women (Martha Whiteley — see Chapter 4 — being the other) who led the fight for the admission of women to the Chemical Society.[27]

Ida Smedley was born on 14 June 1877 in Birmingham, daughter of William T. Smedley, a chartered accountant, businessman, and philanthropist, and Annie E. Duckworth, daughter of a Liverpool coffee merchant. Smedley had an idyllic upbringing as Mary Phillp (Mrs. Epps), a school and university friend observed:

> Ida grew up in a home where the two gifted and far seeing parents
> devoted themselves to the interests of the children, and one may add, to

---

[25] Goodwin, T. W. (1987). *History of the Biochemical Society 1911–1986*. The Biochemical Society, London, pp. 14–15.

[26] Whiteley, M. A. (1946). Ida Smedley Maclean 1877–1944. *Journal of the Chemical Society* 65.

[27] Rayner-Canham, M. and Rayner-Canham, G. (2011). Forgotten pioneers. *Chemistry World* **8**(12): 41.

those of their children's friends. To some of us the house was a second home. Literature, theatricals, music and languages filled up the leisure hours. Independence of thought and action were encouraged, and the whole atmosphere of the home was decades ahead of its time.[28]

Smedley, like Stern, attended KEVI, and it was there that she met Beatrice Thomas (see Chapter 3), the two forming a friendship that endured throughout their lives. Smedley spent 3 years at Newnham College, Cambridge, graduating in 1899. She then became a research student at the Central Technical College, London (later part of Imperial College) with Henry Armstrong,[29] being awarded a D.Sc. by the University of London for her research on benzylaniline sulphonic acids. In 1903, Smedley briefly returned to Newnham as a Demonstrator in Chemistry, resigning in 1904 to take up full-time research at the Davy–Faraday Laboratory of the Royal Institution, London.

Then in 1906, Smedley became the first woman appointed to the Chemistry Department at Victoria University of Manchester, holding the rank of Assistant Lecturer. At Manchester, Smedley continued her research in organic chemistry long into the night, often until about 3 a.m., together with the other "night owls",[30] Robert Robinson and D. L. Chapman.

Smedley embarked on a career change in 1910 that was to gain her more recognition. Awarded a Beit Research Fellowship, she returned to London to take up a position at the Lister Institute of Preventive Medicine where she remained for the rest of her working life. Her field of study was that of fat metabolism and fat synthesis.[31] The value of her research was recognised in 1913, when she received the Ellen Richards Prize[32] of the

---

[28]"M.E. de R.E." [Phillp, M., Mrs. Epps] (January 1945). Ida Smedley Maclean, *Newnham College Roll Letter* 50–51.

[29]Keeble, F. W. (1941). Henry Edward Armstrong, 1848–1937. *Obituary Notices of Fellows of the Royal Society* **3**: 229–245.

[30]Todd, Lord, and Cornforth, J. W. (1976). Robert Robinson. 13 September 1886–8 February 1975. *Biographical Memoirs of Fellows of the Royal Society* **22**: 419.

[31]Nunn, L. C. A. (22 July 1944). Obituary: Dr. Ida Smedley-Maclean. *Nature* **154**: 110.

[32]Ellen Richards had been a leading American woman chemist. See: Rayner-Canham, M. and Rayner-Canham, G. (1998). *Women in Chemistry: Their Changing Roles from Alchemical Times to the Mid-Twentieth Century.* Chemical Heritage Foundation,

294 *Pioneering British Women Chemists: Their Lives and Contributions*

American Association of University Women for the woman making the most outstanding contribution of the year to scientific knowledge. In 1913, Smedley married Hugh Maclean, who became a Professor of Medicine at the University of London and at St. Thomas's Hospital, London and she subsequently had two children, a son in 1914 and a daughter in 1917.

During the First World War, Smedley worked for the Ministry of Munitions and for the Admiralty, her major contribution being the development of the large-scale production of acetone from starch by fermentation (see Footnote 31). After the War, she returned to her work on fats. Her most influential research on fat metabolism came in 1938, when she followed up the classic work of G. O. Burr and M. M. Burr at the University of Berkeley, California, who had described stunted growth and dry scaly skin in young rats totally deprived of fat.

Smedley, and her colleague, Margaret Hume, reproduced the results of the Burrs and showed that the health issues could be cured by either of the two unsaturated fatty acids — linoleic and linolenic acid. Then she isolated and identified arachidonic acid and showed this fatty acid to be as active biologically as linoleic acid.[33] The three acids, linoleic, linolenic and arachidonic, came to be known as "essential fatty acids" because without one or other of them in the diet, the characteristic signs of deficiency appeared. We are now aware of the major importance of these omega fatty acids in human diets. In addition to more than 50 research papers on fats, Smedley authored a monograph, *The Metabolism of Fat.*[34]

Smedley was active in the rights of women outside of her research fields. In particular, she played the leading role in forming the British Federation of University Women.[35] As was noted at the time: "... the

---

Washington, pp. 51–55. The Ellen Richards Prize was terminated in the 1930s as the organizers were convinced that women had finally established themselves in science and that the award was no longer necessary.

[33]Martin, S. A., Brash, A. R. and Murphy, R. C. (2016). The Discovery and Early Structural Studies of Arachidonic Acid. *Journal of Lipid Research* **57**: 1126–1132.

[34]Smedley-Maclean, I. (1943). *The Metabolism of Fat.* Methuen, London.

[35]Dyhouse, C. (1995). The British Federation of University Women and the status of women in universities, 1907–1939. *Women's History Review* **4**: 465–485.

The Professional Societies 295

single gathering sufficed to reveal to several who were present (stimulated and led by Dr. Ida Smedley Maclean) the value and importance of forming an association of University women on a permanent basis, ..."[36] In addition to being a founder, she served in various roles of the organization, including President from 1929 to 1935.

In summing up the later years of Smedley's life, Phillp commented:

> Her marriage was a very happy one, and she showed as much skill in running her home, and bringing up her children, as she had done in other departments of her life. How she managed to hold three threads evenly in her hands, research, social work and home-life, was a constant wonder even to those who knew her best. When for a time serious illness fell upon a member of the family her devotion to the invalid, while all her other work was carried on, was little less than heroic. During the last two years of her life her health was failing; but true to her courageous nature she accomplished what she and in 1943 her last work was published.[37]

Smedley died on 2 March 1944, aged 67.

Harriette Chick

The second of the pioneer women to be admitted to the Biochemical Club was Harriette Chick.[38] Born on 6 January 1875, Harriette Chick was one of the daughters of Samuel Chick, a businessman in the lace trade, and Emma Hooley. All of the daughters attended Notting Hill High School,[39] London, and amazingly for the time, six of the seven completed university degrees.

Harriette Chick attended classes at Bedford College and then entered University College, London (UCL), where she became an outstanding student in botany before shifting her focus to chemistry. Following

---

[36]Tylecote, M. (1941). *The Education of Women at Manchester University*, Manchester University Press, Manchester, p. 63.

[37]Note 28, 'M.E. de R.E', pp. 50–51.

[38]Creese, M. R. S. (1998). *Ladies in the Laboratory? American and British Women in Science, 18001900.*, Scarecrow Press, Lanham, Maryland, pp. 40, 149–150.

[39]Sayers, J. E. (1973). *The Fountain Unsealed: A History of the Notting Hill and Ealing High School.* Broadwater Press, Welwyn Garden City, pp. 39–43.

completion of her B.Sc. (Hons.) in 1896, Chick turned her attention to bacteriology. In 1898, she was awarded an 1851 Exhibition Scholarship which she used to undertake research at the Institute for Hygiene at the University of Vienna and at University College, Liverpool. Then in 1902, Chick was appointed assistant to the Chief Bacteriologist for the Royal Commission on Sewage Disposal. She spent the following year at the University of Munich, then in 1904, she received a D.Sc. for her research into green algae in polluted waters.

In 1905, Chick applied for a Jenner Memorial Research Studentship at the Lister Institute. The application caused a furore as no women had previously been given this award:

> As soon as her application became known, two members of the scientific staff implored the Director not to commit the folly of appointing a woman to the staff. She was, nevertheless, appointed to the Studentship and soon accepted on terms of equality and friendship by the apprehensive males.[40]

Chick's early work at the Lister, undertaken with Charles Martin,[41] was on the chemical kinetics of the disinfection process. Of particular importance, she found that the temperature dependence of the disinfectant action did not follow the Arrhenius rate expression. Instead, the rate increased by as much as seven- or eight-fold for a 10°C increase in temperature, thus showing that warm disinfectant solutions were far better for killing bacteria than cold.[42] This observation led to her being the co-developer of the Chick–Martin Test for the efficacy of a disinfectant.

Following the First World War, Chick was co-leader of a team studying the disease rickets.[43] This research, showing the link between nutrition and the disease, and finding methods of treating and preventing it, resulted

---

[40]Chick, H., Hume, M. and Macfarlane, M. (1971). *War on Disease: A History of the Lister Institute.* Andre Deutsch, London, pp. 87–88.

[41]Chick, D. H. (1956). Charles James Martin, 1866–1955. *Obituary Notices of Fellows of the Royal Society* **2**: 172–208.

[42]Chick, H. (1908). An investigation of the laws of disinfection. *Journal of Hygiene* **8**: 92–158.

[43]Carpenter, K. J. (2008). Harriette Chick and the problem of rickets. *Journal of Nutrition* **138**: 827–832.

in the near-complete disappearance of this debilitating illness. Chick continued in the field of nutrition, particularly the role of vitamins, becoming one of the founder members of the Nutrition Society and its President from 1956 until 1959.[44] Among the many honours she received was a C.B.E. in 1932 and a D.B.E. in 1949. Despite formal retirement in 1945, she kept active in the nutrition field as an honorary staff member until her death on 9 July 1977, aged 103.

The Chemical Society

Though there was initial opposition to the admission of women to every chemical organization, by the early part of the 20th century, there was only one major bastion to be breached — the Chemical Society. The fight for admission to this organization was to be tough and lengthy, spanning a period of 40 years. The most interesting part is the women who spearheaded the attack during the long campaign, together with the men who valiantly championed their cause.

The Chemical Society was founded in 1841, but it was not until 1880 that the question was raised of the admission of women. This convoluted saga has been described in detail by Joan Mason.[45] In the initial discussion, legal opinion was given that, under the Charter of the Society, women were admissible as Fellows. A motion was proposed in 1880 by Augustus G. Vernon Harcourt[46] to clarify the bye-laws, so that any reference to the masculine gender should be assumed to include the feminine gender.[47] The motion was rejected as "not expedient at this time." A similar proposal was put forward by William Ramsay in 1888, but after lengthy discussion, this motion was also withdrawn.[48]

The first attempt by a woman to enter the Society occurred in November 1892. The long controversy started innocuously, as the Minutes

---

[44] Copping, A. M. (1978). Obituary Notices: Dame Harriette Chick (5 January 1875–9 July 1977). *British Journal of Nutrition* **39**: 3–4.

[45] Mason, J. (1991). A forty years' war. *Chemistry in Britain* **27**: 233–238.

[46] "H.B.D." (1920). Obituary Notices of Fellows Deceased. A. G. Vernon Harcourt, 1834–1919. *Proceedings of the Royal Society of London. Series A* **93**: vii–xi.

[47] (1880). *Minutes of the Meetings, IV.* Council of the Chemical Society, pp. 215, 219.

[48] (1888). *Minutes of the Meetings, V.* Council of the Chemical Society, pp. 103, 219.

298   *Pioneering British Women Chemists: Their Lives and Contributions*

of the Council Meeting describes: "The Secretary having read a letter from Prof. [Walter Noel] Hartley suggesting the election of a lady as Associate, Prof. Ramsay gave notice that he would move that women be admitted Fellows of the Society."[49]

The motion, proposed by Ramsay and seconded by William Tilden,[50] came the following January.[51] An amendment was then proposed by William Perkin, Sr,[52] that it was not desirable at that time to amend the bye-laws for the purpose of admitting women. The amendment was defeated by seven votes to six, then curiously, the motion itself was defeated by a margin of eight votes to seven. The Secretary commented:

> Does the Charter of the Society contemplate or permit of the admission of women as Fellows and if so what alteration, if any, is required in the Bye-Laws? The Council were advised that under the Charter women are admissible as Fellows, but when on three occasions (1880, 1888, and 1892) proposals were made to put it into effect no action was taken, the general feeling being that, although there was no objection in principle to the admission of women as Fellows, the case in their favour was not entirely established by any considerable number of applications for the Fellowship and that a change involving so radical an alteration in the policy of the Society, should be recommended by a maximum vote.[53]

So things remained in limbo until 1904, when Marie Curie's name was put forward for Election as a Foreign Fellow.[54] At the following meeting, discussion of her candidacy resulted in a motion to again request the opinion of legal counsel on the eligibility of women for admission as Ordinary

---

[49](17 November 1892). *Minutes of the Meetings, V,* Council of the Chemical Society, pp. 212, 226.

[50]"J.C.P." (1928). Obituary Notices of Fellows Deceased. Sir William Augustus Tilden, 1842–1926. *Proceedings of the Royal Society of London. Series A* **117**: i–v.

[51](19 January 1893). *Minutes of the Meetings.* Council of the Chemical Society, n. pag.

[52]Meldola, R. (1908). Obituary Notices: William Henry Perkin. *Journal of the Chemical Society, Transactions* **93**: 2214–2257.

[53]Anon. (1893). *Proceedings of the Chemical Society, London* **9**: 84.

[54](3 March 1904). *Minutes of the Meetings.* Council of the Chemical Society, n. pag.

Fellows and Foreign Members.[55] Presumably, the opinion of 24 years earlier had either been forgotten or it was hoped that a new counsel would offer a different opinion. This was, in fact, the case. The new counsel, R. I. Parker, reported: "In my opinion married women are not eligible as Fellows of the Chemical Society and I think it extremely doubtful whether the Charter admits of the election of unmarried women as Fellows."[56] Under British common law of the time, a married woman was not an independent person and therefore could not be a Fellow in her own right. However, Parker deemed that as Honorary and Foreign members came with no duties and responsibilities, there was no impediment to the election of Curie. Curie was duly elected.[57]

The 1904 Petition

Emboldened by Curie's success, a memorial (petition) was presented to the council in October 1904 by Ida Smedley and Martha Whiteley requesting admission of women to Fellowship.[58] Nineteen women chemists (see below) signed the petition which stated: "We, the undersigned, representing women engaged in chemical work in this country desire to lay before you an appeal for the admission of women to Fellowship in the Chemical Society."[59]

The women chemist signatories were the following: Lucy Boole, Katherine Burke, Clare de Brereton Evans, E. Eleanor Field, Emily Fortey, Ida Freund, Mildred Gostling (Mrs. Mills), Hilda Hartle, Edith Humphrey, Dorothy Marshall, Margaret Seward (Mrs. McKillop), Ida Smedley, Alice Smith, Millicent Taylor, M. Beatrice Thomas, Grace Toynbee (Mrs. Frankland), Martha Whiteley, Sibyl Widdows, and Katherine Williams.

---

[55](23 March 1904). *Minutes of the Meetings*. Council of the Chemical Society, n. pag.
[56]Note 45, Mason, p. 234.
[57](20 April 1904). *Minutes of the Meetings*. Council of the Chemical Society, n. pag.
[58]Note 45, Mason, p. 234.
[59](21 October 1904). Letter enclosed in: *Minutes of the Meetings*. Council of the Chemical Society, n. pag.

300 *Pioneering British Women Chemists: Their Lives and Contributions*

The petitioners then noted the increasing contributions of women chemists and the willingness of the Chemical Society to publish their results:

> Reference to the publications of the Chemical Society shows that during the last thirty years [1873–1903] the names of about 150 women of different nationalities have appeared there as authors or joint authors of some 300 papers.... Seeing that the Chemical Society recognises the value of the contributions made by women to chemical knowledge by accepting their work for publication, we are encouraged to point out that their work would be greatly facilitated by free access to the chemical literature [in the Society's library] and by the right to attend the meetings of the Society (see Footnote 59).

Following receipt of the petition, the Council at the time which was women-friendly unanimously adopted the proposal to alter the bye-laws, but the changes had to be approved by the body of the organization. Of the over 2700 members, only 45 attended the Extraordinary General Meeting to approve the changes and, of those, 23 voted against. In his 1905 Presidential address, Tilden expressed his displeasure with the outcome:

> The number of women desiring admission is but small, and I fail to see any cogent reason, beside the legal one, for excluding them. Some of them are doing admirable scientific work, and all the memorialists are highly qualified. To deprive them of such advantages as attach to the Fellowship simply on the grounds that they are not men seems to be an unreasoning form of conservatism inconsistent with the principles of a Society which exists for the promotion of science. It seems to be unfortunate that when a subject so important is brought up for the judgement of a body numbering upwards of 2,700 members, the appointed meeting should be attended by no more than 45.[60]

Effect of the Petition

Tilden, the then-President of the Chemical Society, proposed another route to enable the admission of women. He circulated a petition in support of women's admission, signed by 312 of the most distinguished

---

[60]Tilden, W. A. (1905). Presidential Address, delivered at the Annual General Meeting, March 29th, 1905. *Journal of the Chemical Society, Transactions* **87**: 547–548.

The Professional Societies   301

Fellows of the Society.[61] Then in 1908, Tilden co-sponsored with Henry Roscoe[62] a motion that there be a poll of members on the issue.[63] This motion passed, and a ballot was circulated accompanied by a list with six reasons to vote for admission and seven reasons to deny admission.[64]

In the pages of the journal *Nature*, there was an Editorial which expounded the opinions of the journal editors on the issue:

> It cannot be denied that women have contributed their fair share of original communications. Indeed, in proportion to their numbers they have shown themselves to be among the most active and successful of investigators. The society consents to publish their work which redounds to its credit. Why, then, should the drones who never have done, and never will do, a stroke of original work in their lives be preferred to them simply because they wear a distinctive dress and are privileged to grow a moustache?[65]

Of the 2900 Fellows of the Chemical Society, 1094 were in favour of the admission of women with full rights while 642 were opposed. One might have assumed that the battle had been won, but this was not to be the case.

At a special meeting of Council on 3 December 1908, the motion was put forward by Tilden and seconded by Thomas E. Thorpe[66] that:

> The Council having referred the question of the admission of women to the Fellowship of the Society to the whole of the Fellows, and having ascertained that a majority of those who returned an answer are of the opinion that women should be admitted to the Fellowship on the same terms and with the same privileges as men, resolves to give effect to the wishes of the majority, and to take such action as they may be advised

---

[61] Odling, W. *et al.* (1908). Letter from Past Presidents communicating the Memorial to the Fellows. *Nature* **78**: 227–228.

[62] "T.E.T." (1917). Obituary Notices of Fellows Deceased. Sir Henry Roscoe, 1833–1915. *Proceedings of the Royal Society of London. Series A* **93**: i–xxi.

[63] Note 45, Mason, p. 235.

[64] Reprinted (1908). In women and the fellowship of the chemical society. *Nature* **78**: 226–227.

[65] The Editor (1908). Women and the fellowship of the chemical society. *Nature* **78**: 226.

[66] "A.E.H.T." (1925). Obituary Notices of Fellows Deceased. Sir Thomas Edward Thorpe, 1845–1925. *Proceedings of the Royal Society of London. Series A* **109**: xviii–xxiv.

302 *Pioneering British Women Chemists: Their Lives and Contributions*

are necessary in order to permit of the inclusion of women as Fellows of the Society.[67]

The minutes note that a discussion then ensued, but unfortunately, no details are given. It would appear that Armstrong made very persuasive arguments against the motion. He proposed an amendment, seconded by Horace Brown.[68] "That in the opinion of this Council it is desirable that at any time, on the recommendation of three Fellows of the Society, women be accepted as Subscribers to the Society."[67] This new category would allow women to attend ordinary meetings, to use the Society library, and to receive the Society publications. However, women would not be able to attend extraordinary meetings, vote in Society elections, or hold offices in the Society. The amendment passed by 15 votes to seven. The passage of this reversal was prompted by the fear that the Armstrong-led minority would use legal means to block the proposed bye-law.

The rejection by the Council of the Society's own referendum resulted in a stinging editorial in the journal, *Nature* in February 1909:

> No matter what the size of the majority in favour of the admission of women might be, a contumacious and recalcitrant element in the minority — a cabal of London chemists ... set themselves to thwart the wishes of the majority ... The whole business of the referendum was ... deliberately reduced to a fiasco.[69]

In a despairing attempt to right the wrong, at the Annual General Meeting of the Society in 1909, Edward Divers[70] proposed to move the adoption of a new bye-law on the admission of women.[71] However, the move was ruled out of order by reference to the legal opinion of 1898.

---

[67](3 December 1908). *Minutes, VIII*, Extraordinary Meeting of the Council, Chemical Society, n. pag.

[68]"J.B.F." (1925). Obituary Notices of Fellows Deceased. Horace Tabberer Brown, 1848–1925. *Proceedings of the Royal Society of London, Series A* **109**: xxiv–xxvii.

[69]The Editor. (1909). Women and the Fellowship of the Chemical Society, *Nature* 79: 429–431.

[70]"J.M." (1913). Obituary Notices of Fellows Deceased. Edward Divers, 1837–1912. *Proceedings of the Royal Society of London, Series A* **88**: viii–x.

[71]Note 45, Mason, p. 237.

The Professional Societies 303

The 1909 Letter

In 1909, a report was circulated, claiming that the women petitioners were linked to the agitation for the political enfranchisement of women. Astonishing as it may seem today, 31 women chemists, including 14 of the original petitioners, felt it necessary to rebut the accusation that women chemists seeking admission as Fellows were associated with such radical elements of society. In a letter to *Chemical News*, the women noted that the sole bond between them was a common interest in chemistry.[72]

The letter was followed by a statement from the same group of women concerning a "meeting of representative women chemists." In this declaration, the 312 Fellows were thanked for their support in addition; women were urged not to become Subscribers on the grounds that it would prejudice their case for Fellowship of the Chemical Society.

The signatories of the 1909 letter were as follows, where (P) indicates that they were also a signatory of the 1904 petition: Heather H. Beveridge, Mary Boyle, Katherine A. Burke (P), Frances Chick, Louisa Cleaverley, Margaret D. Dougal, Clare de Brereton Evans (P), E. Eleanor Field (P), Emily L. B. Forster, Ida Freund (P), Maud Gazdar, Hilda J. Hartle (P), E. M. Hickmans, Annie Homer, Ida F. Homfray, E. S. Hooper, Edith Humphrey (P), Zelda Kahan, Norah E. Laycock, Elison A. Macadam, Effie G. Marsden, Margaret McKillop (P), Agnes M. Moodie, Nora Renouf, Ida Smedley (P), Alice E. Smith (P), Millicent Taylor (P), M. Beatrice Thomas (P), M. A. Whiteley (P), Sibyl T. Widdows (P), and Katherine I. Williams (P).

Lesser-Known Signatories of the Letter

Most of the signatories of the letter continued with careers in chemistry or related areas, and their biographical details will be found in the most appropriate chapter of this compilation. However, there is little information on four of the signatories; thus, their brief biographies are compiled in this section.

---

[72] Beveridge, H. H. *et al.* (5 February 1909). Women and the fellowship of the Chemical Society. *Chemical News* 70.

Heather Henderson Beveridge

Heather Henderson Beveridge[73] was born on 27 December 1886 at Stow, Midlothian, the daughter of Rev. John Beveridge and Alice Alexandra Henderson. She was educated at Harris Academy, Dundee and then entered University College, Dundee in 1902 at the age of 16. After graduation with a B.Sc. (St. Andrew's), she was appointed a Carnegie Research Scholar at Dundee, working with James Walker. When Walker moved to the University of Edinburgh, Beveridge moved with him continuing research there. She married Richard John Simpson in 1909 in Dundee. Beveridge died on 6 August 1940 at Haymarket, Edinburgh.

Frances Chick (Mrs. Wood)

Frances Chick (Mrs. Wood),[74] one of the Chick sisters (see Harriette Chick), was born on 25 December 1883 and was educated at Notting Hill High School. She obtained a B.Sc. (Hons.) in Chemistry from UCL in 1908 and after graduation carried out postgraduate research in chemistry with William Ramsay at University College and then with Arthur Harden[75] in the Biochemical Department of the Lister Institute.

It was at the Lister Institute that she attended a series of lectures on statistical methods. Chick found her true vocation applying statistics to medical data, especially mortality rates for cancer and for diabetes.[76] In 1911, she married Sydney Herbert Wood, Inspector with the Board of Education. They had one daughter, but unfortunately, 2 weeks after her daughter's birth, Chick died of an infection on 12 October 1919 at the young age of 36.

---

[73] *Student Records*, Archives, University of Dundee.

[74] Cole, T. (2017). The Remarkable Life of Frances Wood. *Significance Magazine* (accessed 10 August 2018).

[75] Hopkins, F. G. and Martin, C. J. (1942). Arthur Harden, 1865–1940. *Obituary Notices of Fellows of the Royal Society* **4**(11), 2–14.

[76] "M. G." [Greenwood, M] (1920). Frances Wood, 1883–1919. *Journal of the Royal Statistical Society* **83**: 178–180.

Louisa Cleaverley (Mrs. Dunstan)

Louisa Cleaverley (Mrs. Dunstan)[77] was born on 10 June 1883 at Plaistow, Essex to Joseph Cleaverley, Police Constable, and Elizabeth Cleaverley. She was a chemistry student at the East Ham Technical College, the Lecturer of Chemistry and Physics being Albert E. Dunstan.[78] In 1907, she co-authored a publication with Dunstan. Cleaverley was an Elementary Assistant Teacher near Chester in 1911 and sometime the same year, she married Dunstan. She died in April 1963, and, deeply affected by her death, Dunstan died 8 months later.

Zelda Kahan (Mrs. Coates)

Zelda Kahan (Mrs. Coates)[79] was born in Russia about 1882 to Isaac Kahan, shipping agent, and Rachel Kahan. Kahan entered Yorkshire College (later University of Leeds) in 1900, graduating in 1905 with a B.Sc. in Chemistry from Victoria University. The same year, she moved to University College, London, where she authored three publications between 1906 and 1908, acknowledging the support of Professor Collie in her research. Kahan was still at UCL in 1912.

Kahan had become involved with left-wing politics, meeting Lenin as a young girl in 1902.[80] She married William P. Coates in 1914, and the two of them, devoted the rest of their lives to the cause of the Communist Party of Great Britain. Kahan died in 1969.

Proponents and Opponents of Women's Admission

Throughout the struggle for the admission of women, there were men who fervently led the battle on behalf of the unfranchised women. Some of the

---

[77]Langton, H. M. (23 May 1964). A. E. Dunstan, D.Sc., F.R.I.C. 1878–1964. *Chemistry and Industry* 883–884.

[78]Sell, G. and Thomas, W. H. (August 1964). Obituary Notices: Albert Ernest Dunstan 1878–1964. *Proceedings of the Chemical Society* 270–271.

[79](a) Personal communication, email, Nick Brewster, 29 August, 2018. Archivist, University of Leeds; (b) *Calendars*, University College, London; and (c) Wikipedia, (9 May 2018). Zelda Kahan (accessed 26 February 2019).

[80]Coates, Z. K. (November 1968). Memories of Lenin. *Labour Monthly* 506–508.

male chemists knew one or more of the women personally; for example, Whiteley was a research student of Tilden.

Ramsay had proposed the 1893 motion for the admission of women that had been seconded by Tilden. In a historical review of the Chemical Society, Tom Moore and James Philip noted:

> This [later] controversy took place during the Presidency of Ramsay [1907–1909], and its result must have been particularly galling to him, as one of the earliest and most consistent advocates of the admission of women.[81]

A feature of all the women's supporters was their affable personality, and this was true of Ramsay as his biographer, Thaddeus Trenn, commented:

> Ramsay, the eternal optimist, whose motto was said to be 'be kind', was above all a highly cultured gentleman. ... He was admired and beloved by almost all who knew him; and his boyish vigour and simple charm, unaffected by the many honours showered upon him, remained with him throughout his life.[82]

Jocelyn Field Thorpe,[83] a later supporter of women in chemistry, also exhibited similar personality traits to Ramsay as Thorpe's protégée and co-author of his obituary, Martha Whiteley, felt important to describe:

> Thorpe owed much of his success, both as a teacher and as a man of affairs, to his personal qualities, his joviality and charm of manner. He had an incorrigible faith in the goodness of human nature and refused to see anything but the best in the people with whom he came in contact.[84]

---

[81] Moore, T. S. and Philip, J. C. (1947). *The Chemical Society, 1841–1941: A Historical Review*. The Chemical Society, London, p. 97.

[82] Trenn, T. J. (1975). William Ramsay. In Gillispie, C. C. (ed.), *Dictionary of Scientific Biography*. Charles Scribner's Sons, New York, Vol. 11, p. 278.

[83] Ingold, C. K. (1941). Jocelyn Field Thorpe. 1872–1939. *Obituary Notices of Fellows of the Royal Society* **3**: 530–544.

[84] Whiteley, M. A. and Kon, G. A. R. (1940). Obituary: Jocelyn Field Thorpe. *The Analyst* **65**: 483–484.

The Professional Societies   307

Against the many supporters, there was one figure who dominated the opposition: Henry Armstrong (see Chapter 10). Moore and Philip commented: "It is agreed by all who have personal recollections of the Society from 1880 onwards that for many years Armstrong's influence was dominating."[85] They added that Armstrong's stubborn refusal to continence admission of women was against the wishes of the majority of Fellows.

Armstrong was a traditionalist rather than a misogynist. In fact, he was the research supervisor of Ida Smedley from 1900 to 1904 (see the respective section), while Clare de Brereton Evans (see Chapter 5) worked under him between 1895 and 1897, synthesizing organonitrogen compounds and obtaining three publications. His opposition was not to women taking chemistry degrees, but rather to women becoming professional chemists instead of producing more "little chemists" as he stated himself:

> If there be any truth in the doctrine of hereditary genius, the very women who have shown their ability as chemists should be withdrawn from the temptation to become absorbed in the work, for fear of sacrificing their womanhood; they are those who should be regarded as chosen people, as destined to be the mothers of future chemists of ability.[86]

Admission at Last

For the 11 years of its existence, only 11 women availed themselves of Subscriber status, thus indicating a strong determination by most women that it was to be full Fellowship or nothing.[87] It was 1919 when everything changed: a Bill was moving through Parliament titled the Sex Disqualification (Removal) Act. Under this Act:

> A person shall not be disqualified by sex or marriage from the exercise of any public function, or from being appointed to or holding any civil or judicial office or post, or from entering or assuming or carrying on

---

[85] Note 81 Moore and Philip, pp. 78–79.

[86] Nye, M. J. (1996). *Before Big Science: The Pursuit of Modern Chemistry and Physics 1800–1940*. Prentice Hall International, London, p. 17.

[87] Note 45, Mason, p. 236.

308   *Pioneering British Women Chemists: Their Lives and Contributions*

any civil profession or vocation, or for admission to any incorporated society (whether incorporated by Royal Charter or otherwise).[88]

Though the law did not receive Royal Assent until 23 December 1919, it would seem more than a coincidence that the sudden revival of the issue of Fellowship for women chemists occurred while the Bill was being debated. At an Extraordinary General Meeting of the Chemical Society on 8 May 1919, a Resolution was proposed by James Philip[89] and seconded by J. W. Leather:[90] "That women should be admitted to the Society on the same terms as men."[91] The motion passed unanimously, and in 1920, the first women were admitted as Fellows.

There were 21 women chemists admitted at that auspicious first election: Agnes Browne, Eunice A. Bucknell, Margaret Carlton, Mary Cunningham, Ellen Field, Mary Frances Hamer, Elizabeth E. Holmes, Mary Johnson, Hilda Mary Judd, May Sybil Leslie, Phyllis V. McKie, Ida Maclean, Frances Micklethwait, Nora Renouf, Marion Crossland Soar, Millicent Taylor, Gartha Thompson, Martha Whiteley, Sibyl Widdows, Florence Mary Wood, and Olive Workman.[92] Among these were four of the original petitioners: Smedley, Taylor, Whiteley, and Widdows. At subsequent meetings of the Society, three other petitioners — Burke, Humphrey, and Thomas — were elected.

Lesser-Known Initial Members

Just as some of the signatories were low-profile individuals, so were some of the initial members; and in this section, we will summarize what we know of them.

---

[88] Creighton, W. B. (1975). Whatever happened to the sex disqualification (Removal) Act? *Industrial Law Journal* **4**(1): 155–167.

[89] Egerton, A. C. (1942). James Charles Phillip, 1873–1941. *Obituary Notices of Fellows of the Royal Society* **4**: 51–62.

[90] Anon. (1934). Obituary: John Walter Leather, *Journal and Proceedings of the Institute of Chemistry* **58**: 457–458.

[91] Extraordinary General Meeting. (Thursday, 8 May 1919). *Proceedings of the Chemical Society* 58–59.

[92] Certificates of Candidates for Election at the Ballot to be Held at the Ordinary Scientific Meeting on Thursday, December 2nd (1920). *Proceedings of the Chemical Society* 82–100.

The Professional Societies 309

Eunice Annie Bucknell

Eunice Annie Bucknell,[93] born on 29 December 1888 at Fulham, was the daughter of Daniel Bucknell, carpenter and joiner, and Anne Marion Read. She attended NLCS and entered Bedford College in 1907. Bucknell completed a B.Sc. there, and in 1911, she was listed in the census as a Chemistry Instructor. By 1920, she was an Analytical and Research Chemist at the South Metropolitan Gas Company. Bucknell died in 1969 at Croydon.

Mary Cunningham

Mary Cunningham[94] was born in Stamford Hill, London, in 1882. She was educated at Ipswich High School for Girls and then completed a B.Sc. (Hons.) in Chemistry in 1907 and an M.Sc. in 1916 from UCL. In addition to becoming one of the first women members of the Chemical Society, she was also a member of the Society for Chemical Industry and of the Society of Dyers and Colourists. Cunningham had seven publications between 1908 and 1910 from the Borough Polytechnic Institute, London. Then, in 1918, she was sole author of two publications from the Chemistry Research Laboratories of the University of St. Andrews, from where she received a D.Sc. In 1920, Cunningham gave her occupation as a Research Chemist with the Fine Cotton Spinners and Doublers Association, Manchester. In 1922 and 1923, she was granted two patents resulting from her work with the Association.

Edna Elizabeth Holmes (Mrs. Taylor)

Edna Elizabeth Holmes (Mrs. Taylor) was born in 1900 at Catford to Ernest Holmes, Manager, Labour Exchange, and Annie Philadelphia Martin. At the time of her admission to the Chemical Society,[95] she listed her occupation as a Chemical Assistant at the Lennox Foundry Company Research Laboratories, New Cross, London, where she had been for the

---

[93] (a) Student Records, Archives, Bedford College; and (b) Note 92, p. 84.
[94] (a) *Student Records*, Archives, University College, London; and (b) Note 92, pp. 85–86.
[95] (a) *Student Records*, Archives, University College, London; and (b) Note 92, p. 90.

310 *Pioneering British Women Chemists: Their Lives and Contributions*

previous 3 years. Holmes married John E. Taylor in 1933. Nothing additional could be found about her.

Olive Workman

Olive Workman[96] was born on 19 January 1893 at Swansea to John Workman, Railway Clerk, and Ellen Ann Slater. She obtained a B.Sc. in Chemistry from UCL in 1916, and then an M.Sc. conjointly from UCL and IC in 1919 undertaking research with J. C. Philip. In 1923, Workman received a Teaching Certificate from the London Day Training College then becoming a school Science Mistress. Workman died on 10 February 1974, aged 81, at Islington, London.

Gartha Thompson

Gartha Thompson[97] was born in 1886 to Daniel Thompson and Lucy Thompson (who remarried to become Lucy Bolwell). She attended KEVI and from there, she went to Wandsworth Technical College, and then to the Royal College of Science, obtaining a B.Sc. (Hons.) in 1910. Thompson's occupation was listed as teacher at her mother's private school, Mill Road Girls' School from 1910 to 1912, after which from 1912 to 1914 she was a chemist to the Polysulphin Company. The following year, Thompson was Secretary and Assistant to the Chief Chemist of British Thomson-Houston Co. Ltd., Rugby.

In 1915, Thompson became Senior Assistant to Samuel Judd Lewis.[98] Lewis was not part of the chemical academic establishment; instead, in 1909 he set up a private laboratory as a consulting and analytical chemist in Holborn. He became convinced that the future of analytical chemistry lay in spectroscopy, a view that was 20 years ahead of its time. After the First World War, Thompson returned to British Thomson-Houston

---

[96](a) Note 92, p. 100; (b) (1951). *Register of Old Students and Staff of the Royal College of Science*, 6th edn., Royal College of Science Association, London.

[97](a) Note 92, p. 98; (b) (1971). Personal News: Obituaries. *Chemistry in Britain* **7**: 35.

[98]Garton, F. W. J. (1960). Obituary Notices: Samuel Judd Lewis, 1869–1959, *Proceedings of the Chemical Society* 156–157.

Co. Ltd. In 1923, she had three sole-author publications listing an address of Rugby. Thompson died on 4 November 1970, aged 84.

Florence Mary Wood

Florence Mary Wood[99] was born on 24 January 1887. She was educated at Bournemouth Collegiate School for Girls and the Municipal Technical College, Bournemouth. She obtained a B.Sc. in Botany from the University of London in 1911 and a B.Sc. in Chemistry from the University of Birmingham in 1912. From 1912 to 1917, Wood was Science Mistress at Hampton School, Malvern, Jamaica, British West Indies. In 1918, she returned to England and undertook research with S. Judd Lewis's group at IC until 1920; thus, it was possible that Wood overlapped with Gartha Thompson (see the preceding text).

Wood received a Ph.D. from Birkbeck College, London, in 1924 and the same year was appointed Headmistress at Kensington High School for Girls where she remained until 1929. After a short period at Twickenham County School, she took a position as Biology Mistress at Chiswick County School (now Chiswick Grammar School) in 1930. In Wood's spare time, she undertook research at Birkbeck College, being the sole author of five publications between 1924 and 1934. Wood was still teaching at Chiswick at the date of her death on 18 November 1948, aged 61.

Women Chemists' Dining Club

For men, there had long been a male-only club where men could socialize among other chemists. The first men's club had been the Chemical Club which lasted only from 1806 to 1828.[100] Then in 1872, the longer-lasting Chemistry Club was formed in London by some of the male members of the Chemical Society (see Footnote 3). The Chemistry Club provided

---

[99] (a) Anon. (1949). Obituary: Florence Mary Wood. *Journal and Proceedings of the Royal Institute of Chemistry* **73**: 62; (b) Note 92, pp. 99–100.

[100] Lacey, A. (2017). The chemical club: An early nineteenth-century scientific dining club. *Ambix* **64**: 263–282.

312 *Pioneering British Women Chemists: Their Lives and Contributions*

(male) chemists the opportunity for socializing and bonding in an exclusively male environment.

Ladies' clubs had been very popular in the late 19th century.[101] However, they tended to be ephemeral and not a function of some common interest or bond, as Eva Anstruther commented on the lack of women's "clubbable" nature:

> A woman uses her club to eat in, or to learn at, or to entertain her personal friends; she does not yet look upon going to it as a means of passing the time in a place which is congenial to her among people who are her very good comrades while she is thrown with them.[102]

In November 1925, Martha Whiteley and Ida Smedley founded the Women Chemists' Dining Club, the first meeting being held in the Lyceum Club, London.[103] The purpose of the Women Chemists' Dining Club was to enable women chemists to have the opportunity to meet and to develop friendships. Three dinners were held each year with an occasional speaker, usually from outside the field of chemistry.

Meetings of the Club were suspended during the Second World War and they resumed about 1946. Mary Truter,[104] former Professor of Crystallography, University College, London, recalled that during 1946:

> My mother, Agnes Jackman, née Browne, took me to a few meetings. At the time the prime mover was Frances Hamer who worked in Kodak. ... Prof. Dame Kathleen Lonsdale was invited to join but refused, not because she was a physicist, but, characteristically, because she did not think it right to have an organisation for eating when there were starving millions in the world.[105]

Some social outings of the Club were arranged: for example, in the summer of 1948, 27 members visited the laboratories and colleges at

---

[101] Anstruther, E. (1899). Ladies' clubs. *The Nineteenth Century* **45**: 598–611.

[102] Anstruther, p. 611.

[103] The Editor. (26 January 1952). For ladies only! *Chemistry and Industry* 71.

[104] Cruickshank, D. (June 2005). Mary Rosaleen Truter, 1925–2004. *Crystallography News* (93): 20.

[105] Personal Communication, e-mail, M. Truter, 15 October 2001.

The Professional Societies 313

Cambridge; in 1949, there was an excursion to Royal Holloway College; while in 1951, there was an outing to Oxford with lunch at St. Hilda's College. Frances Hamer (see Chapter 16) was succeeded as Secretary of the Club by Ellie Knaggs (see Chapter 12).

In the February 1952 issue of *Chemistry and Industry*, there was an Editorial ruminating on the existence of the Club (which by then had been meeting for 27 years):

> It is with much interest that we learned a few weeks ago that women chemists in London had formed a Club. Most men are clubbable, one way or another, but we did not know this was true of women. We wonder if this formation of a Club for women chemists is another sign of female emancipation. We should be glad to think that they mellow over a bottle or two of fine wine. We commend claret — the Queen of wines. Presumably claret attained this title because of its beauty, its grace and its subtlety — admirable qualities which men have always associated with women.[106]

The Editorial on the Club also noted that as of 1952, there were 66 members of the Club. The article continued: "No doubt the ladies of this Club even smoke..." though the Editor recommended that the women consider snuff instead:

> If you give up smoking, ladies, we might permit you a little snuff. Think how beautiful snuff-boxes can be. How lovely they would look in your handbags: so easy to carry, so delightful to toy with. No more trouble filling petrol lighters: and the grace of the gesture, the poise of your arm, the curve of the extended fingers as you delicately administer a little snuff to your nostrils.[106]

The last mention of the Women Chemists' Dining Club was in 1953, when a meeting was held at Queen Elizabeth College, London, the successor to the King's College of Household and Social Science (see Chapter 8). In the report, it was noted that: "After the meal, Mrs. A. Jackman, senior

---

[106]The Editor. (23 February 1952). A plea to the ladies. *Chemistry and Industry* 155.

314 *Pioneering British Women Chemists: Their Lives and Contributions*

lecturer in the Chemistry Department and a keen member of the Club, conducted the party around the College."[107]

No records of the Women Chemists' Dining Club could be found in the Chemical Society holdings. The Club seems to have been very much run by the first generation of women members of the Chemical Society, and it is quite possible that with their demise, particularly Whiteley's death in 1956, the Club ceased to be.

Commentary

The diversity in attitude between the different societies towards the admission of women is striking. It can be interpreted in terms of how much the society was a pure professional society and how much a "men's club." In particular, as the Royal Institute of Chemistry was the accrediting body, the role of the Chemical Society was much more towards the socializing end of the spectrum and therefore more hostile to women's admission. What comes through strongly is the importance of key individuals, particularly the pair of Ida Smedley and Martha Whiteley, who had each established themselves as respected chemists. However, as women were excluded from the governing body, it was supportive male chemists, such as Augustus Vernon Harcourt, William Tilden, and William Ramsay, who had to argue the women's case in the Council of the Society.

---

[107] Anon. (9 May 1953). Meetings, notices, etc.: Women chemists' dining club. *Chemistry and Industry* 465.

Chapter 10

Women Chemistry Teachers

Kathleen Leeds (26 November 1883–11 May 1921)

When pioneering women chemistry students graduated with their B.Sc. degrees, what would they do?[1] We will look at some of the other possibilities in later chapters, but here we will cover one of the popular and

---

[1] Gordon, A. M. (1895). The after careers of university educated women. *Nineteenth Century* **37**: 955–960.

316　*Pioneering British Women Chemists: Their Lives and Contributions*

uncontroversial fields — that of school teaching. With the growing numbers of academic girls' schools across Britain, there was certainly a demand for qualified women chemistry teachers. It was those first generations of women chemistry graduates who would become teachers and enthuse subsequent generations of girls.

In this chapter, we will introduce the available educational pathways to become a chemistry teacher together with some of the challenges the women faced. To provide personalized perspectives, we have chosen a few individual biographical accounts of some women chemistry teachers. Most of these women devoted their lives to the cause of science teaching, particularly chemistry.

Earliest Chemistry Teachers

When academic girls' schools first opened their doors, there were no women with a chemistry background who could become their teachers. Thus, the earliest chemistry teaching in these schools had to be accomplished by men. These were male scientists/science teachers who gave their time and effort, presumably from a belief in the rights of girls to learn science and chemistry in particular. In Chapter 2, we showed that, at Newington Academy for Girls, it had been William Allen; at the Mount School and later at Polam Hall, it had been Edward Grubb; while at North London Collegiate School, it had been Robert Buss.

At Winchester High School (later St. Swithun's School), about the late 1880s and early 1890s, it was Rev. H. Searle who provided chemistry and physics teaching to the students:

> The Rev. H. Searle was asked to give us lessons in Physics and Chemistry. You to whom the possession of a well-equipped laboratory is commonplace can have no idea of the thrill of those early lessons given in the School Hall; ... We had no apparatus and he had to carry everything, flasks, chemicals, heavy iron apparatus, by bicycle to and from the Training College, and yet from the first every lesson was fully illustrated by experiments.[2]

---

[2]Findley, E. (1934). *S. Swithun's School Winchester 1884–1934*. Warren and Son Ltd., Winchester, p. 19.

Though briefly preceded by a woman chemistry teacher, Mary Adamson (see the respective section), George Samuel Newth[3] taught chemistry at about 1890 at Princess Helena's College. Newth was a Demonstrator and Lecturer in Chemistry at the Royal College of Science (later Imperial College) from 1871 to 1909. He wrote several chemistry books, including *Chemical Lecture Experiments*.[4] An issue of the *Princess Helena College Magazine* of 1890 reported that: "Mr G. S. Newth, of South Kensington Science Department, gives the Chemistry Lectures, and makes them most interesting to his class by his numerous and beautiful experiments."[5]

An alternative solution, that of sending the students elsewhere for chemistry, was used at some schools. A documented example is that of Leeds Girls' High School, founded in 1876. The first Head Mistress, Catherine Kennedy was a strong proponent of chemistry as the key science for girls. However, she realized that there were few, if any, women teachers qualified to teach the subject.[6] Instead, she was able to arrange for the students to take their chemistry classes and laboratory sessions at the Yorkshire College of Science (later, the University of Leeds). The Chemistry Professor at the college was T. Edward Thorpe,[7] a strong supporter of women in chemistry.[8] He continued with the practice until 1883 when he began to devote all his time to research.

The girls excelled in chemistry. For example, in the obituary for a former student, Edith Little, in the *Leeds Girls' High School Magazine*, it was noted that:

> ... in 1880 she [Edith Little] gained the Prize and First Certificate for Chemistry of the Non-Metals and Practical Chemistry at the Yorkshire

---

[3] Wikipedia, (29 January 2019). George Samuel Newth (accessed 27 February 2019).

[4] Newth, G. S. (1922). *Chemical Lecture Experiments: Non-Metallic Elements*. Longman, Green and Co., London.

[5] Anon. (Easter 1890). School news. *Princess Helena College Magazine* 9.

[6] Jewell, H. M. (1976). *A School of Unusual Excellence: Leeds Girls' High School 1876–1976*. Leeds Girls' High School, Leeds, pp. 47–48.

[7] "A. E. H. T." (1925). Obituary Notices of Fellows Deceased: Sir Thomas Edward Thorpe, 1845–1925. *Proceedings of the Royal Society. Series A* **109**: xviii–xxiv.

[8] Rayner-Canham, M. and Rayner-Canham, G. (2009). Fight for rights. *Chemistry World* **6**(3): 56–59.

318   *Pioneering British Women Chemists: Their Lives and Contributions*

College, which the older pupils of the school used to attend at that time for instruction in chemistry; ...[9]

It was a neighbouring boys' school which provided the solution for St. Martin-in-the-Fields High School for Girls, then on Charing Cross Road, London. Twice a week, from 1903 until 1916, the girls were walked to the chemistry laboratory of nearby Archbishop Tenison Boys' School. They left their school at 4 p.m. arriving at Tenison's after the boys had left, and returned to St. Martin's at 5 p.m. In her reminiscences, Lydia Mentasti, who started at the school in 1897, recalled: "Girls were chaperoned to Archbishop Tenison Boys' School for chemistry. There were usually no boys about, but it would have been against the rules to speak to them if there were."[10]

Ackworth School, a Quaker school near Pontefract, West Yorkshire, had a boys' school and a girls' school side-by-side. Though the chemistry laboratory was built in the boys' half, girls had a weekly chemistry lesson there with a male chemistry teacher by 1889. Elfrida Vipont notes that in 1893: "... an Old Scholar described his astonishment at seeing "a whole bevy of girls invading the boys wing ... to join a combination chemistry class."[11]

Teachers' Training Colleges

For qualified male chemistry teachers, it was accepted wisdom that a science degree sufficed to become a successful teacher. Their own school teachers had only an academic chemistry background in order to teach. That being the case, why would they, themselves, need any teacher training?[12] Some of the women teachers had a different perspective and wished to gain a knowledge base in the theory of teaching. To satisfy this

---

[9] Kennedy, C. L. (Winter 1908). Edith Little. *Leeds Girls' High School Magazine* **2**(29): 5.

[10] Siddall, M. (1999). *From School to School: Changing Scenes, 1699–1999*. Devonshire House, London, p. 40.

[11] Vipont, E. (1959). *Ackworth School: From Its Foundations in 1779 to the Introduction of Co-Education in 1946*. Lutterworth Press, London. p. 128.

[12] Bottrall, M. (1985). *Hughes Hall 1885–1985*. Cambridge University Press, Cambridge, pp. 4–6.

need, in the late 1800s and the early 1900s, many women-dominated teachers' training colleges and departments were instituted across the country.[13] Among the earliest were the Maria Grey Training College in London (see the following section) founded in 1878 and the Cambridge Training College for Women Teachers founded in 1885 (see the respective section).

For validation, most of the Teacher's Colleges relied upon the examinations offered by the Cambridge Teacher Training Syndicate (CTTS). Thus, a Cambridge Teaching Diploma or Certificate rarely indicated that the woman teacher had actually attended Cambridge University. Cambridge University charged a considerable fee for external students taking the examination. Despite the high fee, taking the examination and obtaining the Diploma/Certificate from the CTTS provided graduates of other teacher colleges with national credibility in their teaching qualification.

Nevertheless, the majority of women chemistry teachers went directly from a university degree to a teaching position. In a biography of Frances Buss, Sophie Bryant remarked:

> ... women educated at the Universities persisted in neglecting professional training. Either they despised it, or they could not afford it, or they did not like it, and could get entrance to [teaching at] the schools without it.[14]

Maria Grey Training College

Maria Grey had played an influential role in the formation of the Girls' Public Day School Company (GPDSC — see Chapter 2). She realized that to sustain academic girls' schools strongly, there was a need for highly trained women teachers.[15] To that end, Grey was one of the leading figures involved in the founding of the Teachers' Training & Registration

---

[13]Edwards, E. (2001). *Women in Teacher Training Colleges, 1900–1960: A Culture of Femininity*. Routledge, London.

[14]Ridley, A. E. (1896). *Frances Mary Buss: and her Work for Education*. Longman's Green & Co., London, p. 283.

[15]Ellsworth, E. E. (1979). *Liberators of the Female Mind: The Shirreff Sisters, Educational Reform and the Women's Movement*. Greenwood Press, Westport.

320 *Pioneering British Women Chemists: Their Lives and Contributions*

Society College which opened in Bishopsgate, London, in 1878. After moving to Fitzroy Square in 1885, the college was renamed the Maria Grey College.[16]

The area of science teaching, and particularly chemistry, seemed to be one specific focus of the college. In the description of Maria Grey college's new building, opened in 1892, it was reported in the *Maria Grey College Magazine* that: "… the laboratory with every possible aid to the study of chemistry, …."[17] There were periodic mentions of the teaching of science at the college. For example, by 1897, chemistry teacher Edna Walter (see the respective section) was giving occasional lectures on "Science Teaching in Girls Schools."[18]

There was a whole paragraph on science teaching in the 1903 *Maria Grey College Annual Report*.

A few Science students also had the very great advantage of going regularly once a week during the last two terms to hear a Course of Science Lessons given by Annie Louise Janau at the Central London Foundation School for Girls.[19]

Cambridge Training College and Homerton College, Cambridge

In the opinion of Frances Buss and Sophie Bryant of North London Collegiate School (NLCS), it made sense to have a Training College located in Cambridge, convenient for the graduates of Girton and Newnham. The new institution was approved and the Cambridge Training College (CTC), later named Hughes Hall, was opened in 1885.[20] In September of 1886, Ida Freund (see Chapter 3) was appointed as Lecturer

---

[16] Lilley, I. M. (1981). *Maria Grey College, 1878–1976*. West London Institute of Education, Twickenham.

[17] Anon. (July 1892). At Brondesbury. *Maria Grey College Magazine* 5–6.

[18] Anon. (November 1897). The college session, 1897–1898. *Maria Grey College Magazine* 2.

[19] Anon. (November 1903). The college year 1902–1903. *Maria Grey College Magazine* 3.

[20] Martin, G. (2011). *Hughes Hall Cambridge 1885–2010*. Third Millennium Publishing Limited and Hughes Hall, London.

in Method,[21] though there is no indication that she taught any chemistry. However, in 1900, one of the science teachers at NLCS, Clotilde van Wyss, was hired as part-time Science Lecturer at CTC.[22]

Nine years later, the CTC was joined by a second women's' teacher training college at Cambridge, that of Homerton College. Homerton Academy had been founded as a boys' school in London in 1695, then re-founded as a teachers' training college in 1850. With the move to Cambridge in 1894, it became a women-only institution.[23]

In the early years, for the women students at Homerton, wandering around the streets of Cambridge could be dangerous as Elizabeth Edwards described:

> Cambridge was a town dominated by the male values of its university, which barely tolerated its own women students, let alone those in an obscure teacher training institution. The university's misogyny had been underlined by a privilege which it had only been forced to relinquish as late as 1894. This was its right to imprison in its own private prison for a period for up to three weeks, women walking in the town-who could well have been Homerton students-whom it suspected of being prostitutes.[24]

Hilda Hartle

Hilda Jane Hartle[25] was the Chemistry Instructor at Homerton College from 1902. She was born on 11 September 1876 in Birmingham, daughter of Edward Hartle, merchant's clerk, and Anne Jane Warillow. Like so

---

[21] Note 12, Bottrall, p. 9.

[22] Note 12, Bottrall, p. 40.

[23] Simms, T. H. (1979). *Homerton College: 1695–1978: From Dissenting Academy to Approved Society in the University of Cambridge*. Trustees of Homerton College, Cambridge.

[24] Edwards, E. (2000). Women Principals, 1900–1960, Gender and Power. *History of Education* **29**: 407.

[25] (a) M.E.G. (1975). Hilda Jane Hartle, 1876–1974. *Newnham College Roll Letter* 43; (b) White, A. D. (ed.), (1979). *Newnham College Register*, Vol. 1, *1871–1923*. Newnham College, Cambridge, p. 141; and (c) Personal correspondence, e-mail, P. M. Warner, 14 May 2007. Senior Tutor, Homerton College.

322 *Pioneering British Women Chemists: Their Lives and Contributions*

many others, she was educated at King Edward VI High School for Girls, Birmingham (KEVI), and then she entered Newnham College in 1897 as a Goldsmiths' Scholar, attaining a First Class in the Natural Science Tripos in 1901.

Awarded a 2-year Gilchrist Research Scholarship by Newnham College, Hartle used it to return to Birmingham, becoming a researcher with Percy Frankland at the University of Birmingham from 1901 to 1903. She was then offered, and accepted, a Lectureship to teach chemistry at Homerton College. Homerton was the only one of the early teachers' training colleges to teach chemistry (see Footnote 13), and this was largely a result of Hartle. While at Homerton, she became a signatory of the 1904 petition for the admission of women to the Chemical Society and the 1909 letter to *Chemical News* (see Chapter 9). It was also at Homerton that Hartle penned her attack in 1911 on the domestic science movement (see Chapter 8). Awarded a Mary Ewart Travelling Scholarship, in 1915–1916, she took a sabbatical year touring educational institutions in the United States. Her studies on innovative educational techniques were published in the *Times Educational Supplement*.

In 1920, Hartle was appointed Principal of the Brighton Municipal Training College for Teachers, a post she occupied until her retirement in 1941. The post-retirement 30 years were spent actively working for numerous women's organizations. Her obituarist, "M.E.G." (probably her long-time colleague at Homerton, Margaret Glennie) remarked: "She maintained a vivid interest in all these many activities almost to the end of her life and she also took an affectionate delight in her extensive correspondence with old pupils in all parts of the world." (see Footnote 25a) Hartle died on 20 May 1974, aged 98.

Life of Women Teachers

The large majority of pioneering women chemistry teachers stayed single. To them, teaching science to girls was their lifelong career path. An Editorial of 1932 in the *Journal of Education* remarked:

> Entering the teaching profession for a woman is equivalent to entering a nunnery; for few marry. Their colleagues look upon those who get

engaged as deserters of the cause; not that teachers have many opportunities of meeting the right sort of men.[26]

In an article in *Girls' Realm*, Dorothea Beale laid out her principles of what it took for a girl to be a teacher:

> ... a teacher must keep in the current of thought; if she does not, her mental powers deteriorate, as her muscular system does when she takes no exercise, and then her lectures become dry, stale and uninteresting.... A teacher must be content to deny herself much that she might do her work without distraction; to those who do this the promise is yet fulfilled:- they have a hundredfold more in this present life — peace, content, sympathy, the joy of feeling that they have not lived in vain, and the hope that they may one day dare to say, "I have finished the work Thou gavest me to do."[27]

The majority of the academic girls' schools were day-schools, and the depressing life of the day-school teacher was described by Beatrice Orange in her discourse on "Teaching as a Profession for Women," in the first volume of *The Woman's Library*. This gloom-filled account included the options for accommodation for the woman teacher:

> Here we must touch on one of the problems in the life of the day-school teacher-where to live, both cheaply and comfortably. Mistresses' houses are almost invariably failures; flats are expensive and a source of anxiety, for the teacher's tenure of office is apt to be dependent on the whim of a head-mistress, and in any case she never reckons on permanently remaining in one place; and so the assistant mistress usually solves the difficulty by choosing the admittedly uncomfortable, but undeniably convenient lodging, which can be left at short notice, and where service, such as it is, is supplied. Nevertheless, to most people, it is a distinct hardship to live for weeks and months together in a small house in a mean street, in an atmosphere of woollen mats and chromo-lithographs.[28]

---

[26] Anon. (1932). Education for women. *Journal of Education* **64**: 560–561.

[27] Beale, D. (1900). Careers for girls, V. — Teaching. *Girls' Realm* **2**: 620–623.

[28] Orange, B. (1903). Teaching as a profession for women. In McKenna, E.M. (ed.), *The Women's Library*, Vol. 1, *Education and the Professions*. Chapman and Hall, Ltd., London, p. 94.

324   *Pioneering British Women Chemists: Their Lives and Contributions*

Then, Orange described the "loneliness and dulness[sic]" of such a life:

> For the teacher labours under the disadvantage of coming into contact with no one but her pupils and her fellow-teachers during the day; thus differing from the doctor or nurse or factory inspector [other careers for women], all of whom meet with a great number of both men and women in the course of their work; and in her spare time she has to choose between the society of her fellow-teachers or her own. For, save in exceptional circumstances, young, single working women are practically excluded from society.[29]

Orange also noted the pressure for career advancement:

> It seems to be the fashion at the present time in most day schools to appoint young head-mistresses; and the assistant-mistress who has not secured a post as head by thirty-five has need to feel anxious. Should she be forced to seek a new post even as assistant-mistress after this age, she would not find the matter very easy; the teaching profession being one where youth counts for more than experience. Therefore, again, I would urge the advisability of emigrating before it is too late to adapt one's self to new conditions, and of seeking in new countries what may still be had for the asking.[30]

For teaching at both day and boarding schools, there could be considerable pressure from the Headmistress not to be emotionally and/or physically exhausted — and this in an age of such virulent diseases as tuberculosis and later, Spanish influenza. As an example, the biographer of Olive Willis, first Headmistress of Downe House, remarked: "...Olive had extraordinary physical stamina, and was apt to expect the same in her colleagues...."[31] Thus, it comes of little surprise to find some of the teachers at this, and other, schools having retired comparatively young due to "ill-health."

---

[29] Note 28, Orange, p. 95.

[30] Note 28, Orange, p. 78.

[31] Ridler, A. L. (1967). *Olive Willis and Downe House: An Adventure in Education.* John Murray, London, p. 104.

The impression is often given that in those earlier times, schoolgirls were placid, obedient, young women. As we have shown elsewhere, this was not the case.[32] Teaching in the classroom and particularly in the laboratory of the time, with the lack of safety precautions, was emotionally stressful in itself. A rhyme, part of which is given subsequently, from Miss Iva's chemistry class at Wimbledon High School for Girls illustrates the point.

The chief defect of VI B Science
Was that, in absolute defiance
Of all Miss Iva's prayers and pleas
(She almost went down on her knees)
They never gave their note books in;
And, adding to this dreadful sin,
In spite of all her earnest pleadings
They never entered up their readings,
Which — mark, all scientists in the making —
Should be put down at time of taking.
They almost always tried to shirk
Preparing any reading work,
And, being peppered with abuse,
Would feebly murmur some excuse.
At practicals they were just terrors,
You've never seen such awful errors!
So huge these were, and so gigantic,
They nearly drove Miss Iva frantic.[33]

Some of the Pioneer Women Chemistry Teachers

There are many accounts of the pioneering Headmistresses, but the teachers — in particular the chemistry teachers — have vanished from the

---

[32] Rayner-Canham, M. and Rayner-Canham, G. (2017). *A Chemical Passion: The Forgotten Story of Chemistry at British Independent Girls' Schools, 1820s–1930s*. Institute of Education Press, London.

[33] Loveless, J. VIB (June 1942). A Warning; or the Sad Story of the Worst Science VIth Miss Iva Ever Had. *Wimbledon High School Magazine* (50): n. pag.

record. Yet, it was these young women who brought their enthusiasm of chemistry to the subsequent generations.

Many of the women chemistry teachers stayed at one school for most of their lives. Others seemed to have "flitted" from school to school. The cause — if there was any general cause — of transient positions is unclear. However, it would seem that, at least during the First World War, there was a significant shortage of women chemistry and physics teachers.

The Headmistress, W. M. Kidd, at the Girls' Grammar School, Maidstone, gave a presentation on the supply of women teachers in secondary schools for girls at the 1916 Conference of the Association of Headmistresses which was reprinted in *School World*. She reported that, whereas there was a glut of women teachers in such subjects as English and history, there was a shortage in science:

> The temporary dearth of [women] teachers of mathematics and science (especially chemistry and physics) is really serious. This, of course, may right itself after the war. But at the present time a great many [physical science] mistresses are entering boys' schools, others are becoming analysts and instrument-makers and taking posts in chemical and physical research laboratories. In such posts the hours are shorter and the evenings not burdened by corrections. Then, again, the war is creating an ever-increasing demand for women doctors.[34]

Mary Adamson

Born on 8 June 1864 in Ealing, Mary Madeline Adamson[35] was the daughter of Frank Adamson, builder, and Elizabeth Adamson. She was educated at Notting Hill High School, matriculating in 1882 and entering Bedford College on an Arnott Scholarship in Physics. Adamson graduated from Bedford College with a B.Sc. (London) in 1885. Her first position was to teach chemistry and physics at Princess Helena College, then located in Ealing.

---

[34] Kidd, W. M. (July 1916). The supply of teachers in secondary schools for girls. *School World* **18**: 267.

[35] Adamson, M. M. (1932). Reminiscences, 1882–1932. *Portsmouth High School Magazine, Jubilee Edition*, 6–8.

In 1889, Adamson obtained an appointment at Bromley High School (her replacement at Princess Helena School being Newth — as discussed earlier). The teaching experience at Bromley came as quite a trauma. Reflecting back on her career following retirement, Adamson recalled in the *Notting Hill and Ealing School Magazine* that at Bromley:

> For Miss Heppel acting on her principle that 'anybody can teach anything if they take the trouble,' had given the Chemistry to her Classical Mistress and given me a big block of Middle School History and Elementary German and Latin.[36]

Fortunately for Adamson, a chance encounter in 1890 changed her life. At the time, all of the London-area GPDSC schools had a common prize-giving held at the Crystal Palace. This event involved 2,000 students and staff and 12,000 parents, relatives, and assorted dignitaries. Adamson continued in her reminiscences:

> Sent to seek a truant student in a distant corridor, I found myself close to Miss Jones [Headmistress at Notting Hill High School], ... "You are at Bromley? Do you teach such and such and such?" I said "Yes" and afterwards had a horribly guilty conscience, realising that these were my subjects [chemistry and physics] during the four years of my first post [Princess Helena College] and not the "mixed bag" of the past term [at Bromley High School]. It proved that they were the exact subjects of her vacant Senior Science post. Two days later I was offered it ...[36]

Though the Headmistress of Notting Hill, Harriet Morant Jones, had been keen to recruit Adamson as a science teacher, Adamson added that Jones was not enthusiastic about science itself:

> She was indifferent to, or even disliked science, and often told me so. Yet my scientific ardour was not lessened, nor my spirits damped. If an interest in science and interesting my students in it was my little best, well, it *was* my best, and therefore commendable.[37]

---

[36] Adamson, M. M. (March 1937). My first and last visits to the Crystal Palace. *Notting Hill and Ealing School Magazine* (52): 45.

[37] Sayers, J. E. (1973). *The Fountain Unsealed: A History of the Notting Hill and Ealing High School*. Broadview Press, Welwyn Garden City, p. 104.

In 1900, Adamson was offered and accepted the Headship at Portsmouth High School. Unfortunately, she had to resign as Headmistress in 1905 due to ill health. She travelled to Australia[38] and seems to have recovered, for she survived until 1955, dying at age 91.

Kathleen Collier

Born on 9 December 1889, Kathleen Mary Collier,[39] daughter of an overseas mining engineer and a Mrs. P. J. Collier, attended Streatham Hill High School for 1 year (1905–1906) and then completed her education at Sutton High School. It is not known what she did during the intervening years, but she graduated with a B.Sc. from Royal Holloway College (RHC) in 1919. She was appointed to the science staff at Bedford High School in 1923. It was noted in 1930 that Collier was also a postgraduate student at IC. She taught at Bedford High School until her retirement in 1956.

Following her retirement, several of her former students reminisced in *The Aquila: Magazine of the Bedford High School*, one noted: "Many generations have been infected by Miss Collier's enthusiasm and love of chemistry and acquired from her the true scientific spirit."[40] While another recalled Collier's attributes in more detail:

> Miss Collier's pride in her subject made her set a very high standard and an "A" was a coveted distinction, rarely achieved. She believed, above all, in laying a firm and solid foundation, and this, combined with her intolerance of slipshod, careless work, led to a high standard in her pupils.... But I think we only appreciated Miss Collier as a teacher when, at the university we found that many of our friends did not know the basic facts. Miss Collier would have been shocked![41]

---

[38] Anon. (March 1906). Changes in Staff and Old Girls. *Notting Hill High School Magazine*, 37.

[39] (a) *Student Records*, Archives, Sutton High School for Girls; and (b) *Staff Records*, Archives, Bedford High School for Girls.

[40] "J.N." (July 1956). Miss Collier. *The Aquila: Magazine of Bedford High School* **7**(6): 108.

[41] "M.W." (July 1956). Miss Collier. *The Aquila: Magazine of Bedford High School* **7**(6): 108.

Daisy Dalston

Born on 3 February 1883 at Nunhead, London, Daisy Florence Dalston[42] was the daughter of William Joshua Dalston, agent to a hat manufacturer, and Emily Susan Dalston. She was educated at Clapham High School, matriculating in 1903 and entering RHC the same year. Dalston completed an B.Sc. (Hons.) in 1906. She returned to Clapham High School in 1906 to train as a teacher, obtaining a Cambridge Teachers Training Certificate in 1907 (Clapham High School was one of many schools to offer teacher training at that time).

Dalston was appointed Science Mistress at Sheffield High School in 1908, then returned south to accept a position as Senior Science Mistress at Streatham Hill High School, a post which she held until her retirement in 1938. Dalston was another woman chemistry teacher to have a "long and trying illness",[43] dying on 29 February 1960 at Croydon.

Like so many of the other women chemistry teachers, she was devoted to her chosen field and to her acolytes, as former student Betty Boyd remembered in an Obituary in the *Streatham and Clapham High School Magazine*:

> From the moment we became LVI Science, our lives were divided into two separate worlds, one in which we were absorbed by the day-to-day business of the school, the other, the world of the "Chemi-Lab", the world of Miss Dalston.... Did she come in the morning, or go at night? We never saw her do so — she was always there, as far as we knew.[44]

Lilian Heather

Like so many of the women chemistry teachers, Lilian Frances Heather devoted her life to her mathematics and science teaching, particularly

---

[42] *Student Records*, Archives, Clapham High School for Girls; *Staff Records*, Archives, Streatham Hill High School for Girls.

[43] Mansfield, R. (née Webster) (July 1960). Miss Dalston. *Streatham and Clapham High School Magazine* (63): 30.

[44] Boyd, B. (July 1960). Miss Dalston. *Streatham and Clapham High School Magazine* (63): 30.

330 *Pioneering British Women Chemists: Their Lives and Contributions*

chemistry. Born on 15 February 1874,[45] Heather's parents were James Heather, a solicitor, and Frances Sherlock. Initially attending Notting Hill High School, her parents transferred her to Wimbledon High School in 1887. According to the obituary in the *Downe House Magazine*, Heather accepted a Mathematical and Science Scholarship for RHC in 1892. She completed a B.Sc. in pure and mixed Mathematics and Chemistry in 1896.[46] During this time, Heather devised and patented a method of food preservation by means of an apparatus which replaced the air above the food by carbon dioxide.[47]

Heather worked in London as Editorial Assistant for the weekly chemistry publication, *Chemical News*, though it is unclear when she ceased holding this position. The Editor of the magazine was the famous chemist, Sir William Crookes. Crookes's biographer, Fournier d'Albe quoted part of a letter Crookes had written in 1882 to Sir William Ramsey, for advice on hiring an assistant:

> I think a woman would be if anything better than a man, always providing the qualifications are equal. A woman is more conscientious than a man in many things, and is not always trying to get another appointment to "better" herself. But on the other hand she goes and gets married, which is quite fatal![48]

From its opening in 1907, Heather taught at Downe House, then located in the village of Downe, Kent (now part of Bromley). She used to drive to the school in her "dog-cart" up from Green Street Green, where she lived with her mother, stabling the pony, and then driving back down the hill each afternoon until her mother died, after which she lived at the school. With the small initial enrolment, Heather taught part-time at Downe House and part-time at Westerham School.[47]

---

[45] *Staff Records*, Archives, Wimbledon High School for Girls.

[46] Anon. (Summer 1943). Lilian Frances Heather, February 15th, 1874–September 16th, 1943. *Downe House Magazine* (87): 1–3.

[47] Note 31, Ridler, p. 88.

[48] D'Albe, E. E. F. (1923). *The Life of Sir William Crookes O.M., F.R.S.* T. Fisher Unwin Ltd., London, p. 373.

It was Heather who organized the School Science Club in 1917.[49] The *Minute Books* of the Science Club have survived, and it is apparent that Heather made sure that her students were enthused by the latest discoveries in chemistry. For example, in the Minute Book for 1923, it was reported that: "Newspaper cuttings [glued into Minute notes] on the element, hafnium, were read by Miss Heather, and the subject of using helium in airships instead of hydrogen was also discussed."[50] While in 1924, the Minutes noted: "Newspaper cutting [glued in] on "Selenium and its Uses" was read by Miss Heather."[51]

Heather was diagnosed with bone cancer, but continued working on school issues until her death on 16 September 1943, aged 67.[52] She was buried on the grounds of the school. Heather's anonymous obituarist reported on a comment by Heather:

> "I am a lucky woman," she said to me, "because my work is just what I should always choose to do if I ever had to decide. It is all so interesting and exciting and one never gets to the end of the possibilities."[53]

Annette Hunt

Annette Dora Hunt was born in 1865 in Waterford, Ireland.[54] When she was 16 years old, her parents, Ambrose Hunt, a surgeon, and Elizabeth Hunt, moved the family to the south of France, where she was educated at the École Normale in Pau.[55] She taught at GPDSC schools throughout her career, beginning at Weymouth High School in 1886 when she was 21 years old. In 1891, she transferred to Croydon High School, and then in 1895 to Sutton High School, where she became Senior Science Mistress. Hunt taught at Sutton High School until her retirement in 1928.

---

[49] Anon. (Michaelmas 1917). Downe House science club, *Downe House Magazine* (24): 8.

[50] Anon. (4 February 1923). *Minute Books, Downe House Science Club.*

[51] Anon. (8 June 1924). *Minute Books, Downe House Science Club.*

[52] Note 31, Ridler, p. 161.

[53] Note 46, Anon., p. 1.

[54] *Staff Records*, Archives, Sutton High School for Girls.

[55] "E.M.L.L." (1932). In Memoriam: Miss A.D. Hunt. *Sutton High School Magazine* **68**: 18.

332 *Pioneering British Women Chemists: Their Lives and Contributions*

One of her former students recalled:

> On a never-to be-forgotten occasion, as she was giving a science lecture while we took notes, she must have seen that our attention was wandering; so, gliding fluently into an obvious error which passed unchallenged, she stopped. 'Parrots! Parrots! The Kindergarten are more alive than you. If I should tell them that I keep a rabbit in this cupboard, do you think they would believe it? They would demand to see it: and if I forbade them, by hook or by crook they would get there and open that door. But *you* — you would swallow any rubbish. When shall I teach you that the whole basis of science is to accept *nothing* without proof?[56]

Hunt died on 26 February 1932 at Hartfield, Sussex.

Kathleen Leeds

Most of the women chemistry teachers survived to an old age, but others died young. One example of an early death was that of Grace Heath (see the respective section) while another was Kathleen Mary Leeds.[57] Leeds returned to her old school for much of her sadly shortened teaching career.

Leeds was born on 26 November 1883, at Forest Hill, Kent. Her parents were Charles Edward Leeds, bank clerk, and Agnes Maria Leeds. She was educated at the Croydon High School from 1894 to 1901. Leeds stayed on at the school as Laboratory Assistant from 1901 to 1903 and then entered the Royal College of Science in 1904 to study chemistry under a Special Studentship given to teachers in training.[58] Upon graduation with a B.Sc. in Chemistry, Leeds was briefly an instructor at Goldsmiths College, New Cross, before being appointed science teacher at Portsmouth High School in 1909.[59]

---

[56] Anon. (1964). *A School Remembers: Sutton High School G.P.D.S.T, 1884–1964*. Croker Brothers Ltd., Merton, p. 65.

[57] *Staff Records*, Archives, Croydon High School for Girls.

[58] Tilden, W. (July 1921). In Memoriam: Kathleen Mary Leeds. *Croydon High School Magazine* (32): 4.

[59] *Staff Records*, Archives, Portsmouth High School for Girls.

In 1911, Leeds had to return to Croydon to live with her aunt — though the precise reason was not given. Fortunately, a vacancy arose for a chemistry teacher at Croydon High School and she was immediately hired. The Headmistress at Croydon High School, Miss Cossey, recalled: "She threw herself whole-heartedly into all the school activities ... Outside the school she worked for the local Women's Suffrage Society. She was always a firm supporter of the women's cause."[60]

Like so many of these early chemistry teachers, Leeds was charismatic. In an obituary, one of her former students remembered:

> Probably one of the things which we felt most strongly about her was her intense vitality.... Instead of ... doing an experiment merely mechanically, one wanted to do it, and do it well, just because she was so keen about it herself.[61]

She added:

> We were a little jealous of them [her suffrage activities] at first and then angry with them, when she would come in to the form room at the end of the morning, looking thoroughly exhausted, and at once give all her energies to settling of various bits of business of the form that came up.[61]

Adding to Leeds workload, in 1917 she was promoted to Second Mistress, essentially deputy head. Leeds encouraged the idea of bright futures for the students, giving weekly inspirational talks. One of the students recalled:

> I remember ... in the Lower Sixth how we used to look forward to Miss Leeds' talk at twenty to one on Wednesdays; and how disappointed we used to be if she were prevented from coming to us. It was when she talked to us about our future careers that she used to show us her own ideals. How awed we were at their height! Then she would tell us of

---

[60] Cossey, A. F. (1921). In Memoriam: Kathleen Mary Leeds. *Croydon High School Magazine* (32): 4.

[61] "By an Old Girl" (1921). In Memoriam: Kathleen Mary Leeds. *Croydon High School Magazine* (32): 5.

334 *Pioneering British Women Chemists: Their Lives and Contributions*

some little thing of her own experience, something so human that we felt that her ideals were not merely dreams but that they were possible for us to reach them if we tried.[62]

Leeds died on 11 May 1921, no cause of death being given. Indicative of her renown, William Tilden, Professor of Chemistry at IC, was one of her obiturists (see Footnote 58). In recognition of school contributions, the Kathleen Mary Leeds Prize was instituted as an annual award "… to a girl in the Sixth Form for Science, especially in Chemistry."[63] Unfortunately, the award no longer exists.

Evelyn MacDonald

Born on 11 April 1869 at Waterford, Ireland, Evelyn (sometimes spelled Eveline or Evelyne) MacDonald was the daughter of George MacDonald and Margaret Smyth. Educated at Queen's School, Chester, she entered Girton College in 1888, completing the Tripos requirements in 1892. As a "steamboat lady," she added an M.A. from the Trinity College, Dublin, in 1905.[64]

For 2 years (1892–1894), MacDonald was the Science Mistress at Wheelwright Grammar School, Dewsbury. Then she was appointed Science Mistress at Oxford High School in 1894. Several reminiscences by former students provide an insight into her character. One of her students from the 1876–1902 period, Dorothy Counsell, remembered: "Miss McDonald was always humorous in her teaching, and we were allowed to speak quietly to one another when weighing and measuring; a most unusual concession in those days."[65]

A student from the 1910s, Marjorie Thekla Cam, later MacDonald's successor as chemistry teacher at Oxford High School, recalled that

---

[62]"By One of the Present Sixth" (1921). In Memoriam: Kathleen Mary Leeds. *Croydon High School Magazine* (32): 6.

[63]Anon. (July 1933). School Notes. *Croydon High School Magazine* (44): 2–3.

[64]Butler, K. T. and McMorran, H. I. (eds.), (1948). *Girton College Register, 1869–1946*. Girton College, Cambridge, Cambridge p. 52.

[65]Counsell, D. (1963). Reminiscences 1876–1902. In V.E. Stack (ed.), *Oxford High School 1875–1960*. Abbey Press, Abingdon p. 61.

MacDonald employed the discovery [heuristic] method (see the section titled "Henry Armstrong and the Heuristic Method"):

> My enthusiasm for Science.....was stimulated and increased by Miss McDonald. She was a delightful person, humorous and friendly. She always made us start our experiments with an open mind, instilled in us a love of discovery and eschewed any kind of text book in the elementary stages. At that time the prevailing attitude in teaching Science was that each pupil should write his own text book, which certainly nurtured a true scientific spirit but made the rate of progress somewhat slow.[66]

McDonald encouraged some of her students, including Muriel Palmer (née King), to study extra science:

> On reaching Form V, I was flattered to receive an invitation to do some extra science with "Smac", as we always called Miss MacDonald. Four close-packed years followed, full of intense interest and wide teaching, such as was then available in very few girls' schools. The one Mistress had to cope with the whole of the teaching of science in the school, and manage the laboratory and keep apparatus in order with no help.[67]

Sadly, a bicycle accident in 1920 ended MacDonald's career at Oxford High School. She died in 1952 at Bournemouth, aged 84.

Dorothy Patterson

Born on 16 February 1889 in Wolverhampton, Dorothy Christina Patterson[68] was the daughter of George Sandford Patterson, inland revenue officer, and Agnes Maria Proffitt. She was educated at KEVI, and then completed a B.Sc. at the University of Glasgow in 1912. Patterson spent the following year at Newnham College as a research student. In 1914, she

---

[66] Cam, M. (1963). Reminiscences 1902–1932. In V.E. Stack (ed.), *Oxford High School 1875–1960*. Abbey Press, Abingdon p. 78.

[67] Palmer, M. (1963). Reminiscences, 1902–1932. In V.E. Stack (ed.), *Oxford High School 1875–1960*. Abbey Press, Abingdon p. 73.

[68] Mayne, D. (June 1943). Dorothy Christina Patterson (An appreciation). *The Persean Magazine* **13**(32): 5–8.

336 *Pioneering British Women Chemists: Their Lives and Contributions*

was appointed Assistant Mistress at Wheelwright Grammar School, then in 1916, accepted a position at Perse School for Girls, Cambridge, where she remained until she retired in 1940.

Patterson was a great believer in taking the chemistry students of the School Science Club to industrial plants. In a 1923 issue of *The Persean Magazine*, a student, H. Chivers reported: "During the Spring Term eight members of the Science Club had the good fortune to be able to visit the Sulphuric Acid Works at Stowmarket, where they spent a very interesting and enjoyable afternoon."[69] While in 1928, M. Skinner described another outing:

> On Tuesday, March 27th, several members made a tour of the local Gas Works, which was most interesting and instructive. First, we saw the coal being shot into the retorts, and then we were shown how the gas was purified and the waste products collected.[70]

Upon her retirement in 1940, one student wrote in *The Persean Magazine*:

> Miss Patterson isn't at Perse any more. How odd that must sound to any girls who during a good many recent years worked their way, painfully sometimes, pleasurably sometimes, but with great thoroughness always, through elementary and advanced chemistry.... What numbers of us she got even into Girton or Newnham when often we were not good material.[71]

Patterson died on 14 February 1943 at Barton-on-Sea, Hampshire.

Lilian Quartly

Lillian Ada Quartly[72] born on 20 October 1881 at Highgate, Middlesex, was the daughter of Francis Alfred Quartly, managing clerk mercantile

---

[69] Chivers, H. (July 1923). Science club notes. *The Persean Magazine* **9**(1): 48.

[70] Skinner, M. (July 1928). Natural science club notes. *The Persean Magazine* **10**(4): 44–45.

[71] Anon. (1940). Miss Patterson. *The Persean Magazine* **13**(28): 10.

[72] *Staff Records*, Archives, Clapham High School for Girls.

and later secretary to a limited company, and Ada Bundy. She had a very "19th century" girl's education at small private schools: from 1887 until 1895, she was a student at Channing House in Highgate; and then at Cecile House School, Crouch End from 1895 until 1899. Quartly spent the period from 1899 until 1902 at the University Tutorial College before entering Northern Polytechnic in 1903, from which she graduated with a B.Sc. (London) in Chemistry and Pure & Applied Mathematics in 1905.

From 1907 until 1908, Quartly was Visiting Mistress in Chemistry at Miss Rice's School in Belsize Gardens. It was in 1915 when she took up an appointment as Chemistry Mistress at Clapham High School, following the departure of Dorothy Marshall for war work (see Chapter 3). From 1925 until 1927, Quartly was a postgraduate student at UCL, then continuing as a chemistry teacher at Clapham High School for the remainder of her career.

Quartly died in December 1959 and in an Obituary in the *Clapham High School Old Girls' Society News Sheet*, one of her former students, Elizabeth Adams (see Chapter 16) provided a detailed account of some of Quartly's contributions:

> It was only after I had left school that I realised not only how well Miss Quartly taught us the fundamentals of chemistry but how many "extras" she provided that other schools did not. Books appeared from Lewis' [Medical and Scientific Lending] Library, ranging from Maurice Traver's "Discovery of the Rare Gases" to a translation of Lucretius' "De Rerum Natura", and one was even taken to evening lectures at the Royal Institution as well as being introduced to the Christmas Lectures there. Practical chemistry was a delight; it was much more exciting to estimate the silver in a threepenny piece than in a silver salt taken out of a bottle and the experiment brought home to us the use of what we had been taught. This greatly appealed to me and perhaps influenced my choice of a career in chemistry applied to food technology and not in pure chemistry.[73]

---

[73] Adams, E. (1960). Miss Quartly. *Clapham High School Old Girls' Society News Sheet* (13): 2.

338 *Pioneering British Women Chemists: Their Lives and Contributions*

Dorothy Rippon

Dorothy May Lyddon Rippon[74] was born on 10 February 1904 at Oxford, daughter of Claude Ripon, journalist, and Belle Wheeler. She entered St. Hugh's College, Oxford, in 1922, obtaining a B.A. in 1925, a B.Sc. (Hons.) in 1928, and an M.A. in 1930. Commencing chemistry research at Oxford in 1927, Ripon was still an active researcher when she accepted a part-time position at Downe House.

The first mention of Rippon in the *Downe House Magazine* was a comment in a Science Club report: "Miss Rippon (new science teacher) told us about her research work, and was most clear and instructive, although she dealt with such alarming substances as hexahydrocarboxylcyano-cyclopentane."[75] This was research which she had accomplished at the Dyson Perrins Laboratory (see Chapter 3), Oxford, working with S. G. P. Plant.[76] In the following issue of the magazine, Miss Rippon was mentioned again, this time for acquiring liquid air, from the Oxford University research laboratories: "… Its success [the experiment] was largely due to the liquid air which Miss Rippon, with great difficulty, procured for us…"[77]

One of the Downe House teachers, Lady de Villiers commented:

> My memory takes me back to 1928 when I myself was teaching at Downe and Miss Rippon, then a young Oxford research chemist, was persuaded to come over from her home at Abingdon to do some part-time teaching for us. Even in those days good science teachers were at a premium and we thought ourselves fortunate to secure her help. To Miss Heather, of course, the appointment of someone of this calibre gave particular satisfaction. But I suppose none of us then thought that the temporary part-time post would develop into a permanent full-time

---

[74] Soutter, A. M. and Clapinson, M. (2011). *St. Hugh's College Register 1886–1959.* St. Hugh's College, Oxford, 76.

[75] "S.L." (1932). D. H. S. C. *Downe House Magazine* (66): 22.

[76] Plant, S. G. P. and Rippon, D. M. L. (1928). The Condensation of hexahydrocarbazole and of tetrahydropentindole with cyclopentanoe cyanohydrin. *Journal of the Chemical Society* 1906–1913.

[77] "S.L." (Summer 1932). D. H. S. C. *Downe House Magazine* (67): 35–36.

one and that Miss Rippon would, so to speak, take root at Downe and play so outstanding a part in building up the science side of the school.[78]

Rippon died on 3 April 1965 after a lengthy illness. In her Will, she left funds for a prize to be awarded for good work in science.

Clara Taylor

A key figure in the Association of Women Science Teachers (AWST, see the respective section) was Clara Millicent Taylor[79] (not to be confused with Millicent Taylor of Cheltenham Ladies' College, see Chapter 6). Clara Taylor was born on 12 December 1885 in Taranaki, New Zealand, attending New Zealand University from 1902 until 1909. She obtained first a B.A. and then an M.A. (Hons.) in Chemistry. A student of hers from Redlands High School recalled that Taylor told them of how she came to England:

> She [Taylor] told [us] a story of how, as a New Zealand Science graduate, determined to do post graduate work at Cambridge, she wrote to Newnham College saying she wished to take a certain course, and would arrive on a certain day! This was typical of Miss Taylor, who always required a strong sense of purpose from the girls.[80]

Taylor attended Newnham College from 1911 to 1912 and then obtained a position teaching part-time at the Clapham High School for the 1912–1913 year. Her biographer, Ian Rae, commented: "At Clapham she worked under Edith Sarah Lees [see the respective section], a stern leader who required of her junior staff that they attend school on Saturday mornings to try out experiments for the week ahead."[81]

---

[78] de Villiers, Lady (1965). Miss Rippon. *Downe House Magazine* 2–3.

[79] (a) *Staff Records*, Archives, Redlands High School for Girls; and (b) White, A. D. (ed.), (1979). Newnham College Register, Vol. 1, 1871–1923. Newnham College, Cambridge, p. 236.

[80] Anon. (Summer 1953). Redland Memories. *The Redland High School Magazine* 17.

[81] Rae, I. D. (1991). Clara Taylor, 1885–1940. *Chemistry in Britain* **27**: 146.

340 *Pioneering British Women Chemists: Their Lives and Contributions*

Taylor was appointed Chief Science Mistress at St. Paul's Girls' School in 1913, leaving there to become Headmistress at Northampton School for Girls in 1921, then Headmistress at Redlands School for Girls in 1926.[82] At Redlands, she continued teaching chemistry as a former student recalled: "I remember the Chemistry lessons we had in the old Upper Lab., where, in spite of frequent interruptions for School business, we knew we were being taught by a scientist who loved her subject."[83] Taylor co-authored the text, *Elementary Chemistry for Students of Hygiene and Housecraft*.[84] She died on 10 January 1940 in Yorkshire but was buried in Taranaki, New Zealand.

Elinor Younie

Elinor M. Younie[85] was born on 19 November 1898 at Cambus, Scotland, daughter of George Younie, inland revenue officer and later excise officer, and Helen Garrow. She graduated from the University of Edinburgh, and then from 1920 to 1921, attended St. George's (Teachers) Training College, Edinburgh, which was associated with the St. George's School for Girls. It was noted in the *St. George's Chronicle* that in 1921, she was teaching at The Park School, Glasgow, but by 1927, she had transferred to the Central Secondary School for Girls, Sheffield.

Younie returned to Scotland in 1928, taking up an appointment to teach Chemistry at St. George's School.[86] During the Second World War when the school buildings were occupied by the British army, the school was evacuated to three houses at Bonchester Bridge, a hamlet in the Scottish Borders. The *St. George's Chronicle* published reminiscences:

> At Bonchester Miss Boyd and she [Younie] did wonders in an impro-vised laboratory they clearly fixed up in the butler's pantry, and her life

---

[82] Allen-Williams, J. (2012). *Redland: Rubra Terra, Redland Court and Redland High School*. Redcliffe Press Ltd., Bristol, p. 80

[83] Milton, D. (April 1940). In Memory. *The Redland High School Magazine* 8–9.

[84] Taylor, C. M. and Thomas, P. K. (1930). *Elementary Chemistry for Students of Hygiene and Housecraft*. John Murray, London.

[85] *Staff Records*, Archives, St. Georges School, Edinburgh.

[86] Anon. (January 1929). Changes in Staff. *St. George's Chronicle* (94): 13.

was complicated there by having to cycle several miles with a small group of girls every night and morning to and from sleeping quarters in a remote farm-house.[87]

Resigning in 1954, Younie emigrated to Adelaide, Australia to marry Bertram Reekie, a chemist. After Reekie's death in 1956, she returned to Scotland. She died on 22 February 1996 in Edinburgh, aged 98.

Henry Armstrong and the Heuristic Method

Many of the pioneering women chemistry teachers had enthusiastically adopted the heuristic method.[88] The heuristic method, teaching chemistry through discovery laboratory work, had been devised by the British chemist, Henry Armstrong, who we mentioned in Chapter 9 in the context of his opposition to the admission of women chemists into the Chemical Society.

In 1894, Armstrong described his vision:

> For the ideal school of the future I picture the teacher no longer giving lessons but quietly moving about among the pupils, all earnestly at work and deeply interested, aiding each to accomplish the allotted task, as far as possible alone.[89]

He was very clear that heurism was the principle of guided inquiry, not simply letting students loose in the laboratory in the hope that they might discover all the basic principles of chemistry by themselves. It was to be the British private girls' schools who truly embraced heurism with enthusiasm.

Armstrong involved himself directly in the improvement of chemistry teaching in schools. For example, in the summer of 1898, a course of

---

[87] Anon. (1953–1954). Miss Younie. *St. George's Chronicle* 4–5.

[88] Rayner-Canham, M. and Rayner-Canham, G. (2015). The heuristic method, precursor of guided-inquiry: Henry Armstrong and British girls' schools, 1890–1920. *Journal of Chemical Education* **92**: 463–466.

[89] Praagh, G. van. (1973). *H. E. Armstrong and Science Education*. John Murray, London, p. 8.

chemistry laboratory work for school science teachers was offered at RHC:

> In Chemistry the course was one prepared by Prof. Armstrong, of the City and Guilds Central Institution, which Miss Whiteley, who took charge of this section, had attended in his laboratory. Prof. Armstrong most kindly came and spent a day in the laboratories, and expressed approval of the work. Miss Thomas has been appointed Assistant Demonstrator for this session.[90]

Grace Heath

It was an article in *Nature* by Grace Heath that first gained a wide readership for Armstrong's heuristic method:

> By this new [heuristic] method the pupils themselves are put into the position of discoverers, they know why they are at work, what it is they want to discover, and as one experiment after another adds a new link to the chain of evidence which is solving their problem, their interest grows so rapidly, that I have seen at a demonstration lesson a whole class rise to their feet with excitement when the final touch was being put to the problem which it had taken them three or four lessons to solve.[91]

Annie Grace Heath was born in 1865 at Wandsworth, daughter of Richard Heath, wood engraver, and Annie Louisa Blakemore. Educated at NLCS,[92] she was a chemistry student at the Central Institution (later City & Guilds College),[93] from 1885 to 1887, while also undertaking research with Henry Armstrong.[94] In 1887, Heath placed an advertisement in the *Journal of Education* for a teaching position:

---

[90] Anon. (December 1898). *College Letter, Royal Holloway College* 7.

[91] Heath, G. (1892). Letters to the Editor: A new course of chemical instruction. *Nature* **46**: 540–541.

[92] *Staff Records*, Archives, North London Collegiate School.

[93] Anon. (n. date). *Women at Guilds 1884–1984*, unknown source.

[94] Eyre, J. V. (1958). *Henry Edward Armstrong. 1848–1937. The Doyen of British Chemists and Pioneer of Technical Education.* Butterworths, London, p. 272.

ANNIE GRACE HEATH, late Student in Chemistry at the City and Guilds of London Technical Institute, seeks an APPOINTMENT as TEACHER of Chemistry and Physics; has had considerable experience in Laboratory work....[95]

Frances Buss, Principal at NLCS (see Chapter 2) hired Heath in 1888 as the School's first qualified chemistry teacher. Heath had a passion for chemistry. One of her former students, Dora Bunting, wrote:

> Chemistry was the subject she devoted most of her time to. She was enthusiastic about it, and tried to make her pupils so, to develop their faculties of thinking and working out everything for themselves. This was especially so in the practical work in the laboratory. She tried to insist on each one doing the work for herself, and understanding each experiment and the reason for it, so that we should not be content with learning what the results should be.[96]

Heath founded the School Science Club in 1890.[97] One of the chemistry students at the time was Edith Humphrey (see Chapter 5). Humphrey chronicled the chemical expedition led by Heath to Daintree & Co's Dye Works at Southwark:

> We saw a vat full of *green* indigo, but we were told that the cloth which was then in it would turn blue when brought under the action of light. We also saw a red dye made by boiling logwood, which was mixed with a soluble ferric salt to produce black. We had specimens of the different aniline dyes given to us, and they may now be seen in our Museum.[98]

Sadly, in 1895, Heath contracted pulmonary tuberculosis. Her successor at NLCS, Edith Aitken, visited Heath, just before Heath died: "She

---

[95] Heath, G. (1887). Advertisements. *Journal of Education* 9: 331.

[96] Bunting, D. E. L. (July 1895). In Memoriam: An address to the science club. *Our Magazine: North London Collegiate School* 28–30, 90.

[97] Heath, G. (March 1890). Club Notices: The science club. *Our Magazine: North London Collegiate School* 42–43.

[98] Humphrey, E, (November 1892). School Societies: Science club. *Our Magazine: North London Collegiate School* 142–143.

344 *Pioneering British Women Chemists: Their Lives and Contributions*

[Heath] sat there so weak yet still so bright, and talked of Dr. Armstrong's chemistry class ..."[99] She died in April 1895 at Le Cannet, France.

Edna Walter

In addition to Heath, another of Armstrong's former women chemistry students, Edna Walter, also pioneered the heuristic method as Armstrong himself described:

> But the most systematic trial given to the method in a girls' school has been that carried out at the Central Foundation School in Bishopsgate, London by Miss Edna Walter, B.Sc. This lady has embodied her experiences in an interesting paper read at the Liverpool meeting of the British Association in 1896, which was afterwards printed in *Education*.[100]

Lavinia Edwardena (Edna) Walter[101] was born on 21 February 1866 at St. Luke, London, daughter of Thomas Robert Walter, pawnbroker and jeweller, and Tabathia Beeston. She was educated at NLCS. Walter entered the Central Institution in 1886, obtaining a B.Sc. in 1889, and like Heath, joined Armstrong's research group (see Footnote 93). In 1889, Walter obtained a teaching position at the Central Foundation Girls' School. Walter was the driving force for chemistry at the school until her departure in 1901 as was reported in the *Central Foundation School Magazine*.

> Miss Walter, B.Sc., to whom is due the credit of organizing the science work of the School, has been appointed an Inspector under the Board of Education (Science and Art Department), being the first woman to hold the position. It is with great regret that we bid her good-bye, after her six years of devoted work with us, but, at the same time, we must congratulate her on obtaining so important a post, and on opening up a new field of work to women.[102]

---

[99] Aitken, E. (July 1895). In Memoriam: An address to the science club. *Our Magazine: North London Collegiate School* 24–28.

[100] Armstrong, H. E. (1903). *The Teaching of Scientific Method and Other Papers on Education*. Macmillan, London, p. 245.

[101] *Student Records*, Archives, North London Collegiate School.

[102] Anon. (1901). The girls' school, miscellanea. *Central Foundation School Magazine* 11–13.

Walter left the school to take up an appointment as an H. M. Inspector of Schools. She stayed in contact with Armstrong, residing with the Armstrong family in the Lake District at the time of the 1911 census. Walter continued to play an active role in chemistry education, giving occasional lectures on science teaching at Maria Grey Training College (see the respective section).

At Colston's Girls' School, Bristol, Walter was an advisor on teaching laboratory design:

> In March 1902 Mr Gough was instructed to draw up plans for a new building adjoining the original building and taking up some of the playground. Later that month Miss E.L. Walter, B.Sc., a Board of Education inspector, was invited to School to give a preliminary course of lessons in physical science and Gough took the opportunity to consult her about current thinking on teaching laboratories ...[103]

Walter was awarded an O.B.E. for her contributions to science education. By 1939, she had retired and was living in Worthing, Sussex. Walter died on 11 January 1962 at Richmond, Surrey.

Spread of the Heuristic Method

It was Heath's successors at NLCS who continued the proselytizing for the heuristic method. The first of these was Edith Aitken. Aitken was invited to address the annual meeting of the Association of Assistant Mistresses on the topic in 1898. She described how the heuristic method was introduced to 12-year-old girls at NLCS:

> We begin with such substances as sand and clay. Each child has her blue pinafore and her hair tied up out of the way of the gas flames. She has her stuff and does what she likes with it, subject, of course, to criticism and advice. She tries to dissolve, she boils it, bakes it, tries the action on it of acids, tries to crystallise it, examines it with a microscope, etc.[104]

---

[103] Dunn, S. (1991). *Colston's Girls' School: The First Hundred Years*. Bristol, Redcliffe Press, p. 46.

[104] Aitken, E. (1898). *Educational Review Reprints: The Teaching of Science in Schools as a Method of Induction from the Concrete*. The Educational Review, London, pp. 8–9.

346 *Pioneering British Women Chemists: Their Lives and Contributions*

Aitken, in turn, was replaced by another supporter of the heuristic method, Rose Stern (see Chapter 9). Stern, together with her friend, Alice Maude Hughes, Science Mistress at Eltham Hill Secondary School for Girls, wrote a laboratory manual along heuristic principles: *A Method of Teaching Chemistry in Schools*. In the Preface, they lay forth their principles:

> ... it is intended that every experiment should be suggested and carried out by the pupils, the part of the teacher being only to guide and supervise. At the same time the teacher must reserve the right of selecting the experiment to be done by the class when several have been suggested, and, in this way, preventing time being wasted in trying experiments which would be of little value to the children and which would break the sequence of their work.[105]

The use of the heuristic method at British girls' schools became accepted practice. For example, in a history of Bedford School for Girls, it is commented: "... when Professor Armstrong and his Heuristic Method ... had caused a good deal of fluttering in the scientific dove-cote, it became absolutely necessary to make some provision for individual practical work."[106]

To use guided inquiry required considerable skill on behalf of the teacher: to guide towards the goal, not to leave the students aimless. However, as Armstrong's heuristic devotees retired, the new generation lacked the training in the proper context and application of the method. As a result, by the 1920s, the heuristic approach was in decline. Dorothy Turner described this decline in *History of Science Teaching in England*:

> Unfortunately the disciples of Armstrong went too far. They regarded practical work in the school laboratory as an end in itself. ...They were afraid to tell their pupils anything, and the unfortunate young

---

[105] Hughes, A. M. and Stern, R. (1906). *A Method of Teaching Chemistry in Schools*. Cambridge University Press, Cambridge.

[106] Westaway, K. M. (ed.), (1932). *A History of Bedford High School for Girls*, F. R. Hockliffe, Bedford, p. 74.

investigators often gained nothing from their work in the laboratory but a marked distaste for the subject. The over emphasis on method and the ignorance of the importance of the content has done much to bring heuristic teaching into disrepute.[107]

Association of Women Science Teachers

Oft-forgotten from educational history, the Association of Women Science Teachers (AWST) played an important role in linking together these pioneering women.[108]

Origins of the AWST

It was the Science Section of the London Branch of the Association of Assistant Mistresses (AAM) from which the AWST was founded. There was a consensus in the Autumn of 1911 that it was important to have an organization to which all women science teachers could belong, not only those who were eligible as members of the AAM. There was deemed to be a particular urgency as, in their view, science teaching was undergoing rapid changes.

Thus, a formal body was needed which could discuss issues and present the conclusions to the educational authorities as the collective viewpoint of its members. Membership would be open to all current and former teachers of science at the secondary or post-secondary level. A six-member committee was struck to draft proposals for such an organisation.[109] Of the six members, three were chemistry teachers: Edith Lees of Clapham High School, Rose Stern of North London Collegiate School (see Chapter 9), and Yolande Raymond of St. Paul's Girls' School.

---

[107]Turner, D. M. (1927). *History of Science Teaching in England.* Arno Press, NY, p. 145.

[108]*Minutes,* Association of Women Science Teachers, Brotherton Library Archives, University of Leeds.

[109]Layton, D. (1984). *Interpreters of Science: History of the Association for Science Education.* John Murray, London, pp. 39–40.

Edith Lees

Edith Sarah Lees[110] was born on 24 August 1865 in Bermondsey, London, the daughter of Joseph Lees, physician, and Jane Eliza Brett. She was privately tutored and then entered Clapham High School in 1881 and matriculating in 1888. Lees was hired as a teacher at the school in 1889, while she continued her education through courses at UCL in the 1890–1891 academic year.

Lees was obviously a powerful voice for the teaching of science in girls' schools. After her retirement in July 1926, Miss Barratt, the Headmistress commented: "She had from the beginning struggled for the teaching of science in girls' schools, and thanks to her efforts and those of like mind the subject now stood where it did."[111] While a former student, Emily Kingston, made a similar comment: "Others will write with more authority about her historic struggle and achievement in making the Sciences an essential subject in the curriculum of Girls' Schools."[112]

Yolande Raymond

Yolande Gabrielle Raymond,[113] the third chemistry teacher on the committee, was born on 19 September 1871 in County Kerry, Ireland. She was educated at the Clergy Daughter's School in Bristol, entering Newnham College in 1890. Raymond completed the Natural Science Tripos in 1893. She subsequently held positions as chemistry teacher at Liverpool (Belvedere) High School (1893–1898),[114] Sydenham High School (1898–1905), and St. Paul's Girls' School (1905–1912).[115] She was appointed Headmistress of Kidderminster High School in 1912, a post she held until 1932. Raymond died on 31 August 1952.

---

[110] *Staff Records*, Archives, Clapham High School for Girls.

[111] Anon. (1927). Changes in Staff. *Clapham High School Magazine* (27): 39.

[112] Kingston, E. (1948). Foreword. *Clapham High School Old Girls' Society News Sheet* 1.

[113] White, A. B. (1979). Y. G. Raymond. *Newnham College Register 1871–1971*, Vol. I, *1871–1923*. Newnham College, Cambridge, p. 107.

[114] Anon. (1898). Editorial. *Liverpool High School Chronicle* (11): 4.

[115] Anon. (1912). Changes in staff. *The Paulina: The Magazine of St Paul's Girls' School* March (23): 2.

Women Chemistry Teachers 349

Meetings of the AWST

Two meetings were held each year: a winter one in London and a summer one elsewhere in the country. During the early years, meetings were often followed by talks or tours. For example, in 1914, the biochemist Ida Smedley Maclean (see Chapter 9) spoke on "The Formation of Fats in Living Organisms."

The first chemistry teacher to occupy a position was Rose Stern as Vice-President from 1912 to 1913. She was followed by Edith Lees as President in 1919.[116] Women chemistry teachers continued to be prominent in the AWST in later decades. For example, the President in 1921 and 1922 was a chemist, M. Beatrice Thomas[117] (see Chapter 3) of Girton College, while Clara Taylor of Redlands High School was President for 1925 and 1926 and Vice-President for 1927.

Of the 465 members of the AWST in 1926, only 10 were married or widowed. David Layton has postulated that the AWST played an important social as well as academic role in the women science teachers' lives:

> For the large proportion of unmarried women members, the Association's social functions were as important as, and probably indivisible from, its professional ones. The 'cozy intimacy' of the meetings in the early years, and the 'jolliness and earnestness' of later gatherings were much valued by the predominantly spinster membership.[118]

In the AWST Annual Report of 1921, the focus was chemistry. Clara Taylor gave a lengthy annual address including a critique of Armstrong's heuristic method:

> We have not made the most of our resources, for so often we have deliberately stood aside and tried to let the facts speak for themselves; and to

---

[116] Anon. (1962). *A Short History with a List of Members in January 1962*. The Association of Women Science Teachers, Archives, Leeds University.

[117] Taylor, P. M. (1954). Miss M. B. Thomas, 1873–1954. *Association of Women Science Teachers, Report for 1954*. Archives, Leeds University, pp. 43–44.

[118] Note 109, Layton, p. 63.

350 *Pioneering British Women Chemists: Their Lives and Contributions*

most of us facts will not speak for themselves; they need an interpreter, and we have not sufficiently realised that our position is to interpret.[119]

Taylor's address was followed by a presentation by Rose Stern on the teaching of chemistry in the middle school. Stern began by giving a general overview of the purpose of the chemistry laboratory as an exercise in developing mental powers:

> A great deal has been accomplished when a whole class of beginners can set up an experiment, bending their own glass tubing, and fitting corks so that it succeeds. How often does one see a girl trying to fit a large cork into a small opening, and how often is there disappointment on finding out that a cork becomes larger when a tube is put through it. These may be small things, but they ensure good training. The class learns that little faults lead to bad endings.[120]

Welsh Branch of the AWST

Autonomous Branches of the AWST were set up across the country, the earliest being North-West (1914), Midlands (1918), Wales (1921), North-East (1926), and Northern (1926).[121] An account of the convoluted history of the North-East (Yorkshire) Branch has been given by Edgar Jenkins.[122] There was a considerable degree of autonomy among the branches of the AWST, in part because local meetings were easier to organise and attend.

However, it was only the Welsh Branch for which we could find any detailed information.[123] According to the *Committee Minute Books*, most of the meetings of the Welsh Branch involved lectures or discussions. For

---

[119] Anon. (1921). *Association of Women Science Teachers, Report for 1921*. Archives, Leeds University, p. 6.

[120] Note 119, Anon., p. 8.

[121] Note 109, Layton, pp. 50–54.

[122] Jenkins, E. (2009). The 'Yorkshire Branch' of the Association of Women Science Teachers, 1926–1963. In *75 Years and More: The Association for Science Education in Yorkshire*. Association for Science Education.

[123] *Minutes: Annual General Meeting*. Welsh Branch, Association of Women Science Teachers. Richard Burton Archives, University of Swansea.

example, on 12 December 1922, a presentation was given by a Miss Abbott at City of Cardiff High School for Girls on "Teaching of Science in Girls Schools." There were also visits to company works, such as the Mellingriffin Tinplate Works in 1927, and to university research laboratories. In addition, the Welsh Branch organised scientific excursions, including a traverse of a South Wales coalfield with a geologist.

As can be inferred from the above, though it was called the Welsh Branch, in fact, for meetings, it was the Cardiff–Swansea Branch. This issue arose at the Annual General Meeting of 1936 when the following Motion was proposed and approved: "...that the scattered members in the schools of North Wales should be invited to meet the members of this branch at some convenient place... Llandrindod Wells or Shrewsbury for the summer meeting."[124] However, there was no indication that such a meeting took place.

The meeting at which the formation of a Welsh Branch was proposed was held at the City of Cardiff High School for Girls (CCHSG), and the subsequent meetings alternated in location between Cardiff and Swansea. During the inter-War period, the attendance was typically between 15 and 20 members. The Welsh Branch continued to be active until the merger of the AWST with the Science Masters' Association in 1962.

Florence Gibson

Florence Gibson, who was appointed Assistant Mistress at CCHSG in 1896, played an active role in the Welsh Branch in the early years. She was born in 1870 at Ecclesfield, Yorkshire, to John Gibson, grocer and druggist, and Ann Chambers. Gibson graduated with an external B.Sc. from the University of London in 1890.

Though Gibson was appointed to teach botany and some mathematics, her main interest was chemistry. When the school building was designed in 1897, a chemistry laboratory was included in the plans. However, the authorities struck out the words "chemistry laboratory" as, in their opinion, chemistry was not required for girls' education. Instead,

---

[124] Anon. (1936). *Minutes: Annual General Meeting*. Welsh Branch, Association of Women Science Teachers. Richard Burton Archives, University of Swansea.

352   *Pioneering British Women Chemists: Their Lives and Contributions*

the space was relabelled as a "sewing room."[125] Gibson was determined to return the room to its original purpose:

> ... it was not long before she was demonstrating the rudiments of Chemistry and teaching General Science in the room labelled 'Sewing Room', and agitating for the equipment that was to turn it into a small cramped laboratory, but one in which the girls could experiment themselves.[126]

Gibson was promoted to Senior Mistress in 1919. She founded the Field Club in 1910 which continued to be active until her retirement in 1924. One of her students wrote:

> Miss Gibson left us last term after a teaching career of — as she put it — 100 terms of which 82, if we reckon rightly, were in the Cardiff High School. Well has she earned her right to far more ease and leisure than she ever allowed herself during all that time. No words can express what the school owes to Miss Gibson's skill, devotion, energy and good will..... May her raspberries always ripen and the potato crop never fail.[127]

Gibson retired to Somerset, dying at Chard, Somerset in 1956, aged 87.

Margaret John

Margaret Elizabeth John[126] was Gibson's successor at CCHSG and John also followed Gibson on the committee of the Welsh Branch. Born at Barry, Glamorgan, on 15 January 1898, she was the daughter of John Harris John, farmer, and Phoebe Harris. John had been admitted to Howell's School, Llandaff, on a scholarship at age 13 and subsequently became Head Girl. She entered Bedford College in 1917, obtaining a

---

[125]"M.C." [Mary Collins] (1924). The Opening. In *City of Cardiff High School for Girls, 1895–1924*. City of Cardiff High School for Girls, Cardiff, p. 18.

[126]Carr, C. (1955). *The Spinning Wheel: City of Cardiff High School for Girls, 1895–1955*. Western Mail and Echo Ltd., Cardiff, p. 123.

[127]Anon. (December 1925). Changes. *City of Cardiff High School for Girls Magazine* 39.

B.Sc. in 1921.[128] John taught at King's High School for Girls, Warwick, from 1921 until 1924, before returning to Bedford College to undertake a M.Sc. in Inorganic Chemistry with James Spencer which she completed in 1926.

In 1926, John obtained her appointment as Chemistry Teacher at CCHSG. She organized an active Science Club which held its first meeting in 1930.[129] A former student, Alicia Lewis reminisced about John:

> I can always see Miss John walking the length of the Chemistry Laboratory, that Palace of Experiment on the top floor of the new building, and seating herself on the dais at the end. Miss John taught us and chaffed us through Chemistry with sureness of method and deftness of touch which makes me try to reflect her lab of those days in my kitchen of these.[130]

John retired in 1946, after teaching 20 years at CCHSG. She moved to Australia and sent back to the Howell's School, Llandaff, detailed accounts of her life in the Antipodes.[131] John must have returned to Wales for she died in 1987 in Radnorshire, Powys, aged 89.

Commentary

Teaching chemistry in a girls' school was one career which was readily available to women chemistry graduates. Yet, for these pioneers it came at a price: an expectation of total dedication of one's life to the teaching career. For many of them, stress resulted in illness and early retirement. However, during the inter-War period, circumstances for women chemistry teachers changed, as we will show in Chapter 16.

---

[128] *Staff Records*, Archives, City of Cardiff High School for Girls.

[129] Note 126, Carr, p. 125.

[130] Note 126, Carr, p. 124.

[131] John, M. (January 1949). In Australia. *Hywelian (Old Girls) Magazine* 30–33.

Chapter 11

Hoppy's "Biochemical Ladies"

Dorothy Jordan Lloyd (1 May 1889–21 November 1946)

During the 1880–1940 period, the University of Cambridge was a contradiction: on the one hand, women were barred from formal undergraduate degrees; on the other hand, women scientists were to be found in many of the university's research laboratories. There were women researchers in

356 *Pioneering British Women Chemists: Their Lives and Contributions*

both the Balfour Biological Laboratory for Women[1] and the Cavendish Laboratory for Physics.[2] Nevertheless, of all the research schools, the biochemical research group of Frederick Gowland Hopkins[3] stands out as exceptional in its high proportion of women researchers.

Women and Biochemistry

The question arises why women were attracted to, and flourished in, the field of biochemistry.[4] It could be argued that biochemistry appealed to women scientists by its relevance to the understanding of living processes; however, there were other fields in which British women scientists clustered and gained recognition, such as crystallography[5] (see Chapter 12) and astronomy.[6]

Why was Biochemistry at Cambridge Women-Friendly?

Margaret Rossiter has proposed that the fields in which women made up a significant proportion were often the new rapid-growth areas where the demands for personnel were so great that there was less strident objection to the hiring of women.[7] Biochemistry certainly fitted this paradigm, as the biochemistry historian Robert Kohler described:

> Biochemistry is one of those fascinating but problematic 'new sciences' that have appeared with some regularity in the history of science. It came

---

[1]Richmond, M. L. (1997). "A lab of one's own": The Balfour Biological Laboratory for women at Cambridge University, 1884–1914. *Isis* **88**: 422–455.

[2]Gould, P. (1997). Women and the culture of university physics in late nineteenth-century Cambridge. *British Journal for the History of Science* **30**: 127–149.

[3]Dale, H. H. (1948/1949). Frederick Gowland Hopkins, 1861–1947. *Obituary Notices of Fellows of the Royal Society* **6**: 115–145.

[4]Long, V., Marland, H. and Freedman, R. B. (2009). Women at the dawn of British biochemistry. *The Biochemist* **31**(4): 50–52.

[5]Ferry, G. (2014). History: Women in crystallography. *Nature* **505**: 609–611.

[6]Brück, M. (2009). *Women in Early British and Irish Astronomy: Stars and Satellites.* Springer, Dortrecht.

[7]Rossiter, M. W. (1980). "Women's Work" in science, 1880–1910. *Isis* **71**: 381–398.

quite suddenly on the scene in the early years of this century, with a new name and intimations of new insights into the nature of life processes.[8]

Mary Creese has expanded upon this contention:

> The entry paths and entry qualifications of its practitioners were not well defined. ...This lack of prestige, due to the slowness of academic chemists to recognize the full power and potential of research in the field, offers one explanation for its [biochemistry] relative openness to women.[9]

Though such explanations have some validity, we have provided evidence that the role of mentor was another important factor.[10] Rossiter has shown that this factor played a role in the high number of women in the Biochemistry Department of Yale University between 1896 and 1935. The Yale women biochemists were all former students of Lafayette Mendel.[11] In fact, of the 124 students who completed doctoral degrees with Mendel, 48 were women. Rossiter concluded that the personality and supportiveness of Mendel was a major factor in this exceptional percentage.

Frederick Gowland Hopkins

In Britain, many women chemists veered towards biochemistry for their careers, joining such organizations as the Lister Institute, London, and the Rowett Institute in Aberdeen (see Chapter 7). Nevertheless, it was the Cambridge group under Hopkins that provided a unique community of talented women biochemists.

---

[8] Kohler, R. E. Jr., (1973). The enzyme theory and the origin of biochemistry. *Isis* **64**: 181–196.

[9] Creese, M. R. S. (1991). British women of the 19th and early 20th centuries who contributed to research in the chemical sciences. *British Journal for the History of Science* **24**: 275–306.

[10] Rayner-Canham, M. and Rayner-Canham, G. (1996). Women's fields of chemistry: 1900–1920. *Journal of Chemical Education* **73**: 136–138.

[11] Rossiter, M. W. (1994). Mendel the mentor: Yale women doctorates in biochemistry, 1898–1937. *Journal of Chemical Education* **71**: 215–219.

358   *Pioneering British Women Chemists: Their Lives and Contributions*

The history of modern biochemistry[12] is inextricably linked with the name of Hopkins, Nobel Laureate, though his encouragement of women to enter biochemical research has been less widely recognised.[13] Creese has been one of the few to note this facet of his character, commenting:

> At a time when there were practically no women research workers in any of the other university departments at Cambridge, Hopkins gave them places in his, despite the criticism which this brought him. Even in the 1920s and 1930s, when, as a Nobel laureate with a world-wide reputation he received hundreds of applications for places in his laboratory, nearly half of the posts in his Department went to women scientists.[14]

But there is more to creating a women-friendly environment than simply opening the doors to women. One of these factors was the personal style and attitude to their research students. In the case of Hopkins, Dorothy Moyle, one of his former students, stated that Hopkins provided valuable moral support and that he regarded his students as fellow researchers rather than as underlings in a research empire.[15] Another of his former students, Malcolm Dixon, explained:

> Hopkins was one of the kindest and most lovable of men. ... He never made an unkind or irritable remark, though he could be critical on occasion, it was always with a courtesy that left no sting. He had great charm of manner and was invariably courteous, even to the least important of us. He was always ready to talk over our work and ideas, and somehow contributed to make us feel by the way that he listened to us that our ideas were extremely interesting and our work important.[16]

---

[12]Teich, M. (1965). On the historical foundations of modern biochemistry. *Clio Medica* **1**: 414–457.

[13]Weatherall, M. and Kamminga, H. (1992). *Dynamic Science: Biochemistry in Cambridge 1898–1949*. Cambridge Wellcome Unit for the History of Medicine, Cambridge.

[14]Note 9, Creese, p. 296.

[15]Needham, J. and Needham, D. M. (1949). Sir F. G. Hopkins' personal influence and characteristics. In Needham, J. and Baldwin, E., (eds.), *Hopkins and Biochemistry*. Cambridge University Press, Cambridge, pp. 113–119.

[16]Dixon, M. (1949). Sir F. Gowland Hopkins, O.M., F.R.S. *Nature* **160**: 44.

Both Mendel and Hopkins engendered a social cohesiveness in their biochemical research students that was remarkable. Students of Hopkins ("Hoppie" or "Hoppy" as he was known) set up "Hoppie Societies" wherever they were and held periodic meetings. For example, the Hoppie Society of America met from 1934 until (at least) sometime in the 1940s, and there was even a Hoppie Club of Leningrad.[17] Not only was there a camaraderie among Hopkins' students *per se* but also there seemed to be a specific friendship amongst the women biochemists. This is particularly apparent from obituaries, where the obituary of one woman biochemist was often written by another woman biochemist.

Contributing to the socialization, Hopkins believed that the tea-room was one of the two most important rooms in the research building (the other being the departmental library).[18] Throughout the 50 years of Hopkins' time at Cambridge, there were weekly or fortnightly tea-room meetings of his group at which research work would be presented.

In addition to Hopkins' encouragement of a social environment, there was a second relevant factor of his personality. This behavioural characteristic was summarised best by Joseph Needham and Dorothy Moyle:

> … he was a living embodiment of the Confucian maxim that one should behave to everybody as one receiving a great guest. The humblest laboratory assistant or the youngest research worker was always sure of a welcome, and a hearing much longer than he was likely to get from any other scientific man of the same standing or generation. Hopkins had faith in people. Colleagues were known to remark lightly, "All Hoppy's geese are swans," but they forget that there is an induction process by which certain geese may be turned into swans if given the hormone of encouragement.[19]

Another aspect of making a group women-friendly was to demonstrate an enthusiasm for the subject. Vivienne Gornick interviewed a large number of women scientists and she commented:

---

[17]Reports of these Societies and Clubs were sent to Hopkins and the items are to be found in Hopkins' Archives, Cambridge University Library, Cambridge.

[18]Note 15, Needham and Needham, p. 114.

[19]Note 15, Needham and Needham, pp. 114–115.

360   *Pioneering British Women Chemists: Their Lives and Contributions*

> Each of them had wanted to know how the physical world worked, and each of them had found that discovering how things worked through the exercise of her own mental powers gave her an intensity of pleasure and purpose, a sense of reality nothing else could match.[20]

This portrays exactly the environment in the Hopkins group as Dixon remarked: "To work there was to feel the thrill and sense of adventure in penetrating into the secrets of living matter, and life was never dull" (see Footnote 19).

Some Women Biochemists at Cambridge

Women undergraduates at Cambridge largely led their separate and sheltered lives in Newnham or Girton (see Chapter 3). However, for graduate work, women had to enter the colder climate of a male-dominated laboratory. The physiologist and biochemist William Bate Hardy[21] recalled the atmosphere in the 1880s and 1890s:

> At that time women were rare in scientific laboratories and their presence by no means generally acceptable-indeed, that is too mild a phrase. Those whose memories go back so far will recollect how unacceptability not infrequently flamed into hostility.[22]

Thus, it is particularly striking that biochemistry at Cambridge attracted a significant number of women graduates as we will show subsequently. In fact, at least 60 women biochemists worked in the biochemistry unit during Hopkins' "reign," including some from overseas, such as Kamala Bhagwat (Mrs. Sohonie).[23] Bhagwat recalled her arrival at Cambridge in 1937:

---

[20] Gornick, V. (2009). *Women in Science: Then and Now*, The Feminist Press at CUNY, New York, p. 15.

[21] "F.G.H." [Hopkins, F. G.] and "F.E.S." (1934). William Bate Hardy, 1864–1933. *Obituary Notices of Fellows of the Royal Society* **1**: 326–333.

[22] Hardy, W. B. (1932). Mrs. G. P. Bidder. *Nature* **130**: 689–690.

[23] Haines, M. C. (2001). Sohonie, Kamala née Bhagwat. *International Women in Science: A Biographical Dictionary to 1950*. ABC-CLIO, Santa Barbara, California, p. 291.

I applied for admission to his [Hopkins'] laboratory, although it was already full. Then the unexpected happened — a kind scientist already working in the laboratory offered me the daytime use of his bench while he would work at night. Professor Hopkins accepted this solution and I was admitted to this great laboratory on 18 December 1937 — the happiest and proudest day of my life.[24]

We have chosen here to document five women biochemists whose contributions, in our view, were particularly interesting.

Edith Willcock

Though Edith Gertrude Willcock had the briefest sojourn at Cambridge of any of the women biochemists, it was the work she performed with Hopkins that was to become a classic in biochemistry.[25] Willcock was born on 7 January 1879 at Albrighton, daughter of Robert A. Willcock, solicitor, and Emma J. Willcock. She was educated at the King Edward VI High School for Girls (KEVI) in Birmingham.[26] Like so many chemistry students at KEVI, Willcock applied to Newnham College for her university education. She entered in 1900, completing Parts I and II of the Natural Science Tripos in 1904.

During her undergraduate studies, Willcock undertook research with Hardy on the oxidation of iodoform, initiated by radiation from radium. Of more significance, on her own, she published a paper on the effects of radium on animal life — probably one of the first studies that showed the damaging effects of exposure to radioactive elements.

Willcock had been inspired by the lectures of Hopkins. She recalled that he did not simply provide facts but left her with: "… a realisation of the existence of vast unexplored tracts and the unfolding of immense

---

[24] Sohonie, K. (1982). Women in Cambridge biochemistry. In Richter, D. (ed.), *Women Scientists: The Road to Liberation*. Macmillan, London, p. 19.

[25] Teich, M. with Needham, D. M. (1992). *A Documentary History of Biochemistry 1770–1940*. Leicester University Press, Leicester, pp. 314–316.

[26] White, A. B. (ed.), (1979). *Newnham College Register, 1871–1971*, Vol. I, *1871–1923*, 2nd edn., Newnham College, Cambridge, p. 33.

362 *Pioneering British Women Chemists: Their Lives and Contributions*

opportunities for research."[27] Resulting from this fervour, Willcock became a Bathurst Research Student (1904–1905) and a Newnham Research Fellow (1905–1909) with Hopkins. It was her discoveries with him between 1905 and 1909 which ensured Willcock a place in the history of biochemistry.

At the time (1906), it was believed that diets were complete as long as they contained the appropriate chemical functional units. In particular, it was known that the indol unit was required for certain biological functions. Willcock and Hopkins were the first to show that diets had to contain specific molecules, in this case, tryptophan, and that other indol-containing compounds would not function in its place.[28] This initial study focussed Hopkins' attention on the essentiality in diets of certain amino acids and led him to formulate his hypothesis of "accessory food factors" later to be named "vitamins."[29] Willcock, another of the steamboat ladies (see Chapter 3), received a D.Sc. from Trinity College, Dublin, on the basis of her research.

At that time, over 60% of the women who studied science at Cambridge remained single,[30] but Willcock was one of the minority who married. In 1909, she married Cambridge zoologist, John Stanley Gardiner.[31] During the First World War, Willcock was a local consultant for the British Ministry of Agriculture on the raising of rabbits and poultry and she wrote leaflets on the subject for public distribution. In addition, she became an advisor on oyster culture. Willcock had several other interests, being a recognized water-colour artist, singer, and the author of a popular children's book, *We Two and Shamus*, published in 1911. She had

---

[27] Stephenson, M. (1949). Sir F. G. Hopkins' Teaching and Scientific Influence. In Needham, J. and Baldwin, E., (eds.), *Hopkins and Biochemistry*, Cambridge University Press, 29–38.

[28] Willcock, E. G. and Hopkins, F. G. (1906). The importance of individual amino-acids in metabolism. *Journal of Physiology* **35**: 88–102.

[29] Kamminga, H. and Weatherall, M. W. (1996). The making of a biochemist I: Frederick Gowland Hopkins' construction of dynamic biochemistry. *Medical History* **40**: 269–292.

[30] MacLeod, R. and Moseley, R. (1979). Fathers and daughters: Reflections on women, science and Victorian Cambridge. *History of Education* **8**: 321–333.

[31] Forster-Cooper, C. (1947). John Stanley Gardiner, 1872–1946. *Obituary Notices of Fellows of the Royal Society* **5**: 541–553.

two daughters and travelled widely with them and her husband. The 1939 wartime registration listed Willcock as married, with domestic duties. She died on 8 October 1953 in Cambridge.

Muriel Wheldale (Mrs. Onslow)

Muriel Wheldale[32] was one of only two genetics researchers in the first decade of the 20th century who performed plant breeding experiments and investigated the corresponding chemistry of flower pigments of the plants (the other being Erwin Bauer in Germany).[33] To indicate the importance of her contributions, J. B. S. Haldane[34] used her research to conclude that genes controlled the formation of large molecules, such as pigment molecules.[35]

Born on 31 March 1880 in Aston, Birmingham, Wheldale was the daughter of John Wheldale, barrister, and Sarah Fannie Hayword.[36] Like Willcock, Wheldale attended KEVI and then went up to Newnham in 1900, where she took First Class in both parts of the Natural Sciences Tripos, specializing in botany. In 1903, Wheldale joined the research group of William Bateson[37] as a Bathurst Research Student (1904–1906) and a Newnham Research Fellow (1909–1914). With the revival of Gregor Mendel's theories of hereditary at the beginning of the 20th century, Bateson had become one of Mendel's most ardent champions. It was Bateson who had devised the term "genetics" to describe the field, and

---

[32]Rayner-Canham, M. and Rayner-Canham, G. (April 2002). Muriel Wheldale Onslow, 1880–1932, Pioneer plant biochemist. *The Biochemist* **24**(2): 49–51.

[33]Olby, R. (1989). Scientists and bureaucrats in the establishment of the John Innes Horticultural Institution under William Bateson. *Annals of Science* **46**: 497–510.

[34]Pirie, N. W. (1966). John Burdon Sanderson Haldane, 1892–1964. *Biographical Memoirs of Fellows of the Royal Society* **12**: 218–249.

[35]Glass, B. (1965). A century of biochemical genetics. *Proceedings of the American Philosophical Society* **109**: 227–236.

[36]Creese, M. W. (2004). Onslow, Muriel Wheldale, 1880–1932. *Oxford Dictionary of National Biography*. Oxford University Press (accessed 16 December 2018).

[37]"J.B.F." (1927). Obituary Notices of Fellows Deceased. William Bateson, 1861–1926. *Proceedings of the Royal Society of London, Series B* **101**: i–v.

364 *Pioneering British Women Chemists: Their Lives and Contributions*

he had surrounded himself with a group of enthusiastic research students, many of them women.[38]

Wheldale's particular niche was the study of the inheritance of flower colour in *Antirrhinums*, her research proving to be among the most widely recognised of Bateson's group.[39] She published a full-factorial analysis of flower colour inheritance in *antirrhinums* in 1907. In a reference for Wheldale's research, Bateson commented:

> The problem of colour inheritance in *Antirrhinum*, which she set out to solve, proved to be far more complex than was expected, and the solution she proposed (*Proc. Roy. Soc.* Vol. 79, B, 1907) is entirely her own work. There is every reason to believe that it is correct and I regard the paper as one of considerable value.[40]

This landmark publication was the first of a flurry of research papers on the linkage between the inheritance of genetic factors and the production of the pigments, the anthocyanins. Her research culminated in the writing of the classic monograph, *The Anthocyanin Pigments of Plants*.[41] In the Preface of the monograph, she commented:

> Herein lies the interest connected with anthocyanin pigments. For we have now, on one hand, satisfactory methods for the isolation, analyses and determination of the constitutional formula of these pigments. On the other hand, we have the Mendelian methods for determining the laws of their inheritance. By a combination of these two methods we are within reasonable distance of being able to express some of the phenomena of inheritance in terms of chemical composition and structure.[42]

---

[38]Richmond, M. L. (2001). Women in the early history of genetics: William Bateson and the Newnham College Mendelians, 1900–1910. *Isis* **92**: 55–90.

[39]Richmond, M. L. (2007). Muriel Wheldale Onslow and early biochemical genetics. *Journal of the History of Biology* **40**: 389–426.

[40]Bateson Letters, Innes Collection, cited in: Note. 38, Richmond, p. 82.

[41]Wheldale, M. (1915). *The Anthocyanin Pigments of Plants*, 1st edn., Cambridge University Press, Cambridge.

[42]Ref. 41, Wheldale, p. v.

Wheldale became convinced that the future lay not with the genetic aspects of botany, but with the biochemical basis of plant pigment synthesis, that is, chemical genetics.[43] To this end, she briefly attended the University of Bristol to strengthen her background in biochemistry. In 1911, Wheldale joined Bateson's new group at the John Innes Horticultural Institution at Merton Park, Surrey.

By 1913, Wheldale's fame was such that she was one of the first three women (the other two being Ida Smedley and Harriette Chick) to be elected to the Biochemical Club (see Chapter 9),[44] the forerunner of the Biochemical Society. However, in 1914, Wheldale decided her future lay in Cambridge and accepted an offer to join Hopkins' group.

With Wheldale's departure, Bateson lost not only one of the most gifted members of his research team but also his leading artist, as he described:

> As an artist in colour she has extraordinary skill. If she leaves us, her loss will be a serious one, for she is the only person I know who can reproduce the colours of flowers in such a way as to be an exact record. In respect of accuracy and appreciation of what scientific colouring should be, the quality of her painting far exceeds that of any professional whose work I know.[45]

In 1916, Wheldale was introduced to Huia Onslow, second son of the 4th Earl of Onslow. As a result of a diving accident, he had been paralysed from the waist down. Nevertheless, he began a research program in chemical genetics from a laboratory constructed in his home. At their first meeting together, they had tea followed by a "certain amount of Mendelian discussion."[46] Over the following years, Wheldale spent much of her time

---

[43] Scott-Moncrieff, R. (1981). The classical period in chemical genetics: Recollections of Muriel Wheldale Onslow, Robert and Gertrude Robinson and J. B. S. Haldane. *Notes and Records of the Royal Society of London* **36**: 125–154.

[44] Goodwin, T. W. (1987). *History of the Biochemical Society 1911–1986*. Biochemical Society, London, p. 15.

[45] Bateson Letters, Innes Collection, cited in: Note. 38, Richmond, pp. 82–83.

[46] Onslow, M. (1924). *Huia Onslow: A Memoir*, Edward Arnold, London, p. 162.

366   *Pioneering British Women Chemists: Their Lives and Contributions*

helping Onslow with his research, particularly that on the origins of the iridescence of some butterflies, moths, and beetles (though she was never acknowledged in his publications). In 1919, they were married and, until his death in 1922, Wheldale became the conduit between Onslow and the Hopkins group.[47]

The same anthocyanin pigments that Wheldale had followed genetically with Bateson, she studied chemically with Hopkins. This fruitful line of research cemented her reputation as a leading chemical geneticist.[48] Following from her biochemical studies, she wrote a second edition of *The Anthocyanin Pigments of Plants*.[49] This was no mere reprint of the first edition, but a complete revision in light of the tremendous biochemical advances of the previous decade. As she noted:

> Since the appearance of the first edition the publications of greatest value on the subject of anthocyanin pigments have been in connection with the chemistry and biochemistry of these substances. This later work has now been included, and the present state of our knowledge of the significance of the pigments in relation to plant metabolism has, as far as possible, been indicated.[50]

The biochemistry group wrote and had published their own magazine, *Brighter Biochemistry*, which contained short stories, pen-and-ink cartoons, and plagiarized poetic verse. One of these modified poems was penned by "D.R.P.M." — Dorothy Moyle (see the respective section) — about Wheldale's work.

> *A weary lot is thine, fair maid*
> *A weary lot is thine,*
> *To search for anthocyanins*
> *In Primula of Sine.*

---

[47]"F.G.H." [Hopkins, F. G.]. (1923). Obituary Notices: Victor Alexander Herbert Huia Onslow. *Biochemical Journal* **17**: 1–4.

[48]Lawrence, W. J. C. (1950). Genetic control of biochemical synthesis as exemplified by plant genetics — flower colours. *Biochemical Symposia* **4**: 3–9.

[49]Onslow, M. W. (1925). *The Anthocyanin Pigments of Plants*, 2nd edn., Cambridge University Press, Cambridge.

[50]Note 49, Onslow, p. vii.

Some picrate crystals on a plate,
A spectrum in the blue —
A row of test-tubes in a rack,
No more of me you knew, My Love!
No more of me you knew.
This morn is merry June, I trow,
The rose is budding fain,
But you must grind her petals bright
In dish of porcelain.
She tipped her beaker as she spake,
It crashed upon the floor:
Far in the distance came a voice,
Said 'Adieu for evermore, My Love!
And adieu for evermore.[51]

Wheldale's devotion to research was balanced by an enthusiasm for teaching. From 1907 until its closure in 1914, she was a Demonstrator in Physiological Botany in the Balfour Biological Laboratory for Women. Then, from 1915 until 1926, she held the position of Assistant in Plant Biochemistry under Hopkins and in 1926, she was promoted to university lecturer — one of the first women to hold this rank at Cambridge. Her pedagogical interest led her to write *Practical Plant Biochemistry*[52] in 1920, followed by the text *Principles of Plant Biochemistry*,[53] the first volume of which was published in 1931.

The latter book was written at Wheldale's house in Norfolk, one of her two favourite places, the other being the Balkans where she spent her holidays. Unfortunately, she died on 19 May 1932 in Cambridge, before completing the second volume, her work on flower pigments and genetics being continued by Rose Scott-Moncrieff (Mrs. O. M. Meares).[54]

---

[51] "D.R.P.M." (May 1931). Ode to a lady researching on anthocyanins. *Brighter Biochemistry* (8): 16.

[52] Onslow, M. W. (1920). *Practical Plant Biochemistry*. 1st edn., Cambridge University Press, Cambridge: 2nd edn., 1923.

[53] Onslow, M. (1931). *The Principles of Plant Biochemistry:* Part I, Cambridge University Press, Cambridge.

[54] Martin C. (April 2016). Rose Scott-Moncrieff and the dawn of (bio)chemical genetics. *The Biochemist* **38**(2): 48–53.

368 *Pioneering British Women Chemists: Their Lives and Contributions*

Marjory Stephenson

Of all Hopkin's women researchers, it is Marjory Stephenson[55] who is most deserving of recognition. As Robert Kohler has shown, the field of bacterial biochemistry was, in large part, defined by the work of Stephenson.[56] Stephenson was born on 24 January 1885 at Burwell, a village near Cambridge, and it was at Cambridge that she was to spend most of her life. Her mother was Elizabeth Rogers, and her father, Robert Stephenson was a farmer.[57]

Stephenson recalled: "I acquired a childish interest in science from my beloved governess and later from my father."[58] She was educated by the governess until the age of 12, at which point she received a scholarship to attend the Berkhamsted High School for Girls. It was Stephenson's mother who insisted that she obtain a university education and that Newnham was the appropriate place. She attended Newnham from 1903 until 1906, taking the Part I Natural Sciences Tripos in chemistry, physiology, and zoology.

After leaving Newnham, Stephenson would have liked to have studied medicine, but lacking the financial resources, she took teaching positions in domestic and household science for the next 5 years, including King's College for Women, in Kensington (see Chapter 4). The *King's College Magazine, Women's Department*, reported:

> The Cookery Side of the Domestic Science staff has a distinct acquisition in Miss Marjory Stevenson [sic] from the Gloucester School of Domestic Science. She has taken the Natural Science tripos (chief subject chemistry) and also a first class Diploma in Cookery; a combination

---

[55] (a) Robertson, M. (1949). Marjory Stephenson, 1885–1948. *Obituary Notices of Fellows of the Royal Society* **6**: 563–577; and (b) Mason, J. (1996). Marjory Stephenson 1885–1948. In Shils, E. and Blacker, C. (eds.), *Cambridge Women: Twelve Portraits*. Cambridge University Press, Cambridge, pp. 113–135.

[56] Kohler, R. E. (1985). Innovation in normal science: Bacterial physiology. *Isis* **76**: 162–181.

[57] Štrbáňová, S. (2016). *Holding Hands with Bacteria: The Life and Work of Marjory Stephenson*. Springer, Berlin.

[58] Stephenson, M. *Personal Records*. Royal Society, cited in: Mason, J. (1992). The admission of the first women to the Royal Society of London. *Notes and Records of the Royal Society of London* **46**(2): 284–285.

of certificates which seems at the present moment to have been achieved only by herself. Unfortunately, owing to present limitations, a good deal of her King's College work will be at the Clapham Housewifery School, though with our students only; she will not be able to spend much time in our own small Kitchen Laboratory.[59]

Her first saviour was Robert Henry Aders Plimmer.[60] In 1911, Plimmer, Assistant Professor in Physiological Chemistry at University College, London, invited Stephenson to teach advanced classes in the biochemistry of nutrition and to join his research group. It was somewhat contradictary, then, that Plimmer, co-founder of the Biochemical Club, proposed that only men should be eligible for membership.

As a result of her research on fat metabolism and on diabetes, Stephenson was awarded a Beit Memorial Fellowship in 1913; however, she relinquished the Fellowship on the outbreak of war, running soup kitchens in France and then supervising a nurses' convalescent home in Salonika. For this war work, she was awarded an M.B.E.

In 1919, Stephenson took up her Beit Fellowship again, moving to Cambridge to work with Hopkins. This move was probably on the advice of Plimmer, as Hopkins and Plimmer had worked together as co-editors on a series of biochemistry monographs. Hopkins was to be the greatest influence on Stephenson's career. After the expiry of the Fellowship, she worked on annual grants from the Medical Research Council (MRC) until 1929, when she obtained a permanent post with the MRC.

It was Hopkins who encouraged Stephenson to develop her own interests, and she chose chemical microbiology. She explained the reasons for her choice in the Preface of her book *Bacterial Metabolism*, first published in 1930:[61]

> Perhaps bacteria may tentatively be regarded as biochemical experimenters; owing to their relatively small size and rapid growth variations

---

[59] Anon. (1910). College notes. *King's College Magazine, Women's Department* (40): 5, K/SER1/170.

[60] Lowndes, J. (2004). Plimmer, Robert Henry Aders (1877–1955). *Oxford Dictionary of National Biography*. Oxford University Press (accessed 31 December 2018).

[61] Stephenson, M. (1930). *Bacterial Metabolism: Monographs on Biochemistry*. Longmans, Green & Co., London.

370   *Pioneering British Women Chemists: Their Lives and Contributions*

must arise very much more frequently than in more differentiated forms of life, and they can in addition afford to occupy more precarious positions in natural economy than larger organisms with more exacting requirements.[62]

Also in the Preface, Stephenson notes the indirect method by which the metabolism of bacterial cells was studied:

> ... we are in much the same position as an observer trying to gain an idea of the life of a household by a careful scrutiny of the persons and material arriving at or leaving the house; we keep accurate records of the foods and commodities left at the door, patiently examine the contents of the dustbin, and endeavour to deduce from such data the events occurring within the closed doors.[63]

The book was highly regarded, with a second edition published in 1938, a third in 1949, and a paperback reprint appearing in 1966.

Stephenson's studies of bacteria were immediately fruitful. At the time, studies had shown that bacteria possessed enzymes similar to those in other organisms. It was Stephenson who first separated out a pure cell-free enzyme from bacteria, in this case, lactic dehydrogenase from *E. coli*. Arguably one of her greatest claims to fame was her research with Leonard Stickland on hydrogenase.[64] Sydney Elsden and Norman Pirie co-authored one of her obituaries and, in it, they explained how the discovery occurred:

> About 1930 the Cambridgeshire Ouse was polluted by waste from a sugar-beet factory to such an extent that an active fermentation could be observed in the river itself. This provided an opportunity for investigating the methane fermentation, using the polluted river as an inoculum. These enrichment cultures, in addition to producing methane from

---

[62]Note 61, Stephenson, p. viii.

[63]Note 61, Stephenson, p. vi.

[64]Stephenson, M. and Stickland, L. H. (1932). Hydrogenase. Bacterial enzymes liberating molecular hydrogen. *Biochemical Journal* **26**: 712–724.

formate, reduced sulphate to hydrogen sulphide and made methane from carbon dioxide and hydrogen.[65]

Despite her inter-War pacifist views, with the onset of the Second World War, Stephenson devoted her time to assisting the war effort. One such avenue was the study of acetone-butyl alcohol fermentation as a means of synthesis of industrial solvents.[66] Of greater importance, however, was her work on pathogenic bacteria, and also her contributions to the MRC Committee on Chemical Microbiology. After the War, she studied the bacterial synthesis of acetylcholine in sauerkraut, while her last years were spent on an investigation of nucleic acids in bacteria and of their breakdown by enzymes within the cells.

Stephenson was largely responsible for the founding of bacterial chemistry as an autonomous branch of biochemistry; hence, it was appropriate that she should have received many honours. The most valued recognition was her election in 1945 as one of the first two woman Fellows of the Royal Society. The first woman proposed for election had been Hertha Ayrton,[67] a physicist, in 1902. Ayrton's nomination was rejected on the grounds of the ineligibility of women for Fellowship.

In 1922 and again in 1925, Caroline Haslett, Secretary of the Women's Engineering Society, wrote to the Royal Society asking for their position concerning the admission of women following the passage of the 1919 Sex Disqualification (Removal) Act. In 1925, the Secretary of the Royal Society, wrote to Haslett that there was now "general opinion" that women were eligible for admission, provided that "their scientific attainments were of the requisite standard."[68]

---

[65] Elsden S. R. and Pirie, N. W. (1949). Obituary Notices: Marjory Stephenson, 1885–1948. *Journal of General Microbiology* **3**(2): 329–339.

[66] Davies, R. and Stephenson, M. (1941). Studies on the acetone-butyl alcohol fermentation. *Biochemical Journal* **35**: 1320–1331.

[67] Mason, J. (1991). Hertha Ayrton (1854–1923) and the admission of women to the Royal Society of London. *Notes and Records of the Royal Society of London* **45**: 208.

[68] Mason, J. (1992). The admission of the first women to the Royal Society of London. *Notes and Records of the Royal Society of London* **46**: 279–300.

372 *Pioneering British Women Chemists: Their Lives and Contributions*

Yet, despite this admission of women's eligibility, nothing happened for nearly 20 years. As Hilary Rose has commented:

> This extraordinary gap suggests at best a collective amnesia — or perhaps a repression of memory — within the Royal Society, in which the fact of legal eligibility and the political likelihood of success become conflated to become an unstated and legally false, but socially powerful, consensus that women were not admissible.... But the pressure which had brought the reluctant admission that women were eligible had weakened. The interwar period, ... meant that the feminist organizations were functioning at just tick-over.[69]

It was not until 1943 that the issue was raised again, this time in an article in the British communist-leaning newspaper, the *Daily Worker*. The author of the article, evolutionary biologist J. B. S. Haldane, noted the "striking omission" of any women's names in the nomination list and added that there were "certainly half a dozen [women]" eligible. Biologist Lancelot Hogben wrote to Haldane on 30 July 1943 to ask Haldane for suitable names of women from the biologically related sciences.[70] Haldane replied on 18 August 1943 supporting the nomination of biochemist Marjory Stephenson:

> I think the strongest claim is that of Dr Marjory Stephenson who was the first person in the world to do work on bacterial metabolism as exact as that on mammalian metabolism, and who has continued to do good work in this field, discovering, for example, a number of new enzymes, in particular those dealing with the production and consumption of hydrogen.[70]

The question of the eligibility of women to become Fellows was raised again in 1944. This time, the Fellows voted by an overwhelming majority to amend the Statutes of the Society formally lifting any restriction on the election of women. Whereas it was common for about 8–10 Fellows to propose a candidate, in Stephenson's nomination, indicative of the breadth

---

[69] Rose, H. (1994). *Love, Power and Knowledge: Towards a Feminist Transformation of the Science*. Polity Press, London, p. 122.

[70] Haldane papers, cited in: Note 68, Mason, pp. 289–290.

of her support, 18 Fellows signed her certificate of application.[71] On 22 March 1945, among the successful candidates elected as Fellows of the Royal Society were the first two women: Marjory Stephenson and crystallographer Kathleen Lonsdale (see Chapter 12).

Stephenson had mixed feelings about being elected Fellow, as Elsden and Pirie explained:

> She [Stephenson] was unsparing in her condemnation of secretiveness, personal vanity and competitiveness in scientists and for this reason jeered at most of the medals and awards that scientists on occasion confer on each other.... Her pleasure [at being made an F.R.S.] was however marred by the realization that she might be accused of inconsistency on what might have been almost a matter of principle.[72]

Stephenson died on 12 December 1948 in Cambridge.

Dorothy Jordan Lloyd

Dorothy Jordan Lloyd[73] had the biological sciences in her genes: her father, George Jordan Lloyd, was a distinguished Professor of Surgery at University of Birmingham, while her paternal grandfather had been a Lecturer in Anatomy at the Old Mason College (the predecessor of the University of Birmingham). She was born on 1 May 1889 in Birmingham, her mother being Marion Annie Hampson Simpson. By the age of 12, Jordan Lloyd had already decided that she wanted to become a scientist. She followed the same path as both Willcock and Wheldale, attending KEVI and then Newnham College, entering the latter in 1908.

In 1912, Jordan Lloyd completed the Part II Tripos in zoology and obtained a B.Sc. (London). However, she was more interested in the functional and dynamic side of biology than in structural studies. Thus, as a Bathurst Scholar from 1912 until 1914, she chose to study the physico-chemical properties of proteins. This research was undertaken with Hardy,

---

[71] Note 69, Rose, p. 128.

[72] Note 65, Elsden and Pirie, p. 337.

[73] Phillips, H. (2004). Lloyd, Dorothy Jordan (1889–1946), *Oxford Dictionary of National Biography*. Oxford University Press (accessed 31 December 2018).

374 *Pioneering British Women Chemists: Their Lives and Contributions*

who had previously been the mentor of Willcock (see the respective section).

Then in 1914, Hopkins invited Jordan Lloyd to join his group as a Newnham Research Fellow. With the outbreak of war, she worked on culture media for meningococcus, one of the anaerobic pathogens involved in trench diseases, and on causes and prevention of "ropiness" in bread. For this research, Jordan Lloyd was awarded a D.Sc. (London) in 1916.

By 1921, her original work on proteins led to an invitation to join the British Leather Manufacturers' Research Association, a move which, according to Marjory Stephenson: "probably robbed this country of a distinguished professor of biochemistry."[74] As a final contribution to her original research field, Jordan Lloyd wrote the classic work *The Chemistry of Proteins*, published in 1926, for which Hopkins wrote the introduction. A second edition, co-authored with Agnes Shore, appeared in 1938.[75]

It was Jordan Lloyd who turned leather manufacture from a craft industry into a scientific process.[76] In 1927, she was appointed Director of the Association, a post in which she served until her death in 1946. For a woman to head such a large scientific organization was an amazing accomplishment for the period. Her predecessor, Robert Pickard,[77] commented in his obituary of her:

> Among scientific women of her time she perhaps was unique in that to the successful management of a large research organization she added the capacity to delegate its routine administration whilst retaining control, and at the same time also to play a leading and personal part in the actual research.[78]

---

[74]"M.S." [Stephenson, M.]. (1947). Dorothy Jordan Lloyd. *Newnham College Roll Letter* 54.

[75]Lloyd, D. Jordan (1926). *The Chemistry of Proteins*. J. & A. Churchill, London.

[76]Theis, E. R. (1947). Dorothy Jordan Lloyd. *Journal of the American Leather Chemists Association* **42**: 40–41.

[77]Kenyon, J. (1950). Robert Howson Pickard, 1874–1949. *Obituary Notices of Fellows of the Royal Society* **7**: 252–263.

[78]Pickard, R. (18 January 1947). Dorothy Jordan Lloyd, *Chemistry and Industry* 47.

In 1939, Jordan Lloyd was awarded the Fraser Muir Moffatt Medal by the Tanners' Council of America for her contributions to leather chemistry. Her change of career direction did not diminish her productivity, and over her lifetime, she authored or co-authored over 100 scientific papers together with planning and contributing to the three-volume *Progress in Leather Science, 1920–1945*, the classic textbook of leather technology.

A keen mountaineer, Jordan Lloyd achieved the distinction of making the first ascent and descent in 1 day of the Mittellgi ridge of the Eiger. Her mountain-climbing exploits led to her election to the Ladies' Alpine Club. Later, she adopted competition horse riding as her main interest, which caused her friend Stephenson to comment: "in both those pursuits she excelled and certainly the element of danger encountered in them was not without its attraction to her" (see Footnote 74). She died of pneumonia on 21 November 1946 at Great Brookham, Surrey.

Dorothy Moyle (Mrs. Needham)

Culminating her career with a study of the biochemistry of muscle contraction, Dorothy Moyle[79] showed that marriage did not have to end a woman scientist's professional life. Daughter of John Moyle, civil service clerk in the Patent Office and Ellen Daves, Dorothy Mary Moyle was born on 22 September 1896 in London. She attended school at Claremont College, Stockport, and then Manchester High School for Girls before being accepted to Girton in 1915. In 1918, as a Fourth Year scholar, Needham started research with Hopkins. Her M.A. followed in 1923, and the next year she received the Gamble Prize for her essay on the correlation of structure, function, and chemical constitution in the different types of muscle.

---

[79](a) Teich, M. (2003). Dorothy Mary Moyle Needham, *Biographical Memoirs of Fellows of the Royal Society* **49**: 353–365; (b) Needham, J. (November 1988). Dorothy Needham, *The Caian*, 128–131; and (c) Coley, N. G. (2004). Needham [née Moyle], Dorothy Mary (1896–1987). *Oxford Dictionary of National Biography.* Oxford University Press (accessed 31 December 2018).

376 *Pioneering British Women Chemists: Their Lives and Contributions*

The year 1924 was also the year of Moyle's marriage to Joseph Needham, and for the remainder of her life, part of her research was done independently and partially with her spouse and others. For example, in 1931, she co-authored a publication with Needham and with Marjory Stephenson.[80] For her contributions to the advancement of knowledge of the biochemistry of cells, Moyle was elected Fellow of the Royal Society in 1948.

There was another side to Moyle which was equally important: her left-wing political activism. During the 1930s, the Cambridge biochemists and crystallographers (see Chapter 12) comprised a high proportion of the Cambridge Scientists' Anti-War Group and Moyle and Needham were among the leaders.[81] Though no list of members has survived, a letter of protest against the militarization of research was signed by 79 faculty, research workers, and students of whom 12 (15%) were women (one of whom was Stephenson) even though women comprised only 7% of the personnel.

As Gary Werskey has remarked:

> Why women scientists were often drawn to the Left is not an easy question. Though they belonged to the first post-suffragette generation, they received little encouragement, either from the Popular Front or from wider social forces to develop an explicitly feminist perspective.[82]

Werskey hypothesized that some women became outraged by the malnutrition issue, and this was certainly true for Moyle, as she had worked hard in the malnutrition campaigns. Moyle's political activities were largely overshadowed by those of Needham. In fact, none of the women scientists were given recognition as Werskey concluded:

> Nevertheless it was the men who publicly dominated the scientific Left, especially as its leading theoreticians. Hence, neither for the first nor the

---

[80]Needham, J., Stephenson, M., and Moyle Needham, D. (1931) The relations between yolk and white in a hen's egg. *Journal of Experimental Biology* **8**: 319–329.

[81]Werskey, G. (1988). *The Visible College: A Collective Biography of British Scientists and Socialists of the 1930s*. Free Association Books, London, pp. 219–231.

[82]Note 81, Werskey, p. 221.

Hoppy's "Biochemical Ladies" 377

last time, the political contributions of such women would be almost completely 'hidden from history' (see Footnote 82).

Despite her 1930s pacifism, at the beginning of the Second World War, Moyle became a research worker for the Ministry of Supply (Chemical Defence), studying the biochemical effects of war gases. Then, in 1943, Needham, whose other main interest was the history of science in China, became Scientific Counsellor of the British Embassy in Chungking.[83] Moyle joined him as Associate Director of the Sino-British Science Cooperation Office. It was in China that she contracted tuberculosis, from which it took her until the 1950s to recover fully.[84]

After the War, Moyle returned to teaching and research at Cambridge, never with a permanent position, just subsisting on research grant after research grant from 1945 until 1962. She later commented on her situation:

> Looking back over my 45 years in research I find it remarkable ... that although a fully qualified and full-time investigator ... I simply existed on one research grant after another, devoid of position, rank, or assured emolument ... [I]t was calmly assumed that married women would be supported financially by their husbands, and if they chose to work in the laboratory all day and half the night, it was their own concern.[85]

After retirement in 1963, she began writing her classic work *Machina Carnis: the Biochemistry of Muscle Contraction in its Historical Development*,[86] which was published in 1971. The book had a dedication page to "F.G.H. mentor and friend." Then for many years she assisted Mikuláš Teich with *A Documentary History of Biochemistry 1770–1940*.

---

[83]Winchester, S. (2009). *The Man who Loved China*. Harper Perennial (reprint edition), New York.

[84]Letter, D. Needham to D. Singer, 17 December 1950. Wellcome Institute Archives.

[85]Needham, D. (1982). Women in Cambridge Biochemistry. In Richter, D. (ed.), *Women Scientists: The Road to Liberation*. Macmillan, London, p. 161.

[86]Needham, D. (1971). *Machina Carnis: the Biochemistry of Muscle Contraction in its Historical Development*. Cambridge University Press, Cambridge.

378 *Pioneering British Women Chemists: Their Lives and Contributions*

Unfortunately, Moyle's health deteriorated, as her former student, Jennifer Williams, described:

> I last saw "Dophi", as she was known at her ninetieth birthday party in Cambridge in 1986. … Although I had worked as her assistant for some years, she did not know who I was, as she was then suffering from Alzheimer's Disease. Only a few years earlier we had played Scrabble in the sunshine of her Grange Road garden. She had beaten us all.[87]

Moyle died on 22 December 1987 in Cambridge.

Commentary

With Hopkins as administrator, the easy-going atmosphere allowed researchers such as Stephenson and Moyle to follow their own intuition and direction. In return, Hopkins was revered throughout the Institute. When Hopkins finally retired in 1943, the "Hoppy regime" came to an end. In a letter written in 1948, Stephenson described how life had changed:

> Gone are the days of Hoppy's 5.30 talks when he used to shut his eyes and draw on his memories of the biochemists belonging to the turn of the century: I enjoyed science then and was ever so happy but at best I could never have stood the pace the modern young scientist must march at.[88]

Mark Weatherall and Harmke Kamminga studied the "Hoppy tradition" which Hopkins' colleagues had enunciated.[89] As much as the loose administrative style, it was also about his warm and supportive personality as we described earlier. In another letter, Stephenson highlighted the difference under the regime which followed Hopkins departure:

---

[87] Williams, J. (11 January 1988). Dr. D. M. Needham. *The Independent.*

[88] Letter, Marjory Stephenson to Sydney Elsden, 9 November 1948. Newnham College Archives.

[89] Weatherall, M. W. and Kamminga, H. (1996). The making of a biochemist II: The construction of Frederick Gowland Hopkins' reputation. *Medical History* **40**: 415–436.

I am worried about the Department; it is beginning to disintegrate owing to the absence of a real Professor. People come here to work and no-one knows they are here or what anyone else is working on or whom to discuss his problems or difficulties with.[90]

The collegial life under the benign influence of Hopkins was no more.

---

[90]Letter, Marjory Stephenson to Sydney Elsden, 13 January 1947. Newnham College Archives.

Chapter 12

Women Crystallographers: The Bragg Descendants

Mary "Polly" Winearls Porter (26 July 1886–25 November 1980)

In the previous chapter, we described how the Cambridge group of F. Gowland Hopkins had been a haven for women biochemists. Maureen Julian was the first to document a similar concentration of women in the field of crystallography.[1] As Georgina Ferry[2] and Michelle Franci[3] have

---

[1] Julian, M. M. (1990). Women in crystallography. In Kass-Simon, G. and Farnes, P. (eds.), *Women of Science: Righting the Record*. Indiana University Press, Bloomington, Indiana, pp. 335–383.
[2] Ferry, G. (2014). History: Women in crystallography. *Nature* **505**: 609–611.
[3] Franci, M. (2014). Seeding crystallography. *Nature Chemistry* **6**: 842–844.

both pointed out, crystallography was essentially a male-preserve except for those women scientists directly or indirectly linked to either W. H. Bragg or W. L. Bragg, hence the title of this chapter. However, there was a significant sociological difference between the women biochemists of Hopkins group and the women crystallographer "descendants" of the Braggs. Probably resulting from the more solitary nature of crystal structure determination, the women crystallographers did not tend to associate or collaborate.

X-ray crystallography did not appear out of nowhere. It built upon centuries of study of the morphology of crystals and the understanding of fundamental principles of the laws of symmetry.[4] To trace the involvement of women, it is crucial to begin this chapter back in the era of "classical" crystallography, a part of mineralogy.

The Early History

The characteristic angles between the faces for each crystalline mineral had been a subject of study in the late 19th century, and one of the more famous researchers in the field was Henry Alexander Miers.[5] Miers, who had been Keeper of Mineralogy at the British Museum, was elected to the Waynflete Chair of Mineralogy at Oxford in 1895, where he set up a laboratory at the Oxford Museum. The Oxford Museum had been built between 1855 and 1860 at the instigation of Henry Acland, who had been concerned that Oxford lacked any facilities for the promotion of the sciences.

In the latter part of the 19th century, the Museum housed astronomy, geometry, experimental physics, mineralogy, chemistry, geology, zoology, anatomy, physiology, and medicine. It was Miers who transformed the mineralogy department into a small but productive centre for research in

---

[4] Burke, J. G. (1966). *Origins of the Science of Crystals*. University of California Press, Berkeley.

[5] (a) Holland, T. H. and Spencer, L. J. (November 1943). Henry Alexander Miers, 1858–1942. *Obituary Notices of Fellows of the Royal Society* 4(12): 368–380; and (b) Porter, M. W. (ed.), (1973). *Diary of Henry Alexander Miers, 1858–1942*. Department of Geology and Mineralogy, Oxford.

classical crystallography.[6] There were women workers at the Museum, as Janet Howarth noted: "Research-oriented professors who were not over-burdened with undergraduate teaching tended to welcome women as collaborators or assistants — over thirty are recorded as working in the Museum Department and Observatory before 1914."[7]

Mary Porter

Of the women workers at the Museum, there were two who contributed directly to the research of Meirs' group: Florence Isaac and Mary (Polly) Winearls Porter.[8] Isaac was Miers' main collaborator on the study of the growth of crystals, co-authoring five publications. However, it was Porter who provided the link to the first generation of women X-ray crystallographers.

Porter was born on 26 July 1886 at King's Lynn, Norfolk. Her parents were Robert Porter, journalist and international correspondent for *The Times* newspaper, and Alice Russell Hobbins. Porter's father believed that education was unnecessary for women, and she was taught simply to read and write.

When Porter was 15, the family were residing in Rome. While there, her mother became ill necessitating a lengthy stay in the city. This stay was to determine Porter's future, for she encountered the archaeologists, Conte Gnoli and Giacomo Boni, who were working on excavations of the Roman Forum. Her curiosity piqued by the different colours of marble used in the Roman buildings, and she commenced collecting specimens of the different types. Teaching herself the fundamentals of geology, Porter compiled records of her marble studies. Porter's brothers were impressed

---

[6]Vincent, E. A. (1994). *Geology and Mineralogy at Oxford 1860–1986: History and Reminiscences*. The Author, Oxford.

[7]Howarth, J. (1984). 'Oxford for Arts': The Natural Sciences, 1880–1914. In Brock, M. G. and Curthoys, M. C. (eds.), *The History of the University of Oxford,* Vol. VII, *Nineteenth Century Oxford,* Part 2, Clarendon Press, Oxford, pp. 457–498.

[8](a) Wikipedia, (11 February 2019). Mary Winearls Porter (accessed 27 February 2019); and (b) "A de V" [De Villiers, A.] (1980). Mary Winearls Porter, 1886–1980. *Somerville College Report and Supplement* 32–33.

384 *Pioneering British Women Chemists: Their Lives and Contributions*

with her potential and implored her father to allow her to have a formal education, but he still refused.

Fortunately for Porter, upon the family's return to England, they settled at Oxford. Porter visited the Oxford Museum and became obsessed with the Corsi Collection of marble specimens. Miers was impressed by the visits of this earnest young woman, and he gave her the task of identifying exhibits and translating the catalogue from Italian. He encouraged Porter to attend Oxford University as a student, but again her parents were opposed. After working with Miers from 1905 to 1907, she authored a book (at age 21) on the subject of Roman marble: *What Rome was Built With*.[9]

For the 1910–1911 year, Porter commenced research in crystallography with Alfred Tutton in London. Then she obtained a 1-year position as mineral cataloguer at the National Museum, Washington, D.C. The following year, the second turning point in Porter's life occurred, being hired as mineral cataloguer at Bryn Mawr College, a women's college in Pennsylvania. Florence Bascom, Professor of Geology, recognized Porter's talent. Bascom wrote to Victor Mordechai Goldschmidt, founder of the Institut für Mineralogie und Kristallographie at the University of Heidelberg:

> I have long had the purpose of writing you to interest you in Miss Porter, who is working this year in my laboratory and whom I hope you will welcome in your laboratory next year. Her heart is set upon the study of crystallography … she should go to the fountainhead of inspiration. Miss Porter's life has been unusual, … she has never been to school or college except for a very brief period. There are therefore great gaps in her education, particularly in chemistry and mathematics, but to offset this I believe that you will find that she has an unusual aptitude for crystal measurement, etc., and certainly an intense love of your subject. I want to see her have the opportunities which have so long been denied her.[10]

---

[9]Porter, M. W. (1907). *What Rome was Built With: A Description of the Stones Employed in Ancient Times for its Building and Decoration*. Henry Frowde, London and Oxford.

[10]Arnold, L. B. (1993). The Bascom–Goldschmidt–Porter Correspondence 1907 to 1922. *Earth Sciences History* **12**: 196–223.

Goldschmidt invited Porter to Heidelberg for the 1914–1915 year, an offer which she accepted. Porter returned to the Oxford Museum in 1916 and, despite a lack of formal education, she was appointed to the Mary Carlisle Fellowship at Somerville College in 1919. Though her career commenced in the time of classical crystallography, she quickly embraced and contributed to the new era of X-ray crystallography in which the atomic arrangements within crystals could be determined.

Porter's research was published in such academically prestigious journals as the *Proceedings of the Royal Society*, *Acta Crystallographica*, and the *Mineralogical Magazine*. She also co-edited all three parts of the classic work *The Barker Index of Crystals*.[11] In 1950, Porter co-authored a publication with Dorothy Crowfoot on crystallographic measurements on the anti-pernicious anaemia factor.[12]

Porter was elected a Member of the Council of the Mineralogical Society of Great Britain from 1918 to 1921 and again from 1929 to 1932. She was a member of the Somerville College Council from 1937 to 1947 and was appointed an Honorary Research Fellow in 1948. During the 1960s, Porter had the opportunity to revisit her decorative stone collection at Oxford, reordering them according to the modern system of classification. She died on 25 November 1980 at Oxford, aged 94.

The Braggs and X-Ray Crystallography

X-ray crystallography is one branch of physical science whose origins can be exactly defined.[13] In 1912, the theoretician Max Laue persuaded the experimental physicists Walter Friedrich and Paul Knipping to test his

---

[11] See: Lewis, J. W. (31 October 1957). A new aid to the study of crystals. *New Scientist* **11**: 13.

[12] Crowfoot H. D. M., Porter, M. W. and Spiller, R. C. (1950). Appendix: Crystallographic measurements on the anti-pernicious anaemia factor. *Proceedings of the Royal Society of London, Series B-Biological Sciences* 136: 609–613.

[13] (a) Ewald, P. P. (1962). The beginnings. In Ewald, P. P. (ed.), *Fifty Years of X-ray Diffraction*. International Union of Crystallography, Utrecht, The Netherlands, pp. 6–80; (b) Gasman, L. D. (1975). Myths and X-rays, *British Journal for the History of Science* **26**: 51–60; and (c) Speakman, J. C. (1980). The discovery of X-ray diffraction. *Journal of Chemical Education* **57**: 489–490.

386   *Pioneering British Women Chemists: Their Lives and Contributions*

hypothesis that crystals diffract X-rays. William Henry (W. H.) Bragg,[14] Professor at the University of Leeds, and his son, William Lawrence (W. L.) Bragg,[15] then at the University of Cambridge, read of the success of Friedrich and Knipping's experiment. The Braggs quickly saw the potential for the determination of crystal structures. W. H. Bragg devised and constructed the Bragg X-ray spectrometer for crystal structure analysis, and shortly afterwards, W. L. Bragg deduced the relationship between the diffraction pattern and the atomic spacing (Bragg's Law). This was the start of X-ray crystallographic methods for the determination of structures.

The First World War interrupted the studies,[16] but with its end, the Braggs decided to divide the crystal world between them. W. H. Bragg chose to work on organic structures and also quartz, while W. L. Bragg was allotted inorganic substances (except quartz). About this time, they both moved: W. H. Bragg to University College, London (UCL), and then to the Royal Institution (RI), London,[17] while W. L. Bragg obtained a faculty appointment at the Victoria University of Manchester, a position that he acquired as successor to Ernest Rutherford, the latter having moved to Cambridge.

The RI seems to have been a particularly welcoming place of research. In his biography of W. T. Astbury, J. D. Bernal commented about the atmosphere at the RI:

> The band of people who had gathered there were diverse but all shared in common a lively enthusiasm for the discovery of the new world of

---

[14](a) Andrade, E. N. da, and Lonsdale, K. (1943). William Henry Bragg, 1862–1942. *Obituary Notices of Fellows of the Royal Society* **4**: 276–300; and (b) Caroe, G. B. (1978). *William Henry Bragg 1862–1942: Man and Scientist.* Cambridge University Press, Cambridge.

[15]Phillips, D. (1979). William Lawrence Bragg. *Biographical Memoirs of Fellows of the Royal Society* **25**: 75–143; and Hunter, G. K. (2004). *Light Is a Messenger: The Life and Science of William Lawrence Bragg.* Oxford University Press, Oxford.

[16]Jenkin, J. (2014). The Braggs, X-ray crystallography, and Lawrence Bragg's sound-ranging in World War I. *Interdisciplinary Science Reviews* **40**: 222–243.

[17]Julian, M. M. (1986). Crystallography at the Royal Institution. *Chemistry in Britain* **22**: 729–732.

Women Crystallographers: The Bragg Descendants 387

crystal structure which they were privileged to share. It was a very happy time: there was no real rivalry because that world was quite big enough for all their work. They were effectively and actually a *band* of research workers, dropping into each other's rooms, discussing informally over lunch and ping-pong and formally in Bragg's colloquia every week.[18]

The Braggs and Women Researchers

The research groups of both Braggs contained a remarkable number of women.[19] Of W. H. Bragg's 18 students, 11 were women: Ellie Knaggs, Grace Mocatta, Kathleen Yardley, Natalie Allen, Thora Marwick, Lucy Pickett, Helen Gilchrist (see Chapter 7), Berta Karlik, M. E. Bowland, Ida Woodward, and Constance Elam.

W. L. Bragg's first research student was Lucy Wilson, from Wellesley College, Massachusetts, and he, too, had other women researchers working with him over the years: Elsie Firth, Helen Scouloudi, and P. Jones. Maureen Julian, herself a former researcher with Kathleen Yardley, has shown that women have contributed throughout the "family tree" of crystallographers, many of the Bragg students (both male and female) themselves taking on women students when they acquired academic positions.[19]

Why was crystallography such an attractive field for women? One viewpoint is that the Braggs and their crystallographic heirs provided a women-friendly environment. Anne Sayre, spouse of crystallographer David Sayre, is convinced that, at least in crystallography, it was the non-aggressive and friendly attitudes of the supervisors that were so vital to the encouragement of women:

> There is something in the ancient history of crystallography that is hard to isolate but nevertheless was there, that I can best describe as modesty. I have often wondered how much the Braggs were responsible for the unaggressive low-key friendly atmosphere that long prevailed in the field (and no longer seems to very much). Somehow the first and second

---

[18]Bernal, J. D. (1963). William Thomas Astbury, 1898–1961. *Biographical Memoirs of Fellows of the Royal Society* **9**: 1–35.

[19]Note 1, Julian, p. 342.

388　*Pioneering British Women Chemists: Their Lives and Contributions*

and a few of the third generation crystallographers consistently conveyed an impression of working for pleasure, for the sheer joy of it — the idea of competition didn't seem to emerge very strongly until the 1960s or so. Uncompetitive societies tend to be good for women.[20]

W. L. Bragg's personality can be gleaned from this comment by one of his former students, W. M. Lomer:

[W. L.] Bragg was a gentle man and a gentleman. He never embarrassed anyone and my every contact with him was a pleasure. When he met me showing my fiancée around the little museum … [he] was so very pleasant that to this day my wife will hear nothing against him. And nor will I.[21]

A less charitable view of why women chose crystallography has been taken by Franklin Portugal and Jack Cohen, who used the laborious nature of crystallography to explain the number of women in the field:

Since the high speed computer had not yet been invented, the business of calculating data was a very laborious occupation and smart fellows who could find other things to do would generally do them, unless they were absolutely dedicated to the business of X-ray crystallography. Is it possible that these first class women got to be X-ray crystallographers because they were willing to do this work.[22]

This parallels the arguments used to account for the very high proportion of women in astronomy during the late 19th and early 20th centuries.[23]

---

[20] Letter, Anne Sayre to Maureen Julian, cited in: note 1, Julian, pp. 339–340.

[21] Lomer, W. M. (1990). Blowing bubbles with Bragg. In Thomas, J. M. and Phillips, D. (eds.), *Selections and Reflections: The Legacy of Sir Lawrence Bragg*. Royal Institution of Great Britain, London, p. 117.

[22] Portugal, F. H. and Cohen, J. S. (1977). *A Century of DNA: A History of the Discovery of the Structure and Function of the Genetic Substance*. MIT Press, Cambridge, Massachusetts, p. 267.

[23] (a) Warner, D. J. (1979). Women astronomers. *Natural History* **88**: 12–26; (b) Dobson, A. and Bracher, K. K. (1992). A historical introduction to women in astronomy. *Mercury* **21**: 4–15; and (c) Fraknoi, A. and Freitag, R. (1992). Women in astronomy: A bibliography. *Mercury* **21**: 46–47.

Margaret Rossiter has shown that the rise in women's participation in astronomy corresponded to the change from the active work of observing (men's work), to the passive role of classifying the thousands of photographic plates (women's work).[24] In fact, it was the first of the women astronomer-assistants, Williamina P. Fleming, who commented that women's superior patience, perseverance, and method made such activities particularly suitable for women in science.[25]

John Lankford and Ricky Slavings, in their studies of women in astronomy, added: "In evaluating women [for astronomy], male scientists tended to focus on their [women's] ability to do routine work. Indeed, some recommendations read as if they were descriptions of machines."[26] Yet, the women in this chapter did not see their crystallography work as dull and routine — to them, the research was the exciting focus of their lives.[27]

Ellie Knaggs

Like Porter, Isabel Ellie Knaggs[28] bridged classical and modern crystallography, but whereas Porter was at Oxford, Knaggs was at Cambridge. Knaggs was born on 2 August 1893 in Durban, South Africa. She was educated at North London Collegiate School (NLCS) and at Bedford College, going up to Girton in 1913, where she completed her studies in 1917. Arthur Hutchinson[29] at the Mineralogical Laboratory, Cambridge, hired her as a research assistant, and she worked with him until 1921.

---

[24] Rossiter, M. W. (1980). 'Women's work' in science, 1880–1910. *Isis* **71**: 381–398; also in Rossiter, M. W. (1982). *Women Scientists in America*. Johns Hopkins University Press, Baltimore, Maryland, pp. 53–57.

[25] Fleming, W. (1893). A field for 'women's work' in astronomy. *Astronomy and Astrophysics* **12**: 688–689.

[26] Lankford, J. and Slavings, R. L. (March 1990). Gender and science: Women in American astronomy, 1859–1940. *Physics Today* **43**: 58–65.

[27] Lonsdale, K. (1970). Women in science: Reminiscences and reflections. *Impact of Science on Society* **20**: 45–59.

[28] Megaw, H. D. (1982). Obituaries: Dr. Isabel Ellie Knaggs. *Girton College Newsletter* 30.

[29] Smith, W. C. (1939). Arthur Hutchinson, 1866–1937. *Obituary Notices of Fellows of the Royal Society* **2**: 483–491.

390 *Pioneering British Women Chemists: Their Lives and Contributions*

That year, she moved to Imperial College (IC), where she undertook research leading to a Ph.D., which she completed in 1923.

The year before she received her Ph.D., Knaggs was appointed Demonstrator in Geology at Bedford College, and the same year, she was elected Fellow of the Geological Society of London.[30] Then, in 1925, she was awarded a 2-year Hertha Ayrton Fellowship, which she chose to take at the Davy Faraday Laboratory of the RI to work under W. H. Bragg. When the Fellowship ended in 1927, she was invited to join the staff of the RI.

Among the many crystal structures she determined was that of the explosive cyanuric triazide, as W. H. Bragg described in a letter that he wrote to the Editor of the journal, *Nature*:

> I send you a short note which I hope you may see fit to publish in Nature. I would like you to know however that my writing it has something to do with an attempt to do a little act of justice to the lady mentioned in the note, Miss Knaggs. She has been working for some time on an extraordinary substance, cyanuric triazide. It is one of the highly explosive nitrogen compounds. In the recent Faraday Society discussion — see Nature, May 26 — reference was made to her preliminary results without mention of the source from which they had come. It was an accident, of course: there is no question of any unfairness. But this is Miss Knaggs' magnum opus so far and she is naturally disappointed. I have thought I might put matters straight by writing you the short note to which I have referred.[31]

However, according to a recent investigation by Bart Kahr, Knaggs should have been given credit for the first definitive structure based upon a tetrahedral carbon atom. Kahr explains:

> Knaggs' most significant body of work, stemming from her doctoral thesis, was the structural determination of symmetrically substituted methane derivatives, and it is here that she should have been recognised

---

[30]Burek, C. V. (2009). The first female Fellows and the status of women in the Geological Society of London. *The Geological Society, London, Special Publications* **317**: 373–407.

[31]Letter, W. H. Bragg to Sir Richard Gregory, 3 July 1934. W. H. Bragg Archives, The Royal Institution. The note was published as: Bragg, W. H. (1934). Structure of the azide group. *Nature* **134**: 138.

for making a lasting contribution. By 1923, both Knaggs and [Isamu] Nitta began assaults on the coordination at carbon in methane derivatives by X-ray diffraction. Knaggs solved the problem and more completely in our estimation.[32]

Helen Megaw commented on Knaggs' personality: "Ellie Knaggs was a kind and gentle person, rather shy. She attended scientific meetings, but did not put herself forward" (see Footnote 28). Knaggs' reticence is probably a major reason why her name is rarely mentioned among the pioneer women of crystallography. She had many publications in a wide range of journals from *Proceedings of the Royal Society* through the *Mineralogical Magazine* to the *Journal of the Chemical Society* and the *Journal of Physical Chemistry*. Of especial note, Knaggs and Austrian physicist, Berta Karlik,[33] co-authored *Tables of Cubic Structures of Elements and Compounds*,[34] with a section on alloys by the metallurgist, Constance Fligg Elam (Mrs. Tipper).[35]

In addition to her research at the Davy Faraday Laboratory, Knaggs was advisor on crystallography to Burroughs Wellcome, the pharmaceutical company. During the Second World War, it was noted in the *Girton Register* that she: "carried out investigations involving knowledge of crystal structure at the request of various government departments."[36] Some of Knaggs' later research involved collaboration with Kathleen Yardley. After retirement, she was elected visiting scientist to the RI for the 1963–1966 and 1974–1977 years. In 1979, Knaggs moved to Australia to join her sister, and it was in Sydney that she died on 29 November 1980, aged 87.

---

[32]Kahr, B. (2015). Broader impacts of women in crystallography. *Crystal Growth & Design* **15**: 4715–4730.

[33]Rentetzi, M. (2011). Berta Karlik (1904–1990) In Apotheka, J. and Sarkadi, L. S. (eds.), *European Women in Chemistry*. Wiley-VCH, Weinheim, Germany, pp. 161–164.

[34]Knaggs, I. E. and Karlik, B. [with a section on alloys by C. F. Elam] (1932). *Tables of Cubic Structures of Elements and Compounds*. A. Hilger Ltd., London.

[35]Ford, A. L. (2004). Tipper [*née* Elam], Constance Fligg (1894–1995). *Oxford Dictionary of National Biography*. Oxford University (accessed 29 December 2018).

[36]Butler, K. T. and McMorran, H. I. (eds.), (1948). *Girton College Register, 1869–1946*, Girton College, Cambridge, p. 680.

Kathleen Yardley (Mrs. Lonsdale)

The most famous woman to work with W. H. Bragg was Kathleen Yardley.[37] Yardley was born on 28 January 1903 at Newbridge, County Kildare, Ireland, to Harry F. Yardley, postmaster, and Jessie Cameron, the youngest of 10 children (though only six survived to adulthood). When 5 years old, her parents separated, and her mother took the four younger children to live in Ilford, Essex. Yardley attended the County High School for Girls, Ilford, but as it offered little science, she attended classes in physics, chemistry, and advanced mathematics at the High School for Boys, Ilford. Though she excelled at almost every academic subject and was a good gymnast, her first love was mathematics.

At the age of 16, Yardley's exceptional talents resulted in the offer of a place at Bedford College to study mathematics.[38] After 1 year, she switched to physics and told the headmistress of her former high school of her change of plan.[39] The headmistress argued that Yardley was a fool to think that she would be able to compete in a "man's field." Yardley, however, had rejected the alternative of a mathematics degree, as she believed mathematics would only lead to a teaching position, while physics offered the opportunity of experimental research.[40]

In 1922, at the age of 19, Yardley was awarded an B.Sc. (Hons.). W. H. Bragg was one of her B.Sc. oral examiners, and, impressed by the fact that she had higher university marks than anyone in the previous 10 years, he offered her a position in his laboratory at UCL. She accepted his offer, commenting: "He inspired me with his own love of pure science and with his enthusiastic spirit of enquiry and at the same time left me entirely free to follow my own line of research."[41] The following year,

---

[37] Julian, M. M. (1981). X-ray crystallography and the work of Dame Kathleen Lonsdale. *Physics Teacher* **3**: 159–165.

[38] Rice-Evans, P. (2001). Physics. In Crook, J. M. (ed.), *Bedford College: Memories of 150 Years,* Royal Holloway and Bedford College, pp. 265–266.

[39] Note 27, Lonsdale, pp. 54–55.

[40] Crowfoot, D. M. C. (1975). Kathleen Lonsdale, *Biographical Memoirs of Fellows Royal Society* **21**: 447–484.

[41] Note 40, Crowfoot, p. 449.

Yardley received an M.Sc., jointly from UCL and Bedford College, and that same year, she moved with Bragg's group to the RI.

In 1924, Yardley constructed a set of 230 space-group tables, mathematical descriptions of the crystal symmetries that became vital tools for crystallographers.[42] Three years later, in 1927, she received a D.Sc. degree for her work on ethane derivatives and that same year she married fellow physics student Thomas Lonsdale. Yardley considered giving up research to become a traditional wife and mother, but Lonsdale argued that he had not married to obtain a free housekeeper and that she should keep on working.[43] They shared the shopping for food, while she specialized in devising dinners that could be prepared in 30 minutes.

The Lonsdale family moved to Leeds in 1927, where Thomas Lonsdale commenced work for the British Silk Research Association. Upon her departure from the RI, she wrote to W. H. Bragg:

> I should like to take this opportunity of thanking you again for all the help you have given me in so many ways. I feel sure that it will be difficult to find a place where I shall be as happy in my work as I was at the Davy Faraday.[44]

W. H. Bragg had held a position at the University of Leeds from 1909 to 1915 and this must have given Yardley credibility, for she was able to set up her own laboratory at the university there.

It was Yardley's research at Leeds between 1927 and 1929 on the structure of benzene[45] which provided her with international recognition. Auguste Kekulé had proposed in 1865 that benzene had a six-member

---

[42]Lonsdale, K. (1936). *Simplified Structure Factor and Electron Density Formulae for the 230 Space-Groups of Mathematical Crystallography*. G. Bell & Sons, London.

[43]Baldwin, M. (2009). 'Where are your intelligent mothers to come from?': Marriage and family in the scientific career of Dame Kathleen Lonsdale FRS (1903–1971). *Notes & Records of the Royal Society* **63**: 81–94.

[44]Letter, K. Lonsdale to W. H. Bragg, 16 November 1927. W. H. Bragg Archives, The Royal Institution.

[45]Julian, M. M. (1981). Kathleen Lonsdale and the planarity of the benzene ring. *Journal of Chemical Education* **58**: 365–366.

ring structure, and this was generally accepted by chemists,[46] but the question remained as to whether the ring was planar or puckered. The Braggs had already shown that the carbon atoms in diamond could be pictured as forming six-member puckered rings, and benzene was expected to possess the same puckered structure. As benzene was a liquid at room temperature, Christopher K. Ingold,[47] Professor of organic chemistry at Leeds, provided Yardley with large single crystals of hexamethylbenzene, a solid at room temperature. Yardley was able to show that hexamethylbenzene was planar, and hence benzene itself was also likely to be planar. Although her results contradicted W. H. Bragg's belief in a puckered ring, he was enthusiastic in his praise of her work.

Between 1929 and 1934, Yardley had three children. This was a difficult time for the family, as she had no formal position and to make matters worse, in 1930 Thomas Lonsdale's job was terminated. The family returned to London, where Thomas Lonsdale obtained an appointment working at the Testing Station of the Ministry of Transport. From 1929 until 1931, Yardley worked at home on the calculations needed to solve her next crystal structure, that of hexachlorobenzene.

In 1931, W. H. Bragg wrote Yardley an enthusiastic letter, telling her that he had acquired funds on her behalf to cover the cost of a full-time home-help so that Yardley could return to the RI. However, when she arrived, she found that all the X-ray equipment was in use. Discovering a large old electromagnet, she changed her focus to that of the diamagnetism of aromatic compounds. Yardley next undertook lattice constant measurements of natural and synthetic diamonds. Her precise work on the diamond structure became so renowned that the discoverers of a rare hexagonal form of diamond announced that it would be named lonsdaleite in her honor.[48]

---

[46](a) Vanderbilt, B. (1975). Kekulé's whirling snake: Fact or fiction. *Journal of Chemical Education* **52**: 709; and (b) Wotiz, J. H. and Rudolfesky, S. (1984). Kekulé's dreams: Fact or fiction? *Chemistry in Britain* **20**: 720–723.

[47]Leffek, K. T. (1996). *Sir Christopher Ingold: A Major Prophet of Organic Chemistry*. Nova Lion Press, Victoria, BC, Canada.

[48](a) Frondel, C. and Marvin, U. B. (1967). Lonsdaleite, a hexagonal polymorph of diamond. *Nature* **214**: 587–589; and (b) Bundy, F. P. and Kasper, J. S. (1967). Hexagonal diamond — A new form of carbon. *Journal of Chemical Physics* **46**: 3437–3446.

W. H. Bragg obtained a grant or fellowship for Yardley each year thereafter until his death in 1942, after which she was appointed a Dewar Fellow at the RI. However, her position was under pharmacologist and physiologist, Henry Dale.[49] No longer part of a crystallography group and feeling isolated, Yardley applied for her first academic appointment. In 1946, at the age of 43, she finally obtained such a position as Reader in Crystallography at UCL, where her career had started almost 25 years earlier. Initially, most of her time was filled in the role of Editor-in-Chief of the *International Tables for X-ray Crystallography*.[50]

Three years later, she was promoted to Professor of Chemistry and Head of the Department of Crystallography. Her reputation continued to grow, and, as we discussed in Chapter 11, she was one of the first two women (the other being biochemist Marjory Stephenson) to be elected a Fellow of the Royal Society, having been nominated by W. L. Bragg.[51] Then in 1966, she was elected the first woman president of the International Union of Crystallography. Among her later work was the study of boron–nitrogen analogues of carbon species, including the identification of a graphite-like form of boron nitride containing alternating boron and nitrogen atoms in arrays of hexagonal rings.

Having lived under the path of air attacks on London in the First World War and having seen the horrific fireball of a bomb-carrying Zeppelin shot down in flames, Yardley became a pacifist. In a letter to Archibald V. Hill,[52] she expanded on her beliefs:

---

[49]Feldberg, W. S. (1970). Henry Hallett Dale, 1875–1968. *Biographical Memoirs of Fellows of the Royal Society* **16**: 77–174.

[50]Henry, N. F. M. and Lonsdale, K. (eds.), (1952). *International Tables for X-ray Crystallography: Symmetry Groups*. Vol. I, Kynoch Press, Birmingham; Kasper, J. and Lonsdale, K. (eds.), (1959). *International Tables for X-ray Crystallography: Mathematical Tables*. Vol. II, Kynoch Press, Birmingham; and Lonsdale, K., MacGillavry, C. H. and Reich, G. D. (eds.), (1962). *International Tables for X-ray Crystallography: Physical and Chemical Tables*. Vol. III, Kynoch Press, Birmingham.

[51]Mason, J. (1992). The admission of the first women to the Royal Society of London. *Notes and Records of the Royal Society* **46**: 279–300.

[52]Katz, B. (1978). Archibald Vivian Hill, 26 September 1886–3 June 1977. *Biographical Memoirs of Fellows of the Royal Society* **24**: 71–149.

396 *Pioneering British Women Chemists: Their Lives and Contributions*

> It was certainly not the imperative of a religious upbringing that made me a pacifist. I like to believe it was commonsense. I came of a military family and have a naturally pugnacious character, the more violent manifestations of which I have to keep continually under control. My religious teaching as a child was of the orthodox kind that rubber-stamps any war that happens to be going. I don't believe in sitting down and being walked on; I believe in non-violent resistance because I believe it to be the only form of resistance that can be really effective, in that it does not perpetuate the evil it aims at eliminating.[53]

Both Kathleen Yardley and Thomas Lonsdale became Quakers, and during the Second World War, she was jailed for a month for refusing to register for war duties. As a result of her personal experiences in jail, she wrote a critique of prison life for women,[54] and as an extension of her beliefs, she became active in prison reform and served on many prison boards. Her indignation at the extensive nuclear testing by the Soviet Union, the United States, and Great Britain, caused her to write the book *Is Peace Possible?*, the foreword of which contains her comment that the book was "written in a personal way because I feel a sense of corporate guilt and responsibility that scientific knowledge should have been so misused."[55]

During the post-war period, Yardley undertook her first teaching, that of a University College graduate course in crystallography, organized jointly with Bernal's crystallography department at Birkbeck College. Teaching was not her forte, as former student Judith Milledge explained:

> Kathleen was not really a good teacher in the generally accepted sense of the word. This was not because she failed to prepare adequate material, but simply because she would not accept the fact that all students did not have her intellectual capacity, and that beyond a certain level, diligence is not a substitute for ability; she tended to believe that

---

[53] Letter, Kathleen Lonsdale to A.V. Hill, 7 June 1953. A.V. Hill collection, Churchill College, Cambridge University.

[54] Lonsdale, K. *et al.* (1943). *Some Account of Life in Holloway Prison for Women*. Prison Medical Reform Council, London.

[55] Lonsdale, K. (1957). *Is Peace Possible?* Penguin, London.

Women Crystallographers: The Bragg Descendants 397

students could master anything if they would only work hard enough, as she herself had always been able to do.[56]

Crystallography was Yardley's life. She began a study of the crystalline nature of kidney, bladder, and gall stones when nearly 60 years old.[57] Yadley developed an interest in stones after the Chief Medical Officer of the Salvation Army had told her of small children in hot, dry, Third World countries suffering from them.[58] When Yardley fell ill, the first medical diagnosis assumed that she had acquired tropical malaria from her world travels in pursuit of stones and their causes. When cancer of the bone marrow was identified and she was told that her remaining time was short, she began 13-hour work days towards a book on human stones. Yardley completed the first draft a few weeks before she died on 1 April 1971 at Bexhill-on-Sea, Sussex, aged 68. Ten years later, in recognition of her contributions to crystallography, the then-chemistry building at UCL was named the Kathleen Lonsdale Building.

In her later years, Yardley relied more and more on Thomas Lonsdale to help with the tremendous amount of correspondence she received. He had always been very supportive of his famous spouse, and Yardley identified the need for such a relationship in her comments on women scientists:

> For a woman, and especially a married woman with children, to become a first class scientist she must first of all choose, or have chosen, the right husband. He must recognise her problems and be willing to share them. If he is really domesticated, so much the better.[59]

Yardley was very fortunate in her choice of husband. After her death, Thomas Lonsdale wrote to W. L. Bragg:

---

[56] Milledge, H. J. (1975). Obituary: Kathleen Lonsdale 28 January 1903–1 April 1971, *Acta Crystallographica* A31: 705–708.

[57] In addition to academic publications on human stones, Lonsdale wrote a popular account, see: Lonsdale, K. (1968). Human Stones. *Science* **159**: 1199–1207.

[58] Letter, Thomas Lonsdale to Mr. Glanville, undated. Lawrence Bragg papers, Royal Institution.

[59] Note 40, Crowfoot, p. 474.

When the apple fell on Newton's head someone gathered it and the other windfalls and made a pie for his dinner, thats [sic] my job now a bit, it always has been. Hilton [a prominent mathematician of the time] told me that a Professor of Maths is lucky if in his life he has one student who can see a whole branch of maths as a structure, they know the text book but they don't need it because they go on to build, where the math doesn't yet exist they invent it, for him, Kathleen was that student. Even before we were married I knew she had one of the most powerful intellects of the time.... Most of my working life has been spent in road engineering, I only know enough about her work to realise its importance and value and how fortunate I have been to have been associated with it, 'in getting Newton's dinner'.[60]

Influence of J. D. Bernal

Though the Braggs had been the "founding fathers" of X-ray crystallography, it was John "Sage" Desmond Bernal[61] who provided the link among the majority of women crystallographers. In addition to a shared passion for crystallography, most of the women espoused the same left-wing political ideals as Bernal[62] and were members of the Cambridge Scientists Anti-War Group (CSAWG). As an example of their political inclinations, both Dorothy Crowfoot and Helen Megaw joined Bernal in signing the protest against the Incitement to Disaffection Bill[63]

---

[60] Letter, Thomas Lonsdale to Sir Lawrence Bragg, 24 May 1971. Lawrence Bragg papers, Royal Institution. For a discussion of the relationship between Kathleen and Thomas Lonsdale, see: Julian, M. M. (1996). Kathleen and Thomas Lonsdale: Forty-Three Years of Spiritual and Scientific Life Together. In Pycior, H., Slack, N. G. and Abir-Am, P. G. (eds.), *Creative Couples in the Sciences*. Rutgers University Press, Brunswick, New Jersey, pp. 170–181.

[61] (a) Brown, A. (2005). *J. D. Bernal: The Sage of Science*. Oxford University Press, Oxford; and (b) Hodgkin, D. M. C. (1980). John Desmond Bernal. *Biographical Memoirs of Fellows of the Royal Society* **26**: 17–84.

[62] Swann, B. and Aprahamian, F. (1999). *J. D. Bernal: A Life in Science and Politics*. Verso, London.

[63] Bell, G. D. H. *et al.* (June 1934). Correspondence: Scientific workers and war. *The Cambridge Review* **56**: 451.

(the biochemists, Marjory Stephenson and Dorothy Moyle, had signed the same protest, see Chapter 11).

Bernal had been a student at Cambridge where he had become fascinated with the different types of spatial arrangements of atoms (see Footnote 61a). In 1923, he showed a theoretical paper he had written during his vacation to Arthur Hutchinson. Hutchinson was so impressed that he extolled Bernal's genius to W. H. Bragg at UCL. Bragg quickly found funding support for Bernal at UCL. Then in 1927, Bernal moved back to Cambridge, taking on Nora Martin, as his first student (who was also a member of the CSAWG).[64]

Dorothy Crowfoot was a member of the Bernal group from 1932 to 1934, and she described the very pleasant working atmosphere, particularly the convivial lunches:

> Every day, one of the group would go and buy fresh bread from Fitzbillies, fruit and cheese from the market, while another made coffee on the gas ring in the corner of the bench. One day there was talk about anaerobic bacteria on the bottom of a lake in Russia and the origin of life, another, about Romanesque architecture in French villages, or Leonardo da Vinci's engines of war or about poetry or printing. We never knew to what enchanted land we would be taken next.[65]

When Hutchinson retired in 1931, he had recommended the foundation of a Chair of Crystallography, obviously with Bernal in mind as an occupant; however, the proposal was rejected. Thus, in 1937, when Bernal was offered the Chair of Physics at Birkbeck College, he accepted and spent the rest of his working life there. Such an environment matched well with his Communist Party affiliation and his interest in the social dimension of science.[66] At Birkbeck, he continued his policy of taking on women

---

[64] Wooster, W. A. (1962). Personal experiences of a crystallographer. In Ewald, P. P. (ed.), *Fifty Years of X-ray Diffraction*. International Union of Crystallography, Utrecht, The Netherlands, pp. 680–684.

[65] Note 61b, Hodgkin, p. 31.

[66] (a) Mackay, A. L. (2003). J. D. Bernal, 1901–1971. In perspective. *Journal of Biosciences* **28**: 539–546; and (b) Muddiman, D. (2003). Red information scientist: The information career of J. D. Bernal. *Journal of Documentation* **59**: 387–409.

400 *Pioneering British Women Chemists: Their Lives and Contributions*

researchers, including Käthe Schiff, Rosalind Franklin, Winifred Booth Wright, and Shirley V. King.

Bernal had been asked whether he ran his laboratory on communist principles, to which he retorted that it was more on feudal lines, where the researchers would spend about half their time on his projects and half on their own.[67] He also continued the social life as his former student, Alan Mackay recalled:

> The lab was very socially and politically conscious, a wide spectrum of views being expressed every teatime. Bernal had also thought about the social organization and had brought the institution of the tea club with him from Cambridge so that everyone met twice a day with considerable regularity.[68]

Helen Megaw

The careers of Bernal and another of his early students, Helen Dick Megaw,[69] crossed twice: Megaw worked with Bernal at Cambridge in the early years and then rejoined him for a short period at Birkbeck. Megaw was born in Dublin, Ireland, on 1 June 1907, to Robert T. Megaw and Annie McElderry.[70] She initially attended the Alexandra School, Dublin, and then was sent as a border to Roedean when her parents returned to Belfast. Initially unable to afford Girton, she spent the 1925–1926 year at Queen's University, Belfast, before transferring to Girton, where she completed her degree requirements in Physics in 1930.

Megaw then commenced research towards a Ph.D. in Mineralogy and Petrology with Bernal on the structure of ice. This topic required growing crystals in narrow capillary tubes at low and very exact temperatures, and then studying the X-ray diffraction patterns.[71] Her pioneer work on the

---

[67] Mackay, A. L. (1995) The lab. *The Chemical Intelligencer* **1**(6): 12–18.

[68] Note 67, Mackay, p. 13.

[69] Glazer, A. M. and Kelsey, C. (2006). Helen Dick Megaw, 1907–2002. In Byers, N. and Williams, G. (eds.), *Out of the Shadows: Contributions of Twentieth-Century Women to Physics*. Cambridge University Press, Cambridge, pp. 213–221.

[70] Glazer, A. M. (2009). Megaw, Helen Dick, 1907–2002. *Oxford Dictionary of National Biography* (accessed 29 December 2018).

[71] Ref. 61b, Hodgkin, p. 33.

Women Crystallographers: The Bragg Descendants 401

structure of ice resulted in the Glaciological Society naming an island in the Antarctic the Helen D. Megaw Island.

After completing her Ph.D. in 1934, Bernal suggested that Megaw should try to solve the crystal structure of a clay mineral, hydrargillite, a form of aluminium hydroxide. Bernal, like the U. S. scientist Linus Pauling, held the theory that life had started on clay minerals, hence the determination of the crystal structure of a clay would advance the study of biogenesis. From this beginning, the crystallographic studies of mineral structures was to be Megaw's research focus for the rest of her active life.

Megaw was awarded a Hertha Ayrton Research Fellowship which enabled her to spend the 1934–1935 year as a postdoctoral researcher with Herman Mark at the University of Vienna and the following year with Francis Simon at the Clarendon Laboratory, Oxford. Between 1936 and 1943, she taught at Bedford High School for Girls and then at Bradford Girls' Grammar School, spending her school holidays doing research back at Cambridge. During the 1943–1945 period, Megaw's contribution to the war effort was as a Crystallographic Scientist in the Materials Research Laboratory, Philips Lamps, Mitcham, London.

Then, in 1945, Megaw was appointed Assistant Director of Research in Crystallography at Birkbeck College. It would seem likely that Bernal had instituted the appointment as a means to get his former protégée back into academic research life. If so, the move succeeded brilliantly, with Megaw being enticed back to Cambridge in 1946 as Assistant for Experimental Research in Crystallography at the Cavendish Laboratory. In addition, she was appointed as Fellow, Lecturer, and Director of Studies in Physical Science at Girton.

This period of her life proved to be the most productive, resulting in a stream of research papers together with several books, including the classic: *Ferroelectricity in Crystals*.[72] Megaw's former postdoctoral student, Michael Glazer (who had previously been a graduate student of Kathleen Yardley) commented:

> Helen's impact on ferroelectricity was profound, especially in the early days. She brought to the subject a visual aspect embodied in the crystal structures of ferroelectrics, thus showing how this important effect

---

[72] Megaw, H. D. (1957). *Ferroelectricity in Crystals*. Methuen, London.

402　*Pioneering British Women Chemists: Their Lives and Contributions*

arose. Her book, *Ferroelectrics in Crystals*, the first book on the subject, was for a long time just about the only book on the subject and became almost a "bible" for those working in ferroelectricity.[73]

Megaw received the Roebling Medal of the Mineralogical Society of America in 1989, the first woman to receive this award. In her award address, Megaw remarked:

> Much has been said about the difficulties of women in science, but I would like to say explicitly that I at least was never or rarely aware of discrimination. Perhaps I was lucky, in that everyone who advised me on my education and guided my career assumed that women should be given the same opportunities as men.[73]

Yet, this belief is contradicted by her early career descent to Assistant Mistress of a High School, for had she been male, an academic position would have followed a postdoctoral. Had Bernal not "rescued" her, she would never have made the major contributions to mineral crystallography that she did, nor garner the fame and recognition she deserved.

Even after formal retirement in 1972, Megaw remained active, dividing her time between her home in Ballycastle, Ireland, and her desk in the Department of Mineralogy and Petrology at Cambridge. During this period, she continued active research and wrote another landmark book: *Crystal Structures: A Working Approach*.[74] She died at Ballycastle on 26 February 2002, aged 94.

Dorothy Crowfoot (Mrs. Hodgkin)

The most famous of all women crystallographers was Dorothy Mary Crowfoot,[75] the Nobel Prize winner, who gained fame for elucidating the

---

[73]Cited in: Crystallography: Helen D. Megaw. Available at: http://cwp.library.ucla.edu/Phase2/Megaw,_Helen@851234567.html (accessed 29 December 2018).

[74]Megaw, H. D. (1973). *Crystal Structures: A Working Approach*, W. B. Saunders Co., Philadelphia.

[75](a) Farago, P. (1977). Impact: Interview with Dorothy Crowfoot Crowfoot. *Journal of Chemical Education* **54**: 214–216; and (b) Perutz, M. F. Dorothy Crowfoot, address delivered at a memorial service in the University Church, Oxford, 4 March 1995. S. Chandrasekhar is thanked for a copy of the address.

crystal structures of penicillin, Vitamin B_{12}, and insulin.[76] Though known particularly for the determination of these three structures, her 180 authored and co-authored publications also covered many other fields of endeavour.[77]

Before we chronologically detail her life and work, it is important to note that social responsibility was one of her guiding forces, as it was of Lonsdale and Bernal. It was no accident that the three crystal structures for which she gained fame were all of great medical importance. In addition, Crowfoot felt a need to share her knowledge with the scientists of the then-developing countries of India[78] and China.[79] As Dong-cai Liang and Chih-chen Wang commented: "Dorothy first came to China in 1959 with her husband Thomas Lionel Hodgkin, cherishing her intrinsic sense of duty for human progress and civilization and her enthusiasm for people and science in developing countries."[80]

Crowfoot was born on 12 May 1910 in Cairo, where her father, John Winter Crowfoot, an Oxford graduate, supervised Egyptian schools and administered ancient monuments for the British government. Her mother, Grace Mary "Molly" Hood, was a self-taught expert in botany who had written a book on the flora of the Sudan. In 1914, with the outbreak of war, Crowfoot and her two younger sisters were left in England in the care of a nursemaid while their parents returned to the Middle East. Crowfoot believed this background led to her quiet, independent character. Even after the end of the First World War, the children stayed in England while their parents returned from the Middle East on annual visits.

---

[76]Dodson, G., Glusker, J., and Sayre, D., (eds.) (1981). *Structural Studies on Molecules of Biological Interest, A Volume in Honour of Professor Dorothy Crowfoot*. Clarendon Press, Oxford.

[77]Kademani, B., Kalyane, V., and Jange, S. (1999). Scientometric portrait of Nobel Laureate Dorothy Crowfoot Hodgkin. *Scientometrics* **45**: 233–250.

[78]Ramaseshan, S. (1996). Reminiscences and discoveries: Dorothy Hodgkin and the Indian connection. *Notes and Records, Royal Society London* **50**: 115–127.

[79](a) Liang, D.-C. and Wang, C.-C. (1997). Dorothy and crystallographic research in China. *Current Science* **72**: 463–465; and (b) Wang, J.-H. (1998). The insulin connection: Dorothy Hodgkin and the Beijing Insulin Group. *Trends in Biochemical Sciences: Reflections* **23**: 497–500.

[80]Note 79(a), p. 463.

404 *Pioneering British Women Chemists: Their Lives and Contributions*

Crowfoot became interested in chemistry at an early age. At 10, her parents enrolled her in the Beccles Parents National Educational Union class. At the school, Crowfoot discovered a book on chemistry which contained instructions on how to grow crystals.[81] Reading about this experiment, led her to try growing crystals of copper(II) sulphate and alum. The next year, she was transferred to Sir John Leman School which had an inspiring woman chemistry teacher, Criss Deeley, a graduate of the University of Birmingham. Then at age 12, she and her younger sister were taken to the Sudan to be reunited with their parents. While there, Crowfoot was introduced to Dr. A. F. Joseph, a government chemist at Khartoum. That summer, and again during her visit to the Sudan 2 years later, he showed Crowfoot how to analyse minerals and he gave her a surveyor's mineral analysis kit.

In England, she set up a small attic laboratory where she continued her chemical analyses. In her notebook, she recorded each of her experiments, including: "I had a violent nosebleed and thought — a pity all this good blood should go to waste so I collected it in a test-tube and used it to make haematoporphyrin."[82] Of particular importance for her future career was the children's book *Concerning the Nature of Things* authored by W. H. Bragg, which discussed crystallography, together with a text on biochemistry, the latter which she read with the help of the *Encyclopedia Britannica*. Miss Deeley arranged for Crowfoot, and another girl in the class, to join the more advanced science courses taken by boys.[83] This experience had the secondary benefit of exposing her to an all-male environment at an early age, so that later in life she seemed comfortable in such circumstances.

In 1926, Crowfoot took the Oxford Senior Local Examination, a national university entrance test and attained the highest grade for any female student that year. A friend of the family, Isobel Fry, upon learning

---

[81] Dodson, G. (2002). Dorothy Mary Crowfoot Hodgkin, O.M. 12 May 1910–29 July 1994. *Biographical Memoirs of Fellows of the Royal Society* **48**: 179–219.

[82] Ferry, G. (1998). *Dorothy Hodgkin: A Life*. Granta Books, London, p. 29.

[83] Hudson, G. (1991). Unfathering the thinkable: Gender, science and pacifism in the 1930s. In Benjamin, M. (ed.), *Science and Sensibility: Gender and Scientific Enquiry, 1780–1945*. Blackwell, Oxford, p. 275.

Women Crystallographers: The Bragg Descendants 405

that Crowfoot wanted to study chemistry remarked: "She wants to do science? Why, they'll practically pay her to come to Somerville."[84] Lacking the Latin knowledge and another science required for admission to Oxford, she had to spend a year studying these subjects before she could be admitted to Somerville.

At this time, her father was a director of the British School of Archaeology in Jerusalem, and in the Summer of 1928, she visited him, becoming entranced by the mosaic floors patterns in the Byzantine churches at Jerash (now part of Jordan). In fact, analysing mosaic patterns had a strong parallel to her later work on crystal symmetry as is apparent from Georgina Ferry's comment.

> The mosaics combined geometrical regularity with exquisite decoration, and while Dorothy clearly enjoyed the decorative qualities, she was intrigued by the geometrical challenges that had faced the craftsman of early Jerash. "They are nearly all geometrical interlacing patterns", she wrote to her grandmother on July 1, 1928. "This one [she drew a sketch] covers the north apse of Bishop Paul's church and has a large key pattern border. I have just shown one repeat ... The north isle is covered with a pattern of repeating octagons."[85]

Crowfoot entered Somerville, in 1928, working towards a B.Sc. (Hons.) in Chemistry. Near the end of her first year, Crowfoot attended the lectures of mineralogist Thomas Vipont Barker[86] in the Crystallography Department, and Barker became an early mentor of Crowfoot until his untimely death in 1931. From then on, her preference was for the connections between chemistry and biochemistry, though she also enjoyed a lecture by J. D. Bernal on the metallic state utilizing X-ray crystallography.

A later research student of Crowfoot, Guy Dodson stated: "... she [Crowfoot] was anxious to return to crystallography. She was greatly

---

[84] Note 81, Dodson, p. 183.

[85] Ferry, G. (2014). Perspectives: The art of medicine. Dorothy Hodgkin: On proteins and patterns. *The Lancet* **384**: 1496–1497.

[86] Miers, H. A. (1931). Obituary Notices: Thomas Vipont Barker, 1881–1931. *Journal of the Chemical Society* 3344–3345.

406 *Pioneering British Women Chemists: Their Lives and Contributions*

encouraged in this by Dr. Polly Porter, a Research Fellow at Somerville."[87] Crowfoot herself recalled:

> Polly Porter advised me to go to Germany, to work for a few months in the laboratory of old Professor Viktor Goldschmidt, a particular friend of hers. He had designed a two-circle goniometer for measuring crystals — Polly bought one of these for Oxford — and also devised a good method for drawing crystals which I learnt. ... — but my time was too short for that.[88]

Her chemistry tutor, Freddy M. Brewer, also encouraged Crowfoot to undertake X-ray crystallography in her final year, suggesting that she work with Herbert Marcus (Tiny) Powell on the crystal structure of dimethyl thallium halides. This she did, growing her own crystals. Powell, using his glassblowing skills, constructed the X-ray tube himself.[89]

Powell and Crowfoot ultimately shared laboratory space for a total of almost 40 years. However, their relationship was strained, initiated by an event back when she was his student as K. A. McLaughlan described in an obituary of Powell:

> Their differences developed from the start when Powell presented Dorothy's Part II work to the male-only Alembic Club [see Chapter 3], a chemistry club within Oxford. She greatly resented not being allowed to do it herself, but this was scarcely Powell's fault, it then being the rule. Indeed, he may well have thought he was honouring her by presenting the work they had done together. 'But', he [Powell] wrote, 'it was cruel and Dorothy was hurt by it.'[90]

Wishing to broaden her experience in crystallography, Crowfoot accepted an offer to join Bernal's group at Cambridge. There she studied early

---

[87] Note 81, Dodson, p. 184.

[88] *Crystallography and Chemistry in the First Hundred Years of Somerville College.* A James Bryce Memorial Lecture delivered in Wolfson Hall, Somerville College on 2 March 1979 by Professor Dorothy Hodgkin, O.M., F.R.S. P. Adams, Archivist, Somerville College is thanked for providing a copy of this unpublished lecture.

[89] McLauchlan, K. A. (2000). Herbert Marcus Powell, 7 August 1906–10 March 1991. *Biographical Memoirs of Fellows of the Royal Society* **46**: 426–442.

[90] Note 89, McLaughlan, p. 437.

Women Crystallographers: The Bragg Descendants 407

preparations of Vitamin B_1, Vitamin D, and several of the sex hormones. Finances were a problem, though, and only a substantial gift of money from her aunt enabled her to survive.

During her first year at Cambridge, Crowfoot was offered a post as Tutor at Somerville. As an enticement, the Oxford University authorities agreed that she could spend a second year at Cambridge before returning to Oxford. Bernal advised Crowfoot to accept the offer, which she did. Once back in Oxford, Crowfoot began an X-ray crystal analysis of cholesterol derivatives, but at every opportunity, she would return to visit the sparkling intellectual life at Cambridge, a contrast to the loneliness she found at Oxford, exacerbated by her exclusion from the Senior Alembic Club.

In 1937, Crowfoot received the Ph.D. degree from Cambridge. That same year, she acquired her first graduate student, Dennis Riley. Riley had invited Crowfoot to address the Junior Alembic Club, and having become excited by her research, insisted that he wished to work with her. His request was not looked upon favourably by fellow chemists. Riley noted:

> This, at the time, was quite revolutionary and several eyebrows were lifted. Here was I, a member of a prestigious college [Christ Church], choosing to do my fourth year's research in a new borderline subject with a young female who held no university appointment but only a fellowship in a women's college.[91]

Also in 1937, Crowfoot married the historian Thomas L. Hodgkin, and over the following years she had three children. Hodgkin decided early in the marriage that Crowfoot was the more creative of the two, and it was he who looked after the children in the evenings while she returned to the laboratory.

To add complication to their later lives, Hodgkin became active in African studies, which required them to have a residence in Ghana as well as England. Despite her work commitments, Crowfoot always found time for her children, being able to switch between her roles as a mother and a scientist with total ease. Fortunately, Crowfoot's social life improved,

---

[91] Riley, D. P. (1981). Oxford: the early years. In Note 76, Dodson, Glusker, and Sayre, p. 17.

408   *Pioneering British Women Chemists: Their Lives and Contributions*

particularly through a friendship with Max Perutz, another of Bernal's former students, who was still at Cambridge.[92] Perutz would bring his wife and children to Oxford, where the two families would walk beside, and sometimes swim in, the Isis (the local name for the Thames).[93]

Throughout her career, Crowfoot picked projects that were always just beyond the currently accepted limits of feasibility, and her research on the structure of cholesterol was one such example. This molecule had long been a puzzle to organic chemists, and it was Crowfoot and another of her students, Harry Carlisle, who determined the actual bonding arrangement — the first time that X-ray crystallography had been used to deduce an organic structure in which the atomic arrangement was unknown. This work was completed in the early years of the war.

In 1942, Crowfoot embarked upon the molecular structure of penicillin. Structural studies were essential to help in the synthesis of penicillin, a task which was of vital importance during the war. Fortunately, she had met biochemist Ernst Chain[94] some years earlier in the streets of Oxford, and he had promised to provide her with crystals of the antibiotic. Even to start the project was difficult, for it was not realized in those early days that penicillin could adopt different packing arrangements, depending upon the conditions of crystallization.

To help in the work, Crowfoot had one graduate student, Barbara Rogers Low. About the middle of the Second World War, they acquired night-time use of an IBM punched card machine in an evacuated government building. Low helped write the first three-dimensional computer program and punched the data onto cards. The structure was finally completed in the summer of 1945, but of equal importance, Crowfoot had attacked the problem by a variety of different crystallographic techniques, broadening the options available for structure determination.

---

[92] Blow, D. M. (2004). Max Ferdinand Perutz, 19 May 1914–6 February 2002. *Biographical Memoirs of Fellows of the Royal Society* **50**: 227–256.

[93] Perutz, M. (1981). Forty years friendship with Dorothy. In Note 76, Dodson, Glusker and Sayre, p. 6.

[94] Abraham, E. (1983). Ernst Boris Chain, 19 June 1906–12 August 1979. *Biographical Memoirs of Fellows of the Royal Society* **29**: 42–91.

Though Crowfoot had made important contributions to science, at the end of the War, her rank at Oxford was still only that of Tutor. Deeply in debt, she realized that most of her male colleagues had university positions, as well as research appointments, so she asked Cyril Hinshelwood, Professor of Physical Chemistry,[95] to help her acquire a better position. With his help, Crowfoot was appointed as a University Demonstrator in Chemical Crystallography in 1946, her first appointment as an Oxford University employee.

Crowfoot's research group steadily expanded in the post-War era, and it included workers from around the world, one undergraduate student being Margaret Roberts (Mrs. Thatcher),[96] who later changed her career direction to become Prime Minister. Crowfoot decided to limit the numbers to about 10 to maintain the interactive environment of the group. A Rockefeller Foundation investigator commented that the lab was:

> ... kept under good strong scientific discipline by their gentle lady boss who can outthink and outguess them on any score. A lovely small show reflecting clearly the quality of its director ... She conducts the affairs of her small laboratory on a most modest, almost self-deprecatory scale.[97]

In 1948, a researcher with the Glaxo pharmaceutical company gave Crowfoot some deep-red crystals of Vitamin B_{12} that he had just obtained. It had been in 1926 that raw liver was shown to cure the disease of pernicious anaemia and the crucial compound had been identified as this vitamin. Before the vitamin could be synthesized, it was essential to determine the chemical structure. With 93 non-hydrogen atoms, many crystallographers regarded the task as impossible. Crowfoot disagreed. Over the following 6 years, the group grew more and larger crystals while they took a total of 2500 X-ray photographs. Having acquired the data, the next task

---

[95]Thompson, H. (1973). Cyril Norman Hinchelwood, 1897–1967. *Biographical Memoirs of Fellows of the Royal Society* **19**: 374–431.

[96]Walters, R. (2014). *Margaret Thatcher and Dorothy Hodgkin: Political Chemistry*, CreateSpace Independent Publishing Platform.

[97]McGrayne, S. B. (1993). *Nobel Prize Women in Science*. Birch Lane Press, New York, p. 250.

410 *Pioneering British Women Chemists: Their Lives and Contributions*

was the analysis, which was far beyond the normal calculating facilities of the time.

A visit to Oxford by Kenneth Trueblood from the University of California at Los Angeles (UCLA) was to make this step possible. At UCLA, Trueblood had programmed one of the first high-speed electric computers for crystallographic calculations, and he had free computing time on the machine. He offered to help, and it was arranged that Crowfoot would send him the data, which he would run through the machine, and mail back the results.[98] In 1956, she finally determined the structure of Vitamin B_{12}.

Crowfoot was not promoted to Reader until 1957, and even then, she was not provided with modern lab facilities until the following year. The academic pinnacle of success, an endowed Chair, was offered to her in 1960, but it was provided by the Royal Society, not the University of Oxford. Worldwide recognition of her work on the determination of the structures of biochemically important molecules came in 1964 with the Nobel Prize in Chemistry. Among the headlines in the British newspapers of 30 October 1964 were: "Nobel prize for British wife" (*Daily Mail*), "Nobel prize for a wife from Oxford" (*Daily Mirror*), and "British woman wins Nobel Prize — £18,750 prize to mother of three" (*Telegraph*).

Some crystallographers felt that Crowfoot should have received the Nobel Prize earlier. For example, Perutz, who was co-recipient of the Nobel Prize for chemistry in 1962, commented:

> I felt embarrassed when I was awarded the Nobel Prize before Dorothy, whose great discoveries had been made with such fantastic skill and chemical insight and had preceded my own. The following summer I said as much to the Swedish crystallographer Gunner Hagg when I ran into him in a tram in Rome. He encouraged me to propose her, even though she had been proposed before. In fact, once there had been a news leak that she was about to receive the Nobel Prize, but it proved false; Dorothy never mentioned that disappointment to me until long after. Anyway, it was easy to make out a good case for her; Bragg [W.L.] and Kendrew [Perutz's Nobel co-recipient] signed it with me, and to my immense pleasure it produced the desired result soon after.[99]

---

[98]Trueblood, K. N. (1981). Structure Analysis by Post and Cable. In Note 76, Dodson, Glusker, and Sayre, p. 87.

[99]Note 93, Perutz, p. 10.

The determination of the structure of the protein, insulin, was Crowfoot's third major project. Crowfoot had always had an interest in poetry and this change in direction was written as a limerick in her honour:

> *There was a chemist named Dorothy,*
> *On crystal structure, an authority.*
> *Into crystals she did delve,*
> *solving vitamin B12,*
> *before insulin became her priority.*[100]

This was the culmination of 30 years of work, since her first X-ray photograph of the compound had been taken in 1935, and as early as 1939, she had published a report of X-ray measurements on wet insulin crystals. As techniques had improved over the years, she had kept returning to this particular molecule, but it was only technical advances in the 1960s that made the solution finally possible. To construct the 3-dimensional representation, the electron-density contours were traced onto a Mylar film which was then stuck onto Perspex sheets. Mamannamana Vijayan explained the scene:

> This method of producing usable electron-density maps obviously involved a great deal of tedious semi-skilled operations. Dorothy often commandeered the help of the wives of research students and postdocs for this purpose. Dorothy's laboratory, therefore, often gave the impression of families working together. For a period of time, an impressive item in the work place was a beautiful playpen for toddlers to play in while their young mothers worked![101]

When the results were published in 1969, the researchers were listed in alphabetical order, showing her willingness to share the credit and her egalitarian attitude towards all the research workers. In addition, she had

---

[100]E-mail communication Rupert Cole, Associate Curator of Chemistry, Science Museum, London. Reproduced by permission.

[101]Vijayan, M. (2002). The story of insulin crystallography. *Current Science* **83**: 1598–1606.

412   *Pioneering British Women Chemists: Their Lives and Contributions*

one of her young postdoctoral fellows give the first lecture on the structure so that the glory was again shared rather than focussed on herself.

Crowfoot continued the philosophy of the Braggs and of Bernal that science was a social activity. One of her later students, John H. Robertson, recalled the family atmosphere of the Crowfoot group and particularly the afternoon tea:

> Each member of the community took his or her turn, weekly, to provide the little cakes that went with the afternoon's cup of tea. When anyone had a birthday, or a new baby, or anything comparable to celebrate, it was, by unwritten rule, that person's duty to provide a large iced cake, free, for that occasion. Each person had his or her own desk, of course, but everyone knew, at least in outline, what everyone else was doing. All the problems, and everything that was going on, were interesting. Mutual assistance was frequent; animated, even heated discussions were normal. The motivation was the interest of the subject. Everyone worked hard. It came naturally to do so.[102]

After the Second World War, Crowfoot joined the Science for Peace organization.[103] Membership in this organization caused her to be denied a visa to attend a meeting in the United States during 1953. For the next 27 years, she had to obtain a special entry permit from the U.S. Attorney General to attend scientific meetings. Only in 1990, at the age of 80, did the State Department relent and approve a visa application.

It is difficult to overstate the challenges that Crowfoot faced in determining the structures of these very complex molecules. As Judith Howard has commented:

> Dorothy belongs to a small group of women who have changed the face of science... In a time when there were no computers and modern diffractometers, her selected goals in structural chemistry (cholesterol, penicillin, vitamin B12 and insulin) must have seemed impossible to attain. However, once she had solved these structures using X-ray

---

[102]Robertson, J. H. (1981). Memories of Dorothy Crowfoot and of the B12 structure in 1951–1954. In Note 76, Dodson, Glusker, and Sayre, p. 73.

[103]Chatterjee, S. (2001). A scientist for peace. *Frontline* **18**(5): 84–86.

crystallography, they revealed the complexities of nature in a beautiful and complete way and this effectively freed chemists from the burden of structure determination.[104]

Crowfoot formally retired in 1977, but she continued to be active in science until her death on 30 July 1994 in the village of Ilmington, Warwickshire.

Rosalind Franklin

Crowfoot's life was one of ultimate success before she died. However, Rosalind Elsie Franklin[105] died without recognition of her contributions to X-ray crystallography and to science as a whole. Only posthumously has fame been granted.[106] Franklin was born on 25 July 1920 in London, to Ellis Arthur Franklin, banker, and Muriel Frances Waley. Though raised in a Jewish family, her parents were "not overly religious"[107] and had family backgrounds as social activists. Many of Franklin's aunts were practicing socialists and women's-rights workers, while one of her uncles had actually been jailed for his suffragette activities.

Franklin's mother concisely described those attributes of her character which were to lead to antagonisms later in life:

> Her affections both in childhood, and in later life, were deep and strong and lasting, but she could never be demonstrably affectionate or readily express her deeper feelings in words. This combination of strong feeling, sensibility and emotional reserve, often complicated by an intense concentration on the matter of the moment, whatever it might be, could provoke either a stony silence or a storm.[108]

---

[104] Howard, J. A. (2003). Dorothy Hodgkin and her contributions to biochemistry. *Nature Reviews Molecular Cell Biology* **4**: 891–896.

[105] (a) Glynn, J. (2012). *My Sister Rosalind Franklin*. Oxford U. P., Oxford; and (b) Maddox, B. (2002). *Rosalind Franklin: The Dark Lady of DNA*. HarperCollins, New York.

[106] Glynn, J. (2008). Rosalind Franklin: 50 years on. *Notes & Records of the Royal Society* **62**: 253–255.

[107] Maddox, B. (Autumn 2002). Mother of DNA. *New Humanist* 24–25.

[108] Franklin, M. (n. d.). *Rosalind*. Butler & Tanner, Frome, p. 5.

414 *Pioneering British Women Chemists: Their Lives and Contributions*

During her early years, Franklin felt discriminated against because she was female and, as she recalled, her childhood was a battle for recognition. She detested dolls, much preferring sewing, carpentry, and Meccano construction sets. She was fortunate to be educated at St. Paul's Girls' School, an academically strong institution that excelled in the teaching of physics and chemistry. Astronomy was her favourite hobby, and by the age of 15 she had decided to become a scientist.

Franklin took and passed the entrance examinations for the University of Cambridge, where she planned to study physical chemistry; however, her father refused to pay for her. He had once wanted to be a scientist, and he insisted that he would have been delighted for a son to follow such a path, but in his view, daughters should only consider volunteer work rather than a full-time career. Outraged at this attitude, Franklin's mother, together with her favourite aunt, offered to pay; in the family crisis that followed, Ellis Franklin finally relented. As Sharon McGrayne commented: "his approval was grudgingly given and resentfully received."[109] With the encouragement of her female relatives, Franklin went up to Cambridge in 1938.

With the start of the Second World War in 1939, the senior scientists were given war-related research and their time with students dwindled. With the lack of supervision, Franklin, working in the Cavendish Laboratory, enjoyed the opportunity for independent work; however, she did not get a first-class degree. Her supervisor in Physical Chemistry, Fred Dainton,[110] provided the reason:

> Miss Franklin was certainly capable of getting a First because she had the qualities of pertinacity and penetration which were rather exceptional. But I did not expect her to do so because she was very selective in the use of her time, going to depth in the things which interested her and neglecting the other parts of the subject.[111]

---

[109] McGrayne, S. B. (1993). *Nobel Prize Women in Science*. Birch Lane Press, New York, NY, p. 308.

[110] Gray, P. and Ivin, K. J. (2000). Frederick Sydney Dainton, Baron Dainton of Hallam Moors, 11 November 1914–5 December 1997. *Biographical Memoirs of Fellows of the Royal Society* **46**: 86–124.

[111] Letter, Lord Dainton to Mansel Davis, 10 March 1982. Cited in: Davis, M. (1990). Notes and discussions: W. T. Astbury, Rosie Franklin, and DNA: A Memoir. *Annals of Science* **47**: 607–618, p. 616.

After graduation in 1941, Franklin accepted a position with the future Nobel Laureate, Ronald Norrish,[112] exploring gas-phase chromatography of organic mixtures. Franklin and Norrish had an abrasive relationship, which was due to personality problems on both sides. She had been used to working very much on her own, rather than having a supervisor telling her what to do, while Norrish objected to Franklin's belief in sexual equality among scientists.[113] Franklin resigned from the position after 1 year in order to contribute to the war effort by doing research for the British Coal Utilization Research Association (BCURA). Her studies at BCURA contributed significantly to the understanding of coal structure and the effect of heating coals.[114] She wrote up some of this work as her Ph.D. thesis, the degree being awarded in 1945.

During the war, Franklin became friends with the French metallurgist, Adrienne R. Weill, a former worker with Marie Curie and Irène Joliot-Curie at the Institut Curie, and after the war, she wrote to Weill: "If ever you hear of anybody anxious for the services of a physical chemist who knows very little about physical chemistry, but quite a lot about the holes in coal, please let me know."[113] Weill suggested Marcel Mathieu, a former student of W. H. Bragg, and Mathieu invited her to France in 1947 to work as a researcher in the Laboratoire Centrale des Services Chimique de l'État. While in Paris, she worked on the growth of graphite crystals with Mathieu's student, Jacques Méring, with whom she had a very positive relationship. Méring introduced her to the crystallographic techniques which were to become her field of expertise.

The time in France seemed to have been the happiest of her life. Franklin took science very seriously, and in the laboratory, she was intense and reserved, but outside she sparkled in the chic social life of Paris. Her co-workers were also friends with whom she had lunch at local bistros, went to dinner parties, and joined for skiing and mountain-climbing vacations. She particularly loved mountains and went on

---

[112]Dainton, F. and Thrush, B. A. (1981). Ronald George Wreyford Norrish, 9 November 1897–7 June 1978. *Biographical Memoirs of Fellows of the Royal Society* **27**: 379–424.

[113]Sayre, A. (1975). *Rosalind Franklin and DNA*. W. W. Norton, New York, NY, p. 58.

[114]Harris, P. J. F. (2001). Rosalind Franklin's work on coal, carbon, and graphite. *Interdisciplinary Science Reviews* **26**: 204–209.

416 *Pioneering British Women Chemists: Their Lives and Contributions*

frequent long hikes and bicycle trips. Holidays, in fact, were one of her greatest joys and they were planned in meticulous detail. As her mother remarked:

> Not for her the lazy holidays basking in the sun. Sunshine and warmth she loved, but her holidays must be full of movement and never idle. Long walks of up to twenty miles a day; climbing, swimming; visits to picture galleries and ancient buildings, these were her delight. She was an eager stimulating companion, but those who travelled with her must possess a zest an energy comparable to her own.[115]

After 3 years in Paris, Franklin accepted an offer in 1951 to work at King's College, London. Physicist John Randall,[116] who had formed an interdisciplinary group of physicists, chemists, and biochemists, needed an expert crystallographer to analyse X-ray photographs of DNA taken by Raymond Gosling, then a graduate student.[117] Having heard of Franklin's skills, Randall offered her the job, noting: "This means that as far as the experimental X-ray effort is concerned, there will be at the moment only yourself and Gosling, together with the temporary assistance of a graduate from Syracuse, Mrs. Heller."[118]

When Franklin arrived, Randall's second-in-command, Maurice Wilkins,[119] was away, and this proved to be a major problem, for Franklin assumed that she was working completely independently while Wilkins, on his return, assumed that Franklin had been hired as a technical assistant to produce data for other members of the structured group to analyse. This was the start of an unhappy relationship between them.

Adding to the source of conflicts, McGrayne stated that the social life at King's was far different to Franklin's previous experience.

---

[115]Note 108, Franklin, p 12.

[116]Wilkins, M. H. F. (1987). John Turton Randall, 23 March 1905–16 June 1984, *Biographical Memoirs of Fellows of the Royal Society* 33: 492–535.

[117]Attar, N. (2013). Raymond Gosling: the man who crystallized genes. *Genome Biology* **14**: 402–414.

[118]Note 109, McGrayne, p. 312.

[119]Struther, A., Kibble, T. W. B. and Shallice, T. (2006). Maurice Hugh Frederick Wilkins, 15 December 1916–5 October 2004. *Biographical Memoirs of Fellows of the Royal Society* **52**: 455–478.

A number of women scientists worked on the staff, but they were not allowed to eat in the men's common room; women ate outside the lab or in the student's cafeteria. After work, the men visited a male-only bar for beer and shoptalk; women were not invited. As a result, the men talked science casually among friends while the women operated in a more formal office atmosphere.[120]

This would not be surprising considering King's reluctance to admit women earlier in the century (see Chapter 4).

This perspective is different to the views of some of the other women who worked with Randall during that time period.[121] For example, Honor Fell,[122] another member of the group, considered that much of the social interaction took place in after-lunch coffee gatherings in one or other research lab. Franklin's "brusque manner" and her "overriding passion" for science meant, according to Fell, that Franklin never developed friendships, even with the eight women of the 31-member team. Whatever the cause, exclusion from social interaction not only meant a loss of companionship but also it prevented her from appreciating the significance of the hierarchical nature of Randall's group.

Brenda Maddox, journalist and biographer, also felt that King's College had been more welcoming than had been painted. To Maddox, a major problem for Franklin was the strong religiousness of the College which had an Anglican chapel built into the core building. She added:

I also found that Franklin felt singularly unhappy at King's not so much because of her gender, but because of her class and religion: a wealthy Anglo-Jew felt out of place in a Church of England setting dominated by swirling cassocks and students studying for the priesthood.[123]

---

[120]Note 109, McGrayne, p. 313.

[121]Judson, H. F. (1986). Annals of science: The legend of Rosalind Franklin. *Science Digest* **94**: 56–59 and 78–83.

[122]Vaughan, J. (1987). Honor Bridget Fell, 22 May 1900–22 April 1986. *Biographical Memoirs of Fellows of the Royal Society* **33**: 236–259.

[123]Maddox, B. (2003). The double helix and the 'wronged heroine'. *Nature* **421**: 407–408.

418   *Pioneering British Women Chemists: Their Lives and Contributions*

Mainstream crystallography was then concerned with finding precise atomic locations in crystals of pure simple substances. However, the purpose of X-ray studies of complex materials, such as coal and DNA, was to gain a more general view of the atomic arrangements. Over the 2 years that she spent at King's, Franklin established that there were two forms of DNA, A and B, the form depending upon the humidity, a point that had confused the other researchers in the field. Franklin developed techniques to produce the most high-resolution X-ray photographs of DNA taken up to that time. She showed that the phosphate groups were on the outside of the DNA molecule and that hydrogen bonding played an important role.

Gosling enjoyed working with Franklin, and he commented on her strong personality and ebullient, argumentative character. These attributes did not endear her to Wilkins, as Wilkins himself was "shy, passive, indirect."[124] As a result, the relationship between Wilkins and Franklin, which had never been positive, deteriorated into active dislike. Delia Simpson (see Chapter 3), who had been one of Franklin's Lecturers at Cambridge, became aware of Franklin's situation as Simpson's obituarist described:

> Delia showed great concern for the well-being of her students ... This solicitude continued after students had left Cambridge. An example of this was the way in which, when Delia became aware of the problems that Rosalind Franklin was having at King's College, she made a special point of calling in for a chat whenever she was in London.[125]

During this period, James Watson and Francis Crick were constructing their models of DNA at Cambridge University. Wilkins developed a friendship with the two, confiding in them about his difficulties with Franklin. Without telling Franklin, Wilkins removed the superb X-ray photograph of the B-form and showed it to Watson and Crick, providing the major experimental evidence for the helical structure of DNA. Franklin had summarized her work in a report to the Science Research

---

[124]Note 121, Judson, p. 79.

[125]Newton, A. A. (1999). Delia Agar, 1912–1998. *Newnham College Roll Letter* 101–103.

Council and this had been distributed to their review committee. A member of this committee passed the report, unethically in the view of many scientists, to Watson and Crick, providing additional essential information.

When Watson and Crick submitted their famous article on DNA structure to the journal *Nature*, the editor contacted both Wilkins and Franklin to ask if they would submit accompanying articles. Franklin had by this time herself deduced the helical nature of DNA but she was not prepared to speculate. As Michelle Gibbons has pointed out: "Rosalind Franklin's research strategy was to avoid exactly the sort of speculation that Watson and Crick freely engaged in."[126] Whether she would have done so in time, or whether Watson and Crick could have solved the structure without Franklin's photograph and report, is still a matter of debate. The three papers appeared sequentially, with the vague comment by Watson and Crick that they had: "been stimulated by a knowledge of the general nature of the unpublished experimental results and ideas of Dr. M. Wilkins, Dr. R.E. Franklin, and their co-workers at King's College, London."[127] Franklin was never aware that they had actually seen her photograph and report.

The protein crystallographer, David Harker, identified the key issue:

> And the real tragedy in this affair is the very shady behaviour by a number of people, as well as a number of unfortunate accidents, which have resulted in the transfer of information in an irregular way.... I would never have consciously become involved in anything like this behaviour, especially the transfer of information through a privileged manuscript. And I think these people are — to the extent that they did these things — outside scientific morals as I know them.[128]

---

[126]Gibbons, M. G. (2012). Reassessing discovery: Rosalind Franklin, scientific visualization, and the structure of DNA. *Philosophy of Science* **79**: 63–80.

[127]Watson, J. D. and Crick, F. H. C. (1953). A structure for deoxyribose nucleic acid. *Nature*, **171**: 737–738. For a more complete account of the DNA saga see: Olby, R. (1974). *The Path to the Double Helix*. Macmillan, London.

[128]Harker, D. Cited in: note 1, Julian, p. 363. See also: Hubbard, R. (1979). Reflections on the story of the double helix, *Women's Studies International Quarterly* **2**: 261–273; and Strasser, J. (September/October 1976). Jungle law: Stealing the double helix, *Science for the People* **8**: 29–31.

By 1953, Franklin felt that relations with Wilkins had deteriorated to the point where she could no longer work at King's. Kathleen Yardley had offered Franklin a position at UCL,[129] but Franklin declined and instead asked Bernal if she could join his group at Birkbeck. Bernal quickly offered an invitation. Randall agreed to the transfer of her Fellowship to Bernal, on condition that she did not continue working on DNA (at that time it was not uncommon in Britain for specific research fields to be the "property" of particular research groups). Bernal arranged for Franklin to be the head of her own research group, and he suggested that she look at the structure of the tobacco mosaic virus, a topic that he had started and abandoned years earlier. By working with both Bernal and Mathieu, Franklin was now doubly scientifically related to the Braggs.

In the 5 years at Birkbeck, Franklin and her research group determined the structural features of the virus, showing it to be hollow-cored, not solid as supposed by microbiologists, and she was able to determine surface features of the virus. During this period, her team was the world leader in the structure of viruses and, for the first time, she was able to develop positive collaborations with other research groups.[130] Of equal or greater importance, Franklin was back in a pleasant working environment. However, there were still moments of frustration, such as the refusal (at the time) of the British Agricultural Research Council to fund any project that had a woman directing it. Fortunately, the U. S. Public Health Service provided her with adequate funding instead.

During 1956, Franklin experienced extreme pain which was diagnosed as ovarian cancer. Three operations and experimental chemotherapy followed, but without success. During these last years, Franklin had become friends with Crick and his French wife, Odile, showing no animosity for the DNA dispute. While convalescencing with them, Crick had suggested that she move her group to Cambridge, and this she decided to do. Realizing that she was terminally ill, Franklin started a dangerous individual project, the structure of the live polio virus. This work was so

---

[129]Letter, K. Lonsdale to R. Franklin, 4 January 1955. Franklin Archives, Churchill College, Cambridge University (CC-CU).

[130]Creager, A. N. and Morgan, G. J. (2008). After the double helix: Rosalind Franklin's research on tobacco mosaic virus. *Isis* **99**: 239–272.

hazardous that after her death the project was halted. She died on 16 April 1958, at age 37.

Four years later, the Nobel Prize was awarded to Crick, Watson, and Wilkins, for the DNA work. The Nobel Prize is only awarded to living scientists and to no more than three people in any category. Having died, Franklin was no longer eligible, but the controversy focussed upon the lack of acknowledgement of her work during their Nobel Lectures in Stockholm.

Many scientists argued that the popular account of the discovery of DNA, *The Double Helix*[131] by Watson minimized Franklin's contributions and painted a very negative picture of her personality and appearance. As her biographer, Anne Sayre, has commented, everyone is entitled to their own perception of another individual, but the picture of Franklin was to generate a totally misleading image which would reinforce the negative stereotype of women scientists. Sayre added:

> "Rosy" [Watson's nickname for Franklin] was less an individual than ... a character in a work of fiction. ... If Rosalind was concealed, the figure which emerged was plain enough. She was one we have all met before, not often in the flesh, but constantly in a certain kind of social mythology. She was the perfect, unadulterated stereotype of the unattractive, dowdy, rigid, aggressive, overbearing, steely, "unfeminine" bluestocking, the female grotesque we have all been taught to fear or to despise.[132]

To which Patricia Fara has added: "Had he [James Watson] been more aware of European fashion, he might have appreciated the care Franklin took to adopt designer Christian Dior's iconic 1947 look ..."[133]

In fact, her closest collaborator at Birkbeck, Aaron Klug, was so incensed at the belittling of her role that he wrote a formal protest,

---

[131] Watson, J. D. (1968). The double helix. In Stent, G. S. (ed.), *The Double Helix: A New Critical Edition.* Weidenfeld and Nicolson, London. This edition includes additional commentaries, several of which contradict Watson's viewpoint.

[132] Note 113, Sayre, p. 18.

[133] Fara, P. (2013). Women in science: Weird sisters? *Nature* **495**: 43–44.

422 *Pioneering British Women Chemists: Their Lives and Contributions*

published in *Nature*.[134] A former colleague of Franklin's, A. J. Caraffi, wrote to thank Klug for his defence of Franklin:

> I would like to say how very much I have appreciated your paper to *Nature* on Rosalind Franklin and DNA. After the journalistic distortions to which her rightful place in the history of science has been subjected, it is most satisfying to one's sense of justice to see published a dignified, objective, and properly documented account of her signal contribution. It brings out her absolute integrity and illustrates admirably the thorough principles applied by great experimentalists to the evaluation of their proofs. There could be no better counter to the titillations of all the offensive "Rosie" nonsense that was bound to attract selection and emphasis by ignorant hacks [journalists].[135]

It is appropriate to give her friend, Klug, the final word on how Franklin should be perceived among the pantheon of scientists. He wrote a letter to P. Siekevitz of the New York Academy of Sciences:

> However, if she is to be honoured, it should not be as a "woman in science" but for her crucial contributions in sorting out the A and B forms, establishing that the phosphates were on the outside and determining the helical parameters which were used by Crick and Watson in their study.... There is also, inevitably, a fair amount of discussion as to whether she would have solved the structure on her own. One can only guess, but my view, as stated, is that she would have done so eventually.[136]

Commentary

Unusual for their time, both Yardley and Crowfoot managed to raise a family as well as devote long hours in the laboratory. This was, to a large

---

[134] (a) Klug, A. (1968). Rosalind Franklin and the discovery of the structure of DNA. *Nature* **219**: 808–810 and 843–844; and (b) Klug, A. (1974). Rosalind Franklin and the double helix. *Nature* **248**: 787.

[135] Letter, A. J. Caraffi to A. Klug, 29 August 1968. Franklin Archives, CC-CU.

[136] A. Klug to P. Siekevitz (New York Academy of Sciences), 14 April 1976. Franklin Archives, CC-CU.

extent, a result of very progressive marriages, with the husbands performing some of the "wifely" duties (whilst hired help undertook other home-making activities). Franklin followed the more common single pattern. Though she expressed a preference for marriage, not wishing to remain a: "spinster professor," she felt that children would have interfered with the research to which she wanted to devote her life. Hilary Rose, sociologist of science, has suggested that the "single and sexually threatening" image contributed to Franklin's problems, compared to the safe and respectable perception of Crowfoot and Yardley as wives and mothers.[137] It can also be argued that Franklin's abrupt manner did not assist with her acceptance into British scientific society.

Crystallography seems to provide one of the clearest examples of the positive and supportive role of supervisors, and particularly female role models. Among the many of subsequent "generations" of crystallographers has been W. H. Bragg's crystallographic "grand-daughter", Mary Truter;[138] W. L. Bragg's "grand-daughter", Caroline MacGillavry;[139] and Crowfoot's "daughter", Jenny Pickworth (Mrs. Glusker).[140] Another of Crowfoot's "descendants", Judith Howard, late crystallographer at the University of Durham, speculated that: "… one of the reasons why women continue to be attracted to crystallography is because so many early women crystallographers were so good."[141] She added that in 1995 about one-third of the British Crystallographic Association's 800 members were women.

---

[137] Rose, H. (1994). Nine decades, nine women, ten Nobel Prizes: Gender at the apex of science. In Rose, H. (ed.), *Love, Power and Knowledge: Towards a Feminist Transformation of the Sciences*. Polity Press, Cambridge, p. 157.

[138] Cruickshank, D. (2005). Milestones: Mary Rosaleen Truter, 1925–2004. *International Union of Crystallography Newsletter* **12**(3): 28.

[139] (a) Haines, M. C. (2001). MacGillavry, Carolina Henriette. *International Women in Science: A Biographical Dictionary to 1950*. ABC-CLIO, Santa Barbara, California, p. 189; (b) Note 1, Julian, pp. 346–347; and (c) Wikipedia (27 February 2018). Carolina Henriette MacGillavry (accessed 31 December 2018).

[140] Rose, R. K. and Glusker, D. L. (1993). Jenny Pickworth Glusker (1931–). In Grinstein, L. S., Rose, R. K., and Rafailovich, M. H. (eds.), *Women in Chemistry and Physics: A Biobibliographic Sourcebook*. Greenwood Press, Westport, Connecticut, pp. 207–217.

[141] O'Driscoll, C. (February 1996). Minorities and mentors: The X-ray visionaries. *Chemistry in Britain* **32**: 5–8.

Chapter 13

Women in Pharmacy

Elsie Hooper (5 September 1879–6 May 1969)

Though pharmacy is regarded as a distinct profession, many academic women chemists saw pharmacy as one of their few employment options. As we will show, several of the women were also associated with the Chemical Society and/or the Institute of Chemistry, and two of them were signatories to the petition for admission of women to the Chemical Society (see Chapter 9).

426 *Pioneering British Women Chemists: Their Lives and Contributions*

Of particular interest, the fight for women's entry to the Pharmaceutical Society shows an interesting similarity with the battle for membership of the Chemical Society. Pharmacy also provides a contrast to chemistry. The National Association of Women Pharmacists became a continuing connection among women pharmacists, whereas the Women Chemists' Dining Club (see Chapter 9) did not survive beyond the 1950s.

Society of Apothecaries

Apothecaries had been involved in the production and dispensing of patent medicines from the Middle Ages.[1] There are records of women being registered as apothecary or "chymist & druggist" as far back as the 17th century.[2] Detailed records exist for three of these women: Widow Wyncke, Susan Reeve Lyon, and Anne Crosse.[3]

Apothecaries were a recognized Guild with its own Court of Examiners to licence those who wished to dispense the herbal remedies of the time. Their first encounter with a woman candidate occurred in 1860.[4] Elizabeth Garrett[5] had been refused by every medical school to which she had applied on the grounds of her gender (including Edinburgh, see Chapter 7). She then focussed her attention on becoming an apothecary. Passing the Apothecaries' examination entitled its holder to a place on the Medical Register, which was Garrett's long-held desire. Having taken private lessons with an apothecary, she enquired about the possibility of being examined at Apothecaries' Hall. This request caused great consternation among the members of the Court of Examiners. Their legal counsel advised them that the Apothecaries' Act of 1815 opened the examinations

---

[1] Hunting, P. (1998). *A History of the Society of Apothecaries*. The Society of Apothecaries, London.

[2] (a) Rawlings, F. H. (1984). Two 17th century women apothecaries. *Pharmaceutical Historian* **14**(3): 7; and (b) Anon. (1999). Information please. *Pharmaceutical Historian* **29**(1): 3.

[3] Woolf, J. S. (Fall 2009). Women's business: 17th century female pharmacists. *Chemical Heritage* 20–25.

[4] Note 1, Hunting, pp. 207–210.

[5] Manton, J. (1965). *Elizabeth Garrett Anderson*, Methuen, London.

to all persons and as, by British law, women were persons, they could not be excluded. Garrett passed the preliminary examination and then both parts of the professional examination, obtaining her diploma in 1865.

The Council realized that Garrett's success would encourage other women to follow suite. In fact, in 1867, three women passed the preliminary examination. The portents of more women in the pipeline persuaded the Council to act. Before admission to the examination, the candidate had to provide a certificate of attendance at the required lectures. As women were excluded from the lectures at public schools, they had to be tutored privately. Then, later in 1867, the Court of Examiners announced that they would no longer accept the validity of certificates of attendance at private lectures. This tactic proved an effective means of circumventing the law and blocking the aspirations of subsequent women candidates.

From the middle of the 19th century, the profession of Apothecary lost ground to that of the more professionally trained pharmacist. Nevertheless, women found a niche as Apothecaries' Assistants. These assistants were required by any general practitioner or dispensary who dispensed their own medicine. In 1887, Fanny Saward became the first woman to gain admission to the Apothecaries' Assistants' examination.

The Assistants' certificate, known as "Apothecaries' Hall" or simply, "Hall," was, as Penelope Hunting remarked: "... a post suited to a young lady willing to work as the handmaiden of the doctor for little pecuniary reward."[6] This occupation became a female ghetto;[7] for example, in 1917, nine men and 233 women registered for the Apothecaries' Assistants' examination. Among those women was an Agatha Mary Clarissa Christie, who later gained fame as a writer of detective stories. As the pharmaceutical industry developed during the 1930s, greater and greater scientific knowledge was demanded of the dispensers of medicines. For this reason, the more basic examinations of the Apothecaries' Hall fell out of favour compared with the more lengthy and rigorous studies demanded of those holding qualifications of the Pharmaceutical Society.

---

[6]Note 1, Hunting, p. 230.
[7]Adams, D. W. (2010). *The Rise and Fall of the Apothecaries' Assistants 1815–1923*. Ph.D. Thesis, U. Hertfordshire. https://core.ac.uk/download/pdf/1641247.pdf.

Pharmaceutical Society

The profession of pharmacy has been controlled by legal statute for over 150 years.[8] There were three significant Pharmacy Acts relevant to our saga — those of 1852, 1864, and 1868.

The Pharmacy Examinations

The Pharmaceutical Society, founded in 1841,[8] was given the authority to organize the required examinations under the terms of the Acts. However, the legal administration of the examinations was quite separate from the regular activities of the Society. Following the 1868 Pharmacy Act, there were three examinations: the Preliminary, the Minor, and the Major.

The Preliminary examination was more of a skills test in Latin, French, and arithmetic, and it entitled the successful candidate to be registered as an apprentice. Passing the Minor resulted in the designation as an assistant to a chemist or druggist, while passing the Major allowed the graduate to call themselves a pharmaceutical chemist. The Major examination was described as "decidedly difficult" and it focussed on advanced chemistry, *materia medica*, and botany. For pharmacy assistants who had actively been involved in the profession for 3 years, passing a Modified examination was all that was necessary.[9]

By the early part of the 19th century, women pharmacists were to be found all over the country.[10] In fact, when the first compulsory Register of all practicing pharmacists was undertaken in 1869, 215 of the 11,638 registered chemists and druggists were women (see Footnote 6). Most of the women were continuing a business that had been started by a father or husband who had subsequently died. For example, in Sheffield, it was reported that Mrs. Sarah Owen (née Haselhurst) had taken over after her

---

[8]Holloway, S. W. F. (1991). *Royal Pharmaceutical Society of Great Britain 1841–1991: A Political and Social History*. The Pharmaceutical Press, London.

[9]Jordan, E. (1998). 'The great principle of English fair-play': Male champions, the English women's movement and the admission of women to the Pharmaceutical Society in 1879. *Women's History Review* **7**(3): 381–409.

[10]Burnby, J. G. L. (1990). Women in pharmacy. *Pharmaceutical Historian* **20**(2): 6.

husband's death in 1837, likewise Mrs. Jane Wilkinson had inherited her husband's shop in 1831.[11]

The first woman to take the examinations required under the 1868 Act was an F. E. Potter, who applied in 1869 to take the Modified examination. It was only when the individual appeared to sit the examination was it realized she was a woman — Frances (Fanny) Elizabeth Potter. As the Act made it clear that the Society had a duty to examine all persons, Potter was allowed to take the exam, which she successfully passed. Potter (Mrs. Deacon) was followed 6 months later by Catherine Hodgson Fisher (see Footnote 8). In 1873, Alice Vickery of Camberwell, Surrey became the first woman to pass the Minor, and then in 1875 Isabella Clarke became the first woman to pass the Major examination and register as a pharmaceutical chemist.

Admission of Women to the Pharmaceutical Society

Though entry to the profession had been easy, admission to the professional body, the Pharmaceutical Society, proved challenging. Even though by law the Society was forced to admit women to its examinations, the Society acted on the premise that pharmacy was a male profession and that the Society itself was a male preserve. In fact, the assumption had been that the practice of female relatives taking over a pharmacy would cease once Registration became law. Thus, they were unprepared for the "women problem" to be a regular business item for the Council of the Pharmaceutical Society for the next decade. However, the fact that, in 1868, *The Englishwomen's Review* had pronounced pharmacy as a suitable profession for a woman[12] should have warned the Society members of what was ahead.

The Pharmaceutical Society had its own School of Pharmacy. In 1861, the discovery that a "lady" had acquired a ticket for admission to the lectures of the school caused a flurry of concern. As she was already in possession of the ticket, the Library, Museum and Laboratory Committee

---

[11] Austen, J. (1961). *Historical Notes on Old Sheffield Druggists*. J. W. Northend Ltd., Sheffield, pp. 30–34.

[12] Bernard, B. (1868). Pharmacy as an Employment for Women. *The Englishwomen's Review* (First Series) **1**: 348.

430 *Pioneering British Women Chemists: Their Lives and Contributions*

of the Society had little choice in the matter, but they added that ladies "must be regarded as attending upon sufferance."[13]

The "lady" seeking admission had been the afore-mentioned Elizabeth Garrett, who had wished to use the lectures of the School of Pharmacy to prepare herself for the examinations of the Society of Apothecaries. A student at the time, Michael Carteighe (later President of the Pharmaceutical Society), recalled:

> I remember about the time I became a student there that we had a lady student in this house. ... I remember the envious eyes with which a number of us regarded her. I do not think we regarded her with envy because she was a lady — in fact, we admired her on that account: but we were conscious that when once a lady comes into a class she means to take prizes, and I am afraid we were selfish enough to think of that rather than anything else.[14]

The following year, the Council of the Pharmaceutical Society rejected a proposal to formalize the admission of women to lectures, thus preventing any other woman from following suite.

The first woman to apply for membership of the Pharmaceutical Society was Elizabeth Leech in 1869.[15] Leech had learned her pharmacy skills from her father, having worked in his shop for 7 years. Following his death, for 6 years she shared the running of the shop with her brother and then on her own for another 9 years. The Lancashire cotton famine had forced her out of business, and she had then become compounder and dispenser of prescriptions at the Munster House Lunatic Asylum, Fulham. Her application noted that she believed that membership in the Pharmaceutical Society would help her resume her business. Fearful that the Council might think she was a trouble-maker, she wrote: "I have no wish upon any occasion to interfere with the Council or its meetings. All I want is the Membership."[13] The Council rejected her request a total of three times, the last being in 1872.

---

[13]Cited in Note 8, Holloway, p. 262.

[14]Hudson, B. (2013). *The School of Pharmacy University of London: Medicines, Science and Society, 1842–2012.* Academic Press, London, p. 63.

[15]Jordan, E. (2001). Admitting a dozen women into the society: The first women members of the British Pharmaceutical Society. *Pharmaceutical Historian* **31**: 18–27.

Robert Hampson and His Crusade

It is often overlooked how much the progress of women relied on sympathetic men,[16] and the cause of women in pharmacy was no exception. The "white knight" in this case was Robert Hampson.[17] In 1872, Hampson was elected to the Council of the Pharmaceutical Society, focussing his activist radicalism on the issue of women's rights. Annie Neve recalled:

> Women pharmacists have good reason to honour the memory of Robert Hampson ... As a member of the [Pharmaceutical] Society's Council he repeatedly pleaded and wrought for the admission of women pharmacists to membership of the Society.[18]

Mary Creese added: "He was willing to put his beliefs into action, employing and training women at his business at St. John Street Road, London."[19]

Hampson was joined in his crusade by other, equally radical, new Members of Council. On 2 October 1872, Hampson proposed to the Council that, since women had been permitted to take the examinations of the Society, they should have been allowed to attend lectures and use the laboratories of the Society's School of Pharmacy. It is interesting that a report of his opening remarks reflected the perceived progress of women that had occurred in the preceding decade:

> In 1862 perhaps the admission of lady students to the classes and laboratory might appear a step fraught with great danger, and tending to revolution; but in the present day, in remembrance of the social and educational changes that had taken place, he could not for a moment

---

[16] Strauss, S. (1982). *Traitors to The Masculine Cause: The Men's Campaigns for Women's Rights*. Greenwood Press, Westport, Connecticut.

[17] Jordan, E. (1998). 'The great principle of English pair-play': Male champions, the English women's movement and the admission of women to the Pharmaceutical Society in 1879. *Women's History Review* **7**(3): 381–410.

[18] Neve, A. (1926). Miss Annie Neve's Reminiscences of Mrs. Clark-Keer. *Pharmaceutical Journal and Pharmacist* **117**: 375.

[19] Creese, M. R. S. (2005). How women pharmacists struggled for recognition before 1905. *Pharmaceutical Journal* **274**: 730–732.

432 *Pioneering British Women Chemists: Their Lives and Contributions*

assume that the present Council, elected on a much broader basis, would endorse the decision of the predecessors, which was, in fact, most arbitrary, unjust and impolitic.[20]

Hampson contended that, by the use of "person" in the Pharmacy Act of 1868, the framers of the Act did not exclude the possibility of women. His eloquent plea — in this instance — was successful. This was Hampson's sole early victory. The Council agreed unanimously to the admission of women to classes; however, as there was limited laboratory space, they were denied access to the laboratories. His proposal in November of that year that Elizabeth Leech be admitted as a member was defeated.[21] In December, his motion that women who took the examinations should be eligible for the Society's prizes, certificates, and fellowships was also defeated by one vote.[22]

At the Council meeting of 5 February 1873, there were names of three women put forward among the 166 candidates for admission as "registered students" of the Society.[23] These were Rose Minshull, Louisa Stammwitz, and Alice Rowland (Mrs. Hart). Minshull had attained the highest mark among the 166 who had passed the Preliminary examination, while Stammwitz was about mid-way in the list. Rowland had a certificate from the Society of Apothecaries *in lieu* of taking the examination.

Role of the Society for Promoting the Employment of Women

Why were these three women's names put forward? Hampson had the active support of one of the women's organizations, the Society for Promoting the Employment of Women (SPEW) (see Footnote 15). Since

---

[20] (5 October 1872). Lady Students (from Meeting of the Council October 2nd, 1872). *Pharmaceutical Journal and Transactions* **3**: 268.

[21] (9 November 1872). Election of Members (from Meeting of the Council November 6th 1872). *Pharmaceutical Journal and Transactions* **3**: 366.

[22] (1872) Lady Students (from Meeting of the Council December 4th 1872). *Pharmaceutical Journal and Transactions* **3**: 456.

[23] Transactions of the Pharmaceutical Society: Meetings of the Council, February 5th, 1872–1873. *Pharmaceutical Journal and Transactions* **3**: 629–631.

its founding in 1859, the Society had been fighting for the opening of male-exclusive trades to women. Its *modus operandum* was to find a few suitable, talented women for sponsorship and find employees who would take them as apprentices or paying students, the Society covering their fees. The Society believed that once a few women had made a success in each particular field, it would be much easier for others to follow in their path. However, the Society seemed to have little interest in supporting the aspirations of comparatively affluent ambitious women seeking careers in the professions — that is, until Hampson entered the cause of women in pharmacy.

In 1866, the year after gaining her medical licence, Elizabeth Garrett opened St. Mary's Dispensary for Women and Children in Marylebone, London. Over the following years, Garrett had willingly accepted students sponsored by SPEW to be trained as dispensers for hospitals and for medical practitioners. Two of the sponsored young women had been Minshull and Stammwitz. After Hampson had persuaded SPEW to take on the cause of women in pharmacy, Minshull and Stammwitz were chosen to lead the assault. Rowland was probably encouraged to apply for admission by her surgeon-husband, Ernest Abraham Hart.[24] Hart subsequently showed his support for women's rights by being a founding member of, and major donor to, the London School of Medicine for Women (LSMW) (see Chapter 5).

Defeat of the Women's Cause

Though Hampson and his allies may have hoped the list would be approved *in toto* without debate as usually happened, this was not to be. The Minutes reported that:

> The above list [of submitted names for approval] originally contained the names of three females, but on the fact being mentioned, it was decided that they should be put separately, in order that the question of

---

[24]Wikipedia. Ernest Hart (medical journalist) (accessed 9 January 2019).

admitting females to connection with the Society might be voted upon separately.[25]

Hampson moved their election, and the President of the Society, George W. Sandford promptly moved an amendment that they "… be not elected …." Heated debate followed. One Member of the Council suggested: "… a compromise, that the ladies should be made honorary associates,"[26] To which the response was given by another Member: "… the ladies did not wish to be admitted on any but equal terms with the opposite sex." The final outcome of events was that women were barred from admission. As discussed in Chapter 9, this same scenario played out decades later in the admission of women chemists to the Chemical Society.

Sandford articulated his position in the Correspondence pages of the *Pharmaceutical Journal*:

> I have always held that the Pharmaceutical Society was intended to be a Society of men, that certain disadvantages would arise from its being a mixed Society of men and women, and that the admission of females as apprentices would be only a stepping-stone to their admission as members, I felt bound to oppose them on the threshold.[27]

The early months of 1873 also saw the first series of letters from readers to the *Pharmaceutical Journal* and the *Chemist and Druggist* on the subject of "Pharmaceutical Women." However, the writers missed the whole point that the 1868 Pharmacy Act already permitted women to become pharmacists: the only issue was whether women were to be admitted to the professional society.

One correspondent stated that the job was unsuitable for women because of "the common occurrence of prescriptions and remedies dealing with maladies of the most revolting nature" and of the necessity to deal with "subjects which possess the power to appal and disgust the

---

[25] Note 23, pp. 629–630.

[26] Note 23, p. 631.

[27] Sandford, G. W. (1 March 1873). Correspondence. *Pharmaceutical Journal and Transactions* **3**: 698.

sternest members of the sterner sex."[28] Another writer thought that modest Englishwomen would not want to see objects: "which are sold generally to men of not the highest moral character."[29] There was even a poem on the topic of which the following is part:

> *I could not bear to see their hands as soft as alabaster*
> *Begrimed all o'er with dirty pill and nasty smelling plaster.*
> *Oh! may I never see them with their chignons in profusion*
> *Attempt to shake the tinctures or prepare the cold infusion.*
> *How could they climb the shaky steps to clean the bottles dusty,*
> *Or go below amongst the wets into the cellar musty?*
> *Their sleek round arms were never made to work the iron mortar.*[30]

Minshull, Stammwitz, and Rowland put forward their case in a letter to the *Pharmaceutical Journal*: "All that we ask is to be allowed the same opportunities for study, the same field for competition and the same honours, if justly won."[31] The issue of women's admission was next raised at the Annual General Meeting in May 1873. Hampson and his friends pressed the case of the women pharmacists. As Holloway commented: "The supporters of Hampson's motion dominated the debate, won all the arguments, and were heavily defeated at the end of the day."[32]

Ellen Jordan noted that, following the defeat, Hampson's crusade changed tack:

> From that point on explicit appeal to nonconformist, radical and feminist principles vanished from their speeches. They seem to have concluded

---

[28]"H.L." (1 March 1873). Correspondence. *Pharmaceutical Journal and Transactions* **3**: 699.

[29]Lancy, M. de. (29 March 1873). Correspondence: Pharmaceutical women. *Pharmaceutical Journal and Transactions* **3**: 780.

[30]"M.P.S." (1 March 1873). Correspondence. *Pharmaceutical Journal and Transactions* p. 698.

[31]Note 8, Holloway, p. 263.

[32]Note 8, Holloway, p. 264.

436   *Pioneering British Women Chemists: Their Lives and Contributions*

> that demonstrating inconsistency with these discourses left most members unmoved, and that the expedient tactic was to show that refusing admission to women was inconsistent with principles to which all members subscribed.[33]

In the following years, Minshull and Stammwitz (Rowland having been admitted to medical school instead) continued their studies by attending the Society's lectures while using the laboratories of the private South London School of Chemistry and Pharmacy.[34] Opened in 1870, the school was run by John Muter,[35] a former public analyst. Annie Neve acknowledged the importance of Muter to the admission of women:

> I believe that Dr. Muter, of Kennington, was the first principal of a school of pharmacy to admit women as students, and the first three or four women who qualified as pharmacists studied at his school (see Footnote 18).

Minshull and Stammwitz were joined in their quest for admission to the Society by Isabella Skinner Clarke. Clarke started her career in Garrett's Dispensary for Women and Children. She passed both the Minor and Major Examinations in 1875 (gaining fourth place in the Major). She became a registered Pharmaceutical Chemist, and early in 1876, she opened her own chemists shop in Paddington, following a stint working in Hampson's shop. Upon passing the Major, Clarke applied for Membership of the Pharmaceutical Society in January 1876. She was refused, many of the Council arguing that they were bound by the previous rejections. Between 1873 and 1879, Hampson re-raised the issue of the admission of women and on each occasion, the motion was defeated.

Success at Last

The issue would not go away. The names of Minshull and Clarke were put forward again at the council meeting of 1 October 1879, as *The Chemist*

---

[33] Note 17, Jordan, p. 397.

[34] Earles, M. P. (1965). The pharmacy schools of the nineteenth century. In Poynter, F. N. L. (ed.), *The Evolution of Pharmacy in Britain*, Charles C. Thomas, Springfield, Illinois, p. 88.

[35] Hehner, O. (1912). Obituary: John Muter. *The Analyst* **37**: 76–80.

and Druggist reported: "Mr. Hampson moved that they should be elected. He ... urged that it was the duty of the Council to elect all eligible persons irrespective of sex."[36]

The outcome this time was very different. As one of the council members, Mr. Bottle, a former strong opponent of women's admission, was quoted as saying:

> He should vote for the motion, not with a view to conceding the ladies what Mr. Hampson asserted was their right, but as a matter of courtesy, which he thought they had well earned by passing the examinations, and also with a view to bringing about a peaceful termination to a question which had formed a bone of contention for some years. A prolonged agitation would be infinitely worse than admitting even a dozen women into the society.[37]

When the vote was called, only Sandford, still the President of the Society, was against their admission. The opposition to women members had collapsed; this particular battle had been won. The *Medical Times and Gazette* applauded the decision:

> We congratulate the Society itself on the settlement of a disturbing question, and Miss Clarke and Miss Minshull on their having at last succeeded in obtaining the right to call themselves members of the Society.[38]

Stammwitz applied and was accepted a year later. Women were now admitted to the Pharmaceutical Society School of Pharmacy.

Lady Pharmacists

Admission to the School of Pharmacy did not end the challenges for women students: they had to use a separate entrance and were often

---

[36] Anon. (15 October 1879). The Pharmaceutical Council: Ladies admitted as members. *Chemist and Druggist* **21**: 422.

[37] (4 October 1879). Transactions of the Pharmaceutical Society: Meeting of the Council, Wednesday, October 1, 1879. *Pharmaceutical Journal and Transactions* **9**: 265.

[38] Female members of the Pharmaceutical Society. (11 October 1879). *Medical Times and Gazette* **2**: 428–429.

438 *Pioneering British Women Chemists: Their Lives and Contributions*

ignored by their male colleagues. Nevertheless, the two journals the *Pharmaceutical Journal* and the *Chemist and Druggist* maintained their strong support for the cause of women in pharmacy.

In 1891, the *Chemist and Druggist* had an editorial on Lady Pharmacists[39] reviewing the history of their cause and suggesting that the defeat of the motion for admission in 1879 was the result of negative votes by young male pharmacists who feared the potential competition from women. In that editorial, the journal asked women pharmacists to report on their experiences, and 6 months later, the journal published a fascinating three and one-half page summary of comments by some of the more prominent women pharmacists of the time, including Louisa Stammwitz, Margaret Buchanan, Annie Neve, and Rose Minshull.[40] Mary Elizabeth Neve gave a typical response: "During the time in Eastbourne I certainly found considerable prejudice against women-chemists (especially among ladies); but as the years went by the prejudice gradually decreased."[41]

Despite the hostile atmosphere at the School of Pharmacy, women performed excellently in their studies (as had been feared by Carteighe). In 1908, the prevalence of female names among the award winners, most notably Gertrude Wren and Grace Neve,[42] was described by a columnist in the *Chemist and Druggist*:

> The Male Intellect is evidently not equal to the contest with feminine rivals in the class rooms and in examinations. Miss Wren annexes three out of four silver medals which the Pharmaceutical Society contributes annually ... and at the same time establishes her claim to the Periera medal. Miss Neve supplements this demonstration of the superiority of the sex by scooping in exactly the same proportion of the bronze medals awarded in the Minor course. This appropriation of the Society's bullion by two very young ladies ... leaves but a scanty "distribution" of honours ... among the masculine majority of competitors. Moreover the ominousness of the event is that it is not merely occasional or accidental.

---

[39] Anon. (12 December 1891). Lady pharmacists. *Chemist and Druggist* **39**: 848.

[40] Anon. (30 July 1892). Lady pharmacists. *Chemist and Druggist* **41**: 143–147.

[41] Neve, M. E. (30 July 1892). Lady pharmacists, letter from Miss M. E. Neve. *Chemist and Druggist* **41**: 146.

[42] Shellard, E. J. (August 1982). Some early women research workers in British Pharmacy. *Pharmaceutical Historian* **12**(2): 2–3.

Women in Pharmacy 439

It is just the climax of a consistent progress which has been noted again and again ... Undoubtedly the average of the academic work of the ladies at schools of pharmacy (and not in this country alone) has been much higher than that of the male students.[43]

Nevertheless, the columnist was reassured that male pharmacists were still in control:

But there remains the consoling fact that though women have been demonstrating their capabilities in pharmacy for thirty years or more, it is still men who teach them, men who examine them, and men who hand them out their medals.[43]

Many male pharmacists still refused to take on women apprentices. One of the major boosts for training women in pharmacy was a single chemists shop: that at number 17, The Pavement, Clapham Common.[44] The pharmacy had been opened by Henry Deane in 1837. Deane had joined the Pharmaceutical Society in 1841, being one of its earliest members, and he was President from 1853 until 1855. The pharmacy was purchased sometime between 1911 and 1914 by Buchanan as a training facility for women. She was joined in 1914 by Agnes Borrowman. One of their student intake of 1915 recalled:

Surely no youngster could have had two more energetic or exacting tutors than the two "Miss B's" — Miss Buchanan and Miss Borrowman — at that time partners in the somewhat decaying business of Deane's of Clapham. I well remember the contagious enthusiasm with which she [Borrowman] tackled the job of putting that old business back on its feet, and how proud we were to be allowed to help and perhaps earn an occasional word of appreciation when we had worked with her from early morning until late into the night.[45]

---

[43]"Xrayser" (4 July 1908). Observations and reflections: The male intellect. *Chemist and Druggist* **73**: 15.

[44]Hudson, B. (September 2004). The 'Petticoat Peril' on the pavement: Women's pharmacy history in Lambeth. *Newsletter, National Association of Women Pharmacists* 8.

[45]Anon. (27 August 1955). Deaths: Miss A.T. Borrowman. *Pharmaceutical Journal* **224**: 155.

440 *Pioneering British Women Chemists: Their Lives and Contributions*

By 1923, of the 15 young women trained at The Pavement who had studied at the Pharmaceutical Society's School of Pharmacy, 14 had taken prizes and scholarships. As *The Pharmaceutical Journal* noted: "Is there another pharmacy in the country that can beat this record?"[46] Borrowman took over sole proprietorship in 1924. The pharmacy was badly damaged by the nearby impact and explosion of a V2 rocket in January 1945. At this point, Borrowman converted the business to a limited company with Hilda Wells as Director. Wells had been an apprentice at The Pavement in 1918. The business ceased to be women-owned and women-run in 1958, and it finally closed its doors in 1984.

Employment of Women Pharmacists

In 1904, a column in the *Pharmaceutical Journal* reported upon a large increase in the number of women students. Nevertheless, the author of the column was very pessimistic on the opportunities for women pharmacists: "There is no sign that any considerable proportion of master-pharmacists will ever be willing to employ women behind their counters as assistants."[47] The article concludes that women of "average intelligence and good health" should seek a calling other than pharmacy. This column provoked a flurry of Letters to the Editor of both journals, debating whether women were capable of being good pharmacists.[48]

---

[46] Anon. (15 December 1923). Personal items: Miss Agnes Thomson Borrowman. *Pharmaceutical Journal* **107**: 625.

[47] Anon. (9 July 1904). Pharmacy for women. *Pharmaceutical Journal* **73**: 84.

[48] (a) Bedell, M. I. (16 July 1904). Women as pharmacists. *Pharmaceutical Journal* **73**: 107; (b) "Homo." (23 July 1904). Women as pharmacists. *Pharmaceutical Journal* **73**: 135; (c) Oliver, E. (30 July 1904). Women as pharmacists, *Pharmaceutical Journal* **73**: 167; (d) Spencer, R. K. (23 July 1904). Women as pharmacists. *Pharmaceutical Journal* **73**: 167; (e) Borrowman, A. T. (23 July 1904). Women as pharmacists. *Pharmaceutical Journal* **73**: 168; (f) "Homo." (7 January 1905). Lady assistants. *The Chemist and Druggist* **66**: 26; (g) "Bismuth." (14 January 1905). Lady assistants. *The Chemist and Druggist* **66**: 64; (h) "Darby." (28 January 1905). Lady assistants. *The Chemist and Druggist* **66**, 181; (i) Weston, J. H. (11 February 1905). Lady assistants. *The Chemist and Druggist* **66**: 258; and (j) One who Has a Sister. (25 February 1905). Lady assistants. *The Chemist and Druggist* **66**: 320.

One of the most prominent schools of pharmacy was the Westminster College of Chemistry, Pharmacy, and Dispensing.[49] J. E. Walden, Professor at the college, wrote an article in the *Girl's Realm Annual* of 1908: "Girls as Chemists: How a Girl May Take up the Work of Chemistry, with a View to Keeping a Pharmacy, or Becoming a Doctor's Dispenser."[50] Despite the title, the author emphasized the route of the Society of Apothecaries examination leading to the role of "lady dispenser." As to becoming a "girl chemist," in view of the "severity" of the Pharmaceutical Society Examination, Walden suggested that girls consider the Apothecary Examination as a stepping-stone. He commented that:

> The girls who take up the profession are principally doctors' daughters, or other relatives, those who have some means, and yet want something to do. A great many nurses also take it up, for the dispensing certificate is a valuable asset to the trained nurse. It is, however, generally suitable for the well-educated of the middle classes.[50]

Women pharmacists usually finished up in the worst-paid and most demanding areas of pharmacy. By 1908, there were still only 160 women registered as pharmacists, of whom only about two-thirds were practicing pharmacists.[51] Of those who had abandoned their career, some had married (and often had been forced to resign), others emigrated, and the remainder had changed careers, for example, to medicine. Of those in the profession of pharmacy, over 60% were working in hospitals and institutions; about 20% were in retail pharmacy, most often as assistants; 12% were dispensers to doctors in private practice; a few were employed by wholesale pharmaceutical houses, and still fewer were analysts, lecturers, or researchers.

Those women in institutions or working for medical practitioners were particularly disadvantaged financially, as they were competing with the lowly paid ranks of the male dispensers who held the Assistants'

---

[49] Kurzer, F. (2007). George S V Wills and the Westminster College of Chemistry and Pharmacy: A chapter in pharmaceutical education in Great Britain. *Medical History* **51**: 477–506.

[50] Walden, J. E. (1908). Girls as chemists. *Girl's Realm Annual* **10**: 395–398.

[51] Note 8, Holloway, p. 265.

442   *Pioneering British Women Chemists: Their Lives and Contributions*

Certificate of the Society of Apothecaries.[52] Women who attempted to practice in the retail trade found that, even as late as 1916, they were not always welcome, as Emily Forster (see Chapter 4), Lecturer at the Westminster College of Pharmacy, described:

> Where to settle! ... The woman pharmacist has something else to weigh besides expense: it is the question of her sex, and the fact that at present she is a pioneer in her profession, and must naturally turn to where she thinks an enterprising woman will be respected and her ability made use of rather than to a locality that appears very "Early Victorian" ... The places to avoid are centres, such as cathedral towns, where anything new is looked upon with suspicion, and must stand the test of time before it can be trusted.[53]

Though the hospital work was simply routine dispensing, the women pharmacists did make their mark, as Margaret Buchanan commented:

> In such a position the orderliness and attention to detail, the tact and desire to please, which are supposed to be natural to most women, are a most necessary part of their stock-in-trade, and there have been not a few instances where women's business capacity and an up-to-date knowledge of drugs and economical methods have led to practical appreciation on the part of committees.[54]

Girls considering careers in pharmacy were also encouraged to look overseas for their future employment. In an 1880 issue of the girls' magazine, *Atalanta*, Isabella Clarke considered India provided the most opportunity:

> One great opening for women Pharmacists is in India, where so many of the women doctors are now settling. Women Pharmacists should

---

[52] Jordan, E. (2002). 'Suitable and Remunerative Employment': The feminisation of hospital dispensing in late-nineteenth century England. *Social History of Medicine* **15**: 429–456.

[53] Forster, E. L. B. (12 August 1916). The ideal neighbourhood for the woman pharmacist. *Pharmaceutical Journal* **97**: 158.

[54] Buchanan, M. E. (23 May 1908). The present position of women in pharmacy. *Pharmaceutical Journal* **80**: 675.

Women in Pharmacy 443

accompany them, for in many instances in consequence of the present want of women dispensers, the doctors are obliged to dispense their own medicines. Dispensing not being the province of medical women nor men.[55]

Thirty years later, in 1910, Elizabeth Stennett in *Girls' Realm*, similarly recommended overseas employment. She suggested mission hospitals in "Africa, India, Korea, etc."[56]

Association of Women Pharmacists

Membership of the Pharmaceutical Society had bought few if any benefits. Women pharmacists found themselves isolated and marginalized. Buchanan and Clarke are said to have discussed the problems in Buchanan's kitchen at her home in Gordon Square, and they concluded that a separate organization for women pharmacists was needed.[57] This discussion led to the founding of the Association of Women Pharmacists on 15 June 1905, with membership confined to those who had passed the Minor or Major examination.[58] A letter in the *Chemist and Druggist* reported on the inaugural meeting, noting:

> It is hoped that all qualified lady chemists will join the Association now formed, which, on the principle that "union is strength," cannot fail to emphasise and bring to the front the latent power of women's work in pharmacy.[59]

The first President was Clarke; and Buchanan was the Vice-President. Fifty women joined immediately, and by 1912 Buchanan, then President,[60]

---

[55] Clarke-Keer, I. S. (1881). Employment for girls: Pharmacy. *Atalanta* **1**: 295.

[56] Stennett, E. (1910–1911). Careers for girls XII: Dispensing. *Girls' Realm* **13**: 439.

[57] Jones, D. M. (10 November 1959). Progress of women in pharmacy. *Chemist and Druggist* **177**: 42.

[58] Symonds, S. (11 June 2005). An event great with possibilities. *Pharmaceutical Journal* **274**: 733.

[59] (24 June 1905). Letters in brief: Women pharmacists. *Chemist and Druggist* **66**: 975.

[60] (6 March 1909). Winter Session of Chemists' Associations: Association Presidents. *Chemist and Druggist* **74**: 377.

proclaimed that "practically every woman practicing pharmacy" was a member.[61] Women were encouraged; to use the Association's employment service; to take up the Association's special insurance and annuity scheme; to participate in the training programme involving the interchange of women apprentices between retail and hospital services; and to start their own business, preferably as a joint venture between "two or three women of congenial tastes."[62]

The first public meeting of the Association of Women Pharmacists was held in London on 17 October 1905. Within the ranks of the Association were a number of "radical" members, and on 17 June 1911, they marched across the city with more than 40,000 other women to show their solidarity with the Suffragette Movement. The *Chemist and Druggist* expressed its approval:

> It was a magnificent demonstration of the organising abilities of women and of the universality of their desire to get their Parliamentary vote. We give two photographs of the small section composed of women pharmacists.[63]

In 1918, branches of the Association were established in other British cities, and the same year, the organization changed its name to the National Association of Women Pharmacists.

With all young able-bodied men called up for military service during the First World War, women were required to take over many positions (see Chapter 15), including that of pharmacist. However, the end of the War brought a flood of returning males expecting (and demanding) to re-occupy their former jobs. Nevertheless, women were able to retain some of the inroads they had made, with 7% of the names on the Pharmaceutical Register being women in 1920.

---

[61] Women and Pharmacy. *Information Sheet*, Museum of the Royal Pharmaceutical Society of Great Britain.

[62] Note 8, Holloway, p. 267.

[63] Anon. (24 June 1911). Women and the Vote. *Chemist and Druggist* **78**: 36.

Pioneer Women Pharmacists

Of the four at the forefront of the battles with the Pharmaceutical Society, Clarke, Minshull, and Stammwitz, all pursued successful careers in pharmacy. Neither Alice Marion Rowland (Mrs. Hart)[64] nor Alice Vickery[65] had any intention of continuing in pharmacy (see Footnote 15). They had both planned to study medicine, but had been thwarted by the collapse in 1872 of a plan to set up a Ladies' Medical College in London. Thus, both women left pharmacy and enrolled with the LSMW upon its founding in 1874. Vickery became a successful medical practitioner and was also active in radical political causes, including the birth control movement.

Isabella Clarke (Mrs. Clarke-Keer)

Isabella Skinner Clarke was born on 29 October 1842 in London, daughter of Edward Clarke, solicitor, and Elizabeth Pemberton. In 1876, Isabella Clarke[66] established her own business at Spring Street, Paddington, London. She provided the dispensing course at her pharmacy for women medical students at the LSMW, as a result, Clarke was appointed Tutor in Pharmacy at the LSMW. Clarke married Thomas Henry Keer in 1883 and taking the family name of Clarke-Keer. She had met him years earlier when they had both been students at the South London School of Chemistry and Pharmacy under Muter.

After marriage, Clarke sold the Spring Street pharmacy and became Keer's partner in a pharmacy in Bruton Street, Berkley Square. She took in women pharmacy students as boarders at her home in Endersleigh Street, which was also used for the early meetings of the Association for Women Pharmacists. Following Keer's death in 1898, Clarke ran a shorthand and typewriting school in Westminster; then during the First World War, in her 70s, she worked daily at the Admiralty. She died in Croydon on 30 July 1926, aged 84.

---

[64] Wikipedia (5 November 2018). Alice Hart (accessed 9 January 2019).

[65] Wikipedia (11 October 2018). Alice Vickery (accessed 21 December 2018).

[66] Buchanan, M. and Neve, A. (1926). The late Mrs. Clarke-Keer. *Pharmaceutical Journal* **110**: 374–375.

446 *Pioneering British Women Chemists: Their Lives and Contributions*

Rose Minshull

Rose Coombs Minshull[67] was born at Birmingham in 1846 to John B. Minshull, merchant, and Elizabeth Allcock. After completing her qualifications, Minshull was appointed the pharmaceutical chemist at the North Eastern Hospital for Children, Bethnal Green, London, while her sister, Flora, was the hospital dispenser at the same hospital. In 1892, she described her life in the hospital:

> As the result of many years' hospital work, I am decidedly of the opinion that, certainly in women's and children's hospitals, a lady dispenser is the right woman in the right place. Of course, she must know her work thoroughly, for she will find it to be more sharply criticised than a man's would be in the same position keep well posted up in the current pharmaceutical literature, so as to have at her fingers' ends when applied to, all that relates to new remedies, etc. (I supplemented my Major work with a course of analysis of food and drugs, and it has been of great value to me in many ways.)[68]

Dying on 9 May 1905 in Hastings, Sussex, at the comparatively young age of 58, Minshull was described in an obituary as: "… not by nature a fighter, but a bright and charming little woman of an affectionate nature."[67] Another obituary commented upon her academic excellence:

> … it is interesting to note that when she passed the Preliminary Examination in 1873, when the list of passes was made out in the order of merit, Miss Minshull's name stood at the head of about three hundred candidates. A few years later she entered the Society's School of Pharmacy, and on passing the Minor, in October 1877, she obtained the highest number of marks of any candidate at that examination.[69]

---

[67] Anon. (20 May 1905). Deaths: Minshull. *Chemist and Druggist* **66**: 786.

[68] Minshull, R. (30 July 1892). Lady pharmacists, letter from Miss Rose Minshull. *Chemist and Druggist* **41**: 145.

[69] Anon. (13 May 1905). Deaths — Minshull. *Pharmaceutical Journal* **20**: 712.

Louisa Stammwitz

Louisa Stammwitz[70] was born about 1849, the daughter of Ernst Stammwitz, tailor, and Hannah Stammwitz. As described in the preceding section, Stammwitz, together with Minshull, were in the forefront of the fight for admission to the Pharmaceutical Society. Stammwitz was another of the women who trained at Muter's South London School of Pharmacy. She spent 9 years as a dispenser at the New Hospital for Women (an expansion of Garrett's original dispensary). In 1890, Stammwitz went into partnership with Annie Neve and they established a successful chemist's shop in Paignton, Devon. She recounted her experiences in a letter to the *Chemist and Druggist* in 1892:

> The business conducted by Miss. A. Neve and myself was very success-ful during the time we had it, and we had no difficulty in disposing of it when, through ill-health, it became necessary for me to give up active work. At first there was some prejudice against us as women chemists, but that almost disappeared after a few months, as far as residents were concerned. Occasionally a visitor would come in and say, 'Is anybody at home?' but when we explained who we were we seldom had the morti-fication of seeing a prescription taken away unprepared.[71]

As described in one of Stammwitz's obituaries, she had major health issues: "... but for many years past she had been a great sufferer and helplessly crippled."[70b] Stammwitz moved back to Sanderstead, Croydon, south London, to be joined later by Neve. Stammwitz died in 1916.

Annie Neve

There were three "pharmaceutical" Neves: Annie Neve[72] and her sister, Mary Elizabeth Beck Neve of St. Leonards-on-Sea, Sussex, and, later, Grace Mary Neve of South Croydon. It seems likely that Grace (1884–1975), who

---

[70](a) Anon. (21 October 1916). Deaths — Stammwitz. *Chemist and Druggist* **88**: 38-39; (b) Anon. (14 October 1916). Deaths — Stammwitz. *Pharmaceutical Journal* **43**: 382.

[71] Stammwitz, L. (30 July 1892). Lady pharmacists, letter from Miss Louisa Stammwitz. *Chemist and Druggist* **41**: 143.

[72] Anon. (26 October 1929). Deaths — Neve. *Pharmaceutical Journal* **117**: 409.

448 *Pioneering British Women Chemists: Their Lives and Contributions*

became a hospital pharmacist, was a relative of Annie and Mary, but we were unable to find a link between the respective family trees.

Annie Neve was born about 1861 at St. Leonards-on-Sea, daughter of Albert Neve, solicitor's managing clerk, and Ann Neve, of private means. She was apprenticed to Clarke in 1879 and registered with the Pharmaceutical Society in 1882, her sister being registered in 1886. They were practicing pharmacists at St. Leonards-on-Sea until Annie moved to Paignton to join Stammwitz. By 1895, Neve had returned to St. Leonards-on-Sea, presumably to work with her sister again until she rejoined Stammwitz in Croydon. Following Stammitz's death, Neve held the position of pharmaceutical chemist at the Royal Eye Hospital, Southwark.

As noted earlier, Neve acknowledged the gratitude of women pharmacists to the efforts of Hampson and Muter. She reiterated the point in a letter to the *Pharmaceutical Journal*. Neve had been upset that the Journal's obituary for Stammwitz had emphasized the male opposition to women's entry into pharmacy without crediting those men who advocated for women. Neve closed her letter with:

> Will you do me the kindness to insert the above [praise of Hampson and Muter], as I think the gratitude of all women should be given to the men who helped us in the early days, when considerable courage and broadness of view were needed to oppose current opinions.[73]

Described as "bright and full of vigour" Neve died in 1929 at the age of 68, her obituarist explaining: "[she] was knocked down and fatally injured by a motor vehicle at Croydon while on her way to work from her home at Sanderstead" (see Footnote 72).

Margaret Buchanan

Over the long-term, the most important pioneer of women in pharmacy, Margaret Elizabeth Buchanan,[74] was born on 26 July 1865 at Clerkenwell,

---

[73] Neve, A. (28 October 1916). The Late Miss Stammwitz. *Pharmaceutical Journal* **43**: 427.

[74] Shellard, E. J. (unpublished work). Chapter II, Some Outstanding Women Pharmacists of the late 19th and the early 20th Century, in *Women in Pharmacy*. We thank the Archivist of the Pharmaceutical Society for a copy of this manuscript.

London. She was the daughter of Albert S. Buchanan, medical doctor, and Elizabeth Anne Brake. Buchanan was educated at North London Collegiate School for Girls before beginning her pharmaceutical training with her father and later with Clarke and Keer. She qualified in 1887 and passed the Major a year later, obtaining second place in the Periera competition.

Until taking over proprietorship of the Deane pharmacy, Buchanan had been a hospital pharmacist. Then, following her transfer of ownership of the Deane pharmacy to Borrowman in 1924, Buchanan founded the Margaret Buchanan School of Pharmacy for Women at Gordon Hall, of which she was Principal for many years. Her links with the Deane pharmacy, however, did not come to an end:

> ... and it was arranged that her pupils, all of whom were women, should attend the Clapham establishment [the Deane pharmacy] — three in the morning and three in the afternoon — in order to gain practical experience (see Footnote 44).

Buchanan also taught pharmacy to women students at the LSMW.

Buchanan was the first woman to be elected to the Council of the Pharmaceutical Society. It was no accident but a well-planned endeavour by members at the Progressive Pharmaceutical Club, as Herbert Skinner described in one of her obituaries:

> Things were not moving nicely and strong resistance from the Pharmaceutical Society at Bloomsbury Square was still strongly manifested.... A few kindred spirits met at the Progressive Pharmaceutical Club including John Humphrey, Hugo Wolff, myself and a few others. We determined to try and bring in one of our club members, Miss Buchanan, and run her as a candidate for the 1918 election. She was a progressive in every sense, but somewhat diffident on the political side. We succeeded in persuading her to stand for the Pharmaceutical Council-the first woman candidate. The consternation that arose in certain quarters is one of those things not easily forgotten.[75]

---

[75] Anon. (13 January 1940). Deaths — Buchanan. *Chemist and Druggist* **132**: 27.

450   *Pioneering British Women Chemists: Their Lives and Contributions*

Perhaps fired by the opposition to the nomination, the Progressive Pharmaceutical Club members threw all their efforts into the campaign to elect Buchanan.

> We progressives had to secure Margaret's return [election]... The *C & D*. [*Chemist and Druggist*] helped us magnificently and for successive nights we addressed envelopes, wrote postcards, sent circulars for Margaret E. Buchanan, the woman progressive pharmacist for the Pharmaceutical Council.... The returns on the election day rewarded us for all we had worked to achieve (see Footnote 75).

Buchanan had made history. Following her election to the Council of the Pharmaceutical Society, though her interest was in education, she was placed on the Benevolent Fund Committee.

During 1922, Buchanan travelled extensively in Canada as the accredited representative of the Council, and on her return she submitted a detailed report on the requirements for qualification, registration, and practice in the different Canadian provinces. One result of this document was a reciprocity agreement between the Society and the Province of Ontario.

Buchanan retired from the Council in 1926 due to ill health. She retired to Devon where she died on 1 January 1940. One of her first students, Elsie Hooper, was quoted in another obituary:

> She was a pioneer of women's work in pharmacy and has trained many of the best women pharmacists. No one who came in contact with her bright intelligent personality could fail to be affected by it. She was a wonderful and inspiring teacher, and laid the foundation of sound pharmaceutical knowledge.[76]

Agnes Borrowman

Born near Melrose, Scotland on 17 October 1881, Agnes Thomson Borrowman[77] was the daughter of Peter Borrowman, shepherd, and

---

[76] Anon. (6 January 1940). Deaths — Buchanan. *Pharmaceutical Journal* **144**: 10.

[77] Anon. (27 August 1955). Obituary: Miss A.T. Borrowman. *Pharmaceutical Journal* **175**: 155.

Women in Pharmacy 451

Margaret Davidson. She spent a 4-year pharmacy apprenticeship before joining the Edinburgh pharmacist, William Lyon, as a junior assistant. During her limited spare time, Borrowman studied for the Minor at the Edinburgh Central School of Pharmacy. She passed the examination in 1903, at the age of 21.

Like elsewhere, woman pharmacists in Scotland often faced discrimination. As was commented by Borrowman's obituarist:

> At that time the Edinburgh public had not outgrown its prejudice against women in medicine and pharmacy, and customers often walked out of the shop rather than let a girl serve them, so that it was difficult for a woman pharmacist to get employment. ... In fact, Lyon would not permit Borrowman to be seen at the front counter. Instead, she had to work in the back-shop where "a large range of galenicals [lead compounds]" were made (see Footnote 77).

For this reason, Borrowman moved south to Runcorn, Cheshire. In addition to pharmacy duties, she undertook her first research. Borrowman presented the results of her research (on an arsenic, iron, and quinine mixture) to an Edinburgh Evening Meeting in 1904. It was while at Runcorn, she penned a spirited letter to *The Pharmaceutical Journal*. In the letter, Borrowman expressed frustration with the timidity of women pharmacists in pressing their claims to equality of remuneration and parity in treatment with male members of the profession.[78]

In 1906, Borrowman moved to Dorking to work for 3 years with J. B. Wilson, an accomplished pharmacist and a cultivator of medicinal plants. She passed the Major examination in 1909, after which she became a research assistant at the Pharmacy School with Henry Greenish, studying components of quinine. Having been recommended by Arthur Crossley,[79] Borrowman subsequently left the school's laboratories and took up a better-paying appointment as Research Chemist in the

---

[78] Borrowman, A. T. (30 July 1904). Letters to the Editor: Women as pharmacists. *Pharmaceutical Journal* **73**: 168.

[79] "W.P.W." [Wynne, W. P.] (1928). Obituary Notices of Fellows Deceased: Arthur William Crossley, 1869–1927. *Proceedings of the Royal Society of London, Series A* **117**: vi–x.

452　*Pioneering British Women Chemists: Their Lives and Contributions*

London laboratory of the Rubber Growers' Association of Malaya and Ceylon.

Borrowman's research activities covered an incredible diversity of topics, as her obituarist in the *Pharmaceutical Journal* noted:

> … the examination of the physical and chemical properties of vulcanized rubber, and of the various processes of vulcanization, with a view to their improvement; soil analysis with the object of increasing the yield of latex, examination of distinctive fungi of the rubber plant, cellulose, paper-making, new processes for production of artificial silk, and examination of possible plant paper-making materials. In this connection Miss Borrowman acquired such a facility in the microscopic examination of fibres that she could tell at a glance the proportion or percentage of different fibres in a given paper. She also experimented extensively with papers, for the detection of forgery (see Footnote 77).

The obituary went on to state that Borrowman also spent four nights per week attending classes at Borough Polytechnic, Chelsea Polytechnic, and the Cass Institute. In her spare time, she read and searched specifications at the Patent Office library.

However, in 1913, following the death of her father, Borrowman had to return to the more remunerative pharmacy trade to support the younger members of her family. She took a year of retail experience at a pharmacy in Slough before joining Buchanan at the Pavement in 1914, becoming sole proprietor in 1924.

Borrowman was active with the Association of Women Pharmacists from its founding. In 1924, she was the first woman to be invited onto the Board of Examiners of the Pharmaceutical Society, an appointment she held until 1937. Borrowman was also elected President of the South-West London Chemists' Association. Among other activities, she helped compile *Pharmacopedia* with Edmund White and John Humphrey (see Footnote 74).

The near-destruction of the Deane pharmacy in 1945 and the years of wartime duties, such as fire-watching, that she added to her load took their toll on Borrowman's health. However, she kept active with the pharmacy until her death on 20 August 1955, aged 74. A correspondent for the *Pharmaceutical Journal*, "An Onlooker," commented: "Miss Borrowman

Women in Pharmacy 453

was a firm believer in the equality of the sexes but this view applied only to other women — she knew that she herself was more than equal to most men."[80]

Elsie Hooper (Mrs. Higgon)

Another pioneer stalwart of the Association of Women Pharmacists was Elsie Seville Hooper.[81] Born on 5 September 1879 at Hackney London she was the daughter of Cleeve Oliver Hooper, leather and factor agent, and Elizabeth Mace. Hooper passed the Major examination in 1902 and was one of the most prominent early women researchers in pharmacy (see Footnote 74). In 1901, Hooper was the first woman to be awarded the Redwood Research Scholarship, while in 1903, she was the first woman recipient of the Burroughs Research Scholarship.

Under Greenish and T. E. Wallis, Hooper performed research into natural products, presenting some of her work at the 1905 British Pharmaceutical Conference. During the same time period, she was a demonstrator at the Pharmaceutical Society's School of Pharmacy, while in the evenings she studied towards a B.Sc. in Botany and Chemistry at Birkbeck College.

Following completion of her degree in 1905, Hooper worked on the publication of the first British *Pharmaceutical Codex*, followed by a year with Alfred Kirby Huntington at King's College (see Chapter 4). Then she joined the staff of E. F. Harrison and worked on the analysis of "secret remedies" for the British Medical Association. Subsequently, for 4.5 years, she held a Lectureship in Chemistry at Portsmouth Municipal College. She was also one of the signatories of the 1909 letter to *Chemical News* (see Chapter 9).

During the First World War, Hooper was an analyst with Ucal at Cheltenham. After the War, she opened a retail pharmacy in Cheltenham, called Ladies Chemists Ltd. In 1920, Hooper left the pharmacy to return to London and assist in the teaching at Margaret Buchanan's School of

---

[80] Anon. (3 September 1955). An Onlooker's Notebook: Miss Borrowman. *Pharmaceutical Journal* **175**: 191.

[81] Anon. (21 June 1969). Deaths — Hooper. *Pharmaceutical Journal* **202**: 714.

454   *Pioneering British Women Chemists: Their Lives and Contributions*

Pharmacy for Women. Kathleen King joined the staff in 1922, and Hooper and King took over ownership of the school from Buchanan in 1925, renaming it the Gordon Hall School of Pharmacy for Women. In view of the continued prejudice against women pharmacists, Hooper purchased two retail pharmacies in northwest London specifically to provide work experience for the women students. She remained owner and main teacher at the school until it closed in 1942. Hooper sold one of the pharmacies in 1945 and the other in 1961 (see Footnote 81).

Hooper was the first Secretary of the Association of Women Pharmacists, serving from 1905 to 1908, and she was elected President for the 1927–1928 year. In 1911, she had been one of the pharmacists demonstrating with the Coronation Suffragette Procession, marching behind the banner: "Women Pharmacists Demand the Vote." There is nothing noted on any later activities, but Hooper did marry a Mr. Higgon. Perhaps the marriage occurred somewhere about the time of the closing of the school. Hooper died in Paignton, Devon on 6 May 1969, aged 89.

Dorothy Bartlett (Mrs. Storey)

Greenish and Crossley seem to have played mentoring roles for the early women research pharmacists. Dorothy Bartlett[82] was another woman pharmacist whose career owed a lot to these two mentors (see Footnote 74).

Bartlett, born on 31 July 1888 at Brixton, London, was the daughter of Willy Hugh Bartlett, clerk to a shipping company, and Emily Sophia Osbourne. Educated at Streatham Hill High School, she passed the Major examination in 1911 and obtained a B. Pharm. from King's College in the same year. Bartlett was another of the brilliant early women students who won a large number of the prizes of the Pharmaceutical Society.

From 1911 until 1912, Bartlett undertook research with Crossley. In 1912, she was awarded the Burroughs Research Scholarship, followed by the Redwood Research Scholarship. The scholarships enabled her to work with Greenish in the Pharmacognosy Research Laboratories. Three publications resulted from her work. Bartlett then became a Research

---

[82] Anon. (1941). Obituary. *Journal of the Royal Institute of Chemistry*, Part 1, 59.

Chemist with Burgoyne Burbridges & Co Ltd. In 1915, she married William Armstrong Storey, a fellow chemist, and apparently terminated her professional activities. Bartlett died 20 January 1941 at Bucklow, Cheshire, aged 53.

Norah Renouf

Norah Renouf (see Footnote 74) was born 31 August 1881 at St. Helier, Jersey in the Channel Islands, daughter of John Renouf, hardware merchant, and Delahay Woods. Norah's twin sister, Lucy, also went into pharmacy. Norah Renouf gained some experience working in a chemist's shop in Jersey and decided to follow a career in pharmacy. She moved to London and studied at the Pharmaceutical Society School of Pharmacy, passing the Minor in 1902 and the Major in 1903. Renouf followed Hooper as recipient of the Redwood Research Scholarship in 1904.

In 1905, Renouf was the first woman and the first pharmacist to be awarded the Salters' Research Scholarship. For the next 2 years, she performed research with Crossley on camphor derivatives, work that resulted in 12 publications. In fact, Renouf was Crossley's leading researcher as his obituary noted: "... in many of those [communications] from the School of Pharmacy, Miss Nora Renouf was his collaborator".[83] She did not take up the third year of the Scholarship.

Renouf was another founder member of the Association of Women Pharmacists, and she was the Treasurer from 1907 until 1916. She, too, was a signatory of the 1909 letter to *Chemical News* (see Chapter 9). Hooper and Renouf overlapped in their time at the School of Pharmacy, Hooper passing the Major the same year that Renouf passed the Minor. Thus, it seems likely that these two sole pharmacy signatories of the letter knew each other.

To contribute to the war effort, in 1916, Renouf went back to the Channel Islands to work in a hospital. Following the end of the War, she returned to London and in 1920 was one of the first cohort of women to be admitted to the Chemical Society (see Chapter 9). In her application

---

[83]Wynne, W. P. (1927). Obituary Notices: Arthur William Crossley. *Journal of the Chemical Society* 3165–3173.

to the Chemical Society, Renouf noted that she was a survey officer at the Fuel Research Board of the Department of Scientific and Industrial Research.[84] Renouf died 12 February 1959 in Jersey.

Hope Winch

The most tragic ending was that of Hope Constance Monica Winch.[85] She was born on 25 August 1894 at Northallerton, Yorkshire, the daughter of George Thomas Winch, Church of England clergyman, and Elizabeth Maud Crofton. Winch had been educated at the Clergy Daughters School, Casterton, Kirby Lonsdale, Westmoreland. After leaving school, she had a year's training in practical pharmacy at the Royal Victoria Infirmary, Newcastle-upon-Tyne, before spending nearly 3 years in dispensing at the hospital. Then Winch moved to London and attended the School of Pharmacy where, among her many honours, she was the fifth woman student to be the Pereira Medallist. She passed her Major in 1917 and obtained her B.Pharm. from King's College in 1918.

Upon graduation, Winch held the appointment of Demonstrator in the School until 1920, while undertaking research in the School's research laboratories as a Redwood Scholar. In 1920, she was appointed Lecturer in Pharmaceutical Subjects and Botany at the Technical College, Sunderland, where from 1930, she was Head of the Pharmacy Department. Winch took an active part in the local branch of the Pharmaceutical Society being its Secretary for 20 years and finally Chairman. Tragically, on 8 April 1944, Winch fell about 150 feet while climbing the peak of Scafell and was killed. At the inquest, it was suggested that: "the fall might have been due to a temporary 'black-out', or to the rock which she was holding disintegrating through frost and crumbling in her hand."[86] Winch was only 49 years old.

---

[84] Certificates of candidates for election at the ballot to be held at the ordinary scientific meeting on Thursday, December 2nd, 1920, *Proceedings of the Chemical Society* **117**: 95.

[85] Anon. (11 May 1918). Personal Items: Miss Hope Constance Monica Winch. *Pharmaceutical Journal* **100**: 227.

[86] Anon. (15 April 1944). Deaths — Winch. *Pharmaceutical Journal* **152**: 161.

Agnes Lothian (Mrs. Short)

We conclude this chapter with an individual whose scholarship bridged pharmacy and scientific librarianship (see Chapter 16). Agnes Edith Lothian,[87] born in 6 July 1903 in Scotland, was the daughter of John Lothian, teacher of pharmacy at both Edinburgh and Glasgow, and Edith Ellen Heard. She took her pharmacy qualification at Herriot-Watt College, Edinburgh, graduating with the Major in 1926. Lothian obtained appointments in retail pharmacies, first in Redhill, Surrey, then with Allen & Hanburys in London. She married Edward Arthur Orme in 1931.

In September 1940, Lothian was appointed as Librarian to the Pharmaceutical Society. Following her appointment, she formally qualified as Librarian, being elected Associate of the Library Association in 1944. Her librarian activities were recalled by Desmond Lewis, a later Secretary and Registrar of the Pharmaceutical Society:

> To a student she was always kind and helpful: to others she could be brusque to the point of abrasion. It took courage to invite an opinion on a personal matter, because she could be devastatingly frank. But the formidable front that she adopted concealed an inner kindness that could burst into ripples of laughter.[88]

Lothian developed a special interest in the history of pharmacy and became a world authority on historical drug jars and mortars. She was the first woman to be elected member of the Academie International d'Histoire de la Pharmacie. Lothian's second marriage was in 1955, to George Reginald Allen Short, Pharmacist, a Fellow of the Pharmaceutical Society. Two years later, the Council of the Pharmaceutical Society bestowed upon her the title of Emeritus Keeper of the Historical Collection. As Lewis added:

> ... she used her combined knowledge and a limited budget to build the Society's library into the finest of its kind in the world. ... Inevitably, she

---

[87] Anon. (1983). Obituary: Lothian Short. *Pharmaceutical Historian* **13**(4): 4.
[88] Anon. (22 October 1983). Deaths — Agnes Lothian Short. *Pharmaceutical Journal* **231**: 475.

will be best remembered for her researches into the history of pharmacy and its artefacts. She was recognised as a world authority on drug jars and mortars, and the collections that she built for the Society and the major acquisitions which she made, often against strong opposition of those who resented the expenditure, will remain as her memorial and our heritage (see Footnote 88).

Lothian retired on 31 December 1967 from her positions as Librarian and Keeper of Historical Records. She died on 13 October 1983, aged 80.

Commentary

In several respects, the fight for the admission of women to the Pharmaceutical Society resembled the later battle for admission to the Chemical Society. The two Societies were not only professional bodies but also served as a "men's club" necessitating all possible measures to exclude the "alien species" known as females. As the Chemical Society had its excluder in Armstrong, so the Pharmaceutical Society had Sandford, with his quote that: "... the Pharmaceutical Society was intended to be a Society of men ..." (see Footnote 27). Likewise, the Chemical Society supporters, particularly Ramsey and Tilden, had their equivalent in the Pharmaceutical Society's Hampson. And, of course, there were a series of young women willing to take on the establishment of the Pharmaceutical Society, just as the 1904 petitioners had taken on the Chemical Society.

There was one notable difference between the two sagas. The Women Chemists' Dining Club, organized by Martha Whiteley and Ida Smedley, never flourished and ceased to exist soon after the death of Whiteley. By contrast, the National Association of Women Pharmacists (NAWP) founded by Buchanan and Clarke lives on until the present day. The key difference was in the purpose of the respective women's organizations. The Women Chemists' Dining Club was solely a social and networking body, while the goals of the Association of Women Pharmacists encompassed the professional rights of women pharmacists. This broader mandate provided the NAWP with a continuing reason for its existence even when the generation of the founders had long passed on.

Chapter 14

Roles of Chemist's Wives

Grace Toynbee (4 December 1858–5 October 1946)

Male chemists usually relied upon women for support, the best example being Frederick Donnan, Professor of Chemistry at University College, London (UCL) from 1913 until 1937.[1] It was Donnan's two sisters, Leonora (Nora) and Jane, who were indispensable to him. Nora kept house for Donnan for 38 years while Jane was his secretary at the UCL.

However, it was more often a wife who provided the support. Some of these wives had no background in science and acted as secretaries and organizers; others had a scientific background, but gave up active

---

[1] Freeth, F. A. (1957). Frederick George Donnan, 1870–1956. *Biographical Memoirs of Fellows of the Royal Society* **3**: 23–39.

460 *Pioneering British Women Chemists: Their Lives and Contributions*

scientific participation upon marriage; while yet others continued activity in chemistry either with their spouse or on their own. Here, we will provide examples of women who filled each of these roles.

Amateur Assistant

In this section, we will consider some examples of wives who lacked any chemistry background. The three Holland sisters became simply helpmates, while Winifred Beilby and Grace Toynbee trained themselves in science and played active roles in the laboratory.

Mina Holland (Mrs. Perkin), Lily Holland (Mrs. Kipping), and Kathleen Holland (Mrs. Lapworth)

The three Holland sisters, Mina, Lily, and Kathleen, daughters of William Thomas Holland and Florence DuVal, exemplified the role of "secretarial wives." In a fictionalized biography, authors Eugene Rochow and Eduard Krahé propose that each sister contributed to the success of her husband.[2] As little documentary evidence survives of the three sisters lives, William Brock has commented on the invented dialogue:

> The imagined conversations are stilted and unnatural, reminding the historian of the improving didactic conversations of early nineteenth-century chemistry popularizers such as Samuel Parkes and Mrs. Alexander Marcet.... despite this negative reaction, the story of the Holland sisters is worth telling. They were surely unique in the annals of science in marrying three brilliant chemists.[3]

The Holland sisters, who lived in Bridgewater, Somerset, were cousins of the silicon chemist, Frederic Stanley Kipping,[4] their mothers being sisters.

---

[2] A fictionalized account of their lives was published as: Rochow, E. G. and Krahé, E. (2001). *The Holland Sisters: Their Influence on the Success of Their Husbands Perkin, Kipping and Lapworth*. Springer, Berlin.

[3] Brock, W. H. (2011). *The Case of the Poisonous Socks*. RSC Publishing, London, p. 215.

[4] Thomas, N. R. (2010). Frederick Stanley Keeping — Pioneer in silicon chemistry: His life & legacy. *Silicon* **2**: 187–193.

Kipping had been friends with Lily from the time that they were both young children as the Holland family often visited the Kipping family in Manchester. In their teenage years, they had often played mixed-doubles tennis as partners. In 1886, Kipping went to the University of Munich[5] to work under William Henry Perkin, Jr.

Perkin was introduced to Mina during a visit to Kipping's home where she was staying with Lily, already Kipping's fiancée.[6] It was the expectation at the time that the eldest daughter would be the first to marry, thus with urging from Kipping, Perkin and Mina were married in 1887, enabling Kipping and Lily to marry in 1888. In the biography of Perkin, Mina's support was acknowledged:

> ... those of us whose privilege it was to know the Perkins more intimately are aware how greatly Perkin was helped by her constant sympathy with his ideals and with his work, and with what unobtrusive care she shielded him as far as possible from the minor worries and irritations of life.[7]

Kipping obtained a position of Chief Demonstrator at the Central Technical College, London, undertaking research with Henry Armstrong, Kipping's assistant being Arthur Lapworth.[8] It was through Kipping and Lily that Lapworth met Kathleen. Lapworth and Kathleen were married in 1900. In 1909, Lapworth was appointed Senior Lecturer in Inorganic and Physical chemistry at Victoria University, Manchester.

The claim by Rochow and Krahé that Kathleen played a significant role in her husband's chemistry life is certainly true. According to Arthur Lapworth's obituarist, Kathleen assisted Lapworth with paperwork, acting as his secretary: "As already mentioned the valuable help of

---

[5]Challenger, F. (1950). Frederick Stanley Kipping, 1863–1949. *Obituary Notices of Fellows of the Royal Society* **7**: 182–219.

[6]"J.F.T." (1931). Obituary Notices: William Henry Perkin — 1860–1929. *Proceedings of the Royal Society of London, Series A* **130**: i–xii.

[7]Greenway, A. G., Thorpe, J. F., and Robinson, R. (1932). *The Life and Work of Professor William Henry Perkin*. The Chemical Society, London, p. 36.

[8]Robinson, R. (1947). Arthur Lapworth, 1872–1941. *Obituary Notices of Fellows of the Royal Society* **5**: 554–572.

462 *Pioneering British Women Chemists: Their Lives and Contributions*

Mrs. Lapworth lightened the load of routine for several years."[9] However, there is no documentary evidence that Mina and Lily played any similar role in the lives of their husbands.

Lapworth was joined in 1922 by Robert Robinson to form what was known as the Lapworth–Robinson "golden age" of organic chemistry at Manchester. In a review of the Chemistry Department, George Burhardt noted that the spouses of Lapworth and Robinson played a prominent role:

> An unusual feature of the life of the School of Chemistry at this time was the presence in it of the wives of both professors. Mrs. Lapworth as her husband's secretary helped him greatly with the detail of the heavy administrative responsibilities in the department. Mrs. (later Lady) Robinson [see the respective section], as an Honorary Research Fellow, worked on long-chain acids in the professor's laboratory. Both took a kindly and active interest in staff and students.[10]

Winifred Beilby (Mrs. Soddy)

An example of an untrained spouse performing laboratory work is provided by Winifred Moller Beilby,[11] spouse of Frederick Soddy (see Chapter 7). Born on 1 March 1885 in Edinburgh, Scotland, she was the daughter of George Thomas Beilby, an industrial chemist. When she was 14 years old, the family moved to Glasgow where she was educated at home by a governess.

It was through her father that she met Soddy, for George Beilby had helped finance Soddy's research work. This assistance included a campaign in 1904 to raise money for equipment needed for radium research. Soddy and Beilby became engaged in 1906 and married in 1908.[12] Beilby

---

[9] Note 8, Robinson, p. 566.

[10] Burkhardt, G. N. (1954). Schools of Chemistry in Great Britain and Ireland–XIII The University of Manchester (Faculty of Science). *Journal of the Royal Institute of Chemistry* **78**: 448–460.

[11] Letter, Hilda M. Beilby (spouse of Hubert Beilby, Winifred's brother) to Muriel Howorth, 30 October 1957. Soddy collection, Oxford University Archives (OUA).

[12] Howorth, M. (1958). Chapter 15: The Perfect Marriage. *Pioneer Research on the Atom: The Life Story of Frederick Soddy*. New World Publications, London.

wanted a quiet, early morning wedding so that they could depart quickly on their mountain-climbing honeymoon.

Beilby became interested in Soddy's work and helped him considerably, her work on the gamma rays emitted by radioactive atoms being published in 1910. Soddy himself commented: "She did quite a bit of work for me at Glasgow. You can read it all in my communications. It was a tedious investigation and she stuck at it, like Marie Curie, I used to say."[13]

Upon Soddy's move to Oxford, Beilby seems to have given up research activity to devote her time to the required social life and spending her free moments painting and gardening. Towards the end of 1935, Beilby fell ill and she died suddenly in August 1936. Quite possibly, her death was as a result of exposure to lengthy doses of gamma rays during her research work. Her passing came as a grievous shock to Frederick Soddy, and it was the cause of his resignation from his position at Oxford.

Grace Toynbee (Mrs. Frankland)

The most acknowledged of the amateur assistant spouses was Grace Coleridge Toynbee.[14] Daughter of Joseph Toynbee, pioneer ear specialist, and Harriet Holmes, Toynbee was born on 4 December 1858 in Wimbledon. Initially home schooled, she also studied in Germany, and then spent 1 year at Bedford College although little else is known of her early life. In 1882, she married Percy Frankland,[15] who was then a Lecturer in Chemistry at the Normal School of Science, South Kensington (later part of Imperial College, see Chapter 4); subsequently, they had a son.

---

[13] Note 12, Howorth, p. 168.

[14] (a) Cohen, S. L. (2004). Frankland [née Toynbee], Grace Coleridge. *Oxford Dictionary of National Biography.* Oxford University Press (accessed 27 December 2018); and (b) Creese, M. R. S. (1998). *Ladies in the Laboratory: American and British Women in Science 1800–1900.* Scarecrow Press, Lanham, Maryland, pp. 150–151.

[15] Garner, W. E. (1945–1948). Percy Faraday Frankland. *Obituary Notices of Fellows of the Royal Society* **5**: 697.

464 *Pioneering British Women Chemists: Their Lives and Contributions*

Percy Frankland, together with his father, Edward Frankland, had set up a private analytical laboratory in London, and it was here that Toynbee commenced her scientific career. Though both father and son were chemists, they had a strong interest in bacteriological problems, particularly those relating to human health. Toynbee's first publication, co-authored with Percy Frankland, was on microorganisms in air.

Percy Frankland was appointed in 1888 as Professor of Chemistry at University College, Dundee, and the institution's magazine, *The College*, reported:

> Any notice of Dr. Frankland would be incomplete without some reference to Mrs. Frankland, who has worthily aided and seconded him in his scientific career, and whose achievements are about as well known in the world of science as are those of her husband.[16]

At Dundee, Toynbee continued her research which included studies of the reactions involved in bacteriological fermentation as a means of synthesizing chemical compounds. In 1894, they moved to Birmingham, where Frankland had accepted a position as Professor of Chemistry. It was during her time at Birmingham that Toynbee added her signature to the 1904 petition for the admission of women to the Chemical Society (see Chapter 9).

In 1894, the Franklands co-authored a book, *Micro Organisms in Water: Their Significance, Identification and Removal*,[17] and then Toynbee, on her own, wrote a more popular book *Bacteria in Daily Life*[18] published in 1903. It seemed that after the move to Birmingham, Toynbee focussed more on science journalism than laboratory research. She was elected a Fellow of the Royal Microscopical Society in 1900 and was one of the first 12 women scientists admitted to the Linnean Society in 1904.

---

[16]Anon. (16 March 1889). Professor Percy F. Frankland. *The College, Magazine of University College, Dundee* **1**(3): 87.

[17]Frankland, P. and Frankland, P., Mrs. (1894). *Micro Organisms in Water: their Significance, Identification and Removal*. Longmans, Green & Co., London.

[18]Frankland, P., Mrs. (1903). *Bacteria in Daily Life*. Longmans, Green & Co., London.

With Frankland's retirement in 1919, they moved to Scotland, where Toynbee died on 5 October 1946, Frankland dying 3 weeks later. Frankland's obituarist, William Garner, noted:

> Probably in few cases have husband and wife collaborated so effectively and enthusiastically in both research and professional work. On one occasion it was said, "Many women in the past have helped their husbands, but Percy Frankland is the first man who had the chivalry to admit it."[19]

Woman Chemist as Professor's Wife

In the socially restricted world of academic research, university-educated women tended to marry university-educated men — particularly their supervisors or colleagues. Some of them gave up any academic aspirations in favour of the expected domesticity. Louisa Cleaverley (see Chapter 9) and Elison Macadam (see Chapter 4) were two of the women chemists who terminated their careers upon marriage to a chemistry professor. Such a fate was mourned by Flora Garry, a graduate of the King's College, University of Aberdeen, in her poem, *The Professor's Wife*, in which the final stanza of the poem reads:

> *'Learnin's the thing', they wid say,*
> *'To gie ye a hyste up in life.'*
> *I wis eence a student at King's.*
> *Noo I'm jist a professor's wife.*[20]

A small, but slowly increasing number of these women continued with their research after marriage, joining the ranks of the "academic couples."[21] However, the women usually found themselves given a secondary status and were overshadowed by their male partners.

---

[19] Note 15, Garner, p. 700.

[20] Garry, F. (1995). *Collected Poems*. Gordon Wright Publishing, Edinburgh, pp. 16–17.

[21] Dyhouse, C. (1998). *No Distinction of Sex? Women in British Universities, 1870–1939*. UCL Press, London, pp. 162–163.

466　*Pioneering British Women Chemists: Their Lives and Contributions*

The writer Sharon McGrayne has commented on the prevalence of the spousal problem:

> The academic countryside was littered with scientific couples studying botany, genetics, chemistry, and other sciences. Professor husbands and their low-ranking, low-paid [or unpaid] wives often worked together for decades ... the women were generally low-level instructors, lecturers, or research assistants while their male partners were professors with tenure.[22]

May Badger (Mrs. Craven) and Louise Badger (Mrs. Sinnatt)

The Badger sisters, daughters of Frank Badger, salesman, and Annie Midgley provide an interesting contrast of marital outcomes. May Badger (see Chapter 6) continued with her chemistry career after marriage to William Craven, working with Frank Sinnatt at the Faculty of Technology, Victoria University of Manchester.

May Badger,[23] born on 18 February 1887, attended Ardwick Higher Grade School. Admitted to the Faculty of Technology of Victoria University in 1904,[24] Badger received a B.Sc. Tech. (Hons.) in Applied Chemistry together with a Diploma of the School of Technology in 1907. On a postgraduate scholarship, she worked with William Pope[25] towards an M.Sc. Tech., which she completed in the Spring of 1908. During the 1907–1908 year, Badger was also a Demonstrator in Chemistry at the School of Domestic Economy. For the remainder of 1908, she had a temporary position as an assistant in the laboratory, and she also taught chemistry from September to December 1908 at the Manchester High School for Girls (MHSG).[26]

---

[22]McGrayne, S. (1993). *Nobel Prize Women in Science: Their Lives, Struggles, and Momentous Discoveries*. Birch Lane Press, New York.

[23]Anon. (1954). Obituary Notes: May Badger Craven. *Journal of the Royal Institute of Chemistry* **78**: 105.

[24]Wood, J. K. (1954). Schools of Chemistry in Great Britain and Ireland–XXXII The Manchester College of Science and Technology. *Journal of the Royal Institute of Chemistry* **78**: 755–762.

[25]Gibson, C. S. (1941). Sir William Jackson Pope, 1870–1939. *Obituary Notices of Fellows of the Royal Society* **3**: 291–324.

[26]*Staff Records*, May Badger, Archives, Manchester High School for Girls.

In 1909, May Badger accepted a position as a Chemist with the Pilkington Tile and Pottery Co., carrying out research on glass and pottery. Then the following year, she became Chemist to the Clifton and Kersley Coal Co. That same year, she married Percival William Craven. In 1916, Badger returned to the University of Manchester, Faculty of Technology as a Senior Demonstrator in Chemistry. She continued her interest in coal chemistry, co-authoring (as M. B. Craven) with Frank Sinnatt[27] a bulletin on the heat content of coal in 1921 and with Sinnatt and others a book *Coal and Allied Subjects* in 1922.[28]

May Badger remained at the Victoria University of Manchester until her retirement in 1952, having risen to be Head of the Inorganic Chemistry Laboratories. She died on 24 November 1953, her obituarist noting:

> Mrs Craven had a flair for teaching the methods of inorganic analysis and her enthusiastic approach was always stimulating to her students. She will be affectionately remembered by several generations of Manchester chemists who owe to her their early training in this branch and many of whom will recollect with pleasure the friendly hospitality of the Craven household (see Footnote 23).

Unfortunately, we know much less about her sister, Louise Midgely Badger. Louise Badger had followed in her sister's path, also completed a B.Sc. Tech. in Applied Chemistry from the Faculty of Technology, but in 1912. She became a researcher in the Chemistry Department at the Victoria University, though it is not known if Louise, too, worked with Sinnatt. Louise married Sinnatt in 1921. Unlike May, Louise apparently discontinued research upon marriage, the sole mention of her in Sinnatt's obituary being: "Sinnatt left a widow, Louise Midgley Badger, who, herself a University trained chemist, had been his devoted companion for over 20 years, ..."[29]

---

[27] Egerton, A. C. (1943). Frank Sturdy Sinnatt, 1880–1943. *Obituary Notices of Fellows of the Royal Society* **4**: 429–445.

[28] Sinnatt, F. S. *et al.* (1922). *Coal and Allied Subjects*. H. F. & G. Witherby, London.

[29] Note 27, Egerton, p. 441.

Marjory Wilson-Smith (Mrs. Farmer)

Once married, some of the women chemists are simply noted as having assisted their husbands, but without having any formal position. The career of Marjory Jennet Wilson-Smith[30] provides an example. Wilson-Smith was born on 19 May 1899 at Bath, daughter of Thomas William Wilson-Smith, physician, and Alice Maud Wilks. She was educated at Bath High School and Cheltenham Ladies' College (CLC) before entering the Royal Holloway College (RHC) in 1918, graduating with a B.Sc. in Chemistry in 1921.

From 1921 until 1925, Wilson-Smith was a Demonstrator in Chemistry at the London School of Medicine for Women, though she spent the 1922–1923 year as a research student in the organic chemistry department of Imperial College (IC). In 1925, she accepted an appointment with RHC, but resigned in 1930, as the *College Letter* reported:

> The last loss to be recorded is of Miss M. J. Wilson-Smith (now Mrs Farmer) for two years Assistant Lecturer and Demonstrator in the Chemistry Department. Shortly after the end of term we heard of her engagement to a fellow-scientist at Imperial College. Her marriage followed within a few weeks. Those who remember her record for industry will note, perhaps only with faint surprise, that, according to Rumour, Mrs Farmer has for the last two months done "not a stroke" of scientific work.[31]

According to the obituarist of Ernest Farmer, Wilson-Smith worked with Farmer after marriage: "During this period, in 1930, he married Marjorie Wilson-Smith, she was one of his research students and continued for many years to assist him with his researches."[32] However, none of Farmer's later publications list Wilson-Smith as a co-author. We will never know how many more women chemists continued to work upon marriage, but without acknowledgement or recognition. Helena Pycior, Nancy

---

[30] *Student Records*, Archives, Royal Holloway College.

[31] Anon. (November 1930). *College Letter*, *Royal Holloway College Association* 16.

[32] Gee, G. (1952). Ernest Harold Farmer. 1890–1952. *Obituary Notices of Fellows of the Royal Society* **8**: 159–169.

Slack, and Pnina Abir-Am have named such women the "invisible" assistants.[33]

Collaborative Couples

A small number of married women chemists forged a path independent of their non-chemist spouse. In Chapter 4, we described the life and work of Margaret Seward, who continued university teaching even after marriage and the birth of her son. In crystallography (see Chapter 12), Dorothy Crowfoot and Kathleen Yardley were both very fortunate in their choice of academic husbands who took over much of the parenting role to allow their famous wives to continue research. In biochemistry, Helen Archbold (see Chapter 4) was also a true "independent."

The research relationship between the married biochemists Dorothy Moyle (see Chapter 12) and Joseph Needham could better be described as "autonomous" rather than independent as their research overlapped to a significant extent. We will see in the respective section that the descriptor "autonomous" also fits the research profile of Gertrude Walsh with Robert Robinson. Finally, Muriel Wheldale's collaboration with Huia Onslow (see Chapter 11) might be considered farther along the collaborative spectrum as "semi-autonomous." Nevertheless, most active women spouse-chemists were part of a collaborative couple. Here, we will provide five examples of such collaborative couples.

Anne Sedgwick (Lady Walker)

Anne (Annie) Purcell Sedgwick,[34] daughter of Lieutenant-Colonel Sedgwick and Mary Purcell White, was born at Queenstown, Cork on 4 December 1871.[35] She completed the Science Tripos at Girton in 1893,

---

[33] Pycior, H. M., Slack, N. G., and Abir-Am, P. G. (1987). Introduction. In Pycior, H. M., Slack, N. G., and Abir-Am, P. G. (eds.), *Creative Couples in the Sciences*. Rutgers University Press, New Brunswick, New Jersey, p. 12.

[34] Melville, H. W. (2004). Walker, Sir James (1863–1935). *Oxford Dictionary of National Biography*, Oxford University Press (accessed 28 December 2018).

[35] Butler, K. T. and McMorran, H. I. (1948). Teaching and Administrative Staff, 1895. *Girton College Register 1869–1946*, Girton College, Cambridge, pp. 637–638.

470 *Pioneering British Women Chemists: Their Lives and Contributions*

and then joined Norman Collie[36] at UCL as a research student in chemistry.[36,37] It was at UCL that Sedgwick met James Walker, who she was to later marry.[38]

It is documented that Sedgwick undertook research at Caius College, Cambridge, but no details could be discovered. In 1895, Sedgwick was appointed as assistant to Ida Freund (see Chapter 3) at Newnham College, while also being a Resident Lecturer in Natural Science, both posts which she held until 1897.

In 1894, Walker had been appointed Professor of Chemistry at the University College, Dundee. Thus, Sedgwick's departure from Newnham in 1897 coincided with her move to Scotland to marry Walker in the same year. After marriage, Sedgwick commenced research with her husband, the last co-authored publication appearing in 1905.[39] Her termination of research may have corresponded with the birth of their only child, Frederick. Walker accepted a Professorship at the University of Edinburgh in 1908, and it was in Edinburgh that Sedgwick died on 7 September 1950.

Mildred Gostling (Mrs. Mills)

Daughter of George James Gostling, dental surgeon and pharmaceutical chemist, and Sarah Abicail Aldrich, Mildred May Gostling[40] was born on 15 December 1873 at Stowmarket, Suffolk. She entered RHC in 1893 obtaining a B.Sc. (Hons.) in Chemistry in 1896. Gostling spent 1898–1900

---

[36] Baly, E. C. C. (1943) John Norman Collie, 1859–1942. *Obituary Notices of Fellows of the Royal Society* **4**: 329–356.

[37] Sedgwick, A. P. and Collie, N. (1895). XLIV — Some oxypyridine derivatives. *Journal of the Chemical Society, Transactions* **67**: 399–413.

[38] Kendal, J. (1932–1935). Sir James Walker. *Obituary Notices of Fellows of the Royal Society* **1**: 537–549.

[39] Walker, J. and Walker, A. P. (1905). CI. — Tetrethylsuccinic acid. *Journal of the Chemical Society, Transactions* **87**: 961–967.

[40] (a) White, A. B. (ed.), (1979). *Newnham College Register, 1871–1971*, Vol. I, *1871–1923*. 2nd edn., Newnham College, Cambridge, p. 146; (b) Creese, M. R. S. (1998). *Ladies in the Laboratory: American and British Women in Science 1800–1900*. Scarecrow Press, Lanham, Maryland, p. 268; and (c) Mann, F. G., revised by Shorter, J. (2011). Mildred May Gostling (1873–1962). In Mills, William Hobson (1873–1959). *Oxford Dictionary of National Biography*. Oxford University Press (accessed 7 January 2018).

at the Newnham College as a Bathurst Scholar, working with Henry Fenton[41] on carbohydrate chemistry and co-authoring four publications. In Fenton's obituary, one of Gostling's contributions was described:

> With Miss M. M. Gostling he found that various carbohydrates, in particular fructose, gave a purple colour when dissolved in ether and treated with hydrogen bromide, and this proved to be due to an oxonium salt of a yellow crystalline compound which could be thus obtained in considerable quantity and was shown to be ω-bromomethylfurfuraldehyde.[42]

Returning to RHC in 1901, Gostling was appointed Demonstrator in Chemistry. During her time at Newnham, she had met William Mills.[43] In 1902, Mills was appointed as Head of the Chemistry Department of Northern Polytechnic Institute in North London. His research group included Alice Bain (see Chapter 7) and Sibyl Widdows (see Chapter 5).

Gostling and Mills must have remained in contact, for in 1903, Gostling resigned her position at RHC to marry Mills. She joined Mills' group. Her extensive research on dinaphanthracene, co-authored with Mills, was published in 1912.[44] That year, Mills accepted a position at Cambridge, and it seems Gostling ceased research at that time. She had one son and three daughters; two of the daughters, Mildred Marjorie Mills and Sylvia Margaret Mills, became students at Newnham, while the third, Judith Isobel Emily Mills became, later in life, a staff member at Newnham. Gostling died on 19 February 1962.

Ellen Field (Mrs. Stedman)

Not to be confused with E. Eleanor Field, Ellen Field together with her spouse Edgar Stedman formed the most equal chemistry partnership of those we have studied. Of course, in those days, equal work was not

---

[41]"W.H.M." [Mills, W. H.] (1930). Obituary Notices: Henry John Horstman Fenton, *Proceedings of the Royal Society of London. Series A* **127**: i–v.

[42]Note 41, "W.H.M.", p. v.

[43]Mann, F. G. (1960). William Hobson Mills. 1873–1959. *Biographical Memoirs of Fellows of the Royal Society* **6**: 200–225.

[44]Mills, W. H. and Mills, M. (1912). The synthetical production of derivatives of dinaphthanthracene. *Journal of the Chemical Society, Transactions* **101**: 2194–2208.

472   *Pioneering British Women Chemists: Their Lives and Contributions*

reflected in equal status. Field was born on 29 October 1883 at Greenwich, Kent, the daughter of William Frederick Field, labourer, and Ellen Bobey.[45]

Field studied towards a chemistry degree at Goldsmiths' College. Goldsmiths' College, New Cross, London had originally been the Royal Naval School. Purchased by the Worshipful Company of Goldsmiths in 1889, it was opened as the Goldsmiths' Company's Technical and Recreative Institute in 1891. The intention of the Institute was the "promotion of the individual skill, general knowledge, health and well being of young men and women belonging to the industrial, working and poorer classes," and between 1907 and 1915 included evening classes towards a B.Sc. (London).[46]

After completing her B.Sc. degree in 1912, Field was hired as a Demonstrator in Chemistry at Goldsmiths'. In her spare time, she commenced research with George Barger,[47] Head of the Department of Chemistry. Her work in 1912 on the blue compounds of iodine formed with starch and other substances provided her first publication. While she was a demonstrator, a young undergraduate student, Edgar Stedman, regularly visited her house for help with his chemistry studies.

Field followed Barger to Royal Holloway College (see Chapter 5). Then the following year, when Barger was appointed to the Department of Biochemistry and Pharmacology of the Lister Institute, Field undertook research at Bedford College, resulting in an M.Sc. in 1916 jointly from Bedford and Goldsmiths'. In 1919, when Barger accepted the Chair of Chemistry in Relation to Medicine at the University of Edinburgh, he invited Field to join him, which she did. Barger found the workload of the new department overwhelming, and Field proposed that he hire the newly graduated Stedman as Lecturer. Barger concurred and hired Stedman. Field, equally qualified as Stedman, was not appointed Lecturer for another 8 years.

---

[45] Cruft, H. J. (1976). Edgar Stedman. 12 July 1890–8 May 1975. *Biographical Memoirs of Fellows of the Royal Society* **22**: 528–553.

[46] Firth, A. E. (1991). *Goldsmiths' College: A Centenary Account*. Athlone Press, London.

[47] Dale, H. H. (1940). George Barger, 1878–1939. *Obituary Notices of Fellows of the Royal Society* **3**: 63–85.

Roles of Chemist's Wives 473

With Stedman having been offered a position, he and Field married. This was to prove an ideal match, as Stedman's obituarist, H. J. Cruft, commented:

> She [Field] had a stimulating, often mildly caustic, sense of humour and was never to be overawed by Edgar with his single-mindedness. She never hesitated to express her views, even if they might be unpopular — an attribute Edgar admired. They enjoyed many activities together such as hill walking, and later gardening and above all the sense of achievement when experiments went well and produced a positive result in the laboratory.[48]

While Stedman worked towards a Ph.D., Field continued research with Barger, studying plant alkaloids and their derivatives, resulting in two more publications — one in 1923 and the other in 1925. With Stedman's Ph.D. granted, Field shifted to research with her husband, studying haemocyanin between 1925 and 1928 and co-authoring five publications on the subject. As Cruft noted: "This work was primarily carried out by Ellen and for which she had received several grants."[49]

In the 1930s, Field and Stedman undertook research on acetylcholine and choline-esterase, co-authoring 10 publications. Then, in the 1940s, they entered a more controversial study: that of the chemical composition of the cell nucleus, arguing in a series of papers that DNA was not a major constituent of chromosomes. By the 1950s, Field and Stedman had become isolated from the life of the Department, as Cruft described:

> He [Stedman] and Ellen at this stage, to a certain extent, retired from the banter of the department coffee room and had their lunch of fruit and sandwiches in two leather chairs in Edgar's almost Dickensian office overlooking the Medical Quadrangle.[50]

Sadly, during the summer of 1953, Field suffered a massive coronary thrombosis and was severely ill for some months. For the rest of her life

---

[48] Note 45, Cruft, p. 533.
[49] Note 45, Cruft, pp. 535–536.
[50] Note 45, Cruft, p. 543.

she was essentially bed-ridden, boredom being a major problem for her. She died on 22 September 1962.

Mary Laing (Mrs. McBain)

Mary Evelyn Laing, born on 17 October 1891 at Alcester, Warwickshire, was the daughter of Scottish parents, David Laing and Jeannie McGeogh.[51] Laing was educated at Redland High School for Girls, Bristol. After matriculating in July 1911, she accepted a teaching position in France.[51] Returning to Britain before the First World War, Laing entered the University of Bristol, completing a B.Sc. in 1915 and an M.Sc. in 1919.[52] During the War, she undertook some unspecified research for the British Government.

In 1919, she was hired as a "scientific collaborator"[51] with the physical chemist James McBain. Her research resulted in nine publications on the structure of soap solutions — five on her own and the remainder with McBain (one also being co-authored with Millicent Taylor, see Chapter 6). Laing was awarded a D.Sc. in 1924 on the basis of her research.[53]

McBain accepted a position at Stanford University, California in 1926, and Laing moved with him, becoming a Research Associate in Chemistry. They married in 1929 and had one son. As Hugh Taylor, one of McBain's obituarists commented:

> His second marriage on New Year's Day 1929, to Mary Evelyn Laing, a scientific collaborator, member of the Science Faculty at Bristol, and Research Associate in the earliest years at Stanford, was a model of scientific and social harmony. [54]

---

[51] Rideal, E. K. (1952–1953). James William McBain 1883–1953. *Obituary Notices of Fellows of the Royal Society* **8**: 529–547.

[52] Anon. [with extracts written by E. McBain] (Summer 1955). India turns to Science. *Redland High School Magazine* 37–39.

[53] Anon. (May 1926). News of Old Girls. *Redland High School Magazine* 36.

[54] Taylor, H. (1956). Obituary notices: James William McBain, 1883–1953. *Journal of the Chemical Society* 1918–1920.

Laing authored eight papers from Stanford, four under her name alone, which was now given as M. E. L. McBain. In 1932, she was appointed Assistant Professor in Physical Chemistry.[55]

In 1937, McBain and Laing went to Russia where they both lectured at the Physical Chemical Institute in Leningrad (see Footnote 51). During the Second World War, Laing was active with a British War relief unit, reporting back to her high school:

> For Britishers in California the war is not something remote, but is something which enters into our daily lives in many ways. We either lack news and worry, or we get censored news. The activities of the British War Relief Units are many and varied. We all knit; we hold food sales; we get money by organising concerts and lectures; we collect, mend and make garments, and finally we pay visits and talk in order to procure foster parents for the British children who were not able to be convoyed safely across the Atlantic.[56]

In 1949, following the independence of India, McBain was asked to develop and direct the National Physical and Chemical Laboratories as part of the planned industrialisation (see Footnote 52). Laing described some of the challenges:

> We felt very early in our stay that the chemists working in different sections met very rarely. To remedy this we planned a self-service cafeteria where all groups could get together. Many of the older members of the laboratory said that our plan would never work, since no Indian with any self respect would carry his soiled dishes anywhere. However, on opening day, Dr. McBain and I fetched our coffee and returned our cups, and after that no one questioned the procedure. India for many centuries has had a strict caste system. ... It was our aim to have no regard for caste in our laboratory. We succeeded in this endeavour chiefly because we inaugurated the National Laboratory Club to take care of the leisure hours of our workers.[57]

---

[55] Anon. (December 1932). News of Old Girls. *Redland High School Magazine* 30.

[56] McBain, J. W., Mrs. (née E. Laing) (June 1941). News from other parts of the world. *Redland High School Magazine* 33.

[57] Note 52, Anon., p. 38.

476　*Pioneering British Women Chemists: Their Lives and Contributions*

McBain and Laing returned to California in 1952, McBain dying in 1953. Laing continued to be active, having research published until 1957 and being the author of a monograph, *Solubilization and related phenomena.*[58] Laing died on 25 June 1981 at Palo Alto, California.

Catherine Tideman (Mrs. LeFèvre)

The highest profile spouse-collaborator was Catherine (Cathie) Gunn Tideman. Born in Glasgow, 1 November 1909, she went to UCL to study chemistry,[59] and it was there that the physical organic chemist Raymond LeFèvre noticed her:

> Among those taking organic chemistry during 1928–29 was one who seemed always cheerful, lively, ready with relevant comments on current or local affairs, full of conversational topics, of repartee, energy and vigour; a hockey player of enthusiasm (who had once knocked unconscious an opponent through an accidental head-to-head collision), a tireless dancer, a rider of horses (her grandmother had owned a riding school in Glasgow), not excessively teetotal but convivial with most of the women and men contemporary with her at U.C.L.[60]

A romance soon developed, with the marriage taking place in 1931. Tideman became LeFèvre's long-term research colleague, co-authoring papers and reviews under the name of C. G. LeFèvre. Much of the careful laboratory work was performed by Tideman, such as that which led to the validation of the molar Kerr constant (the constant is a numerical criterion for the discrimination between often closely related molecular configurations).[61]

---

[58] McBain, M. E. L. (1955). *Solubilization and related phenomena.* Academic Press, New York.

[59] Aroney, M. J. and Buckingham, A. D. (1988). Raymond James Wood Le Fèvre. *Biographical Memoirs of Fellows of the Royal Society* **34**: 375–403.

[60] Autobiographical notes by R. J. W. Le Fèvre. Cited in: Note 59, Aroney and Buckingham, p. 383.

[61] Davies, M. (1987). R. J. W. Le Fèvre 1905–1986. *Chemistry in Britain* **23**: 564.

In 1938, Tideman's son, Ian, was born. The birth did not stop her academic life and shortly afterwards, Tideman was appointed Demonstrator in Chemistry at UCL and she also taught chemistry at Queen's College, Harley Street (see Chapter 2). She had a second child, Nicolette, in 1940. As a result of Tideman's research contributions at UCL, she was awarded a D.Sc.

In 1946, LeFèvre accepted the offer of the position of Director of Chemistry at the University of Sydney, Australia. Upon arrival, Tideman and LeFèvre continued their research partnership. In addition to continuing work on the Kerr effect, during the 1960s Tideman became active in social issues, particularly drug dependency and the drug problem in New South Wales. Subsequently, she developed an interest in forensic science, becoming the first woman elected to the Council of the Australian Academy of Forensic Sciences.[62] Tideman was also heavily involved in programmes to encourage women into science. She died in Sydney on 9 March 1998.

A Contrast in Women Organic Chemists

For much of their later lives, organic chemists Robert Robinson[63] of Oxford and Christopher Ingold[64] of UCL were in professional competition. The dispute, and the resulting enmity between them, have been described elsewhere[65]; what is usually overlooked is that they each had a wife who continued research after marriage.

Their spouses, Gertrude Walsh (Mrs. Robinson) and Hilda Usherwood (Mrs. Ingold), combined a dedication to organic chemistry with marriage and motherhood. However, the nature of the research relationships with their spouses were quite different. Walsh, outgoing and strong willed,

---

[62] Walker, R. (5 March 2018). Le Fèvre, Catherine Gunn (1909–1998). *Encyclopedia of Australian Science* (accessed 08 January 2019).

[63] Todd, Lord, and Cornforth, J. W. (1976). Robert Robinson. 13 September 1886–8 February 1975. *Biographical Memoirs of Fellows of the Royal Society* **22**: 414–527.

[64] Shoppee, C. W. (1972). Christopher Kelk Ingold, 1893–1970. *Biographical Memoirs of Fellows of the Royal Society* **18**: 348–411.

[65] Saltzman, M. D. (1980). The Robinson-Ingold controversy: Precedence in the electronic theory of organic reactions. *Journal of Chemical Education* **57**: 484–488.

478   *Pioneering British Women Chemists: Their Lives and Contributions*

developed her own clear research field, even though she was involved in some joint projects. Usherwood, on the contrary, "painfully shy," became very much a collaborator of Ingold. Her contributions merged into Ingold's productivity, and this contributed to a lower profile of Usherwood compared with Walsh.

Both Usherwood and Walsh, however, faced the common problem for married women chemists at the time: that of being perceived as append-ages of their husbands with no opportunity or expectation of an independ-ent scientific career. It was their shared love of organic chemistry which allowed them to spend countless hours at the research bench, unpaid for their efforts, while each raised a family.

Gertrude "Gertie" Walsh (Mrs. Robinson)

Of all the chemist spouses, Gertrude Maud Walsh[66] was the most renowned, being deemed worthy of obituaries in two scientific journals. Born on 6 February 1896 in Winsford, Cheshire, Walsh was the daughter of Thomas Makinson Walsh, a representative of a coal merchant, and Mary Emily Crosbie. She was educated at Verdin Secondary School, Winsford, and then obtained her B.Sc. degree in 1907 and M.Sc. in 1908 from Victoria University of Manchester. After graduation, she undertook organic chemistry research with Chaim Weizmann (later President of Israel) and taught Chemistry at MHSG.

Robinson, an organic chemist at Manchester, organized a walk for the "Tea Club" researchers through Cheedale and Cowdale in Derbyshire; among the participants were Ida Smedley (see Chapter 11) and Walsh. Torrential rain discouraged the participants, except Walsh, from Robinson's subsequent invitations to hikes. As Robinson recalled: "The mass invasion of the Derbyshire Dales was not repeated, but Miss Walsh and I devoted many weekends to rock climbing on the millstone grit edges of the Peak District."[67]

A common bond in mountaineering turned into a personal relation-ship, and in 1912, they were married. They had two children, a daughter,

---

[66]Baker, W. (1954). Lady Robinson (Obituary). *Nature* **173**: 566–567.

[67]Robinson, R. (1976). *Memoirs of a Minor Prophet: 70 Years of Organic Chemistry*, Vol. 1, Elsevier, Amsterdam, p. 45.

Marion (born 1921) and a son, Michael (born 1926). Robinson's first post was at the University of Sydney, Australia, where Walsh was employed as an unpaid Demonstrator in Organic Chemistry, the university banning the paid employment of married women. Walsh was very popular among the students as this commentary notes:

> Mrs Professor Robinson, who acts as a honorary demonstrator for her husband, and says the things to students that he is not game to say, is an amazing amateur fire brigade, all by herself. Fires are of frequent occurrence in organic chemistry and a rug is specially provided with which to extinguish them. When a blaze arises Mrs Robby hurls herself into the air, grabbing the rug as she flies, falls upon the conflagration, puts it out regardless of singed hair and eyelashes, and without even waiting to regain her breath, rounds upon the luckless student until he wishes the fire had consumed him too![68]

During her years in Sydney, Walsh developed elegant methods of synthesising both saturated and unsaturated fatty acids. However, her most prolific period was during the 1930s, after Robinson was appointed Professor of Chemistry at the University of Oxford. It was here that she began studying chemical genetics, specifically plant pigments (see Muriel Wheldale, Chapter 11).[69] J. C. Smith, who chronicled life in the Oxford chemistry department commented:

> R.R.[Robinson] took over Perkin's office and laboratory, wisely allotting two benches to his wife. Mrs Robinson was probably the hardest worker of us all. She arrived at 9 am...., left at 1 pm and then put in a 2.15-7 pm afternoon.[70]

Walsh drove around Oxford in her Standard 12 car, as Smith recalled:

---

[68] Williams, T. I. (1990). *Robert Robinson: Chemist Extraordinary*. Clarendon Press, Oxford, pp. 34–35.

[69] Scott-Moncrieff, R. (1981). The classical period in chemical genetics: Recollections of Muriel Wheldale Onslow, Robert and Gertrude Robinson and J. B. S. Haldane. *Notes and Records of the Royal Society of London* **36**: 125–154.

[70] Smith, J. C. (n.d.). *The Development of Organic Chemistry at Oxford*. Part II, *the Robinson Era 1930–1955*. unpublished, p. 5. H. Anderson, Memorial University, is thanked for a copy of this manuscript.

480 *Pioneering British Women Chemists: Their Lives and Contributions*

> Mrs. (Gertrude) Robinson was not a good driver; she not only forgot to wear her glasses, but she sat so low that she peered through the steering-wheel. One was often surprised at the positions in which her car stood outside the D.P. [Dyson Perrins Laboratory].[71]

As Walsh and Robinson did not have a crushing machine to extract the flower pigments, they put the plants in a pile, covered them with boards, and then drove one of their cars backwards and forwards over them.[72]

The extractions were performed using ether (ethoxyethane) as a solvent, which was often contaminated by highly explosive peroxo-compounds. One particular explosion involved Walsh as Smith remembered:

> In 1938 there was a shattering explosion on the top floor while I was demonstrating below. Senior assistant ("Gertie") Miller and I rushed upstairs and found Mrs. Robinson on the floor, her face covered with blood and pieces of glass. Miller, the unflappable, quickly did all that we dared and quickly took her to the Oxford Eye Hospital. Fortunately there was no glass in the eyes, but many pieces were removed from the face.[73]

Walsh was one of the few wife-assistants to gain some recognition for her work. Robinson, in his autobiography, gave her due credit:

> Further reference to the work of G.M.R. and to incidents in her later life will be found in Volume II of these Memoirs [which was never published]. Nevertheless, I cannot postpone an acknowledgement of the very great help which she gave me at all stages of my career. Looking back, I can see how she subordinated her interests to mine, was always such a ready collaborator in scientific work, and cheerfully followed my chief vacation activity, namely mountaineering.[74]

---

[71] Note 70, Smith, p. 46.

[72] Williams, T. I. (1990). *Robert Robinson: Chemist Extraordinary*. Clarendon Press, Oxford, p. 67.

[73] Note 70, Smith, p. 43.

[74] Note 67, Robinson, pp. 55–56.

Robinson's support for Walsh was very public, as Smith recalled in the context of Walsh's pioneering and often overlooked synthesis of oleic acid[75] and higher fatty acids:

> R.R. admired her confident and cheerful attitude to research-work (and to the rather frequent breakages of glassware). He became embarrassingly enthusiastic over her successes. ... R.R. startled one of his demonstrators dining with him in Magdelen College, by suddenly asking, "Have you heard about my wife's fatty acids?"[76]

During the Second World War, Walsh became involved in antibiotic synthesis, and she was the first chemist to synthesise a penicillin analogue with antibiotic character. Unfortunately, little was published on her research in this field.[75]

As a woman, Walsh was excluded from the meetings of the Alembic Club — the chemistry society at Oxford (see Chapter 3). So it was not surprising that she exhibited a determination for sexual equality, illustrated by the following account from Louis Hunter of the University of Leicester:

> A memorable event in this period was the visit of the British Association for the Advancement of Science to Leicester in 1933.... It was the custom at that time for the Sectional Dinner to be open only to male members, and this custom fell particularly unfairly on women chemists. With characteristic energy and hospitality, Mrs (later Lady) Robinson arranged a dinner party at the same time as the Sectional Dinner, in the same hotel and with the same menu, to which she invited other women chemists as well as wives of the sectional officers and of other prominent members. This bold action finally broke down the practice of restricting the dinner to men, and at all meetings of the British Association subsequent to 1933 the dinners of Section B have been graced by the presence of ladies.[77]

---

[75] Simonsen, J. L. (1954). Obituary Notices: Gertrude Maud Robinson 1886–1954. *Journal of the Chemical Society* 2667–2668.

[76] Note 70, Smith, p. 8.

[77] Hunter, L. (1955). Schools of chemistry in Great Britain and Ireland–XV The University College of Leicester. *Journal of the Royal Institute of Chemistry* **79**: 15–16.

482 *Pioneering British Women Chemists: Their Lives and Contributions*

In 1953, the University of Oxford conferred on Walsh an honorary M.A. degree. Tragically, she died of a heart attack on 1 March 1954.

Edith Usherwood (Mrs. Ingold)

Whereas Walsh's productivity was maintained throughout her career, that of Edith Hilda Usherwood diminished over time.[78] Usherwood, the daughter of Thomas Scriven Usherwood, teacher of engineering, and Edith Howarth, showed academic brilliance from an early age, winning a scholarship to North London Collegiate School.[78] She graduated from RHC in 1920 with a First-Class Honours degree in Chemistry, and then undertook postgraduate research with Whiteley (see Chapter 4) at IC, from where she obtained a Ph.D. in 1923.

It was at IC that she met Christopher Ingold.[79] One of Ingold's co-workers, Frank Dickens,[80] recalled that, in 1923, Ingold gassed himself with phosgene, and it was Usherwood who first rushed to the rescue.[81] Dickens added: "As far as we were concerned that was the first intimation that we had that an engagement was in the offing and, of course, they were married shortly afterwards."[82] Their honeymoon was spent in Snowdonia, though Usherwood preferred walking to climbing.

Marriage did not initially diminish Usherwood's research, and she received the D.Sc. in 1925, from IC and RHC, resulting from a series of papers on tautomerism, published under her name alone. However, a move by Ingold in 1924 did affect her work. Ingold was appointed Professor of Organic Chemistry at University of Leeds, while Usherwood served as unpaid demonstrator, in addition to undertaking research. At the time, socialising was the major occupation of the other faculty wives,

---

[78]Rayner-Canham, M. and Rayner-Canham, G. (1999). A tale of two spouses. *Chemistry in Britain* **35**: 45–46.

[79]Leffek, K. T. (2004). In Ingold, Sir Christopher Kelk (1893–1970). *Oxford Dictionary of National Biography.* Oxford University Press (accessed 9 January 2018).

[80]Thompson, R. H. S. and Campbell, P. N. (1987). Frank Dickens. 15 December 1899–25 June 1986. *Biographical Memoirs of Fellows of the Royal Society* **33**: 188–210.

[81]Leffek, K. T. (1997). *Sir Christopher Ingold: A Major Prophet of Organic Chemistry.* Nova Lion Press, Victoria, British Columbia, p. 55.

[82]Note 81, Leffek, pp. 100–101.

particularly morning visits to each other, a pastime the shy Usherwood avoided when possible.

From 1925 onwards, most of Usherwood's publications were joint with Ingold. An incident in 1926 illustrates how closely Usherwood and Ingold worked as a team. Robinson's group had repeated Ingold's research on the nitration of S-methylthioguaiacol, obtaining very different results. On being told this, Usherwood and Ingold went into the laboratory and worked for 72 hours without a break, repeating the disputed reactions, and rapidly submitting a follow-up paper declaring the correctness of Robinson's findings (see Footnote 82).

Usherwood's productivity declined with the birth of their first child, Sylvia, in 1927 (followed by Keith in 1929 and Dilys in 1932), though she did co-author a paper on the first synthesis of benzyl fluoride. In 1930, Ingold was appointed as Professor of Chemistry at UCL. Usherwood continued to participate in research work at UCL as honorary Research Associate, hiring a "mother's helper" and a cleaner to help her to cope with family responsibilities. During the 1930s, she took on the additional task of aiding Jewish refugees from Germany and Austria.

In 1939, UCL's chemistry department was evacuated, mostly to Bangor, and the remainder — including Ingold and Usherwood — to Aberystwyth. No secretarial staff was posted to Aberystwyth, so Usherwood became *de facto* Administrative Assistant, and, after teaching herself typing, proceeded to become Departmental Secretary as well. At Ingold's request, Usherwood finally received a salary — for secretarial work.

Following their return to London in 1946, Usherwood was formally appointed as Administrative Assistant to UCL's Chemistry Department, with her last research publication appearing in 1947. The chemistry historian, Joan Mason, who overlapped with Usherwood at UCL from 1951 to 1956 recalled:

> I found her not especially friendly to me, and realized after a while that she and her friends were anti-Oxbridge as centres of privilege, and extended this to people like me! I counted myself as politically on the left as she was; also we had been to the same school (North London Collegiate, where I overlapped with her daughter Sylvia). But I can well understand the resentment of non-Oxbridge people, particularly high

484 *Pioneering British Women Chemists: Their Lives and Contributions*

calibre ones such as Ingold, to effortless assumptions of superiority by Oxbridge.[83]

After Ingold's death in 1971, Usherwood continued to be active at UCL, including serving as President of the UCL Chemical and Physical Society during the 1976–1977 academic year. She died on 8 August 1988 at Chelmsford, aged 90, leaving her body to the anatomy department of University College, London. The science historian, William Brock, summed up Usherwood's situation:

> By changing her name to Ingold and forgoing productive research in kinetics during the 1930s while she raised a family, her extraordinarily productive husband subsumed her identity.... to be fair to Christopher Ingold, however, he never failed to acknowledge his wife's help and assistance. Moreover, he always recognized Hilda's contribution to theoretical chemistry.[84]

Commentary

For those women chemists who married chemists, the wife often continued to participate in some way, for example, as a secretarial assistant or a laboratory worker. There were very few "independents," while "autonomous" and "semi-autonomous" relationships, where the spouses worked on related or overlapping areas, seemed more common.

Discovering co-worker spouses is challenging. In some cases, the wives were given co-author status, and thus, they could be identified by perusing lists of the husband's publications. For many others, the wife's contribution was found only as a footnote or as an acknowledgement at the end of a publication. In a few cases, the husband's obituarist acknowledged the importance of the spousal role. However, one wonders how many more spousal women chemists were "invisibles" — women who contributed significantly to their husband's chemical career, but whose name was neither acknowledged in publications nor by the husband's obituarist.

---

[83] Personal communication, e-mail, J. Mason, 14 October 1999.
[84] Note 3, Brock, p. 230.

Chapter 15

Women Chemists and the First World War

Martha Whiteley (11 November 1866–24 May 1956)

The First World War changed everything for women in terms of employment opportunities, especially for women with scientific training.[1] In addition, universities lost most of their male staff and students, resulting in women

---

[1] Fara, P. (2018). *A Lab of One's Own: Science and Suffrage in the First World War*. Oxford University Press, Oxford.

486   *Pioneering British Women Chemists: Their Lives and Contributions*

being hired as lecturers and leaving women students in the majority. A Newnham student, M. G. Woods (Mrs. Waterhouse) remarked: "the preponderance of women in the classrooms made the salutation of 'Gentlemen' more ridiculous than ever."[2]

The War and Young Women

Up until 1914, women had entered university as part of the "New Woman" movement.[3] Marriage and child-raising were still the preference and expectation of the majority, while the unmarried could still be ladies of leisure. Not only was the War opening up opportunities of employment, but the option of marriage was disappearing for many. A career for a middle-class girl as a necessity was a new dimension. In 1915, The Editor of *Girl's Realm* authored an incredibly honest article:

> In the old Victorian days the woman who did not marry was supposed in some way or other to find a niche in the family, where she made herself "generally useful" but did not work, where she was dependent on the money left her by her father, or the generosity of her relations. Those days are passing. Every woman, as well as every man, has in times like these to show "the reason of her existence."[4]

The Editor pointed out that the enormous number of deaths on the battlefields, particularly among the middle-class,[5] made the possibility of marriage by the teen girl readers an unlikely prospect: "I address these remarks principally to girls who are facing life at the opening of their careers. The war has made us all face reality, and, not least of all, the women of the world."[4] The article ended by promising that

---

[2] Brooke, C. N. L. (1993). *A History of the University of Cambridge,* Vol. IV, *1870–1990*. Cambridge University Press, Cambridge, p. 332.

[3] Jordan, E. (1999). *The Women's Movement and Women's Employment in Nineteenth Century Britain*. Routledge, London.

[4] The Editor. (1915). The war and women: The influence of the world-conflict on women's status and work. *Girl's Realm* **17**: 45–46.

[5] Winter, J. M. (1989). *The Experience of World War I*. Oxford University Press, Oxford.

Girl's Realm would continue to provide information on careers open to young women.

Duty of Women University Students

At university, women students were uncertain of what was expected of them in wartime. At Armstrong College, Newcastle, the women students heard of the need for women munitions workers:

> Perhaps the most thrilling event of the term from the feminine point of view has been the appeal for women munition-makers. After twenty-four distracted and heart-searching hours; after some of us had obtained permission to go and were feeling heroic; after some of us had decided to stay and were feeling "desper'te mean"; and after some of us had defied our obdurate parents and guardians in the names of Patriotism and Legal Majority, and were feeling Shelleyesque, we were informed that our highest patriotic duty was to complete our education![6]

The call for women students to continue their education had come from H. A. L. Fisher in early 1917, who represented the Board of Education:

> My own view is that for the present women students at the universities should continue their academic courses until such time as they may be called up by the branch of the National Service Department presided over by Mrs. Tennant.[7]

The fear of being "called up," as Fisher had indicated in his communication was very real for the Newcastle women students:

> Nowadays the most serenely stay-at-home and selfish of us have been under Zeppelin fire two or three times. When Professor Mawer announced that he had a communication from the Government to read to us the involuntary start and look of dismay by which everyone betrayed the impression that conscription for women had come at last, and the general relief which succeeded his explanation that the Government only meant to warn

---

[6]"De Virginibus." (1916). *The Northerner: The Magazine of Armstrong College* **16**(2): 39 (by permission, Archives, Newcastle University).
[7]Fisher, H. A. L. (8 March 1917). Duty of women students. *The Serpent* **6**: 52.

us not to repeat military information. Yet even among the women, numbers have made it already in the interests of their country. The latest to go are Miss Oliver, Miss Clayburn and Miss Cummings, late candidates for Honours in Chemistry, now chemists to munition works at Middlesboro'.[8]

WWI Women's Work Collection

In fact, as we will discuss in the following sections, women chemists made up a significant proportion of the scientific workforce. Fortunately, the Women's Work Collection of the Imperial War Museum (IWM) has a significant amount of documentary evidence on the wartime women scientists. This useful material was compiled in 1919 by Agnes Ethel Conway of the Women's Work Sub-Committee of the IWM. Conway circulated a questionnaire to universities and industries informing them that the committee was compiling a historical record of war work performed by women for the National Archives. In particular, Conway added: "they [the Sub-Committee] are anxious that women's share in scientific research and in routine work should not be overlooked."[9] Enough replies were received to provide a sense of the breadth of employment of scientifically trained women during the War.

Had it not been for the significant increase in the number of women taking chemistry degrees during the first 15 years of the 20th century, it is apparent that the British war machine would have faced a significant shortage of chemists. Fortunately, there was a pool of qualified women chemists ready and willing to do their part towards the war effort. As illustration, Kennedy Orton, the Professor of Chemistry at the University College of Wales, Bangor (see Chapter 7), commented in a report that "The demand for young women who have received a training in Chemistry, both for educational and professional work, has increased greatly during the past Session. The demand is far in excess of the supply."[10]

---

[8]"De Virginibus." (1917). Correspondence. *The Northerner: The Magazine of Armstrong College* **17**(2): 29 (by permission, Archives, Newcastle University).

[9]Conway A. E. Women's War Work Collection, Imperial War Museum, London (WWWC-IWM) (by permission, Archives, Newcastle University).

[10]Orton, K. J. P. (1918). *Annual Reports of the Heads of Department, 1917–1918.* University College of Wales, Bangor, pp. 8–9. T. Roberts, Archivist, University of Wales, Bangor, is thanked for supplying this information.

Responding to the Need

Some women chemistry students did interrupt their studies in order to contribute to the war effort. For example, a note in *Our Magazine*, the student magazine of North London Collegiate School for Girls (NLCS), reported: "Gwendolen A. James is working in an analytical laboratory for a year before going on with her Science Degree work. Margarethe Mautner is analysing drugs in a Chemical Laboratory before taking up her medical studies."[11]

Two graduates of the Edinburgh Ladies' College deferred their university studies as recounted in the school magazine, *The Merchant Maiden*:

> Maisie B. Robertson has for the present forsaken her scientific studies at the University here and is engaged meantime in an explosives factory in the South of England in chemical analytical work, the nature of which, the Censor says, may not be divulged.

> Helen Lumsdaine writes:- "I have been working since September as an analytical chemist in the Cement Works at Cousland. This may not sound very attractive, but I have found the work quite interesting. The employment of lady chemists is an experiment which one or two firms are trying on account of the war — another occupation thrown open to women."[12]

The War also impacted those who had just completed their university studies. For example, Somerville College recorded the war work of its former students, three of whom were undertaking chemistry-related duties:[13]

1. **Arning, Dorothy**: Asst. Loading Chemist, No. 21 Natl. Filling Factory, Coventry.

---

[11] Anon. (December 1916). Pupils leaving. *Our Magazine: North London Collegiate School for Girls* 93.

[12] Anon. (1916). Former pupil notes. *The Merchant Maiden* **9**(1): 24–26.

[13] Anon. (November 1917). *Somerville Students' Association Thirtieth Annual Report and Oxford Letter* 2–33.

2. **Cudworth, Mary**: Analytical Chemist for Mssrs. Richardson & Co., manufacturers of agricultural fertilizers.
3. **Markham, Claudia E.**: Science Mistress, Bude County School, substitute for a man. Then Chemist to British Resorcin Manufacturing Co.

Women Chemists in Organic Synthesis

At the start of the First World War, nearly all fine chemicals, such as pharmaceuticals and anaesthetics, were produced in Germany and Austria. Thus, it was of utmost priority to generate a home-grown organic chemical industry. This task took time. Arthur Schuster, Secretary to the Royal Society, wrote to all university chemistry departments:

> It has been thought desirable to enlist the voluntary services of the many chemical laboratories connected with the educational institutions of the country, to meet the urgent demand for the immediate supply of certain of these drugs, mainly organic products.[14]

The academic chemical laboratories were turned into miniature factories. It was noted in a Royal Society report of 21 May 1917 that the narcotic β-eucaine was being produced by the University of Glasgow, Bedford College, and Imperial College (IC); arabinose by the University of St. Andrews, University of Leeds, IC, and Borough Polytechnic Institute; and atropine by the University of Liverpool and the University of Dundee.[15] In a more detailed report of December 1917, the production of β-eucaine by Millicent Taylor (see Chapter 6) and her students at Cheltenham Ladies' College (CLC) was noted.[16]

The Chemistry Department of the University of Sheffield was one of the major participants in organic synthesis (see Chapter 6). William

---

[14]Draft Letter, Arthur Schuster, Royal Society, to all U.K. Universities, November 1914, Royal Society Archives, Cmb 28.

[15]Statement regarding position of drug work, 21 May 1917, Royal Society Archives.

[16]Report presented to the Royal Society Council, Sectional Chemical Committee, December 1917, Royal Society Archives.

Palmer Wynne assembled a team of six women chemists, including Emily Turner, Dorothy Bennett, and Annie Mathews to synthesize β-eucaine.[17]

The group of organic chemists producing pharmaceuticals at Imperial College was organized by Martha Whiteley (see Chapter 4).[18] Whiteley's seven assistants were all women: Dorothy Haynes, Winifred Hurst, Hilda Judd, Frances Micklethwait (see Chapter 4), Sibyl Widdows (see Chapter 5), and two graduate students, Louise Woll and Ethel Thomas. Micklethwait received an M.B.E. for her contributions to the war effort.[19]

The women chemists at St. Andrews worked on the production of synthetic drugs and bacteriological sugars, research on explosives and poison gases, and the improvement of industrial processes. The individuals who undertook this research sacrificed their own career advancement as this report sent to Conway noted:

> It should be stated that the whole of this work was unpaid from Government sources, the workers receiving only their University salaries, in cases where they were members of staff, or the value of their Scholarships, if they held any such distinctions. Not only so, but the demand for chemists throughout the war was continuous, so that the workers who remained with me gave up many opportunities for professional advancement. I mention these facts as an index of public spirit with which these women gave their services, services which have not received any public recognition.[20]

All these contributions by women chemists were "invisible." No mention was made of them in an article of 1917 by the Institute of Chemistry on the work accomplished by wartime [male] chemists.[21]

---

[17]Chapman, A. W. (1957–1958). The early days of the chemistry department. *By Product (Journal of the University of Sheffield Chemistry Department)* **11**: 2–5.

[18]Letter, M. A. Whiteley to A. E. Conway, 3 October 1919, WWWC-IWM.

[19]Burstall, F. H. (1952). Frances Mary Gore Micklethwaite (1868–1950). *Journal of the Chemical Society* 2946–2947.

[20]Letter, (Illegible signature) to War Committee, 23 August 1919, Royal Society Archives.

[21]The Registrar (1917). Chemists in war. *Proceedings of the Institute of Chemistry of Great Britain and Ireland* 29–34.

Margaret K. Turner

Yet Margaret Kathleen Turner[22] of the University College of Wales, Aberystwyth (UCWA), did participate in the war effort. Turner, born on 14 September 1880 at Newcastle-under-Lyme, was the daughter of William Turner, solicitor, and Emma Sutton. She was educated at Orme Girls' School, Newcastle-under-Lyme, after which she took a governess position. Turner entered UCWA in her mid-20s completing a B.Sc. (Hons.) in Chemistry in 1909.

After graduation, Turner stayed on at UCWA as a student assistant in the Chemistry Department. Then she held the position of Senior Mathematics and Chemistry Mistress at City and County School, Chester, from 1910 to 1913, followed by 1 year as Assistant Lecturer in Chemistry at Bedford College. Turner was appointed as Senior Science Mistress, County Secondary School, Streatham, in 1914. It was the War which provided her with a university appointment as was noted in the history of the chemistry department of UCWA:

> The advent of war in 1914 caused considerable disruption.... Miss Margaret Turner and, later, J. B. Whitworth took up posts as demonstrators and under the direction of the professor carried out large-scale preparations of intermediates for certain essential drugs.[23]

In 1915, Turner wrote a stirring letter to the War Committee, volunteering for additional duties:

> I was one of the workers in the preparation of diethylamine some weeks ago and should be very glad to hear of any further help I could give. I can put all my time and energy at your service for the next 6 weeks, and am anxious to know whether the few helpers down here could not be allowed to contribute further to the needs of the country? I should be much obliged if you would inform me whether there is any other

---

[22] *Staff Records*, Archives, Pontypool County School.

[23] James, T. C. and Davis, C. W. (1956). Schools of Chemistry in Great Britain and Ireland–XXVII The University College of Wales, Aberystwyth. *Journal of the Royal Institute of Chemistry* **80**: 568–574.

preparations we can make, as I, for one, am willing and eager to give up all ideas of holiday while there remains so much to be done.[24]

Presumably, her entreaty was ignored, for in a lengthy review of the mobilization of civilian chemists in the First World War, it was noted that: "And so, too, left without an offer from the [Reserved Occupations] Committee, the only woman of whom we have record, Margaret Turner, of the Chemical Laboratories at Aberystwyth."[25]

In January 1916, Turner was appointed Chemistry Mistress at Pontypool County School for Girls (see Footnote 22), being promoted to Senior Science Mistress in 1934. She retired in 1945 and died in 1961. In the obituary in the school magazine, *The Dragon*, it was commented: "She was an enthusiast in the teaching of her subject [chemistry] and introduced us to many interesting 'sidelines,' such as the crystal-growing competition in the Eistedfod."[26]

Phyllis McKie

Of all the women chemists, Phyllis Violet McKie[27] of the University College of Wales, Bangor, one of Orton's protégées, seems to have been the most productive during the war period. McKie was part of the team at Bangor producing paraldehyde.[28] In addition, she devised a new method for the preparation of the explosive tetranitromethane for the Ministry of Munitions. She also studied methods of preparation of saccharin and vanillin for war purposes.[29] Orton reported back to the War Committee:

> Miss McKie's investigations have been mainly concerned with a study of the methods of preparation of 'materials offensive and defensive' for

---

[24] Letter, M. K. Turner to War Committee, 31 August 1915, Royal Society Archives.

[25] MacLeod, R. (1993). The chemists go to war: The mobilization of civilian chemists and the British war effort, 1914–1918. *Annals of Science* **50**: 455–481.

[26] Wilks, F. E. (July 1961). Miss M. K. Turner. *The Dragon Magazine of the Pontypool County School for Girls* 6.

[27] *Student Records*, Archives, University of Wales, Bangor. E. W. Thomas, Archivist, is thanked for supplying a copy of this document.

[28] Letter, K. J. P. Orton to War Committee, 4 December 1915, Royal Society Archives.

[29] Certificates of Candidates for Election at the Ballot to be held at the Ordinary Scientific Meeting on Thursday, December 2nd (1920). *Journal of the Chemical Society* **117**: 92.

494 *Pioneering British Women Chemists: Their Lives and Contributions*

the Department of Explosives Supply. Her work had been particularly successful as in the two cases investigated, new and far superior methods of preparation have been discovered and examined (see Footnote 28).

McKie was born 18 July 1893 at Bangor, the daughter of William McKie, clerk at the Penryhyn Quarry Office and Edith McKie. She was educated at the County School for Girls, Bangor and entered University College of Wales, Bangor, in 1912. McKie completed her B.Sc. in 1916 and was awarded an M.Sc. by research on the basis of her war work which had resulted in 12 publications.

By 1920, McKie had moved to University College, London (UCL). Then in 1921, she was appointed Demonstrator at Bedford College, the same year that she was awarded a Ph.D. in chemistry. McKie moved to Lady Margaret Hall, Oxford, on a Research Fellowship for the 1925–1926 academic year.

In 1926, McKie was appointed Head of the Chemistry Department at Maria Grey Training College (see Chapter 10), where another Bangor alumna, Alice Smith, had been a Lecturer from 1914 until 1917 (see Chapter 7). In 1929, McKie made yet another move, this time to become a Lecturer in Chemistry at Westfield College, London. The biographer of Westfield College, Janet Sondheimer, noted:

> In 1929 when chemistry teaching started the only place suitable was the attic… however in 1935 the lecturer Dr. Phyllis McKie, was able to transfer her department to the hut, left vacant by the removal of botany to the new laboratory.[30]

When McKie left Westfield College in 1943, chemistry lapsed for several years. Her new appointment was Principal of St. Gabriel's Training College, a post she held until her retirement in 1956. McKie died in 1991 at Chichester, aged 97.

---

[30]Sondheimer, J. (1983). *Castle Adamant in Hampstead: A History of Westfield College 1882–1982*. Westfield College, University of London, London, p. 102.

Women Chemists and the First World War 495

Ruth King

Picric acid was another explosive used in the War, and from 1917 to 1919, Ruth King[31] was assigned as a Wartime Research Worker at Chiswick Laboratory, Department of Explosive Supply, Ministry of Munitions, to research the optimum conditions for its synthesis.

King was born on 13 May 1894 at Tottenham, daughter of Alfred Wilborn King, grocer, and Agnes Elizabeth Matthews, and was educated at Uxbridge County School. King graduated with a B.Sc. (Hons.) in Chemistry from East London College in 1915 and was granted a M.Sc. in 1918, like McKie, on the basis of her wartime research. The title of her Thesis was "Production of Picric Acid from the Sulphonic Acids of Phenol."

At the conclusion of the war, King was appointed Lecturer in Organic Chemistry at University College of South West England (later the University of Exeter). In addition, she was made Warden of Hope Hall, the women's residence.[32] Like other women chemists given academic positions, King was assigned a high proportion of the teaching duties. In fact, from 1919 to 1945, she was the only organic chemist in the department at Exeter.

King remained at Exeter until 1955 when she took early retirement to move to Canada to help care for her aged mother. Her mother, however, died while King was crossing the Atlantic, but, undaunted, she continued to Vancouver. She obtained a post there as Lecturer at the University of British Columbia, retiring in 1961. She died on 28 February 1988 at Vancouver, Canada, aged 93.

---

[31] We thank the following for information on King: A. Nye, Archivist, Queen Mary College, London; Personal communication, letter, K. Schofield, University of Exeter, 17 February 1998; Personal communication, letter, Erwin Wodarczak, Records Analyst/ Archivist, University Archives, University of British Columbia, Vancouver, Canada, 3 March 1998; and (1922). Certificates of Candidates for Election at the Ballot to be held at the Ordinary Scientific Meeting on December 15th, 1921. *Journal of the Chemical Society* **118**: 103.

[32] Britton, H. T. S. (1956). Schools of Chemistry in Great Britain and Ireland–XXVIII The University of Exeter. *Journal of the Royal Institute of Chemistry* **80**: 617–623.

Women Chemists and the War Gases

We mentioned that Grace Leitch of Armstrong College (see Chapter 6) and Helen Gilchrist of the University of St. Andrew's (see Chapter 7) had undertaken research on mustard gas. They were not the only woman chemists to work in this hazardous field. In addition to having pioneered the university production of pharmaceuticals, Martha Whiteley also performed research on lachrymatory war gases.

Whiteley recounted her wartime experiences in a speech at a luncheon at Royal Holloway College:

> Those were very exciting days, for our laboratories were requisitioned by the Ministry of Munitions who kept us busy analysing and reporting on small samples collected from the battlefields or from bombed areas at home. These included flares, explosives and poison gases. It was my privilege to examine the first sample of a new gas, used to such effect on the front that our troops had to evacuate from Armentières as it was reputed to cause blisters; it was called Mustard Gas. I naturally tested this property by applying a tiny smear to my arm and for nearly three months suffered great discomfort from the widespread open wound it caused in the bend of the elbow, and of which I still carry the scar. Incidentally, when shortly afterwards we were carrying on a research for a method of manufacturing the gas, my arm was always in requisition for the final test.[33]

Winifred Hurst (Mrs. Wright)

Winifred Grace Hurst[34] was one of Whiteley's team producing pharmaceuticals at IC; however, her particular claim to fame is her subsequent work on devising protection against the German war gases. She was born on 6 June 1891 in Clapham, daughter of Herbert Edward Hurst

---

[33] (1953). Speech by Whiteley at the Summer Luncheon of the Royal Holloway College Association Summer Luncheon, 1953. *Royal Holloway College Association, College Letter* 48–49, RHC AS/902/96.

[34] (a) Mercury Reporter. (1978). Top Natal Scientist Dies at 87. *Natal Mercury* Durban, South Africa. Dr. Michael Laing (late) is thanked for providing a copy of the article; and (b) Anon. (1979). News Review: Obituaries. *Chemistry in Britain* **15**: 5.

commercial traveller, tea trade, and Clara Cecilia Constant. Hurst was educated at King Edward VI School for Girls, Birmingham (KEVI), before moving to Streatham Hill High School in 1905. She entered Bedford College in 1909, obtaining a B.Sc. in 1912.

Hurst started her career at Berkhamsted School for Girls, teaching chemistry, physics, and mathematics from 1912 to 1913. She then took a position at Lichfield High School for 1 year, before moving to IC in 1914. Between 1914 and 1918, Hurst undertook research at IC on synthetic pharmaceuticals for the Navy, work for which she received the Diploma of Imperial College (DIC, similar to an M.Sc.).

Hurst married Albert John Wright, civil and electrical engineer in 1915 and subsequently had three daughters. In 1918, she was attached to the Anti-Gas Department of the Ministry of Munitions where she worked on gases for wartime defence. Together with the rest of the staff, Hurst had no alternative other than to test the gases on herself. Her superior, Colonel Harrison, died as a result of such experiments. Fortunately, she survived.

From 1919 until 1949, Hurst was affiliated to the Chemistry Department at Battersea Polytechnic. During the Second World War, when bombing prevented her reaching her laboratory, she set up her chemical equipment on the kitchen table of her home in Esher, Surrey. She commented: "after that, I kept one table for cooking and one for chemistry."[35]

In 1949, Hurst emigrated to South Africa to rejoin her husband, who was a consultant on high-tension electrical lines. He suggested that she keep a record of the wild flowers destroyed by the erection of the pylons, this new direction leading to her book on the *Wildflowers of Southern Africa*.[36] Hurst started research on South African plants at the University of Natal, receiving her Ph.D. in 1954 which resulted in newspaper headlines as a "degreed grandmother,"[35] of 62. The studies of plants led her back into the field of chemistry, this time an analysis of alkaloids

---

[35] Anon. (27 January 1964). Woman Ph.D. at 62 now adds authorship to achievements. *The Daily News* Durban, South Africa. Dr. Michael Laing (late) is thanked for providing a copy of this article.

[36] Wright, W.G. (1963). *The Wild Flowers of Southern Africa: A Ramblers' Pocket Guide*. Nelson, Natal, S.A.

obtainable from various plants. Hurst died on 8 September 1978 at Durban, South Africa, aged 87.

Women as Industrial Analytical Chemists

The majority of women chemists entered the analytical field. Women were probably more accepted for this work as repetitious and exacting analyses was considered compatible with women's talents. One such task was the determination of the purities of pharmaceuticals.[37]

Analytical Chemistry in the Explosives Industry

The vast majority of women in the wartime chemical industry were unskilled, simply working at specific synthesis tasks and following exact recipes to produce the enormous quantities of TNT, nitro-glycerine, ammonium nitrate, and ammonium perchlorate that were required by the explosives industry.[38] In fact, the proportion of women in chemical factories was as high as 88%. The work was hard, often very dangerous, and it led frequently to debilitating effects from the toxic chemicals.

TNT poisoning was among the worst health problems, the sufferers being called "canary girls" as a result of the yellow colour of their skin.[39] Many of these factories were enormous. The largest of all was the Gretna Explosives Factory in Scotland, a 10-mile-long complex for the synthesis of cordite, which was largely operated by women workers. The working conditions were extremely hazardous, some of the women developed lung damage from working with open vats of concentrated mineral acids, while others became comatose on a daily basis from ether vapour, having to be dragged outside until they recovered.[40]

---

[37] *Home Office Report: Substitution of Women in Non-munitions Factories during the War.* His Majesty's Stationery Office, London, 1919.

[38] Woollacott, A. (1994). *On Her Their Lives Depend: Munition Workers in the Great War.* University of California Press, Berkeley, California.

[39] Thom, D. (1998). *Nice Girls and Rude Girls: Women Workers in World War I.* I.B. Tauris Publishers, London, pp. 122–143.

[40] Rayner-Canham, M. and Rayner-Canham, G. (1996). The Gretna Garrison. *Chemistry in Britain* **32**: 37–41.

A crucial part of explosives manufacturing was the quality control or chemical analysis laboratory, and the factory at Gretna was no exception. A poem written by one of the women analysts has survived. This poem followed the style of Kipling's poem "If" which Claire Culleton has documented as being one of the favourite formats for women workers poems[41] ("C-in-C" was the abbreviation for chemist-in-charge while "Spee Gee" refers to measurement of specific gravity, the term for density measurements by comparison to that of water, and therefore unitless).

> *And if when Solvent figures turn out 'wonky,'*
> *And 'C-in-C's' most violent rage display,*
> *If you can simply hustle out your hankie,*
> *And wipe the glistening tear away;*
> *If you can fill each tiny minute*
> *With millions of titrations and 'Spee Gee's'*
> *Then you will truly be the limit and we will give you CROWDS*
> *of laurel trees.*[42]

Analytical Chemistry in the Iron and Steel Industry

A crucial task was to analyse samples of the iron and steel used in the production of, for example, ships and tanks. It is not surprising, then, that Sheffield, the centre of the British steel industry, became the focus for training of women in chemical analysis. In a response to Conway's questionnaire, Fred K. Knowles of the Faculty of Metallurgy at the University of Sheffield noted that when the War started, the men in the analytical and research laboratories of the industry were barred from joining the armed forces, due to the essential nature of their occupation.[43]

By the Autumn of 1916, however, the demand for "cannon fodder" became so great that even these individuals were drafted. With the

---

[41] Culleton, C.A. (1995). Working class women's service newspapers and the First World War. *Imperial War Museum Review* (10): 4–12.

[42] "D.M.E.," (1919). *Mossband Farewell Magazine*. 1916–1919, 43.

[43] Letter, F. K. Knowles to A. E. Conway, 21 August 1919, WWWC-IWM.

500   *Pioneering British Women Chemists: Their Lives and Contributions*

situation drastically changed, women now became essential. Knowles continued:

> In these laboratories there is a large amount of routine repetition work which can be carried out by semi-trained assistants, as distinct from chemists and physicists. To meet this emergency, special 1 month Intensive Courses for Women were started in the Metallurgical Department of the Faculty of Applied Science, University of Sheffield: the aim being to give a training in accurate weighing, filtration, titration, general manipulation and calculations. At the end of the Course those students who passed an Examination in the rapid determination of the elements:- carbon, silicon, manganese, sulphur, phosphorus, readily found remunerative employment. The Classes commenced on the 6th November 1916, and continued practically for 2 full University years: during this time 96 women students entered for this work (see Footnote 43).

Sheffield also provided specialized courses in other areas. In another case, six women were trained as analysts for coke oven laboratories.[44]

Some of the steel companies welcomed the women analysts. The Chief Supervisor of the Women's Welfare Department of Thos. Frith and Sons Ltd. of Sheffield, J. H. A. Turner, wrote to Conway to inform her that four women had worked in the research laboratory and 16 in the general laboratory at the company, primarily on the analysis of iron and steels and in microphotography. He added: "I understand that this Firm was one of the first (if not the first) in the Country to employ women at such work and the results have been quite satisfactory to the Heads of the two Laboratories."[45]

Not all companies were effusive in their praise of women chemists. William Rintoul of Nobel Explosives Company in Ayrshire reported to Conway: "Only routine work was entrusted to women. Our experience agrees with the generally accepted view that, in the main, women are unsuitable for the control and carrying out of research work unless under strict [male] supervision."[46]

---

[44] Letter, L. T. O'Shea to W. M. Gibbons, 21 August 1919, WWWC-IWM.

[45] Letter, J. H. A. Turner to A. E. Conway, 6 October 1919, WWWC-IWM.

[46] Letter, W. Rintoul to A. E. Conway, undated, WWWC-IWM.

Women chemists were employed as analysts at the National Physical Laboratory (NPL). Their work, too, was mainly in the analysis of iron and steel samples for the Admiralty. It is noticeable, though, that 10 of the 12 Junior Assistants at the NPL were female, while all of the Assistants, the Senior Assistants, and the Supervisor, were male.[47] The reports listing women's contributions, such as that of the NPL, provide only names and assigned duties. Two Glasgow women chemists were employed directly by the Admiralty as analytical chemists. Ada Hitchens (see Chapter 7) was assigned to work in the Admiralty Steel Analysis Laboratories while Ruth Pirret (see Chapter 7) became a wartime researcher on marine engine boiler corrosion for the British Admiralty.[48]

Analytical Chemistry in the Chemical Industry

Set up in 1915, the chemical factory of Chance & Hunt in Oldbury had a Ladies' Laboratory, staffed and run solely by women. In the *Edgbaston High School Magazine*, H. F. Fry described the facility:

> A works laboratory which is run entirely by women may, perhaps, be considered one of the most advanced products of present-day civilisation. ... The staff consists of eight girls of ages ranging from fifteen to eighteen known technically as "testers," and two University women with science degrees to organise the work and train and superintend the girls. The latter were, with one exception, entirely without chemical knowledge in the beginning ... However, they were most anxious to learn and, though the task of training them must have been arduous at first, they rapidly gained a working knowledge of beakers, burettes, Bunsens, and so forth. ... one did the acidity in the caustic liquors and other tests in connection with the alkali plant, another estimated the moisture and ash in the different kinds of fuel used on the works; the various ores from the copper process kept three girls busy, the zinc plant two, and so on.[49]

---

[47] *The National Physical Laboratory: Report for the Year 1917–1918*, HMSO, London, 1918.

[48] See, for example, Bengough, G. D., May, R., and Pirret, R. (2 November 1923). The cause of rapid corrosion of condenser tubes. *Engineering* 572–576.

[49] Fry, H. F. (1918). A works laboratory. *Laurel Leaves. Edgbaston High School Magazine* 33–34.

502 *Pioneering British Women Chemists: Their Lives and Contributions*

Women in Biochemistry

The biochemist Dorothy Jordan Lloyd, researcher with F. Gowland Hopkins at Cambridge (see Chapter 11), was also given a specific task. On the outbreak of war, the Medical Research Committee assigned her to study culture media for meningococcus, one of the anaerobic pathogens involved in trench diseases, and on causes and prevention of "ropiness" in bread.[50] Jordan Lloyd was one of several women with a background in biochemistry who were enlisted in the war effort.

Annie Homer

The expertise of Annie Homer[51] was considered so vital to the war effort that she was bought back from Canada. Born on 3 December 1882 at West Bromwich, daughter of Joseph Homer, tax collector, and Keziah Skidmore, Homer was educated at KEVI. She studied chemistry at Newnham from 1902 until 1905, and like so many other women, obtained a "mailboat" B.Sc. from Dublin (see Chapter 3).

From 1907 to 1910, Homer was an Assistant Lecturer and Demonstrator in Physical Science at Newnham, then Demonstrator in Chemistry from 1910 to 1914. She was also a signatory of the letter to *Chemical News* in 1909 (see Chapter 9). Between 1907 and 1913, she authored 13 papers on organic synthesis, many involving the Friedel–Crafts reaction. As a result of her research, Homer was awarded a D.Sc. (Dublin) in 1913. In 1914, Homer moved to Canada, becoming a Medical Research Fellow and Demonstrator in Biochemistry at the University of Toronto and an Assistant Chemist at the Dominion Experimental Farm, Ottawa.

---

[50] Bate-Smith, E. C. (1947). Obituary Notice: Dorothy Jordan Lloyd. *Biochemical Journal* **41**: 481–482.

[51] (a) White, A. B. (ed.), (1979). *Newnham College Register, 1871–1971*, Vol. I, 1871–1923. 2nd ed., Newnham College, Cambridge, p. 33; (b) (1953). Obituary: Annie Homer. *Journal of the Royal Institute of Chemistry* **77**: 369; (c) Creese, M. R. S. (1998). *Ladies in the Laboratory: American and British Women in Science 1800–1900*. Scarecrow Press, Lanham, Maryland, p. 279; and (d) Ogilvie, M. and Harvey, J. (eds.), (2000). *The Biographical Dictionary of Women in Science. Pioneering Lives from Ancient Times to the mid 20th Century*. Routledge, New York, pp. 613–614.

At the beginning of the First World War, Homer was promoted to Assistant Director of the Antitoxin Laboratories, Toronto, and it was because of her expertise that she was brought back to Britain for the special work of reorganising the commercial production of antitoxins to meet war demands. In fact, her methods for the manufacture of high-grade commercial antitoxins were adopted in many parts of the world. Her appointment was as assistant at the Lister Institute, though much of her research was accomplished at the Physiological Institute, University College, London. Despite completely changing her field from organic chemistry to biochemical toxicology, her publication rate never diminished, with 24 research papers appearing under her name between 1914 and 1920.

In the 1920s, Homer abandoned this part of her career, devoting her life instead to securing the development of oil and potash and other mineral resources in Palestine. Homer died on 1 January 1953.

Other War Work by Women Chemists

We have already discussed the war work of some other women chemists. May Leslie worked on the improvement of nitric acid synthesis (see Chapter 6), first at H. M. Factory, Litherland, Liverpool, then at H. M. Factory in Penrhyndeudraeth, North Wales.[52] Isabel Hadfield was consigned to research chemical problems relating to aeronautics (see Chapter 9). In addition, we know that towards the end of the War, Millicent Taylor was appointed to H. M. Factory, Oldbury, as a research chemist.[53] It seems likely that other explosives factories also employed women chemists, but that they were not documented.

Other aspects of the war effort involved women scientists. For example, the women science students at King's College for Women undertook research on the manufacture of optical and laboratory glass.[54] This was

---

[52]Rayner-Canham, M. and Rayner-Canham, G. (1993). A chemist of some repute. *Chemistry in Britain* **29**: 206–208.

[53]Baker, W. (1962). Millicent Taylor 1871–1960. *Proceedings of the Chemical Society* 94.

[54]Marsh, N. (1986). *The History of Queen Elizabeth College: One Hundred Years of Education in Kensington*. King's College, London, p. 125.

504 *Pioneering British Women Chemists: Their Lives and Contributions*

carried out on behalf of the Glass Research Committee for the Ministry of Munitions.

Lovelyn Eustice (Mrs. Bickerstaffe)

One of the women chemists with the widest government employment was Lovelyn Elaine Eustice.[55] Eustice was born on 6 August 1894 in Southampton, daughter of John Eustice, Professor of Engineering and Vice-Principal of University College, Southampton, and Evelyn Margaret Gay. She was educated privately and then went to Girton and to University College, Southampton, being granted a B.Sc. (London) in 1917.

During the war years, Eustice was first a Chemist at the Government Rolling Mills, Southampton, then moved to the Air Ministry in London in the same position in 1918. From 1919 to 1922, she was a Research Chemist for the Admiralty, Royal Naval College, Greenwich before becoming a Science Mistress at Dudley High School in 1923. The same year, she married chemist Robert Bickerstaffe, and they spent the years 1927–1934 living in New York. Eustice became Director of the Research Bureau, Encyclopedia Britannica Co. in 1936, a position she held to 1941. She died on 10 August 1943.

Hilda Judd

Another former colleague of Martha Whiteley at IC, Hilda Mary Judd[56] undertook wartime research for the Silk Association for which she was awarded a D.I.C. in Biochemistry in 1918. Judd was born on 26 June 1982 at Richmond, daughter of John Wesley Judd, Professor of Geology at the Royal College of Science (later part of IC), and Jeannie Frances Jeyes. She obtained her B.Sc. at the Royal College of Science (RCS) in 1904.

---

[55] Butler, K. T. and McMorran, H. I. (eds.), (1948). *Girton College Register, 1869–1946.* Girton College, Cambridge, p. 253.

[56] Certificates of Candidates for Election at the Ballot to be held at the Ordinary Scientific Meeting on Thursday, December 2nd (1920). *Journal of the Chemical Society* **117**: 90–91; and (1951). *Old Students and Staff of the Royal College of Science.* 6th edn., Royal College of Science, London, p. 37.

Judd stayed on at RCS as a Research Chemist until 1906, co-authoring a series of publications on the synthesis of nitrogen-containing organic compounds. During this period, she was the College's representative on an expedition to Africa, as the College magazine, *The Phoenix*, reported:

> It is with considerable relief that we welcome back to the College Miss Judd, our representative in the British Association Expedition to South Africa. Considering the dangers and difficulties through which she has passed, and the arduous work of attending all the meetings, she is looking remarkably well, and appears much better for the change.[57]

In 1906, Judd was appointed as Lecturer in Chemistry and Physics at Goldsmiths' College. She relinquished the position in 1916, returning to IC to undertake war work for the Silk Institute. Silk was a fibre of strategic importance in the War, particularly for the envelopes of observation balloons. With the War ended, Judd obtained a post as researcher for the Food Investigation Board of the Department for Scientific and Industrial Research (DSIR). Two publications resulted from her studies, one in 1919, the other in 1920. Judd died in December 1943 in Surrey.

Dorothea Hoffert (Mrs. Bedson)

Another researcher with the Food Investigation Board of DSIR was Dorothea Annie Hoffert.[58] Hoffert was born on 29 January 1893 at Ealing, daughter of Hermann Henry Hoffert, H.M. Inspector of Secondary and Technical Schools, and Annie Ward. She attended Croydon High School, then Manchester High School for Girls before entering Girton in 1910 to study chemistry. Hoffert passed the first part of the Natural Science Tripos in 1913, before transferring to Victoria University of Manchester, where she received a Diploma of Education in 1914.

---

[57] Anon. (1905–1906). College news from a lady correspondent. *The Phoenix* **18**: 18.
[58] Butler, K. T. and McMorran, H. I. (eds.), (1948). *Girton College Register, 1869–1946*. Girton College, Cambridge, p. 214; and (1922). Certificates of Candidates for Election at the Ballot to be held at the Ordinary Scientific Meeting on Thursday, June 15th. *Journal of the Chemical Society* **119**: 46.

506 *Pioneering British Women Chemists: Their Lives and Contributions*

In 1914, Hoffert was appointed a Junior Science Mistress at Bede School for Girls, Sunderland; then, in 1916, she was "requisitioned" for the war effort as a Research Worker under the Food Investigation Board of the DSIR at the City and Guilds of London Institute, Department of Technology. At the same time, she was engaged in research on dopes and varnishes for aeroplanes. She continued working in these roles after the War, receiving the silver medal in her final exam on Painters' Oils, Colours and Varnishes in 1920. That year, with a grant from DSIR, she returned to Cambridge to continue her chemistry studies.

From 1922 to 1926, Hoffert was a Research Assistant to Ida Smedley (see Chapter 9) at the Lister Institute. During that 4-year period, she co-authored four publications with Smedley on carbohydrate and fat metabolism and was awarded a Ph.D. (London) on the basis of her research. In addition, she published two papers under her own name, one in *Chemistry and Industry* on an empirical rule for substitution in benzene derivatives. It was at the Lister Institute that Hoffert met Samuel Phillips Bedson, another researcher at the Lister.[59] They were married in 1926 and they had three children, which ended her research career. Hoffert died in 1969 at Brighton, Sussex.

Traditional Roles for Women

There were still the traditional women's roles needing to be filled — even by chemistry schoolgirls. The chemistry students at Mary Erskine School, Edinburgh, were required to knit socks for soldiers. With each pair, they enclosed a note:

> Dear Somebody, I suppose you are up to your necks in H_2O, and I suppose your feet will show traces of moisture. Doubtless in the circumstances, you will welcome the SOK_2S which I enclose. Also the X_3 which go along with them to speed them on. Ever yours, ...[60]

---

[59] Downie, A. W. (1970). Samuel Phillips Bedson. 1886–1969. *Biographical Memoirs of Fellows of the Royal Society* **16**: 15–35.

[60] Paterson, N. (December 1915). Letter Writing. *Merchant Maiden Magazine* **8**(1): 33.

The bacterial biochemist, Marjory Stephenson (see Chapter 11), spent the War as a nurse.[61] She left her research position at UCL to join the British Red Cross in France and then Salonika. In Salonika, she was in charge of a nurses' convalescent home and also had responsibilities for invalid diets. She was mentioned in dispatches in 1917 and awarded an M.B.E. for her war work.

Jesse Slater (*Mrs. Baily*)

Another women scientist initially given nursing duties was Jesse Mable Wilkins Slater.[62] She was born on 24 February 1879 at Hampstead, daughter of John Slater, architect, and Mary Emily Wilkins. Slater matriculated from South Hampstead High School and then entered Bedford College. She then transferred to Newnham as a Gilchrist Scholar in 1899, excelling in both chemistry and physics, returning to Bedford to complete a B.Sc. (Hons.) in 1902.

Slater was invited to work with J. J. Thompson[63] at the Cavendish Laboratory in Cambridge, where she studied the decay products of thorium from 1903 to 1905. For her research with Thompson, she was awarded a London D.Sc. That year, Slater accepted a position as science teacher at KEVI, staying there until 1909, when she took up an offer of science teacher at CLC. In 1913, she returned to Newnham, this time as Assistant Lecturer in Physics and Chemistry, being promoted to Lecturer in 1914.

Slater obtained leave from Newnham to undertake war-related duties, and for the first 3 years, she was a part-time nurse. She was then called for full-time duty as a radiographer at British military hospitals in France and later held the rank of Officier de l'Instruction Publique with the French

---

[61] Robertson, M. (1949). Marjory Stephenson 1885–1948. *Obituary Notices Fellows of the Royal Society* **6**: 563–577.

[62] We thank Ann Phillips, Archivist, Newnham College, for biographical information on Slater.

[63] Rayleigh, Lord (1941). Joseph John Thomson. 1856–1940. *Obituary Notices of Fellows of the Royal Society* **3**: 586–609.

army.[64] At the end of the War, Slater resumed her position at Newnham where she stayed until 1926, marrying Walter Harold Baily that year and presumably ending her scientific career. Slater died on 25 December 1961 at Hampstead.

Commentary

With few exceptions, the end of the war resulted in the termination of employment for women chemists. The government closed the explosives factories, while the male chemists returned from their war duties and reoccupied their former faculty and research positions. The respondent to Agnes Conway from the Sheffield Steel Company of Thos. Frith noted:

> On the signing of the Armistice most of the women were replaced by returning soldiers, but two [of 16] in the General Laboratory have become so proficient that their services have been retained (see Footnote 45).

The women chemists with specialized training stood the best chance of survival, for example, according to Knowles' letter to Conway, the graduates of the metallurgical analysis course at Sheffield seemed to survive:

> That women have been an undoubted success in this branch of industry, is proved by the fact that notwithstanding so many of the men (who are now demobilized) have resumed duty, a large proportion of the women who desired to stay on have retained their positions to the present time (see Footnote 43).

As part of the final chapter, we will revisit the issue of the employment of women chemists in the post-First World War era.

---

[64]White, A. B. (ed.), (1979). *Newnham College Register, 1871–1971*, Vol. I, *1871–1923*. 2nd edn., Newnham College, Cambridge, pp. 11–12.

Chapter 16

Women Chemists and the Inter-War Period

Frances Hamer (14 October 1894–29 April 1980)

As we have documented in previous chapters, the story of women in chemistry up to the First World War had been one of tremendous advancement. The inter-War period brought an end to many of the hopes. In this chapter, we will document individual stories of success, but they must be seen through the lens of a changed world. The words of feminist journalist, Cicely Mary Hamilton, summed up the viewpoint of many women of the period.

> Today, in a good many quarters of the field, the battle we had thought won is going badly against us — we are retreating where once we

510   *Pioneering British Women Chemists: Their Lives and Contributions*

> advanced; in the eyes of certain modern statesmen women are not personalities — they are reproductive faculty personified.[1]

A vivid depiction of the change in attitudes was expressed through one of the many subversive novels for school girl readership by Angela Brazil. In *The Madcap of the School*, Miss Gibbs, the science teacher, was of the suffragette era, while her young students saw things in a very different light:

> She [Miss Gibbs] was determined … to turn her pupils out into the world, a little band of ardent thinkers, keen-witted, self-sacrificing, logical, anxious for the development of their sex, yearning for careers, in fact the vanguard of a new womanhood. … They listened to her impassioned addresses on women's suffrage without a spark of animation, and sat stolidly while she descanted upon the bad conditions of labour among munition girls, and the need for lady welfare workers. The fact was that her pupils did not care an atom about the position of their sex, a half-holiday was more to them than the vote and their own grievances loomed larger than those of factory hands.[2]

Chemistry Education for Girls

During the 1880s to 1910s, in academically focussed girls' schools, chemistry was seen as essential to enable their graduates to be admitted to university science degree programs. The girls' scientific education had to be at least as good as that of the boys if young women were to take their rightful place in the sciences at university.

The 1920s and 1930s saw this vision as becoming marginalised. Gillian Avery, in her history of girls' independent schools, commented on the inter-War decline in academic focus, citing Cheltenham Ladies' College (CLC) as an example: "Cheltenham, for instance, had started out with high intellectual ideals, but by the 1920s these had waned."[3]

---

[1] Hamilton, C. (1935). *Life Errant*. J. M. Dent, London, p. 251.

[2] Brazil, A. (1917). *The Madcap of the School*. Blackie & Sons, London, pp. 59–60.

[3] Avery, G. (1991). *The Best Type of Girl: A History of Girls' Independent Schools*. London: Deutsch, p. 12.

Of all the influences, a key role must have been played by the Hadow Report of 1923, or to give its full name, *Report of the Consultative Committee on Differentiation of the Curriculum for Boys and Girls Respectively in Secondary Schools*,[4] which enunciated a very different vision for girls' education. University was no longer the primary goal: equipping girls for marriage and motherhood was to be the new, very different focus.

The Hadow Report of 1923

By the very title of the report, "... Differentiation of the Curriculum for Girls and Boys ...", the outcome was predetermined. There were 21 people on the committee, of whom only 4 were women. Each woman was appointed to represent a specific constituency: women elementary teachers; girls' secondary schools; teachers' training colleges; and women at university.[5]

The committee laid great store upon the medical evidence of the time which "proved" that girls were fundamentally weaker than boys. For example, the report noted that the blood of adult males had higher haemoglobin content than that of females:

> The materially lessened amount of hæmoglobin in the woman's blood after puberty is significant; hæmoglobin is the agent of internal respiration, the oxygen carrier of the system; and oxygen is the great liberator of energy. It is therefore evident that the male is the better prepared for a more abundant liberation of energy with less exhaustion or fatigue.[6]

Following from this, it was concluded that a lesser quantity and rate of academic education was preferable for girls. Having given the "scientific"

---

[4] Anon. (1923). *Report of the Consultative Committee on Differentiation of the Curriculum for Boys and Girls Respectively in Secondary Schools.* H. M. Stationery Office, London.

[5] Harrop, S. (2000). Committee women: Women on the consultative committee of the Board of Education, 1900–1944. In Goodman, J. and Harrop, S. (eds.), *Women, Educational Policy-Making and Administration in England: Authoritative women since 1880.* Routledge, London, pp. 156–174.

[6] Note 4, Anon., p. 82.

512   *Pioneering British Women Chemists: Their Lives and Contributions*

basis for their contention of female physical inferiority, the writers of the report continued:

> It appears to be generally recognised that girls in general are not so strong physically as boys and are more highly strung and liable to nervous strain. Moreover, medical statistics seem to indicate that there is a higher percentage among girl pupils of cases of Anæmia, spinal curvature, defective eyesight, and minor physical defects. It should be added that these defects are sometimes caused and often accentuated by sedentary occupations such as needlework.[7]

The Commission then addressed the mental inferiority of women, including in their discourse that:

> It is significant from the psychological standpoint that up to the present, despite ample opportunities, no first class genius on the creative side ... has appeared among women ... Again, in science very few women have attained to the first rank ...[8]

Each subject area is discussed in the report. Under *Science*, the Commission recommended that biology, not chemistry or physics, was the most appropriate science for girls. In their explanation, they stated:

> A special reason is the comparative lack in girls of an attitude of scepticism and curiosity which gives the best approach to Natural Science. Girls have, however, an aptitude for the Biological Sciences, in which they are helped by their greater diligence and neatness; they excel in subjects which require descriptive powers and a capacity for comprehending elaborate classification.[9]

Changed Societal Atmosphere

The Committee Report reflected the view of many new teachers. As Felicity Hunt has commented:

---

[7] Note 4, Anon., p. 85.

[8] Note 4, Anon., p. 86.

[9] Note 4, Anon., p. 104.

In the 1920s and 1930s it was fashionable to accuse girls' secondary schools of neglecting the 'feminine' side of their pupils' development. The Victorian pioneers (and Miss Buss and Miss Beale were frequently cited on these occasions) were supposed to have adopted a model of 'liberal education' and in doing so had 'assimilated' the 'boys' curriculum' and ignored the needs of femininity in their schools. The result, said the accusers, was that girls' education was a 'slavish imitation' of boys' and by definition, therefore, inappropriate for girls.[10]

Even among the science teachers at the schools of the Girls' Public Day School Trust, formerly proponents of equal education for girls, the Hadow philosophy prevailed:

> Miss Esdaile urged that Biology should be a compulsory subject, especially as such a small percentage of girls went on to Universities. Everyone should know the natural laws governing plants and animals. ... There was no practical difficulty in keeping certain live things in schools, such as Bees, Ants' Nests, etc. Miss Haig Brown agreed that Biology was the best subject for girls not going to a University: ... Miss M. E. Lewis and Miss Cossey said that Biology developed thought along interesting lines, made girls healthy and natural, and fitted them for public health work and social life.[11]

Women at University in the Inter-War Period

Carol Dyhouse has succinctly identified this difference between modern historical perception and reality in the context of university education for women.

> Most of the historians who have considered the impact of women's admission to the universities between 1880 and 1939 have drawn a distinction between the early pioneers, who established a right to entry, and

---

[10]Hunt, F. (1987). Divided aims: The educational implications of opposing ideologies in girls' secondary schooling, 1850–1940. In Hunt, F. (ed.), *Lessons for Life: the Schooling of Girls and Women, 1850–1950*. Blackwell, Oxford, pp. 3–21.

[11]Anon. (1930). *Minutes of the Council and Committees, Reports of Examiners, &c. for 1930*. 4 July, Archives, Girls' Public Day School Trust, p. 136.

514 *Pioneering British Women Chemists: Their Lives and Contributions*

a later period of acceptance and integration. ... any tendency towards a narrative of steady progress makes it more difficult to account for the conflicts of the 1920s and 1930s, except in simple terms of backlash or stagnation.[12]

As the returning servicemen flooded back into the educational system, women at the co-educational universities went rapidly from confident majority to discrete minority. Ina Brooksbank, a student at St. Hugh's, Oxford in 1917, vividly remembered the resentment displayed by male students who returned to the university after the First World War and felt that a "regiment of women" had taken over "their" university. This attitude was shared by some of the dons:

> We went down that term expecting only the lifting of a few restrictions, but on our return we found a different world. The city was full of men, bicycles and motor bicycles, often ridden in carpet slippers. We went to our usual lecture at Magdalen and found the hall full of men, seated, and women standing or sitting on the floor. Professor Raleigh entered, saw the situation and postponed the lecture at once. ... Another [don] announced that he didn't lecture to women, so out they had to go.[13]

The academic women of the inter-War period lived in a different society from that of their mothers and, even more so, from that of their grand-mothers. Margaret Tuke, Principal of Bedford, articulated the difference, looking back from the vantage of 1928 at the bygone era:

> Not a few students of that time [the 1880s] worked ten to twelve hours a day and put into those hours an intensity of concentration difficult to realize in our own more easy-going times. ... upon her success, and that of her fellow-students, depended not their own reputation only but that of women as a whole.[14]

---

[12]Dyhouse, C. (1995). *No Distinction of Sex?: Women in British Universities, 1870–1939.* UCL Press, London, p. 189.

[13]Brooksbank, I. (1991). Bingles and bicycles. *Oxford Today* **3**(2): 35.

[14]Tuke, M. J. (January/June 1928). Women students in the universities: Fifty years ago and to-day. *Contemporary Review* **133**: 71–77.

Tuke noted that the loss of "singleness of purpose" had to be balanced by the floodgates having opened, giving university access to the "average woman." She added:

> To-day the majority of students enter university as a necessity — sometimes and unwelcome necessity — at the wish of their parent, or because they realise that in doing so they will have a better start in life (see Footnote 14).

As a percentage, women's participation at university declined during the inter-War period.[15] The decline had been noticed as early as 1933 by Doreen Whiteley.[16] Whiteley found that the percentage of women students at English universities had declined from 31% in 1924–1925 to 27% in 1930–1931. She attributed the decline in part to the difficulty of girls obtaining scholarships, and in part to families putting a lower priority on their daughters' education rather than their sons' at a time of economic recession. Carol Dyhouse has shown that the decline continued to 23% in 1934–1935 and to 22% in 1937–1938.[17]

At the University of Manchester, Mabel Tylecote reported that the decline was particularly apparent in the sciences, adding: "An increased number of women entered the chemistry department in the early post-war years, but declined again afterwards."[18] She also noted increased hostility towards the women students:

> Criticism of women students by no means abated and members of the Men's Union expressed themselves freely. ... Women were said to be seen at lectures taking down every word and to attend meetings which

---

[15] Rayner-Canham, M. and Rayner-Canham, G. (1996). Women in chemistry: Participation during the early twentieth century. *Journal of Chemical Education* **73**: 203–205.

[16] Whiteley, L. D. (1925). *The Poor Student and the University: A Report on the Scholarship System with Particular Reference to Awards Made by Local Educational Authorities*. London, pp. 23–25.

[17] Dyhouse, C. (2002). Going to university in England between the wars: Access and funding. *History of Education* **31**: 1–14.

[18] Tylecote, M (1941). *The Education of Women at Manchester University 1883 to 1933*. Manchester University Press, Manchester, pp. 120–121.

516   *Pioneering British Women Chemists: Their Lives and Contributions*

the men ignored. ... Antagonism was sharpened by the economic problems which had arisen (see Footnote 18).

Opposition to Women in Medicine and Pharmacy

During the closing decades of the 19th century, we have shown that medicine (see Chapter 5) and pharmacy (see Chapter 13) became career options for young women who had taken chemistry at school. With the commencement of war, enrolment in the male-only London medical schools plummeted and admission of women became a necessity for the survival of many of them.[19] At Charing Cross Hospital, the historian, R. J. Minney commented: "They came swarming in. Within a few weeks the male students were mere dots amid the fluttering skirts and flowing hair in the lecture theatre."[20]

This victory of the opening of many of the men-only medical schools to women in the 1910s proved short-lived. One-by-one the bar to women was reinstated in the 1920s. For example, in the case of Charing Cross Hospital, Minney noted that with the end of the War and the return of the soldiers plus new male school graduates: "The situation had begun to adjust itself and not long afterwards women were again barred from the School. It was not until 1948 on the University's insistence, that they were readmitted."[21] By the end of the 1920s, the London School of Medicine for Women (LSMW), now-named the London (Royal Free Hospital) School of Medicine was the sole institution in the capital to freely admit women to medical studies.

Seizing the example of the medical schools, unsuccessful attempts were made to similarly reverse the inroads of women into pharmacy. For example, in 1922, the *Pharmaceutical Journal* contained a letter from 'A Pharmacist':

---

[19]Dyhouse, C. (1998). Women students and the London medical schools, 1914–1939: The anatomy of a masculine culture. *Gender and History* **10**(1): 110–132.

[20]Minney, R. J. (1967). *The Two Pillars of Charing Cross: The Story of a Famous Hospital*. Cassell, London, p. 153.

[21]Note 20, Minney, p. 154.

It has recently been reported to the Press that in future, the London Hospital intends to restrict its students to men only. It occurs to me that this may be a good lead to the Pharmaceutical Society and also to Colleges of Pharmacy generally. While there are so many male chemists and chemists' assistants unemployed at the present time, it seems the limit of absurdity to flood the business, or profession, with a motley horde of untrained and incompetent surplus females. Is it not practicable to eliminate this undesirable element all together?[22]

Women's Employment in the Inter-War Period

The mood had changed towards women workers. From saviours of the nation, they were now job-stealers from men, the breadwinners.[23] The women who had entered employment during the War as heroines — particularly the munitions workers — were vilified in the post-War era, as Elizabeth Roberts commented:

> They [women] were often regarded with open hostility by men who had realised for the first time that women were fully capable of carrying out jobs previously perceived as men's, thus presenting a real challenge.[24]

The feminist author, Irene Clephane, writing in 1935, described how public sentiment towards working women had changed dramatically and rapidly with the return of the job-hungry ex-servicemen:

> From being the saviours of the nation, women in employment were degraded in the public press to the position of ruthless self-seekers depriving men and their dependents of a livelihood. The woman who had no one to support her, the woman who herself had dependents, the

---

[22] "Pharmacist, A." (11 March 1922). Letters to the Editor: Women in pharmacy. *Pharmaceutical Journal* **108**: 208.

[23] Beddoe, D. (1989). *Women Between the Wars, 1918–1939: Back to Home and Duty.* Pandora, London, p. 82.

[24] Roberts, E. (1988). *Women's Work 1840–1940.* Macmillan Education, London, p. 67.

518   *Pioneering British Women Chemists: Their Lives and Contributions*

woman who had no necessity, save that of the urge to personal independence and integrity, to earn: all of them became, in many people's minds, objects of opprobrium.[25]

It was not just a change in attitude to the employment of women: it was as if the "New Woman" of the 1880s had never been, as Susan Kent commented: "The post-war backlash against feminism extended beyond the question of women's employment: a *Kinder, Küche, Kirche* ideology stressing traditional femininity and motherhood permeated British culture."[26] Adrian Bingham has pointed out that the view of young women in the inter-War popular press was actually more complex and nuanced. He contended though it was certainly true that there was "a fear and dislike of the 'surplus' woman, who threatened the basis of political and social stability"[27] at the same time the "modern young woman" was being encouraged to grasp new opportunities.

Marriage — "matrimonial mortality" — marked the end of a woman's employment, especially in the teaching profession. Pat Thane described the problem in the context of Girton women graduates:

This 'marriage bar', as it was known, spread especially fast during the inter-war years … Even childless women were not permitted to work, though widows were permitted to return to employment.… some women evaded it. One Girtonian schoolteacher described how she was able to hide the fact that she was married because she lived in a different part of London from the school where she worked. Another commented on the surprising number of her colleagues who appeared to have married over-night when the bar was abolished in London. In smaller towns it was harder to hide. The bar more or less disappeared, and then was formally abolished in most occupations in World War Two due to a shortage of labour.[28]

---

[25] Clephane, I. (1935). *Towards Sex Freedom: A History of the Women's Movement*. John Lane at the Bodley Head, London, pp. 200–201.

[26] Kent, S. K. (1988). The politics of sexual difference: World War I and the demise of British feminism. *Journal of British Studies* **27**: 232–253.

[27] Bingham, A. (2004). *Gender, Modernity, and the Popular Press in Inter-War Britain*, Clarendon Press, Oxford, p. 7.

[28] Thane, P. (2004). The careers of female graduates of Cambridge University, 1920s–1970s. In Mitch, D, Brown, J., and van Leeuwen, M. H. D. (eds.), *Origins of the Modern Career*.

But marriage and family was the natural goal of a woman, not that of being a spinster teacher, proclaimed an Editorial of 1932 in the *Journal of Education*:

> How often do we hear it said, "We had high hopes of her — but she married." Nevertheless women are beginning to see how empty are the Pankhurstian victories. They are weary of ambition and office life in continually inferior positions; what they really need is marriage, a home, and children. And that is right, for it is their natural function.[29]

Opportunities for Employment

There were more young women seeking careers in the inter-War period — and more than the employment opportunities available. The Headmistress at Howell's School, Llandaff, addressed the issue in her school report of 1926 in *Howell's School Magazine*. Interestingly, one of the new careers she identified later in the report was that of "Chemical Research":

> It is increasingly difficult to find Careers for our girls. One notable difference between the present times and the days of my youth is the fact that every girl now sets forth to have a Career, whereas formerly those girls who took up Medicine or Nursing or Teaching were the exceptions; as a result, all the professions open to women are full.... There are, however, new avenues of work for the girl who has a strong and willing heart, a trained mind and a desire to follow the ancient precept "whatsoever thy hand findeth to do, do it with all thy might."[30]

In 1927, the Institute of Chemistry issued a book, *The Profession of Chemistry*, which included a chapter on Women in Professional Chemistry. The author, Richard Pilcher, reviewed the options open to women chemists, noting that, though in theory any position available to men were

---

Ashgate, Aldershot, UK, p. 214.

[29] Anon. (1932). Education for women. *Journal of Education* **64**: 560–561.

[30] Trotter, E. (1926). Miss Trotter's letter. *Howell's School Magazine* 6–7.

equally available to women, women sometimes were left taking any vacancy that they could find:

> Some turn to secretarial work or to scientific journalism, but the majority take up teaching, for which they are often particularly suited.... For the trained woman chemist who has no vocation for teaching, appointments are occasionally offered by the Research Associations ... In industry, and particularly in those industries where a large number of women are employed — such as food, margarine and jam factories — women are not infrequently engaged in analytical and research laboratories; but in other industries prospects for women are limited, because the higher positions call for experience in dealing with workmen, and moreover employers realize that, after a year or two, when the experience which a woman has gained in her work is becoming most valuable, her professional career may be terminated by marriage.[31]

The Woman "Super-Chemist" as Negative Role Model

Some women chemists stood out as exemplars, but, in a way, they served as a discouragement. Margaret Rossiter described the effect of Marie Curie's visit to the United States as raising the bar for women chemists to unattainable levels:

> Before long most professors and department chairmen were ... expecting that every female aspirant for a faculty position must be a budding Marie Curie. They routinely compared American women scientists of all ages to Curie, and finding them wanting, justified not hiring them on the unreasonable grounds that they were not as good as she, twice a Nobel Laureate![32]

The same phenomenon occurred in Britain. Ida Smedley (see Chapter 9) was used as a benchmark for women applicants for the Position of Reader in Chemistry at King's College of Household and Social Science

---

[31] Pilcher, R. (1927). *The Profession of Chemistry*. Institute of Chemistry of Great Britain and Ireland, London, p. 92.

[32] Rossiter, M. (1982). *Women Scientists in America: Struggles and Strategies to 1940*. Johns Hopkins Press, Baltimore, Maryland, p. 127.

Women Chemists and the Inter-War Period 521

(see Chapter 8): "... it would be of great value to the Department to secure the services of a woman with the high scientific standing and personality of Dr. Ida [Smedley] Maclean."[33]

Smedley was not the only one to be chosen as the expectation for a woman chemist or biochemist. A 1929 article in the *Journal of Careers*[34] held up Martha Whiteley (see Chapter 4) as a role model, while an article in the same journal in 1938[35] extolled Ida Smedley (see Chapter 9), Marjory Stephenson (see Chapter 11), Kathleen Coward (see the respective section), and particularly Dorothy Jordan Lloyd (see Chapter 11) as the heights of careers to which women chemists and biochemists could aspire — but only those who were exceptional.

Women Chemists in Teaching Careers

In the inter-War years, teaching was the most common career for women science graduates.[36] Dyhouse commented: "Teaching, then, be it a vocation, the only realistic option or a last resort, remained the fate of the majority of women graduates in this period [pre-1939]."[37] There was a second reason that women became teachers: from Dyhouse's statistics, about 30% of the women attending university in the inter-War period were only able to afford to do so on grants for teacher training. This commitment was spelled out very clearly at some institutions, as Kathleen Uzzell recalled:

> When we were first at University we were called into a room where we were told we had to swear an oath to teach for five years, but it was

---

[33] Marsh, N. (1986). *The History of Queen Elizabeth College: One Hundred Years of Education in Kensington*. King's College, London, p. 124.

[34] Anon. (1929). Prospects for Women in Science. *Journal of Careers* **9**: 18–19.

[35] Anon. (1938). Prospects of employment for women science graduates: Part I of a Survey of opportunities in government, industrial and other research laboratories. *Journal of Careers* **17**: 88–93.

[36] Anon. (June 1936). Demand for science graduates maintained: Women graduates still mainly teachers. *Journal of Careers* **15**: 37–38.

[37] Dyhouse, C. (1997). Signing the pledge? Women's investment in university education and teacher training before 1939. *History of Education* **26**: 207–223.

522 *Pioneering British Women Chemists: Their Lives and Contributions*

pointed out it was a 'moral not a legal' oath. ... The promise to teach for five years meant a promise not to marry as there were no married female teachers except war widows.[38]

Nearly all of the women school teachers remained single throughout their careers. Some women academics saw themselves following the equivalent to a religious vocation as Elsie Phare recounted about her German Tutor at Newnham: "Miss Paues told me that she was married to scholarship (she wore a wedding ring), with the implication that that was the state to be wished."[39] Though scholarly celibacy might have suited a few, E. H. Neville in an article in 1933 titled "This Misdemeanor of Marriage" abhorred the fact that for women to have an academic career, they were required to be: "enforced celibates, predestined spinsters, and women cunning enough to maintain complete secrecy in their sexual relations."[40]

Hazel Reason

One of the later women chemists who entered into school teaching was Hazel Alden Reason.[41] Born on 24 February 1901 at Friern Barnet, her parents were William Reason, Congregational minister, and Kate Alden. Both of her parents were university graduates. Reason attended Milton Mount College, Gravesend and then entered Bedford College, graduating in 1924 with a B.Sc. in chemistry.[42] She then obtained a position as Senior Science Mistress at the County School for Girls, Guildford. During her free time, Reason worked towards an M.Sc. on the history of science, titled "Historical survey of theories concerning the elementary nature of matter, with special reference to the four element theory and its overthrow by modern scientific work," which she completed in 1936.

---

[38] Uzzell, K. cited in: Note 37, Dyhouse, p. 217.

[39] Phare, E. (1982). From Devon to Cambridge. *Cambridge Review* 149.

[40] Neville, E. H. (1933). This misdemeanor of marriage. *Universities Review* **6**: 5–8.

[41] Wikipedia, (13 June 2017). Hazel Alden Reason. On-line (accessed 17 January 2019).

[42] Forms of Recommendation for Fellowship. The Ballot will be Held at the Ordinary Scientific Meeting on Thursday, February 20th (1936). *Proceedings of the Chemical Society* 7.

Continuing her interest in the history of science, Reason authored the book *The Road to Modern Science*,[43] which was first published in 1936. A second edition of the book appeared in 1939 (reprinted four times) and indicating an amazing longevity. She authored a third, revised edition in 1959. In the Foreword, she commented: "The primary object in writing this book was to present the story of scientific discovery in a form which would appeal to intelligent boys and girls." She added that she did not approve of the "great scientist" approach to the history of science, rather her book covered "... the broad view of scientific discovery."[43] Reason was still teaching at Guildford in 1939. Living for much of her life with her sister, Joyce Reason, Hazel Reason died on 13 April 1976.

Peggy Lunam (*Mrs. Edge*)

Technical Colleges also employed women chemists as lecturers. Peggy Lunam[44] was born on 20 May 1909 at Ormsby, North Yorkshire, to Nevison John Lunam, structural engineer, and Emily Cartwright. She attended Armstrong College where she was Student Treasurer of the Bedson [Chemistry] Club from 1929 to 1930. Lunam earned a B.Sc. (Durham) in 1930 and an M.Sc. (Durham) in 1932. In 1935 she was a part-time Lecturer and Researcher in the Department of Chemistry at Constantine College, Middlesbrough.

Constantine Technical College[45] had been founded in 1930 to support Middlesbrough's engineering, bridge, and shipbuilding industries. At first, Constantine College concentrated on metallurgy, engineering, and chemistry, offering courses leading to University of London degrees, only later broadening its offerings to the Arts.

---

[43] Reason, H. A. (1936). *The Road to Modern Science*. 1st edn., G. Bell and Sons Ltd., London; (1940), 2nd edn.; (1959), 3rd edn.

[44] *Student Records*, Archives, University of Newcastle-upon-Tyne; Armstrong College Old Student Association. *Year Book*, 1933, 1935; Kings Old Student Association. *Year Book*, 1938; *Register*, Royal Institute of Chemistry, 1948; and Anon. (1983). Personal News: Deaths. *Chemistry in Britain* **19**: 990.

[45] Leonard, J. W. (1981). *Constantine College*. Teeside Polytechnic, Middlesbrough.

524 Pioneering British Women Chemists: Their Lives and Contributions

By 1938, Lunam had been promoted to Lecturer in Chemistry. In 1942, she married Herbert Allan Edge of the Research and Development Department, Imperial Chemical Industries (ICI) Agricultural Division, Billingham. Lunam was noted as still being Lecturer at Constantine College in 1948, but later became Head of the Chemistry Department at Kirby Grammar School. She died in 1983 at Cleveland, Yorkshire.

Elfreida Cornish (Mrs. Venn, Mrs. Mattick)

For women chemists, there seemed to be mobility between the different types of academic institutions as is illustrated by Elfreida (Freida) Constance Victoria Cornish[46] who taught at a secondary school, a teachers' training college, and finally became a researcher at a university.

Born in Somerset on 11 September 1887, she was the daughter of Thomas James Cornish and Louisa Ann Allard. Cornish attended Colston's Girls' School, Bristol, from 1900 to 1904. After passing the London University matriculation examination, she attended Merchant Venturers' Technical College from 1904 until 1909, then taught Chemistry at Colston's while completing her B.Sc. at the University of Bristol from 1909 to 1910. News of her degree success — and of her subsequent activities — appeared in *Colston's Girls' School Magazine*: "We were very proud that one of our Old Girls, Freida Cornish (B.Sc. Hons.), was the very first woman graduate to be presented at the first 'Degree Day' of our new University."[47]

On graduation, Cornish was appointed Lecturer in Science at Fishponds Training College.[48] Located at Fishponds, Bristol, and founded in 1853 as the Gloucester and Bristol Diocesan Training College for School Mistresses, the college graduated qualified teachers for the Church of England elementary schools.

---

[46] Kay, H. D. (1944). Obituary Notice: Elfreida Constance Victoria Mattick (1887–1943). *Biochemistry Journal* **38**: 1.

[47] Anon. (1910). President's notes. *Colston's Girls' School Magazine* **9**(3): 54.

[48] Anon. (1910). Honours gained by old pupils at the school. *Colston's Girls' School Magazine* **9**(3): 30.

Concurrently with her teaching duties at Fishponds, Cornish undertook research at Bristol with J. W. McBain. As a result of her studies on the colloidal properties of soap solutions, she was awarded an M.Sc., the first woman to gain an M.Sc. from the Bristol University.[49]

In 1913, Cornish received a Board of Education Agricultural Scholarship, tenable at the newly formed Institute of Research in Dairying at University College, Reading. This was the first Ministry of Agriculture scholarship to be awarded to a woman (see Footnote 46). However, like that of so many other women chemists, her career was temporarily suspended by the First World War, as the school magazine described:

> Elfreida Cornish MSc who has been holding a Research Scholarship at University College, Reading, has been allowed to postpone her Scholarship work for a time in order that she may devote her time to work in the connection with the production of munitions.[50]

Cornish married Bertram Joseph Venn, Assistant Examiner in H. M. Patent Office. At that date, he was in the Army, stationed at Bristol. Sadly, the following month he died, probably as a result of war injuries. In 1920, Cornish married again, this time to Alexander Torovil Robert Mattick,[51] bacteriologist, and later Head of Bacteriology at the Institute.

The most crucial research undertaken by Cornish was on the cause of discolouration of Stilton cheese, as one of her obituarists, H. D. Kay described:

> Her study of this problem led her almost immediately into the fundamental question of the mode of degradation of proteins by micro-organisms.... and she spent some months in Hopkin's laboratory in Cambridge getting to closer grips with the biochemical changes involved (see Footnote 46).

Her research resulted in the award of a Ph.D. by the University of Bristol in 1923. Cornish continued her research on biochemical problems in the dairy industry until her death on 19 July 1943.

---

[49] Anon. (1912). News of old friends. *Colston's Girls' School Magazine* **11**(3): 7.

[50] Anon. (Summer 1915). News of old girls. *Colston's Girls' School Magazine* **14**: 11.

[51] Scott Blair, G. W. (1943). Obituary: Dr. E. C. V. Mattick. *Nature* **152**: 183.

Women in Chemical Industry

A reply from Dorothy Adams to a questionnaire on Women's War work (see Chapter 15) pointed to bleak post-First World War opportunities for women chemists in industry:

> With regard to the prospects of scientifically trained women after the [First World] war my experience has led me to the conclusion that there will be practically no scope for them in industry. There is, and will continue to be for some time, a far larger supply of male Chemists than will be needed. Under such circumstances women with the same qualifications will stand the poorest chances of employment. As teachers and lecturers there is still some demand for such women, but in industry there is next to none. I have been led to this conclusion by my experience in endeavouring to obtain a fresh post myself.[52]

Employability of Women Chemists

The author of the series of articles on the prospects for employment for women science graduates in the *Journal of Careers* was blunt about potential marriage being a significant deterrent to hiring women in industry:

> This question of marriage is undeniably a deterring factor in the employment of women scientists in industry. Firm after firm, among the large number which the Journal of Careers has consulted, raises it as an objection. Even a woman who did brilliant work for some years, of a quality which is still remembered by men colleagues in terms of highest praise, apparently closed the door to other women in that particular firm, for it is recorded "but she left to get married and we haven't employed a woman since."[53]

A second reason for not hiring women, or restricting them to routine work, was the claim that women did not have a mental aptitude for

---

[52]Letter, D. Adams to A. E. Conway, 24 December 1918, Women's War Work Collection, Imperial War Museum, London.

[53]Anon. (1938). Prospects of employment for women science graduates: Part III, Industrial research laboratories, *Journal of Careers* **17**: 289–296.

research or that imaginative women were far rarer than imaginative men. The same author also commented on this perspective:

> That women lack the research type of mind and are therefore not suited to industrial research is a generalisation often advanced by firms which do not employ them. It is also advanced by firms which do employ them and which therefore restrict them to analytical work and to routine work as technicians.[54]

The advantage of men was that they would "take their job home with them"[54] and that it was during evenings and nights at home that inspiration would strike, whereas women, when leaving work, would have their minds stray onto other matters.

Women had one employment advantage: they were inexpensive to hire. It was common for the salary paid to a woman to be much less than that paid to a man for the same position (see Culhane's experience). In his opening remark, the author recounted one "typical" experience:

> Some months ago a leading firm telephoned the Secretary of one of the Colleges of London University and asked her to recommend an experienced woman chemist. "We have an immediate vacancy," said the voice at the other end of the telephone. "We should like to appoint a man, but we are only offering £150 a year, and we'll never get a man for that".[55]

In an interview in the *Journal of Careers* on the prospects for women in science in 1929, Jordan Lloyd made clear her beliefs that only the very best should consider an industrial career. She also described the perceived shortcomings of her own gender:

> Again, there are a number of industrial chemistry posts for which second or third rate ability is sufficient, and generally speaking a man of second or third-grade is to be preferred to a woman of that mental calibre, because he can be used for a wider range of duties and he is not usually,

---

[54]Note 53, Anon., p. 291.
[55]Note 35, Anon., p. 88.

528 *Pioneering British Women Chemists: Their Lives and Contributions*

though he may be, quite as inert mentally. Women are sometimes taken on for routine posts because they can be offered a lower salary than men and because they sometimes show a placid contentment in routine posts and do not crave for responsibility or any duties beyond those for which they are specifically appointed.[56]

Nevertheless, despite the gloomy forecasts, Sally Horrocks has shown that, during the inter-War period, many women chemists did find employment in industry, particularly the food, pharmaceutical, cosmetics, textiles, and photographic industries.[57] We will conclude this section with one case study, the life of Kathleen Culhane. For so many of the forgotten women chemists, scanty information remains on their life and work, but for Culhane, we have a rich narrative which epitomises the struggle of women seeking an industrial chemistry career during the inter-War period.

Kathleen Culhane (Mrs. Lathbury)

Kathleen Culhane[58] was born on 14 January 1900 at Hastings, daughter of Frederick William Slater Culhane, medical doctor, and Lucie "Minnie" Dann. Her father insisted on equal opportunity for his daughters and his sons, and, as a result, he became Culhane's idol. She attended Hastings and St. Leonards College, Sussex, where the only science at the time was botany. Culhane then attended the Hastings School of Science, before entering Royal Holloway College (RHC) in 1918.[59] It was at RHC that she discovered that chemistry was her real interest, and she graduated in 1922 with a B.Sc. (Hons.) in chemistry.

Wanting to enter the chemical industry, Culhane was extremely frustrated that employers would not take an attractive young woman seriously

---

[56] Jordan Lloyd, D. cited in: Anon. (1929). Prospects for Women in Science. *Journal of Careers* **9**: 18–20.

[57] Horrocks, S. M. (2000). A promising pioneer profession? Women in industrial chemistry in inter-war Britain. *British Journal for the History of Science* **33**: 351–367.

[58] Bramley, R. (unpublished manuscript). Mrs. Kathleen Lathbury B.Sc., C.Chem., F.R.S.C.

[59] *Student records*, Archives, Royal Holloway College.

for chemist positions.[60] In fact, she was only considered for interviews when she signed her applications as "K. Culhane" rather than "Kathleen Culhane." However, once her gender became apparent at the interviews, she failed to obtain any of the positions (the ruse was more successful for her chemist daughter in the 1950s). Marriage was scarcely a survival option, for as we mentioned in Chapter 15, a significant proportion of middle-class single males of her generation had been killed in the War. Finally, she obtained work as a school teacher and later, a private tutor.

Joining the Institute of Chemistry proved to be the turning point in her fortunes. Through the Institute, Culhane met John R. Marrack of the Hale Clinical Laboratory of the London Hospital. Marrack allowed her to gain experience of medicinal chemistry by permitting her to do emergency blood sugar determinations in her free time, but without pay. After 2 years of combining teaching with unpaid analytical work, Culhane obtained an industrial chemistry position with Neocellon, Wandsworth, a manufacturer of lacquers and enamels. However, her delight was diminished after being told by the company that the only reason for hiring her was that they could not afford the salary of a male chemist.

The completion of a study of enamel coatings for light bulbs coincided with an offer of a job back at the clinical laboratory, this time as a paid chemical advisor and insulin tester. Not long after, Culhane accepted a position in the physiology department of the large chemical and pharmaceutical company, British Drug Houses (BDH). However, as time progressed, her initial enthusiasm waned:

> I gradually discovered that it was not the intention to employ me as a chemist but as a woman chemist ... I was expected to do all the boring, routine jobs ... while anything interesting was handed to one of the men ... The routine work increased enormously in quantity and I took pride in perfecting my technique ... thinking it must surely win promotion that way. This did not materialize so, by superhuman efforts and late work, I got some research done which was successful and I was allowed to publish it ... The problems I worked on were of my own finding ... I managed to avoid being disliked and was merely regarded as eccentric.[61]

---

[60]Bramley, R. (1991). Kathleen Culhane Lathbury. *Chemistry in Britain* **27**: 428–431.

[61]Cited in Note 60, Bramley, p. 428.

530 *Pioneering British Women Chemists: Their Lives and Contributions*

As well, the senior staff lunch room was male-only and Culhane recalled how she had to eat her lunch with the women cleaners and clerks (see Footnote 61).

At the time, there were no chemical tests for insulin, and Culhane significantly improved the physiological testing procedures. Four researchers were chosen by the League of Nations Health Organization Committee, Culhane being one (though against the wishes of some committee members), to compare independently the physiological activity of amorphous and crystalline insulin. Her results had significant differences from those of the other three researchers, and the committee concluded that hers must have been in error. Culhane was asked to withdraw them, but this she refused to do, convinced of their correctness. Only later was it shown that her results were indeed the more accurate (see Footnote 60).

Culhane commenced a study of vitamins in 1933, though she continued with her work on insulin as well. As part of her diverse research studies, she gave a presentation on the need to standardize products containing added vitamins, arguing particularly for enhanced levels of vitamins in margarine. A British newspaper reported on the meeting, describing her as "Miss Kitty Culhane, the Girl Pied Piper of Science" as Culhane had mentioned the use of mice in vitamin research, and the newspaper added how an "abstruse lecture on vitamins" had been delivered by "a pretty girl with blue eyes and bobbed hair."[62]

That same year, Culhane married Major G. P. Lathbury. Having mentioned her intent to marry to her supervisor, she was amazed that the directors had to give special approval for the employment of a woman after marriage. The approval was granted in her case due to the importance of her work. Culhane resigned her position as senior chemist in 1935 due to pregnancy. Her inexperienced male successor was given a higher starting salary than what she had been earning at the time of her resignation.[62]

With the arrival of war in 1939, Culhane wanted to contribute to the war effort, having sent her small daughter to the country. As she later wrote: "Although it was often publicly stated that industry was short of scientists the Appointments Board were unable to tell me of a single

---

[62] Note 60, Bramley, p. 430.

opening for which a woman would be considered" (see Footnote 62). Persistence finally resulted in a position as an assistant wages clerk, where the senior clerk offered to teach her percentage calculations.

After more badgering of officials, Culhane was appointed manager of a statistical quality control department at a Royal Ordnance Factory, and on the basis of her work, she was made a Fellow of the Royal Statistical Society in 1943. Because salaries determined travel status, earning less than half the salary paid to males in the same position had a secondary effect. That was, on train journeys to London, her male colleagues travelled first class, while Culhane had to sit alone in a third-class compartment. Amazingly, she was not embittered by her experiences, instead regarding them as a source of amusement.

After the war, Culhane retired from science and took up a second career as an artist, becoming a member of the Haslemere and Farnham Art Society and producing paintings that were included in professional exhibitions. It was of great joy to her that her daughter and one of her grandsons became graduate chemists. Culhane died on 9 May 1993 at Hindhead, Surrey, aged 93.

Women Chemists in the Food Industry

A significant number of women chemists found employment in the food industry, as Horrocks has reported:

> The food industry offered the largest proportion of posts, and in Lyons provided the firm employing the greatest number of women chemists. Many of them were graduates of King's College of Household and Social Science (KCHSS). Indeed, eighty-five KCHSS graduates were employed by eighteen different food firms and four analytical chemists' practices during the period 1910 to 1949. Over half of these — forty-four — worked for Lyons. Other food manufacturers known to have employed women chemists included Glaxo, United Dairies, Chivers, CWS, Lever Brothers, Fullers, Peek Frean, Robertson's, Schweppes and Vitamins Ltd.[63]

---

[63] Note 57, Horrocks, p. 356.

532 *Pioneering British Women Chemists: Their Lives and Contributions*

We have chosen four individuals in this category: Ethel Beeching (Mars Confections Ltd.), Mamie Olliver (Chivers and Co.), Elizabeth Adams (Horlicks Ltd.), and Enid Bradford (Marmite Food Extract Company, Ltd. and J. Lyons & Co. Ltd.).

Ethel Beeching

Born on 29 June 1900 in Islington, London, Ethel Irene Beeching[64] was the daughter of Charles Lionel Thomas Beeching, Secretary of the Institute of Certified Grocers, and Annie Louise Beeching. She was educated at James Allens' School for Girls, Dulwich, and then studied at Bedford, obtaining a B.Sc. in chemistry in 1923.

After graduation, Beeching initially became a school teacher. In 1924, she received a University Postgraduate travelling studentship, which she used to return to Bedford to undertake part-time research with James Spencer towards an M.Sc. degree. Beeching was also part-time teacher at St. Philomena's College, Carshalton, a convent boarding school. She completed an M.Sc. in 1925 on the topic of the magnetic susceptibility of some metals and their oxides.

In 1926, Beeching accepted a position in the laboratories of the British Association for Research in the Cocoa, Chocolate, Confectionary and Jam Trades. Then in January 1927, she joined Alfred Hughes and Sons Ltd. of Birmingham as a Works Chemist. At this date, Beeching was also elected a Member of the Analytical Society. Subsequently, she changed employer to that of Mars Confections Ltd., Slough, initially being the Analytical and Control Chemist and then promoted to Chief Chemist. Beeching stayed at Mars until her retirement. In 1931, following in her father's footsteps, she became a Director of the Institute of Grocers. Beeching also joined the Women's Chemists' Dining Club (see Chapter 9), being a committee member in 1953. She died on 28 July 1967 at Reading, aged 67.

---

[64] (a) (1945). List of applications for Fellowship, *Proceedings of the Chemical Society* 81; (b) Anon. (October 1953). Analytical Society, obituary, 165; (c) *Student Records*, Archives, Bedford College.

Mamie Olliver

In the inter-War period, Mamie Olliver,[65] a researcher in the food industry, had a high profile among women chemists. She was also one of the first women to gain success and recognition in the field of industrial chemistry. Olliver was born on 10 April 1905 in Coventry, daughter of Harry Olliver, secretary to a motor manufacturer, and Mary Olliver. She attended Barr's Hill Secondary School, then entered King's College of Household and Social Science in 1923, completing a B.Sc. in Household and Social Science in 1926.[66] On graduation, Olliver was hired by the Excel Co. Ltd. of London as a Chemical Analyst. Simultaneously, she worked towards a B.Sc. in Chemistry at Birkbeck College, which she completed in 1928, followed by an M.Sc. in Biochemistry in 1930.

In 1930, Olliver joined Chivers and Co. at their research facilities at Histon, near Cambridge, being promoted to their Chief Chemist in the same year. During her time there, she designed and organised the central research and quality control laboratories. Olliver was an active researcher, authoring 13 publications, many related to vitamin C content of foods. Her most important research was the discovery of blackcurrant juice as a valuable source of vitamin C. During the Second World War, Olliver was a member of several committees of the Ministries of Food and Health.

Olliver took her professional responsibilities seriously. She was a member of the Biochemical Society, Nutrition Society, Society of Chemical Industry, Society for Public Analysts, and also Fellow of the Chemical Society. Elected Member of Council of the Society for Analytical Chemistry, Olliver had the third highest vote among the eight candidates for the six positions.[67]

Appointed in 1948 as the first woman member of the Council of the Royal Institute of Chemistry (RIC), Olliver served on many committees of the RIC and later, on those of its successor, the RSC. In 1951, she was elected Vice-President of the RIC. Olliver was particularly determined

---

[65]McCombie, T. (1995). Mamie Olliver (1905–1995). *Chemistry in Britain* **31**: 569.

[66]Note 33, Marsh, p. 268.

[67]Anon. (1955). Proceedings of the Society for Analytical Chemistry. Annual General Meeting, *The Analyst* **80**: 325.

534 *Pioneering British Women Chemists: Their Lives and Contributions*

to promote women in chemistry.[68] Her long-time friend, Doris Kett, recalled:

> I first met Mamie on the education committee of RSC when I was the association of science education representative, several years before I became a member myself. We were the only two women on the committee.... When the RSC moved to Cambridge she frequently had lunch there as they thought a lot of her ... she once lectured at the Royal Institution in Albermarle St which was a rare honour for a women in those days.[69]

Following the merger of Chivers and Schweppes in 1959, Olliver was appointed Research Investigator and Consultant for the Schweppes Group of Companies. Forced into retirement in 1965 at age 60, Olliver retained a keen interest in all aspects of food technology. She died on 17 January 1995 at Cambridge, just a few days before her 90th birthday.

Elizabeth Adams

The first woman chemist hired by Horlicks Ltd. was Winifred Elizabeth Adams.[70] Adams, born on 10 November 1909, at Wandsworth, London, was the daughter of Herbert Edward Adams, a Teacher at Dulwich College, and Winifred Rackham, a Preparatory School Mistress. She was educated at Clapham High School, then studied at the Chelsea Polytechnic before entering Newnham to study biochemistry in 1929.[71]

Graduating in 1933, Adams was hired by Horlicks Ltd. of Slough as an Assistant Chemist. She was the first woman to be appointed to their Research Staff and "an object of some suspicion to her male colleagues in the early years."[70] Promoted to Senior Chemist in 1940, Adams was

---

[68] Olliver, M. (1955). Women in Chemistry. *Journal of the Royal Institute of Chemistry* **79**: 413–420.

[69] Personal communication, letters, Doris Kett, 6 April 2000 and 14 June 2000.

[70] Richards, A. (1995). Elizabeth Adams 1909–1994, N.C. 1929. *Newnham College Roll Letter* 60.

[71] White, A. B. (ed.) (1981). *Newnham College Register, 1871–1971*, Vol. II, *1924–1950*. 2nd edn., Newnham College, Cambridge, p. 60.

made Chief Chemist in 1948, the same year that she was elected Fellow of the Royal Institute of Chemistry. In 1960, it was reported in the *Clapham High School Old Girls' Society News Sheet* that: "Elizabeth Adams is still continuing as Joint Secretary of the Food Group of the Society of Chemical Industry (being the first woman to hold this office)."[72]

Outside of her professional life, Adams was a good musician, and she formed a quartet with her parents and with her only sister (a professional cellist). She was an avid traveller, particularly to Canada and the Shetland Islands. Adams had to retire in 1964, the mandatory retirement age for women staff then being 55. After retirement, she moved to Brighton and became involved with conservation groups, dying on 15 July 1994, aged 84.

Enid Bradford (*Mrs. Bentley*)

One of the many women scientists who worked at J. C. Lyons & Co. was Enid Agnes Margaret Bradford.[73] Bradford was born on 5 May 1902 at Battersea, daughter of Percy James Bradford, mercantile clerk, and Ellen Ada Furness. She was educated at Clapham High School for Girls, where she was taught by Lilian Quartly (see Chapter 10). Elizabeth Adams commented that she thought: "Enid Bradford is probably the most enterprising and professionally successful woman chemist started on her career by Miss Quartly."[74] Bradford obtained a B.Sc. (Hons.) at University College, London, (UCL) in 1926.

An article in the *Clapham High School Magazine* described her early career:

> E. B. after taking honours degree in psychology with organic chemistry subsidiary, and doing some research in pharmacology and the bacteriology course for the post graduate medical diploma of public health, has

---

[72] Anon. (1960). *Clapham High School Old Girls' Society News Sheet* 4.

[73] (1941). List of Applications for Fellowship. *Proceedings of the Chemical Society* 62; and Anon. (1981). Personal News, Deaths. *Chemistry in Britain* **17**: 556.

[74] Anon. (1968). *Clapham High School Old Girls' Society News Sheet* 10.

536　*Pioneering British Women Chemists: Their Lives and Contributions*

just gone to California to continue her research work. She expects to go round with the doctors in the clinic and follow up the biochemistry of their patients by various pathological tests eg blood pressure. She is also going to try to find the chemical formula of a substance that has recently been isolated from the liver, and which produces a marked depressor effect when injected into the most cases of high blood pressure."[75]

In 1928, Bradford wrote a lengthy account of her experiences in the U.S., which was published in the *School Magazine*.[76]

Returning to Britain, in 1931, Bradford was noted in the *School Magazine* as being employed in the laboratories of Bexmax, then in 1936 as Assistant Works Chemist at King's Langley, Herts. She completed a B.Sc. in Chemistry from King's College for Household and Social Science in 1939. In 1939, a research paper was authored by her on the analysis of fat in flours and pastries, giving her address as the Research Department of the Metal Box Company, Acton, London. For the 1939–1940 year, Bradford was the only woman on the Council of the Society of Public Analysts. Then by 1941, she held a position as a Research Chemist to the Marmite Food Extract Company, Ltd.

Later in the Second World War, Bradford worked at the Charterhouse Rheumatism Clinic, London, from where she co-authored papers on winter sources of vitamin C and on the determination of riboflavin in blood. After the War, she was employed by J. C. Lyons & Co. Ltd. from where she authored four papers for *The Analyst*, two on riboflavin in tea, one on the microbiological assay of vitamins, and one describing the use of a single-tap source to simultaneously run a vacuum distillation, a condenser, and a constant-level water bath.

Then in 1948, the *Clapham High School Old Girls' Society News Sheet* reported:

E. B. is working under the Medical Research Council doing research on rheumatism at the British Post-Graduate Medical School. She is fitting up a basement room in her house as a laboratory and means to set up as

---

[75] Anon. (1927). *Clapham High School Magazine* 64.
[76] Bradford, E. (1928). *Clapham High School Magazine* 64–67.

a Consulting Chemist, specializing in micro-biological assay of vitamins and food technology.[77]

The *News Sheet* continued to carry information on Bradford's career. In 1959, it provided an update:

> E B B.Sc. F.R.I.C. M.I.Biol is the only consulting Research Chemist specializing in food technology. She has a growing practice and has had several clients recently recommended by the Royal Institute of Chemistry. She is the only woman registered as a consultant Biologist by the Institute of Biology and on the whole she has found her men colleagues very co-operative and treat her like a brother.[78]

Later in life Bradford married Mr. Bentley. She died on 20 August 1981 at Folkestone, Kent, aged 79.

Women Chemists in Biomedical Laboratories

In her article titled "Biochemistry as a Career for Women" in the *Journal of Careers*, Dorothy Jordan Lloyd made it clear that biochemical sciences were among the most demanding in preparation:

> Biochemistry at most of the universities in this country is not included in the syllabus of any degree course in science, but must be studied as a post-graduate subject for a further one or two years. The preliminary training for a career in biochemistry, therefore, usually calls for four or five years of hard work. Even after this, two years spent working on a research problem in a first-class laboratory and the attainment of a Ph.D. are a very desirable supplementary training.[79]

Katherine Coward

In Chapter 8, we discussed the biochemists at Cambridge with Hopkins, but in addition, there were women who entered biochemistry through

---

[77] Anon. (1948). *Clapham High School Old Girls' Society News Sheet* 4.

[78] Anon. (1968). *Clapham High School Old Girls' Society News Sheet* 10.

[79] Jordan Lloyd, D. (1933). Biochemistry as a career for women. *Journal of Careers* **12**: 20–22.

538   *Pioneering British Women Chemists: Their Lives and Contributions*

other paths. One of these women biochemists was Katherine Hope Coward.[80] Born on 2 July 1885 at Blackburn, Lancashire, she was the daughter of Edward Coward, headmaster of a public (private) school, and Jane Hall. Unlike other biochemists who started with chemistry, Coward studied botany at the Victoria University of Manchester, obtaining a B.Sc. in 1906 and an M.Sc. in 1908.

In 1920, Coward entered UCL to study biochemistry. Awarded a Beit Fellowship, she undertook research on vitamin A with Jack Drummond,[81] a total of 22 publications being authored or co-authored from her work. She was elected Fellow of the Chemical Society in 1923. Receiving her D.Sc. in biochemistry in 1924, Coward travelled to the United States on a Rockefeller Travelling Scholarship to continue her studies on vitamin A at the Department of Agricultural Chemistry of the University of Wisconsin at Madison.

Returning to Britain in 1926, Coward took charge of the newly formed vitamin-testing department of the Pharmaceutical Society's Pharmacological Laboratories, and she was also appointed Reader in Biochemistry at the University of London in 1933.

Coward's research broadened to include the study of each of the vitamins, resulting in a total of 62 additional papers. One of her obituarists "J.H.B." described the crucial role of her contributions:

> The work she did in when in charge of the vitamin-testing Department of the Pharmaceutical Society's Pharmacological Laboratories was of outstanding importance. There were already qualitative methods for various vitamins, but there was a grave lack of methods which were quantitative. Dr Coward was a mathematician with a knowledge of, and a liking for, statistical methods, and was therefore ideally suited for the work of defining such quantitative methods as were needed. Between 1926 and 1938, she and her colleagues completed a series of papers,

---

[80] (a) *Student Records*, Archives, University of Manchester; (b) *Annual Report. 1925–1926.* University College, University of London; (c) Anon. (8 May 1937). A new honorary member. *Pharmaceutical Journal* **138**: 484; (d) Anon. (12 August 1978). Deaths: Coward. *Pharmaceutical Journal* **221**: 134.

[81] Young, F. G. (1954). Jack Cecil Drummond, 1891–1952. *Obituary Notices of Fellows of the Royal Society* **9**: 99–129.

which were of great value for the growing number throughout the country who were concerned with these problems.[82]

Coward was a member of the Vitamin Committee of the British Pharmacopoeia Commission from 1933 to 1953 and a member of the Committee of the Biochemical Society from 1932 until 1936. In 1937, she was elected an Honorary Member of the Pharmaceutical Society. Her advice was sought by committees of the League of Nations and the World Health Organisation.

Retiring in 1950, Coward continued to be active as G. S. Cox, another of her obituarists, recalled:

> After she retired in 1950 she was only to happy to be involved in assisting with problems in statistics, and she was a frequent visitor to the pharmacology laboratories at No 17, giving freely of her time to help research workers. She had a great gift with figures and could perform the most complicated calculations in her head. As a retirement gift she was given a hand-operated calculating machine and, some months after her retirement, when I asked her how she liked it her comment was "It's all right, but it is far too slow; I can do the calculations quicker in my head." And I am sure she could![83]

Coward died on 8 July 1978 at Sidmouth, Devon, aged 93.

Winifred "Freda" Wright

In her short life, Winifred Mary Wright[84] became a highly published biochemist. Wright was born on 1 January 1900 in Brighton, the daughter of Arthur Wright, consulting electrical engineer, and Edith M. Wassel. She was educated at West Hill House, Eastbourne, then attended Bedford College for 1 year before entering Girton in 1920. Completing the

---

[82]"J.H.B." (26 August 1978). Deaths: Coward. *Pharmaceutical Journal* **221**: 177.

[83]Cox, G. S. (2 September 1978). Deaths: Coward. *Pharmaceutical Journal* **221**: 195.

[84](a) Rideal, E. K. (1932). Winifred Mary Wright. *Journal of the Chemical Society* 2998–2999; and (b) Butler, K. T. and McMorran, H. I. (eds.), (1948). *Girton College Register, 1869–1946*, Girton College, Cambridge, p. 321.

540 *Pioneering British Women Chemists: Their Lives and Contributions*

requirements for a chemistry degree in 1924, Wright was awarded a Yarrow Studentship, allowing her to undertake research at Cambridge. Initially, her research was with Eric Rideal,[85] then on her own, on the topics of low-temperature oxidation at charcoal surfaces and on the decomposition of hydrogen peroxide under different conditions. The research resulted in a total of eight publications and provided the basis of her Ph.D. in 1928.

Wright worked for a year with C. G. L. Wolf at Addenbrooke's Hospital, Cambridge, her research culminating in a lengthy paper on the serological diagnosis of cancer. In 1930, she was appointed assistant in the Pharmacology Department at UCL. In addition to lecturing, Wright collaborated with Harry Ing,[86] in a study of the curariform action of quaternary ammonium salts.

One of Wright's obituarists, "F.M.H." (probably her contemporary at Cambridge, Frances Mary Hamer), commented:

> Her specialisation in the field of physical chemistry did not prevent her from throwing herself whole-heartedly into those different problems involving animal tissues, and it is perhaps typical that, only shortly before her death, she embarked upon a medical training.[87]

Her death came on 21 June 1932 in London, at the early age of 32.

Mollie Barr

The Wellcome Physiological Research Laboratories, Beckenham, Kent, hired a significant number of women chemists and biochemists and, as an example, we have chosen Mollie Barr.[88] Barr was born on 11 June 1906 at Wentworth, Yorkshire, the daughter of Horace Carlos Barr, surgeon, and Edith Annie Kingston, and educated at several schools, including a year at

---

[85]Eley, D. D. (1976). Eric Keightley Rideal. 11 April 1890–25 September 1974. *Biographical Memoirs of Fellows of the Royal Society* 22: 381–413.

[86]Schild, H. O. and Rose, F. L. (1976). Harry Raymond Ing. 31 July 1899–23 September 1974. *Biographical Memoirs of Fellows of the Royal Society* 22: 239–255.

[87]"F.M.H." (Lent Term 1933). Winifred Mary Wright. *The Girton Review* 5–6.

[88]*Student Records*, Archives, Bedford College.

Polam Hall, before completing her school studies at Blackheath High School, London.

Barr entered Bedford College, graduating with a B.Sc. in Chemistry in 1927. She was then awarded a 3-year grant from the Department of Scientific and Industrial Research (DSIR) to work towards her Ph.D. on molten salt electrolysis under James Spencer. However, Barr terminated her research in 1928 with a M.Sc. and instead accepted a position at Wellcome as a biochemist working with Alexander Glenny.[89] Most of her research was on immunology, particularly diphtheria, and between 1931 and 1955, Barr authored or co-authored a total of 33 publications. She died in 1975 at Hitchen, aged 69.

Elsie Widdowson

Elsie May Widdowson[90] became a pioneer in the scientific analysis of food and on the importance of diet in infant development. She was born on 21 October 1906 in Wallington, Surrey, daughter of Thomas Henry Widdowson, shop assistant, grocer, and Rose Hannah Elphick. Widdowson attended Sydenham High School which had a tradition for its matriculants to enter either Bedford or RHC. However, she had heard good reports of Imperial College (IC) from three girls who had entered a year ahead of her, so she decided to follow them.

Widdowson completed a B.Sc. (Hons.) in Chemistry, during which time she had commenced research with Samuel Schryver[91] in the biochemistry laboratory, separating amino acids from plant and animal sources. While at IC, she was offered a 3-year research position working with Helen Archbold (see Chapter 4) on reducing sugars in apples. Her interest was more in human biochemistry, thus when the grant expired, she enrolled in a 1-year postgraduate diploma course in dietetics at King's College of Household and Social Science.

---

[89] Oakley, C. L. (1966). Alexander Thomas Glenny. 1882-1965, *Biographical Memoirs of Fellows of the Royal Society* **12**: 163–180.

[90] Ashwell, M. (2002). Elsie May Widdowson, 21 October 1906–14 June 2000. *Biographical Memoirs of Fellows of the Royal Society* **48**: 483–506.

[91] "V.H.B." (1932). Obituary Notices. Samuel Barnett Schryver — 1869–1929. *Proceedings of the Royal Society, Series B* **110**: xxii–xxiv.

542 *Pioneering British Women Chemists: Their Lives and Contributions*

As a preliminary to the course, Widdowson was sent to the main kitchen at King's College Hospital to learn about large-scale food preparation. It was there that she met Robert McCance[92] who was studying the loss of nutrients from food during cooking. From her experience with apples, Widdowson challenged McCance's figures for carbohydrates, proposing that some had been lost by hydrolysis. Impressed by the correctness of her arguments, McCance obtained a grant for her.

Widdowson realized that there was a need for tables of the chemical content of foods. She persuaded McCance to allow her to determine the food composition for those items not previously analysed. With the results of these analyses, the two of them compiled tables, noting values before and after cooking, and published them as *The Chemical Composition of Foods.*[93] This monograph has been in print to the present day, the current, seventh edition, titled *McCance and Widdowson's The Composition of Foods*[94] being published in 2014.

During Widdowson's life, she authored or co-authored over 600 publications, mainly with McCance; in fact, an account of their 60-year collaboration has been published.[95] Here we can only give a superficial account of the many avenues of food chemistry research in which she worked. For example, their studies together between 1934 and 1938 were on salt deficiency in humans, the absorption and excretion of iron, diet variations between individuals, and how kidney function differed between babies and adults.

In 1938, McCance was invited to become Reader of Medicine at the University of Cambridge. At his request, the Medical Research Council agreed to Widdowson moving with him. During their first year at Cambridge, they studied strontium absorption by means of injecting each

---

[92]Petch, N. J. and Barnard, L. (1984). Andrew McCance. 30 March 1889–11 June 1983. *Biographical Memoirs of Fellows of Royal Society* **30**: 388–405.

[93]McCance, R. A. and Widdowson, E. M. (1940). *Chemical Composition of Foods.* Medical Research Council Special Report Series No. 235, HMSO, London.

[94]Food Standards Agency (2014). *McCance and Widdowson's The Composition of Foods.* Seventh Summary Edition, Public Health England.

[95]Ashwell, M. A. (ed.), (1993). *McCance and Widdowson — A Scientific Partnership of 60 Years.* British Nutrition Foundation, London.

other with larger and larger doses and determining the fraction excreted through the kidneys versus the bowels. Unfortunately, the last batch of strontium lactate was contaminated by bacteria and they both fell seriously ill in the laboratory. As the previous doses had had no significant physiological effect, they did not have their usual observer with them, but by good fortune, someone came by and called for help. Even during their sickness, they continued to collect their urine and faeces samples for later analysis.

When the Second World War started, Widdowson and McCance realized that dietary aspects of food rationing would become of high importance. Again they experimented upon themselves (and some of their colleagues), going on a diet which, at the time, was thought to be far too little to maintain health. They also identified the importance of the addition of a calcium supplement to the diet, particularly through incorporation into bread.

After the War, Widdowson worked for the next 20 years on animal nutrition. In 1968, she finally received the formal recognition she deserved, being appointed Head of the Infant Nutrition Research Centre at the Medical Research Council's Dunn Nutrition Laboratory. Retiring in 1973, Widdowson then accepted a research opportunity at the Department of Investigative Medicine at Addenbrooke's Hospital, where she continued working until she was 82. She believed in "active" retirement, including Presidency of the British Nutrition Foundation from 1986 until 1996. Among the many honours Widdowson received was election as Fellow of the Royal Society in 1976 and appointment as C.B.E. in 1979. She died on 14 June 2000 at Cambridge, aged 93.

Women Chemists in the Photographic Industry

As Horrocks has noted,[96] the two leading photographic companies of the time, Ilford and Kodak, had positive attitudes to the employment of women chemists, and as a result, gained two outstanding women researchers, Frances Hamer and Nellie Fisher.

---

[96] Note 57, Horrocks, p. 358.

Frances Hamer

Frances Mary Hamer,[97] born on 14 October 1894 at Kentish Town, London, was the daughter of William Heaton Hamer, Medical Officer of Health, and Agnes Conan. Her mother, together with all her aunts on both sides of the family, had attended North London Collegiate School (NLCS), and she was named after Frances Mary Buss (the founder of NLCS, see Chapter 2), who became Hamer's godmother.[98] Hamer, herself, was enrolled at NLCS, matriculating in 1916, then entered Girton later that year.

While still an undergraduate student at Girton, Hamer joined the research group of William Pope.[99] As a matter of wartime urgency, Pope had been asked to investigate the structure of, and reliable synthetic method for, photographic sensitizers.[100] In 1905, a German company had synthesized a dyestuff, pinacyanol, which, when incorporated into photographic plates, improved sensitivity towards the red end of the visible spectrum. This discovery had become of vital military importance as the air reconnaissance in the latter part of the First World War was undertaken at dawn, when the light had a strong reddish bias. As a result, photos of the German battlefronts taken by British planes were far inferior to those taken of the Allied battlefronts by German planes. In Pope's group, Hamer worked with William Mills (see Chapter 14) to determine the structure of pinacyanol and find a reliable method of synthesis. She was successful in both ventures, and nearly all the pinacyanol used in the new British panchromatic film came from the Cambridge laboratories. The vastly improved dawn images played a major role in enabling British

---

[97] (a) Delius, P. (1963). Dr. Francis Mary Hamer. *British Journal of Photography* **110**: 260; (b) Jeffreys, R. Q. and Gauntlett, M. D. (1981). Frances Mary Hamer 1894–1980. *Chemistry in Britain* **17**: 31; and (c) Butler, K. T. and McMorran, H. I. (eds.), (1948). *Girton College Register, 1869–1946*. Girton College, Cambridge, p. 678.

[98] Anon. (n.d.) Frances Mary Hamer. Unpublished manuscript Archives, North London Collegiate School.

[99] Gibson, C. S. (1941). Sir William Jackson Pope, 1870–1939. *Obituary Notices of Fellows of the Royal Society* **3**: 291–324.

[100] Mann, F. G. (Michaelmas Term 1964). Book reviews: The chemistry of photographic sensitisers. *Girton Review* 9–11.

intelligence to spot German troop movements. Following the end of the War, Mills and Hamer were permitted to publish their synthetic procedure.

An essay on her pinacyanol work gained her the Gamble Prize in 1921, and the same year she was awarded the Yarrow Scientific Research Fellowship to continue her work in Cambridge, where she remained until 1924. She worked for a few months in the Davy Faraday Research Laboratory of the Royal Institution and then was hired by the photographic company of Ilford Ltd.[101] The 6 years she spent at Ilford were marred by difficult relations with the Research Director, Frank Forster Renwick. Matters came to a head in 1930, when Hamer complained that her new laboratory was so hot in summer that it was impossible to work. Renwick refused to take any action and Hamer resigned.

By good fortune, Kodak Ltd. was looking for an organic chemist, as was recounted in the Harrow Research Laboratory (Kodak Ltd.) Album:

> I [the unknown author] wanted a first-class organic chemist, and heard that Dr. Frances Mary Hamer, well-known for her work on sensitizing dyes at Cambridge with Pope and Mills, and who had been at the Ilford Research laboratories, in 1930 had fallen out with her directors over the matter of ventilation in her laboratory. I wrote and told [C. E. Kenneth] Mees. He wrote back and said: "Take her out to tea and sound her out." So I did, and she came, and Dr. Renwick accused me of pinching one of his people. Well, first of all Dr. Hamer was unpinchable, and in any case the whole thing was perfectly ethical.[102]

Thus, Hamer became Kodak's Head of the Organic Chemistry Research Department.

During her years at Ilford and Kodak, Hamer authored over 70 research publications. She was also responsible for a large number of patents, having made major contributions to the synthesis of cyanine dyes and discovering new classes of sensitizers, including some extremely effective

---

[101] Harrison, G. B. (1954). The laboratories of Ilford Limited. *Proceedings of the Royal Society of London, Series B* **142**: 9–20.

[102] *Harrow Research Laboratory Album, 1928–1976*, unpublished manuscript, pp. 14–15, Archives, North London Collegiate School.

546　*Pioneering British Women Chemists: Their Lives and Contributions*

in the infrared region. Hamer had a valuable assistant, Nellie Fisher, while between 1938 and 1942, they were joined at Kodak by Hamer's old friend, Enid Pope, who worked as Journal Abstractor and Indexer.

Hamer served on the Councils of the Chemical Society, the Royal Society of Chemistry, and the Royal Photographic Society (RPS). It was the RPS which recognised her achievements: first, in 1948, the Henderson Award, then in 1963, the RPS Progress Medal, and also election to an Honorary Fellowship of the RPS. Despite her major contributions to the chemistry of photography, and to her own disappointment, she was never elected Fellow of the Royal Society. With hindsight, this was not surprising, for in addition to the handicap of being a woman, Hamer's applied research was outside of mainstream academic chemical circles, and thus, she would have had few champions among the Fellows of the time.

In 1945, Hamer returned to academic research and became an Honorary Lecturer at IC while still being a Research Chemist and Consultant at Kodak, finally retiring in 1959. Upon retirement, she undertook the full-time work of writing the definitive monograph on the cyanine dyes, *The Cyanine Dyes and Related Compounds*.[103]

Hamer was an enthusiastic walker, but a serious accident in 1964 and another in 1971, meant that her mobility was reduced to local walks with the aid of a walking stick. However, she was able to maintain her other interest, gardening, until shortly before her death on 29 April 1980 at Hastings, aged 85.

Nellie Fisher

Nellie Ivy Fisher,[104] initially an assistant to Hamer, became a well-known researcher in her own right following her move to Australia.[105] Fisher was

---

[103] Hamer, F. M. (1964). *The Cyanine Dyes and Related Compounds*. Interscience Publishers, New York.

[104] (a) (1930). Forms of Recommendation for Fellowship. The Ballot will be Held at the Ordinary Scientific Meeting on Thursday, February 20th, *Proceedings of the Chemical Society*, 8; and (b) (1951). *Old Students and Staff of the Royal College of Science*, 6th edn., Royal College of Science, London.

[105] Walker, R. (2 March 2018). Fisher, Nellie (1907–1995). *Encyclopedia of Australian Science* (accessed 10 February 2019).

born on 15 October 1907, at Paddington, London, daughter of Francis Frederick Fisher, jeweller, and Mary Jane Davis. She entered IC in 1926, completing a B.Sc. in Chemistry in 1929. Fisher spent 1 year as a researcher at IC before joining Hamer as a Research Assistant at Ilford Research Laboratories. She subsequently received a Ph.D. for her research on the synthesis of isocyanine dyes.

Fisher followed Hamer to Kodak in 1934, and they worked together co-authoring a total of seven publications and several patents. In 1939, Fisher was asked to transfer to Kodak's laboratories in Australia (see Footnote 105). Kodak(Australia) needed a specialist organic research chemist to provide expertise in preparing emergency quantities of vital spectral sensitizers in case supplies were restricted during the Second World War.

Arriving in Melbourne in late 1939, Fisher became actively involved in many manufacturing problems using her knowledge of dyes as filters. In 1945 she was placed in charge of the Manufacturing Analytical and Services Laboratory, where she stayed until her retirement in 1962. Her former colleague, Nigel Beale, recalled:

> She inspired the laboratory staff in all areas and provided encourage-ment and enthusiasm to young trainees who served part of their program time in the control area. Successful careers were launched as a result of a positive atmosphere and some graduates aspired to management status. Dr Fisher was a great walker and enjoyed activities with the Melbourne Walking Club.[106]

Fisher died on 10 August 1995 at Mont Albert, Victoria, Australia.

Women Chemists in Other Research Laboratories

We have shown in the preceding sections that certain companies were receptive to the hiring of women, but the majority were not. The *Journal of Careers* article on "Careers for Women Scientists" in 1938 reported that many of the large companies requiring chemists did not employ women

---

[106] Personal communication, letter, Nigel Beale, Melbourne, Australia, March 2000.

548   *Pioneering British Women Chemists: Their Lives and Contributions*

and were very open about the fact, one firm replying that: "all applications from women scientists are 'automatically ruled through'."[107] An Institute of Chemistry survey showed that of 963 available positions during 1935 and 1936, only 80 were open to women.[108] In this section, we will provide a selection of other research career pathways followed by women chemists.

Rona Robinson

The earliest industrial chemist we could identify was Rona Robinson.[109] Robinson was born in Manchester on 26 June 1884 to Alfred (Fred) Robinson, cotton goods traveller, and Jessie Robinson, lodging housekeeper. She attended Manchester Central School for Girls then entered Owen's College in 1902. Rona completed a B.Sc. (Hons.) in Chemistry in 1905, being awarded the LeBlanc Medal and also a Mercer Scholarship for the best final year student entering research. Her subsequent research on hydroxyphthalic and methoxyphthalic acids resulted in an M.Sc. in 1907.

Robinson accepted a position as a teacher at Altrincham Pupil-Teacher Centre. Developing a passion for women's suffrage, she left her teaching position to work with the Women's Social and Political Union. Robinson was arrested several times in 1909 for her suffragette activities and twice went on hunger strikes.

During this period, Robinson also undertook research on dyes in her own private laboratory at Moseley Villa, Metford Road, Withington, Manchester, together with some research at the Royal Institution Laboratory, Manchester.[110] In 1912, she was a Gilchrist Scholar in Home Science and Economics at King's College for Women. It was while at King's College for Women that she wrote a scathing attack on the Domestic Science degree program (see Chapter 8).

---

[107] Note 53, Anon., p. 294.

[108] Anon. (1937). Salaries and employment in chemistry. *Journal of Careers* **16**: 169.

[109] Wikipedia, (23 February 2019). Rona Robinson (accessed 26 February 2019).

[110] Anon. (1963). Obituary: Rona Robinson. *Journal of the Royal Institute of Chemistry* **87**: 66.

In 1915, Robinson joined J.B. & W.R. Sharpe Ltd. as an Analytical and Research Chemist. In addition, she was the Works Chemist responsible for transferring chemical reactions that she had devised to large-scale production. The following year, she was promoted to Chief Chemist. Robinson left in 1920 to become Chief Chemist to Clayton Aniline Co. Ltd., where three patents were issued with her as inventor, two of which were on aldehyde-amino condensation products. She held the post at Clayton Aniline until her retirement. She died on 7 April 1962, age 77. Initiated by funds in her will, the University of Manchester has the Rona Robinson scholarship for the support of female graduate students in chemistry.

Edith Pawsey (Mrs. Murch)

The life history of Edith Hilda Pawsey[111] is particularly interesting in that she worked for industry, then government, before entering the "women's role" of scientific librarian. Pawsey was born on 12 January 1897 in Manchester, daughter of George B. Pawsey, bedding manufacturer, and Rachel Pawsey, who helped in the business. Pawsey studied at Portsmouth Municipal College. Despite being assigned a tutor who refused to teach women, she passed the examinations and was awarded a B.Sc. in Chemistry by private study in 1916. Then in 1917, she earned a B.Sc. (Hons.) in Chemistry.

From 1918 to 1933, Pawsey worked as an Assistant Research Chemist at the South Metropolitan Gas Company, her research on coal gas having earned her an M.Sc. (London) in 1924. Pawsey was at the Gas Company the same time as Eunice Bucknell (see Chapter 9), and Bucknell was the first-named nominator of Pawsey for her admission to the Chemical Society.

Upon marriage to chemist William Owen Murch, Pawsey was required to resign her position. Over the subsequent years she had two daughters, but the exigencies of war overruled marital status, and during

---

[111]Truter, M. (1995). Edith Hilda Murch 1897–1995. *Chemistry in Britain* **31**: 902; and Anon. (1996). Corrigendum. *Chemistry in Britain* **32**: 55.

550   *Pioneering British Women Chemists: Their Lives and Contributions*

the Second World War, she was employed in the Armament Research Establishment, Fort Halstead.

In 1946, Pawsey joined the organic chemistry section of the Chemical Research Laboratory at the DSIR. When, in 1949, she was asked to take over the chemistry library temporarily, she did the job so well that she was appointed the permanent Librarian, resulting in her becoming an Associate of the Library Association. Pawsey first retired in 1962, but was soon rehired by the National Physical Laboratory to help in the library, retiring for a second and final time in 1968. She died on 14 March 1995 at Helston, Cornwall, aged 98.

Women Chemists as Scientific Librarians and Indexers

Science librarian or archivist work was an avenue of employment for women chemists suggested by an article in the *Journal of Careers*:

> Technical librarianship is another opening for the woman science graduate, or an appointment in a university science library or technical college library, or the post of information officer in a large firm where there is much circulation of technical information from scientific periodicals between heads of departments.[112]

Later in the article, the pay range specifically for women was provided.

Many women science graduates did indeed proceed into technical librarianship and information work as Helen Plant has shown.[113] In Chapter 13, we have already described how Agnes Lothian became librarian and historian in pharmacy. Here we will provide some examples of women chemists who filled this role in chemistry.

---

[112]Anon. (1938). Prospects of employment for women science graduates: Part IV, Careers in which science is useful though not the main subject. *Journal of Careers* **17**: 421–425, p. 423. Margaret Rossiter has described this avenue of employment in the American context: Rossiter, M. W. (1996). Chemical librarianship: A kind of "women's work" in America. *Ambix* **43**: 46–58.

[113]Plant, H. (2005). Women scientists in British industry: Technical library and information workers, c.1918–1960. *Women's History Review* **14**: 301–321.

Enid Pope (*Mrs. Hulsken*)

Enid Marian Pope[114] proceeded directly from university to library work. Pope was born on 19 November 1911 at West Bromwich, daughter of Thomas Henry Pope, research chemist, and Florence George. She was educated at Wallasey High School in Cheshire and then Croydon High School before entering Newnham in 1930. Pope completed the requirements for a degree in chemistry in 1933 and received a B.Sc. (London) in 1934.

That year, Pope accepted a post as an Abstractor and Editor of Abstract Journals for the Research Association of British Rubber Manufacturers. She left the position in 1938, to join Kodak as Journal Abstractor and Indexer at their Research Laboratories in Harrow, joining her long-time friend, Hamer. Pope was then appointed as Research Librarian at the Distillers Research Department, Epsom in 1944. In the same year, she married Johannus Martinus Hulsken, a Chief Officer in the Dutch Merchant Navy. Pope continued working after marriage, but in 1948, she resigned her position prior to the birth of her son. In later years, she was a self-employed Abstractor and Indexer. Pope died on 25 October 1987 at Market Harborough, Leicestershire.

Margaret Robertson (*Mrs. Dougal, Mrs. Chaplin*)

Though belonging to an earlier generation, the life and work of Margaret Douie Robertson[115] fits more appropriately here. Born in 1858 or 1859 in Straits Settlements, Singapore, she was the daughter of J. H. Robertson, a medical doctor. Nothing can be found on her early adult life or her education.

After marriage to William Dougal, Robertson's first publications were a series of four articles on the teaching of inorganic chemistry, which appeared in *Chemical News* in 1893.[116] Then in 1894, she commenced research with Thomas Edward Thorpe at the Royal College of Science,

---

[114]White, A. B. (ed.), (1981). *Newnham College Register, 1871–1971*, Vol. II, *1924–1950*. 2nd edn., Newnham College, Cambridge, p. 81.

[115]Wikipedia (15 October 2018). Margaret Douie Dougal (accessed 8 February 2019).

[116]Creese, M. R. S. (1998). *Ladies in the Laboratory: American and British Women in Science 1800–1900*. Scarecrow Press, Lanham, Maryland, p. 266.

552 *Pioneering British Women Chemists: Their Lives and Contributions*

the research being published in 1896.[117] Thorpe believed that there was a need for an index to the publications of the Chemical Society, and for 15 years, until 1909, Dougal worked as Compiler and Indexer, preparing *A Collective Index of the Transactions, Proceedings and Abstracts of the Chemical Society*. The first two volumes, covering the 20 years from 1873 to 1892, took Dougal over 5 years to produce.

After Dougal had completed this labour, James Dewar made a point of congratulating her in his 1899 Presidential Address to the Chemical Society.[118] Dewar emphasized the tremendous usefulness of the indexes to members. He went on to note that although the task had been of unexpected length and complexity because of lack of system in the previous annual indexing, Margaret Dougal's compilation was "an example of thoroughness and accuracy to her successors."[118] After the death of Dougal, in 1909, Robertson remarried, this time to Arnold Chaplin, M.D. She died on 9 November 1938 at the age of 79.

Margaret Le Pla

Dougal's replacement in 1910 was Margaret Le Pla.[119] Le Pla, born on 17 April 1885 at Southall, Middlesex, was the daughter of Henry Le Pla, Nonconformist Minister, and Sarah Elizabeth Islip. She was educated at South Hampstead High School and then entered Bedford College. After graduation with a B.Sc. in 1906, she was appointed Demonstrator in Chemistry and also Research Assistant to James Spencer.

In 1910, Le Pla accepted a position with the Chemical Society as an indexer, particularly of the *Journal of the Chemical Society*.[119] In the early part of the 20th century, the *Journal of the Chemical Society* carried an Abstracts supplement and when the publication of the supplement was transferred to the Bureau of Chemical Abstracts, Le Pla indexed these also, herself moving to the Bureau offices.

---

[117]Dougal, M. D. (1896). Effect of heat on aqueous solutions of chrome alum. *Journal of the Chemical Society* **69**: 1526–1530.

[118]Dewar, Sir J. (1899). Presidential Address. *Journal of the Chemical Society* **75**: 1168.

[119]Cummings, A. E. (1953). Margaret Le Pla: 1885–1953. *Journal of the Chemical Society* 3335–3336.

When Le Pla retired from indexing the Abstracts, she continued to index the *Journal of the Chemical Society* and *Annual Reports of the Chemical Society*. In addition, she indexed Thorpe's *Dictionary of Applied Chemistry*. Amazingly, these tasks left her with idle time and, her obituarist reported, Le Pla endeavoured to fill it:

> Her energies, however, sought further outlets, and she took over the re-organization of the Research Library of Messrs. C.C. Wakefield & Co., and the care of this library was placed in her hands. Although in her later years she did not give herself much time for recreation, Margaret Le Pla loved music, and for many years she sang in the Old Vic Opera Company. Latterly, she took great interest in the functions of the Women Chemists Dining Club [see Chapter 2], of which she was a member.[120]

She died on 26 January 1953 at Chelsea, London, still working hard.

Margaret Dampier Whetham (Mrs. Anderson)

Both Dougal and Le Pla had been indexers of chemical publications, but Margaret Dampier-Whetham[121] left her biochemical background to index publications across many disciplines. Born on 21 April 1900 in Cambridge, Whetham was a daughter of William Cecil Dampier Whetham (later, Sir William Dampier), Lecturer in Physics at the University of Cambridge, and Catherine Durning Holt, former science student at Newnham (see Chapter 3), and author of several books.

Whetham was home-schooled and then attended University College, Exeter, before entering Newnham in 1918, where she studied science for 3 years (one of her sisters, Edith Holt Whetham also attended Newnham).[122] In 1920, while still a student, she commenced research work with Marjory Stephenson (see Chapter 11) undertaking pioneering work on the washed cell suspensions technique for analysing cells. During her 7 years working

---

[120]Cummings, A. E. (7 February 1953). Margaret Le Pla, *Chemistry and Industry* 130–131.

[121]Wikipedia (8 August 2018). Margaret Anderson (indexer) (accessed 8 February 2019).

[122]White, A. B. (ed.), (1979). *Newnham College Register, 1871–1971*, Vol. I, *1871–1923*. 2nd edn., Newnham College, Cambridge, p. 38.

with Stephenson, Whetham co-authored four research papers. She was also the co-founder of the Cambridge in-house biochemistry magazine, *Brighter Biochemistry*, which lasted from 1923 until 1930 and served to give insights into life under 'Hoppy' (see Chapter 11). Whetham, herself, penned the following verse:

> *A monograph by MS*
> *Would do much to relieve the distress*
> *Caused by all these inferior*
> *Books on bacteria.*[123]

In 1927, Whetham married Alan Bruce Anderson, a clinical pathologist, and gave up her research work. Over the ensuing years, Whetham had five children. She returned to the workforce in 1948 as Abstractor for *British Chemical Abstracts*, then in 1950, she switched to abstracting for *Food Science Abstracts* for 7 years. From being an abstractor, Whetham became an indexer, and over her remaining career she compiled indexes for a total of 567 books. She became a Member of the Society of Indexers, finally becoming Vice-President. Whetham wrote numerous journal articles, authored *Book Indexing*,[124] and co-authored with her father a compilation: *Readings in the Literature of Science*.[125] Whetham died in 1997.

Women Chemists as Factory Inspectors

Becoming a Lady Factory Inspector was another path of employment for women graduates — particularly those with a scientific background.

The first Lady Factory Inspectors had been appointed by the Home Secretary in 1893.[126] It was a tough task for a middle-class woman,

---

[123] Anderson, M. D. (February 1929). *Brighter Biochemistry* **6**: 47.

[124] Anderson, M. D. (1971). *Book Indexing*. Cambridge University Press, Cambridge.

[125] Whetham, W. C. D. and Whetham, M. D. (1924). *Cambridge Readings in the Literature of Science*. Cambridge University Press, Cambridge.

[126] (a) Jones, H. (1988). Women health workers: The case of the first women factory inspectors in Britain. *Social History of Medicine* **1**: 165–181; and (b) McFeeley, M. D. (1988). *Lady Inspectors: The Campaign for a Better Workplace, 1893–1921*. Blackwell, London.

requiring her to venture alone into unsavoury neighbourhoods, cope with aggressive plant managers, and attempt to offer some specific recommendations to alleviate the plight of the working poor. As Adelaide Anderson,[127] Head of the Lady Inspectors, wrote in 1905 to the Chief Inspector:

> A woman, as a Factory Inspector, in an Industrial district away from her own family and social surroundings, as well as her women colleagues, can find no normal associates in, or through, her work. Her work compels her to lead a life that is quite different from that of other women, and the slightest deviation from extreme caution and prudence may subject her to injurious criticism.[128]

The male Inspectors were often less than supportive, disliking the zeal of these well-qualified and determined women.

Dorothy Fox (Mrs. Richards)

One of the later Lady Factory Inspectors was Dorothy Lilian Fox.[129] Born on 24 March 1905, Fox was the daughter of George Frederick Fox, a manufacturer of motor accessories in Birmingham, and Alice Lilian Fox. She was educated at KEVI and then entered Bedford in 1924, having applied late because her father would not permit her to leave home. In addition to academic success, Fox was a gifted hockey player and sculler for the University of London women's teams. After obtaining a B.Sc. in Physiology in 1927 and a B.Sc. in Chemistry in 1928, she looked for employment, but was unsuccessful.

Fox stayed on at Bedford where she undertook research with Eustace Turner, completing a Ph.D. in 1930 on diaryl ethers. Having heard from her father about Lady Factory Inspectors, she obtained a position as

---

[127]Zimmeck, M. (23 September 2004). Anderson, Dame Adelaide Mary (1863–1936). *Oxford Dictionary of National Biography*. Oxford University Press (accessed 8 February 2019).

[128]Note 126a, Jones, p. 175.

[129]*Student Records*, Archives, Bedford College; *Hampshire Chronicle*, February 9 1963.; *Lincolnshire Echo*, January 17 1984; and Anon. (1984). Personal news: Deaths. *Chemistry in Britain* **30**: 304.

556   *Pioneering British Women Chemists: Their Lives and Contributions*

Factory Inspector in Reading. She continued as a Factory Inspector until 1939, being the first woman to inspect large chemical works, such as ICI.

It was in 1939 that Fox married H. G. H. Richards, a consultant pathologist, and she changed her career direction, assisting Richards with his medical research work. Then, with the start of the Second World War, she took a course in building construction and spent the War working on the design and construction of factory air raid shelters and canteens. Following the War, Fox became active in local politics, being first elected as a Lincolnshire County Councillor in 1967. She remained on the Council until 1982, dying 2 years later on 14 January 1984 at Lincoln, aged 78.

The Late- and After-Career of Some Women Chemists

In the late 19th and the first part of the 20th centuries, university authorities considered that women students were in particular a moral danger and needed an authority figure to oversee them during their free hours.[130] A unique role for older woman academics was therefore created, that of Lady Superintendent or Warden of women's halls of residence. Of the women chemists discussed in this book, nine held appointments of this type.

The role was also found in the United States, as Margaret Rossiter had noted, the incumbent often a senior woman scientist, one difference being that the position was called Dean of Women.[131] The duties of Warden of University Hall, University of St. Andrews make it clear that the incumbent had to be an academic:

> For the proper discharge of all these advisory duties, as well as her administrative duties as head of a woman's College Hall, the Warden must not only have academic experience, but must also keep herself constantly in touch with educational movements and ideas.[132]

---

[130]Dyhouse, C. (1995). The British Federation of University Women and the status of women in universities, 1907–1939. *Women's History Review* **4**: 465–485.

[131]Rossiter, M. W. (1980). "Women's work" in science, 1880–1910. *Isis* **71**: 381–398.

[132]Dyhouse, C. (1998). *No Distinction of Sex? Women in British Universities*, 1870–1939. UCL Press, London, p. 254.

At Bedford, Mary Crewdson (see Chapter 5) was appointed Warden of Northcutt House and then Lindsell Hall, retiring in 1954, and being succeeded by Mary Lesslie (see Chapter 5) who was given the title of Dean of Lindsell Hall, herself retiring in 1968. Kathleen Balls (see Chapter 4) at Queen Mary College was appointed Lady Superintendent in 1924. At the University of Sheffield, Dorothy Bennett (see Chapter 6) was Tutor for Women from 1926 and Warden of University Hall for Women (later Fairfax House) from 1934, resigning from both in 1947.

Ruth King (see Chapter 15) was appointed Warden of Hope Hall at University College of South West England (later the University of Exeter) in 1918, while Ettie Steele (Chapter 7) was Warden of Chatten House (later McIntosh Hall), University of St. Andrews from 1930 until 1959. Two of the women chemists became sub-Wardens: Grace Leitch (see Chapter 6) at Easton Hall, Armstrong College, from 1921 to 1941, and May Leslie (see Chapter 4) at Weetwood Hall, University of Leeds, from 1935 to 1937. Finally, Millicent Taylor (see Chapter 6) was acting Warden during Spring and Summer Terms 1934 and 1935 at Clifton Hill House, University of Bristol.

Though the role of Warden could be interpreted in a negative light, at the time, these single women, who had little opportunity for academic advancement, at least had a position and an income for their later years.

Commentary

The 20th century had started so promisingly for women: educational opportunities were taken for granted. The First World War had given access to a wide range of job opportunities for academic women, particularly those with a chemical background. Women's university enrolments — including chemistry — continued to rise into the 1920s. Then came the reversal. Marjorie Nicholson, a 1914 graduate of the University of Michigan, eloquently captured the sentiment of feminist academic women on both sides of the Atlantic:

> We of the pre-[First World] war generation used to pride ourselves sentimentally on being the "lost generation," used to think that, because war cut across the stable path on which our feet were set, we were an

558 Pioneering British Women Chemists: Their Lives and Contributions

unfortunate generation. But as I look back upon the records, I find myself wondering whether our generation was not the only generation of women which ever found itself. We came late enough to escape the self-consciousness and belligerence of the pioneers, to take education for granted. We came early enough to take equally for granted professional positions in which we could make full use of our training. This was our double glory. Positions were everywhere open to us; it never occurred to us at the time that we were taken only because men were not available.... The millennium had come; it did not occur to us that life could be different. Within a decade shades of the prison house began to close, not upon the growing boy, but upon the emancipated girls.[133]

For women chemists there were gains and losses. As we showed in Chapter 9, women chemists were finally admitted to the Chemical Society. On the contrary, women's positions as university and college teaching staff ceased to exist as each of the original incumbents retired. Nevertheless, the 1920s and 1930s did offer opportunities in industrial chemistry for women graduates but only with certain companies or organizations. Whatever the organization, there was always the assumption that "matrimonial mortality" made training women a waste of time.

In the latter-half of the 20th century, it seems as though women chemists had to start their crusade for acceptance all over again. The progress they had made between 1880 and 1925 had been forgotten. Even as late as 1958, a woman student, J. Lemon, at IC lamented:

> From the staff, the attitude [towards women students] is gentle contempt or amusement. From the [male] students, we are an object of amazement, except when they are in difficulty (such as ironing shirts).[134]

But the Hadow Report of 1923 was a creature of its times. It was part of a wider cultural phenomenon of the reversal in fortunes for women's rights and expectations. This problem was succinctly expressed by

---

[133] Nicholson, M. (1938). The rights and privileges pertaining thereto. *Journal of the American Association of University Women* **31**(3): 136.

[134] Lemon, J. (Summer 1958). A woman's point of view. *The Phoenix* (new ser.) **73**: 10–11.

Johanna Alberti in her account of the feminist and suffragist, Elizabeth Haldane:

> But without the binding force of a single unifying issue such as suffrage, feminists in the 1920s had to return to the sisyphean task of making change at a time when political life mediated against change.[135]

---

[135] Alberti, J. (1990). Inside out: Elizabeth Haldane as a women's suffrage survivor in the 1920s and 1930s. *Women's Studies International Forum* **13**(1/2): 117–125.

Index

A

Aberdeen Ladies' Educational Association, 235

Academic Successes, 46

Ackworth School, 318

Adams, Elizabeth, 337, 532, 534–535

Adamson, Mary, 317, 326–328

Admiralty Steel Analysis Laboratories, 501

Aitken, Edith, 343, 345–346

Alcock, Lucy, 125

Alice Ottley School, Worcester, 41

Allen, William, 23–24, 316

American Association of University Women, 294

Analytical Society, 184, 532

Anderson College, *see* Royal Technical College, Glasgow, 227

Andrew, Gertrude, 183–184

Andross, Mary, 233–234

Annual Science Exhibitions, 57

Apothecaries' Act of 1815, 426

Apothecaries' Hall, 426–427

Archbold, Helen (Mrs. Porter), 130–132, 469, 541

Armstrong College, *see* University of Newcastle, 209–212, 487, 496, 523

Armstrong, Henry, 36, 164, 261, 293, 302, 307, 335, 341–342, 344–346, 349, 461

Arthur Sanderson & Sons, 144

Association for Promoting the Education of Girls in Wales, 32

Association for the Higher Education of Women, Glasgow, 227

Association for Women Pharmacists, 443–445, 452–455

Association of Women Science Teachers, 339, 347

Association of Women Science Teachers, Meetings of, 349

Association of Women Science Teachers, Origins of, 347

Association of Women Science Teachers, Welsh Branch, 350

Aston, Emily, 112–113

Avery Hill Training College, 78, 262

562 *Index*

B

Badger, Louise (Mrs. Sinnatt), 466–467

Badger, May (Mrs. Craven), 177, 466–467

Bain, Alice, 83, 237, 471

Balls, Kathleen (Mrs. Stratton), 132–134, 557

Baly, Edward C. C., 114, 116

Barger, George, 158, 472

Barr, Mollie, 540–541

Bartlett, Dorothy (Mrs. Storey), 454–455

Bascom, Florence, 384

Bateson, William, 239, 363–366

Battersea Polytechnic Institute, 275, 282

Battersea Polytechnic Institute, Department of Women's Subjects, 276

Battersea Polytechnic, London, 134–135, 497

Baylis, Dorothy, 181, 183–184

Beale, Dorothea, 28, 31, 37, 60, 89, 112, 323

Bede School for Girls, Sunderland, 81, 212

Bedford College, Cambridge, 389, 532

Bedford College, London, 78, 84, 102, 104, 112, 114, 130, 138–139, 143–149, 151–153, 160, 200, 214, 252, 274–275, 309, 326, 352–353, 392–393, 472, 490, 492, 494, 507, 522, 539, 541, 552, 555

Bedford College, London, Chemical Society, 142

Bedford College, London, Founding of Chemistry Department, 139

Bedford College, London, Life of Women Chemistry Students, 140

Bedford College, London, Women Chemistry Staff, 145

Bedford High School, 38, 48, 328, 401

Bedson, Samuel Phillips, 506

Beeching, Ethel, 532

Beibly, Winifred (Mrs. Soddy), 460, 462–463

Belvedere School, 41–42, 55

Benenden School, Kent, 99

Bengough, Guy Dunstan, 232

Bennett, Dorothy (Mrs. Leighton), 193, 195, 491, 557

Berkhamsted High School for Girls, 368, 497

Bernal, J. D., 386, 396, 398–402, 405–408, 412, 420

Berridge House, Hampstead, 264

Besant, Annie, 151

Beulah House High School, Balham, 135

Beveridge, Heather, 304

Bhagwat, Kamala (Mrs. Sohonie), 360

Bickerstaffe, Robert, 504

Biochemical Club, *see* Biochemical Society, 292, 295, 365, 369

Biochemical Society, 292, 365, 533, 539

Birkbeck College, London, 111, 114, 128, 131, 311, 396, 399, 401, 420–421, 453, 533

Birkbeck, George, 111, 287

Birkbeck Literary and Scientific Institution, *see* Birkbeck College, London, 111

Blackheath High School, 52, 541
Bodichon, Barbara Leigh Smith, 66–67
Boole, Lucy, 162–164, 290
Borough Polytechnic Institute, London, 309, 490
Borrowman, Agnes, 439–440, 450–452
Boyle, Mary, 133, 156–157
Bradford, Enid (Mrs. Bentley), 532, 535–537
Bradford Girl's Grammar School, 119–120, 401
Bragg, W. H., 226, 386–387, 390, 392–395, 399, 404, 415
Bragg, W. L., 386–388, 395, 397
Bremner, Christina, 269
Brighton and Hove High School, 166
Brighton Municipal Training College for Teachers, 322
Briscoe, Vincent, 166
Bristol College, *see* University of Bristol, 112
British Agricultural Research Council, 420
British Association for the Advancement of Science, 10, 481
British Coal Utilization Research Association, 415
British Cotton Industry Research Association, 178
British Federation of University Women, 294
British Leather Manufacturers Research Association, 135, 374
British Medical Association, 453
British Museum, 382
Bromley High School, 327

Browne, Agnes (Mrs. Jackman), 272–274, 312–313
Bryant, Sophie, 139, 320
Bryn Mawr College, Pennsylvania, 384
Buchanan, Margaret, 438–439, 442–443, 448–450, 452, 454
Bucknell, Eunice, 309, 549
Burke, Katherine (Kate), 114–116
Burroughs Wellcome, 391
Buss, Frances, 28, 30, 112, 320, 343
Buss, Robert, 316

C
Cambridge Scientists Anti-War Group, 398–399
Cambridge Teachers Training Certificate, 329
Cambridge Teacher Training Syndicate, 319
Cambridge Training College for Women, *see* Hughes Hall, 76, 320
Campbell, Ishbel, 213–215
Carlton, Margaret, 128–129
Cavendish, Georgiana (*née* Spencer), Duchess of Devonshire, 5–6
Cavendish, Margaret, Duchess of Newcastle, 4–5
Central Foundation Girls' School, 320, 344
Central Technical College, *see* Imperial College, London, 293
Charles, Jessie (Mrs. White), 198
Cheltenham Ladies' College (CLC), 28, 31–32, 34, 46, 60, 89, 94, 164, 207–208, 262, 281, 291, 468, 490, 507, 510

564   *Index*

Chemical Passion, 22, 25

Chemical Society, 62, 144, 292, 297, 299–300, 302–303, 306–309, 314, 322, 341, 434, 455, 464, 533, 538, 546, 549, 552

Chemical Society, 1880 Motion to Admit Women, 297

Chemical Society, 1888 Motion to Admit Women, 297

Chemical Society, 1892 Motion to Admit Women, 297

Chemical Society, 1904 Admission of Marie Curie, 299

Chemical Society, 1904 Discussion of Admission of Women, 298

Chemical Society, 1904 Petition, 299

Chemical Society, 1904 Women Chemist Signatories, 299

Chemical Society, 1905 Presidential Address, 300

Chemical Society, 1908 Motion to Admit Women, 301

Chemical Society, 1908 Poll of Members, 301

Chemical Society, 1908 Women Subscribers, 302

Chemical Society, 1909 Letter, 303

Chemical Society, 1909 Proposed Bye-Law on Admission of Women, 302

Chemical Society, 1909 Women Chemist Signatories, 303

Chemical Society, 1920 Admission at Last, 307

Chemical Society, 1920 Women Chemist Admissions, 308

Chemical Society, Effect of the 1904 Petition, 300

Chemistry Club, 311

Chemistry Courses for Girls' Schools, 36

Chemistry Laboratories at Girls' Schools, 37

Chemistry Laboratories at Girls' Schools, Basement and Attic, 40

Chemistry Laboratories at Girls' Schools, Construction of, 39

Chemistry Laboratories at Girls' Schools, Custom-Built of the 1890s–1910s, 41

Chemistry Laboratories at Girls' Schools, Custom-Built of the 1920s to 1930s, 42

Chemistry Laboratories at Girls' Schools, in Wales and Scotland, 42

Chemistry Teachers, Earliest, 316

Chick, Frances (Mrs. Wood), 304

Chick, Harriette, 292, 295–296, 365

Choice of University, 62

Christie, Agatha, 427

City and Guilds College, *see* Imperial College, London, 123, 342–344

City of Cardiff High School for Girls, 33, 351, 353

City of Cardiff High School for Girls, Field Club, 352

City of Cardiff High School for Girls, Science Club, 353

City of London School for Girls, 53, 132

Clapham High School, 55, 78, 329, 337, 339, 347–348, 534–536

Clarke, Isabella (Mrs. Clarke-Keer), 429, 436–437, 442–443, 445, 449

Cleaverley, Louisa (Mrs. Dunstan), 305, 465

Index   565

Clemo, George, 211
Cliffford, Lady Margaret, 3
Clifton Association for the Higher
  Education of Women, 201
Clifton High School, Bristol, 130,
  205
Clough, Anne Jemima, 67–68, 72
College of Chemistry, Liverpool, 179
College of Perceptors, 46
College of Physical Science,
  Newcastle, *see* University of
  Newcastle, 210
College of Science and Arts, *see*
  Royal Technical College, Glasgow,
  227
Collie, J. Norman, 113, 305, 470
Collier, Kathleen, 328
Colston's Girls' School, Bristol, 56,
  345, 524
Constantine College, Middlesbrough,
  523
Conversations on Chemistry, 13,
  15–19, 24
Cooper, Yvonne, 168
Corner, Mary, 135
Cornish, Elfreida (Mrs. Venn,
  Mrs. Mattick), 240, 524
County Girls' School, Pontypool, 33,
  43, 493
Coward, Katherine, 537–539
Cox, Claudia (Mrs. McPherson),
  279–280
Cranston, John A., 238
Crewdson, Mary, 147, 150, 557
Crick, Francis, 418–422
Croham Hurst School, Croydon, 102
Crompton, Holland, 139–140,
  145–146, 149
Crossley, Arthur, 451, 454–455

Crowfoot, Dorothy (Mrs. Hodgkin),
  96, 101, 385, 398–399, 402–413,
  469
Croydon High School, 44, 128, 237,
  331–333, 505, 551
Crum Brown, Alexander, 243–244
Culhane, Kathleen (Mrs. Lathbury),
  528–531
Cunningham, Mary, 309
Curie, Marie, 188, 298, 520

D

Dalston, Daisy, 329
Davies, Dilys (Mrs. Glynne Jones),
  32
Davies, Emily, 66–67
Davy Faraday Laboratory, *see* Royal
  Institution, 390–391, 545
Davy, Humphry, 10, 12, 16, 23
Dawson, Harry M., 188–190
Deane, Henry, 439
Deane Pharmacy, *see* Pharmacy, 17,
  The Pavement, 449, 452
Degrees for Cambridge Women,
  1920s Battle, 71
Degrees for Cambridge Women, First
  Defeat, 68
Degrees for Cambridge Women, First
  Victory, 68
Degrees for Cambridge Women,
  Second Defeat, 71
Degrees for Cambridge Women,
  "Steamboat Ladies", 69, 334, 362,
  502
Degrees for Cambridge women,
  Success at Last, 72
Demonstrations by Students, 50
de Morgan, Sophia Elizabeth
  (*née* Frend), 24

566   *Index*

Department for Scientific and Industrial Research, 98, 226, 288, 456, 505–506, 541, 550
Desch, Cecil Henry, 121
Dixon, Harold Baily, 92–93
Dodson, Guy, 405
Domestic Chemistry, 264–265, 267, 275
Domestic Science Controversy, 260
Domestic Science Controversy, Consultative Committee 1913, 260
Domestic Science Controversy, Report on Housecraft 1911, 260
Dominion Experimental Farm, Ottawa, 502
Donnan, Frederick, 114, 273
Dove, Frances, 37–38, 45
Downe House, 32, 52–53, 85, 324, 330, 338
Downe House, Science Club, 331, 338
Downing College, Cambridge, 85
Drury, Dorothy (*née* Moore), 3
Dunstan, Albert, 305
Dunstan, Wyndham, 163
Durham College of Physical Science, *see* University of Newcastle, 209

E

East Ham Technical College, 305
East London College, *see* Queen Mary College, London, 132–133, 288, 495
Edgbaston High School for Girls, 56
Edgeworth, Maria, 18
Edinburgh Central School of Pharmacy, 451
Edinburgh Ladies' College, *see* Mary Erskine School, 489

Edinburgh Seven, 218, 243, 245
Edinburgh Seven, Irish Brigade, 220
Edinburgh Seven, Riot of 18 November, 218
Elam, Constance (Mrs. Tipper), 387, 391
Essay on Combustion, 7, 16
Eustice, Lovelyn (Mrs. Bickerstaffe), 504
Evans, Clare de Brereton, 37, 163–165, 168, 307
Eves, Florence, 110
Ewbank, Elinor, 97–98
Examiner's Reports, 44
Experimental Basis of Chemistry, 77, 79

F

Faithful, Lilian, 60, 262
Faraday, Michael, 5, 11, 17
Farmer, Ernest, 468
Farrow, Elizabeth, 101–102
Fawcett, Millicent Garrett, 67
Federal University of Yorkshire, 191
Field, Eleanor, 155–157
Field, Ellen (Mrs. Stedman), 471–473
Findlay, Alexander, 200
Finsbury Technical College, London, 123
First World War, Chemistry Women's War Work, 488
First World War, Duty of Women University Students, 487
First World War, Responding to the Need, 489
First World War, Women Chemists and other War Work, 503
First World War, Women Chemists and War Gases, 496

First World War, Women Chemists in Organic Synthesis, 490

First World War, Women Chemists in the Chemical Industry, 501

First World War, Women Chemists in the Explosives Industry, 498

First World War, Women Chemists in the Iron and Steel Industry, 499

First World War, Women Chemists Traditional Roles, 506

First World War, Women in Biochemistry, 502

First World War, Womens contributions, Ida Smedley, 294

First World War, Young Women, 486

Fisher, Catherine Hodgson, 429

Fisher, Nellie, 543, 546–547

Fishponds Training College, Bristol, 524

Forster, Emily, 122, 442

Fortey, Emily, 205–206

Fox, Caroline, 11

Fox, Dorothy (Mrs. Richards), 555–556

Frankland, Percy, 322, 463–465

Franklin, Rosalind, 118, 400, 413–422

Freund, Ida, 36, 75–79, 96, 255, 261, 320, 470

Friend, J. Newton, 190

Fulhame, Elizabeth, 6–10, 14, 16, 287

G

Garbutt, Phyllis, 272, 280–283

Garrett, Elizabeth, 109, 218, 245, 426–427, 430, 433, 436

Gazdar, Maud (Mrs. Taylor), 116

Geology at Bedford College, 390

Gibson, Florence, 351–352

Gilchrist, Helen (Mrs. Childs), 223, 225–226, 387, 496

Girls' Boarding Schools, 31

Girls' Day Schools, 30

Girls' Public Day School Company, 30–31, 33–34, 36, 39, 44, 138, 319, 327, 331, 513

Girls' Schools in Scotland, 33

Girls' Schools in Wales, 32

Girton College, Cambridge, 59, 67–75, 78–80, 86, 89–90, 104, 320, 334, 349, 360, 375, 389, 400–401, 469, 504–505, 518, 539, 544

Girton College, Cambridge, Chemistry Laboratory, 77

Glasgow and West Scotland College of Domestic Science, 233

Glasgow Mechanics' Institution, see Royal Technical College, Glasgow, 227

Glenny, Alexander, 541

Gloucestershire Training College of Domestic Science, 263

Goldschmidt, V. M., 384–385, 406

Goldsmith's College, London, 158, 332, 472, 505

Gordon Hall School of Pharmacy for Women, see Margaret Buchanan School of Pharmacy for Women, 454

Gosling, Raymond, 416, 418

Gostling, Mildred (Mrs. Mills), 156, 470–471

Grantham Ladies' College, 95

Granville School, Leicester, 95

Greenish, Henry, 451, 453–454

Gretna Explosives Factory, Scotland, 498

Grubb, Edward, 25–27, 316

Guye, P. A., 113, 115

568   *Index*

H

Haberdashers' Aske's Hatcham Girls'
  School, 84, 94
Hack, Maria, 22
Hadfield, Isabel, 288, 503
Haldane, J. B. S., 363, 372
Hamer, Frances, 313, 540, 543–547,
  551
Hampson, Robert, 431–437, 448
Harcourt, Augustus Vernon, 90, 93,
  153, 297
Hardy, William Bate, 360, 373
Hartle, Hilda, 263, 321
Hartley University College, *see*
  University of Southampton, 213
Hatfield, Isabel, 136
Haworth, Norman, 159, 211–212, 248
Haynes, Dorothy, 131
Heather, Lilian, 329–331
Heath, Grace, 332, 342
Heaton, Charles William, 161–163
Henderson, George Gerald, 232–233
Herbert, Mary Sydney, Countess of
  Pembroke, 2
Heriot-Watt College, *see* University
  of Edinburgh, 246–247, 457
Heuristic Method, 335, 341–342,
  345–346, 349
Hewitt, John T., 133
Hickmans, Evelyn, 200
Higginbotham, Lucy, 178
Higher Education of Women, 67
Highfield School, Hendon, 98
Hitchins, Ada (Mrs. Stephens), 97,
  237–239, 501
Hoffert, Dorothea (Mrs. Bedson),
  505–506
Holland, Kathleen (Mrs. Lapworth),
  460–462

Holland, Lily (Mrs. Kipping),
  460–462
Holland, Mina (Mrs. Perkin),
  460–462
Holmes, Edna (Mrs. Taylor), 309
Homer, Annie, 502–503
Homerton College, Cambridge, 263,
  320–322
Homfray, Ida, 115
Hooper, Elsie (Mrs. Higgon), 450,
  453–455
Hope Scholarship, 243–244
Hope, Thomas Charles, 242–243
Hopkins, Frederick Gowland, 71, 131,
  357–362, 365–369, 374, 502, 537
Hoppie Societies, 359
Howell's School, Denbigh, 33, 82
Howell's School, Llandaff, 32, 42,
  95, 352–353, 519
Huddersfield Municipal High School,
  78
Hughes Hall, Cambridge, 76
Humphrey, Edith, 143–144, 343
Hunt, Annette, 331–332
Huntington, Alfred Kirby, 120–122,
  453
Hurst, Winifred (Mrs. Wright), 491,
  496–498

I

Imperial Chemical Industries, 82, 99
Imperial College, London, 123–125,
  128, 131, 166, 181, 232, 254, 334,
  390, 468, 482, 490, 496–497, 504,
  541, 546–547
Imperial College, London, Chemical
  Society, 124–125
Imperial College, London, Women
  Chemistry Students, 124

Imperial College, London, Women's Association, 126

Imperial War Museum, Women's Work Collection, 488

Ingold, Christopher, 394, 477–478, 482–484

Institute of Chemistry, 62, 83, 163, 198, 289–290, 491, 519, 529, 533, 535, 537, 548

Institute of Food Science Technology, 234

Inter-War Period, Biology for Girls, 513

Inter-War Period, Changed Societal Atmosphere, 512

Inter-War Period, Chemistry Education for Girls, 510

Inter-War Period, Difference for Academic Women, 514

Inter-War Period, Employability of Women Chemists, 526

Inter-War Period, Hadow Report of 1923, 511

Inter-War Period, Hostility Towards Women Students, 515

Inter-War Period, Marriage Bar, 518

Inter-War Period, Opportunities for Employment, 519

Inter-War Period, Opposition to Women in Medicine and Pharmacy, 516

Inter-War Period, Returning Servicemen, 514

Inter-War Period, Salary Inequity for Women, 527

Inter-War Period, Woman "Super-Chemist" as Negative Role Model, 520

Inter-War Period, Women at University, 513

Inter-War Period, Women Chemists After-Careers, 556

Inter-War Period, Women Chemists as Factory Inspectors, 554

Inter-War Period, Women Chemists as Indexer, 550, 552–554

Inter-War Period, Women Chemists as Scientific Librarians, 550–551

Inter-War Period, Women Chemists in Biomedical Laboratories, 537

Inter-War Period, Women Chemists in Dye Industry, 548

Inter-War Period, Women Chemists in Food Industry, 531

Inter-War Period, Women Chemists in Gas Industry, 549

Inter-War Period, Women Chemists in Photographic Industry, 543

Inter-War Period, Women Chemists in Teaching Careers, 521

Inter-War Period, Women in Chemical Industry, 526

Inter-War Period, Women's Employment, 517

Inter-War Period, Women's Participation at University, 515

Ionian Society, 48

Ipswich High School, 39, 59, 309

Irvine, James Colquhoun, 222–225

J

James Allen's Girls' School, 151, 274, 532

Jamison, Margaret (Mrs. Harris), 151–152

Jex-Blake, Sophia, 218, 220, 244

570   *Index*

John Innes Horticultural Institution, 239, 365
John, Margaret, 352–353
Johnson, Mary (Mrs. Clark), 81
Jones, Katherine, Lady Ranelagh (*née* Boyle), 3
Jordan Lloyd, Dorothy, 373–375, 502, 521, 527, 537
Judd, Hilda, 491, 504–505

K

Kahan, Zelda (Mrs. Coates), 305
Kathleen, Coward, 521
Kempson, Elizabeth (Betty) (Mrs. Percival and Mrs. McDowell), 248–249
Kenner, James, 194, 196
Kensington High School, 116, 126, 311
King, Annie Millicent, 208–209
King Edward VI High School for Girls, 31, 78–79, 203, 291, 293, 310, 322, 335, 361, 363, 373, 497, 502, 507, 555
King, Ruth, 495, 557
King's College for Household and Social Science, 536
King's College for Women, London, 118–120, 207, 228, 267–268, 270–272, 275, 281–282, 368, 503, 548
King's College for Women, London, Chemistry Department, 272
King's College, London, 53, 108, 117–118, 120–121, 123, 138, 416–419, 420, 453–454, 456
King's College, London, Household Chemistry Controversy, 269

King's College, London, Laboratory of Professor Huntington, 120
King's College of Household and Social Science, 271, 273, 313, 520, 531, 533, 541
King's High School, Warwick, 103
Kipping, Frederick Stanley, 460–461
Kletz, Leonore (Mrs. Pearson), 177–178
Klug, Aaron, 421–422
Knaggs, Ellie, 313, 387, 389–391
Kuroda, Chika, 100–101

L

Laboratoire Centrale des Services Chimique de l'État, 415
La Chymie charitable et facile, 13
Ladies Chemistry, 2
Ladies' College, Bedford Square, *see* Bedford College, London, 138
Ladies' Medical College, 445
Lady Hardinge Medical College for Women, 82–83, 237
Lady Margaret Hall, Oxford, 85, 88–90, 98, 103, 494
Lady Pharmacists, 437–438
Laing, Mary (Mrs. McBain), 208, 474–476
Lapworth, Arthur, 177–178, 461–462
Lavington, Elizabeth (Mrs. Hedley), 99
Laycock, Norah, 168
Lectures with Experiments, 52
Leech, Elizabeth, 430, 432
Leeds Girls' High School, 200, 317
Leeds, Kathleen, 332–334
Lees, Edith, 339, 347–349

LeFevre, Raymond, 476–477
Leicester College of Technology, 183
Leishman, Margaret, 101–102
Leitch, Grace, 211–212, 496, 557
Le Pla, Margaret, 552–553
Leslie, May (Mrs. Burr), 187–189, 256, 503, 557
Lesslie, Mary, 148–150, 250, 557
Lewis, Ida, 51
Lewis, Samuel Judd, 310–311
Liquid Air Demonstrations, 53
Lister Institute of Preventive Medicine, 293, 296, 304, 357, 503, 506
Liverpool High School, *see* Belvedere School, 41, 348
Lloyd, Emily, 198, 289
London Chemical Society, 286–287
London Day Training College, 288, 310
London Matriculation; the Department of Science and Art, 46
London School of Medicine for Women, 116, 145, 160–169, 220, 245–246, 433, 445, 449, 468, 516
London School of Medicine for Women, Chemistry at, 161
lonsdaleite, 394
Lonsdale, Thomas, 393–394, 396–397
Lothian, Agnes (Mrs. Orme, Mrs. Short), 457–458, 550
Luis, Ethel, 250, 253–254
Lunam, Peggy (Mrs. Edge), 523–524

M

Macadam, Elison (Mrs. Desch), 120–121, 465
MacDonald, Evelyn, 334–335

Mallen, Catherine, 212–213
Malvern Girls' College, 49, 182
Manchester and Salford College for Women, 173
Manchester High School for Girls, 31, 44, 63, 70, 110, 173, 177–178, 186, 261, 375, 466, 505
Manchester University, 173, 177–178, 549
Manchester University, Chemical Society, 175–177
Manchester University, Women Students, 174
Marcet, Jane (*née* Haldimand), 13–18, 29, 287
Margaret Buchanan School of Pharmacy for Women, 449, 453
Maria Grey College, London, 256, 320, 494
Maria Grey Training College, *see* Maria Grey College, 319, 345
Marsden, Effie (Mrs. Soloman), 116
Marshall, Dorothy, 77–79, 155, 337
Martin, Charles, 296
Mary Datchelor School, 31, 44
Mary Erskine School, 506
Mason College of Science, *see* University of Birmingham, 196–197, 199, 289–291, 373
Masters, Helen, 272–274, 280–282
Mathews, Annie M. (Mrs. Kenner), 193, 195, 491
Mathieu, Marcel, 415, 420
McBain, James William, 208, 474–476, 525
McCance, Robert, 542–543
McKenzie, Alexander, 148, 249, 254

572   *Index*

McKie, Phyllis, 255, 493–494
Medical Research Council, 369, 371, 542–543
Megaw, Helen, 391, 398, 400–402
Mendel, Lafayette, 357
Men's Chemistry Club, London, 311
Method of Teaching Chemistry in Schools, 346
Meudrac, Marie, 13
Mexborough Secondary School, Yorkshire, 82
Micklethwait, Frances, 129–130, 240, 491
Microchemical Club, 136, 289
Miers, Henry Alexander, 382–384
Miller, Christina, 246–248
Mills, William, 237, 471, 544–545
Milton Mount College, 31, 39, 43, 522
Mineralogical Society of Great Britain, 385
Minshull, Rose, 432–433, 435–438, 445–446
Moodie, Agnes, 223–224
Moore, T. S., 127, 158
Morgan, Gilbert T., 129
Morrison, Edith (Mrs. Corran), 181–182
Mount School, 24–25, 59
Moyle, Dorothy (Mrs. Needham), 358–359, 366, 375–378, 399, 469
Murray, Alice Rosemary, 85–87
Muter, John, 445, 447–448

N
Nadon, Constance, 198–200
National Chemical Laboratory, 135
National Physical Laboratory, 78, 288, 501, 550

National Society for Promoting Religious Education, 264
National Union for Improving the Education of Women of all Classes, 30
Necker, Anne, Madame de Stael, 18
Necker de Saussure, Albertine, 18
Needham, Joseph, 359, 376–377, 469
Neve, Annie, 431, 436, 438, 447–448
Neve, Grace, 438, 447
Neve, Mary, 438, 447
New College, Oxford, 85
New Girls' Schools, 27
New Hall, Cambridge, 86–87
Newington Academy for Girls, 23
Newnham College, Cambridge, 36, 68–72, 74–85, 89–90, 110, 114, 155, 198, 212, 219, 261, 293, 320, 322, 335, 339, 348, 360–361, 363, 368, 373, 470–471, 502, 507–508, 522, 534, 551, 553
Newth, George Samuel, 317
Newton, Lucy, 113–114
New Woman, 518
Northern Polytechnic, London, 237
Northern Universities, 172
North London Collegiate School (NLCS), 28, 30–32, 39, 58, 63, 90, 107, 110, 139, 143, 145, 147, 291, 309, 320–321, 342–345, 347, 389, 449, 482, 489, 544
North London Collegiate School (NLCS), Science Club, 343
Notting Hill High School, 36, 59, 148, 295, 304, 326–327, 330
Nutrition Society, 297, 533

Index 573

O

Ochanomizu University, Japan, 101
Olliver, Mamie, 532–534
Onslow, Huia, 365–366, 469
Orme Girls' School, 492
Orton, Kennedy Joseph Previte, 255–256, 488, 493
Owens College, *see* Manchester University, 172–175, 185–186, 256, 548
Oxford and Cambridge Schools Examination Board, 46
Oxford High School, 101, 334–335
Oxford Local Examinations Delegacy, 46
Oxford Museum, 382, 384–385

P

Park School, Glasgow, 34, 43, 340
Partington, James Riddick, 133
Patterson, Dorothy, 335–336
Pawsey, Edith, 549–550
Pease, Marian F., 202
Pechey, Edith (Mrs. Pechey-Phipson), 218–219, 243–244, 246
People's Palace Technical Schools, *see* Queen Mary College, London, 132
Perkin, William Henry Jr, 100, 461
Perkin, William Sr., 298
Perse School for Girls, Cambridge, 336
Perutz, Max, 408, 410
Pharmaceutical Society, 163, 427–434, 436–437, 439, 441, 443, 445, 447–448, 450, 452, 454, 456–457, 538–539

Pharmaceutical Society, Admission of Women, 429
Pharmaceutical Society, Defeat of the Women's Cause, 433
Pharmaceutical Society, Examinations, 428
Pharmaceutical Society, Pharmacy Acts, 428
Pharmaceutical Society, School of Pharmacy, 429, 431, 437–438, 440, 446, 451, 453, 455–456
Pharmaceutical Society, Success for Women at Last, 436
Pharmacy, 17, The Pavement, 438, 440
Philip, James, 308
Pirret, Ruth, 230–232, 501
Plant, Millicent, 159–160
Plant, Sydney, 103, 338
Plimmer, Robert H. A., 369
Poetic verse, 25, 34, 107, 142, 161, 175, 199, 244, 278, 325, 366, 411, 435, 465, 499, 554
Polam Hall, 26–27, 541
Pope, Enid (Mrs. Hulsken), 546, 551
Pope, F. G., 133
Pope, William, 71, 466, 544
Porter, Mary (Polly) Winearls, 383–385, 389, 406
Portsmouth High School, 328, 332
Potter, Frances (Fanny) Elizabeth (Mrs. Deacon), 429
Presentations on Contemporary Topics, 50
Presentations on the History of Chemistry, 49

574 *Index*

Princess Helena College, 317, 326–327
Progressive Pharmaceutical Club, 449–450

Q
Quaker schools, 22, 27
Quartly, Lilian, 336–337, 535
Queen Elizabeth College, London, 271, 274, 313
Queen Margaret College, Glasgow, 227–229
Queen Margaret College, *see* University of Glasgow, 221
Queen Mary College, London, 95, 104, 132–133
Queen's College for the Education of Ladies, Glasgow, 227
Queen's College, Harley Street, 28–29, 112, 477
Queen's University, Belfast, 400

R
Radium Demonstrations, 53
Raisin, Catherine, 113
Ramsey, William, 78, 112–115, 164, 203–204, 297, 304, 306, 330
Ram, Sosheila, 82, 237
Randall, John, 416–417, 420
Ratcliffe, Anne, 167–168
Raymond, Yolande, 347–348
Reading Agricultural Institute, 240
Reason, Hazel, 522–523
Redland High School, 208, 339–340, 349, 474
Reid, David Boswell, 34
Renouf, Norah, 455–456
Research Association of British Rubber Manufacturers, 551

Richards, Marion, 240–241
Rich, Mary, 94–95
Riley, Dennis, 407
Rippon, Dorothy, 338–339
Roberts, Margaret (Mrs. Thatcher), 409
Roberts, Muriel, 183–184
Robertson, Mary W., 154–155
Robinson, Margaret (Mrs. Dougal, Mrs. Chaplin), 551–552
Robinson, Robert, 97, 103, 224, 293, 462, 469, 477–481, 483
Robinson, Rona, 270, 548–549
Roedean School, 32, 37, 40–41, 48, 52, 58, 95, 119, 275, 400
Rogers, A. M. A. H., 88, 91
Rogers, Ivy, 148
Rogers, Kathleen (Mrs. Penfold), 98–99
Roscoe, Henry, 301
Ross, Katherine (Mrs. Wilson), 99
Rowett Research Institute, Aberdeen, 239–242, 357
Rowland, Alice (Mrs. Hart), 432–433, 435–436, 445
Royal College of Chemistry, London, 123
Royal College of Science, Dublin, 273
Royal College of Science, London, 123, 125–126, 129, 156, 253, 310, 317, 332, 504–505, 551
Royal Free Hospital, *see* London School of Medicine for Women, 160–161
Royal Holloway College, London, 61, 79, 85, 99–100, 119, 126, 133, 139, 152–159, 249, 253, 328, 330, 342, 468, 470–472, 482, 496, 528

Royal Holloway College, London, Life of Women Chemistry Students, 153
Royal Holloway College, London, Succession Controversy, 157
Royal Holloway College, London, Women Chemistry Staff, 154
Royal Institute of Chemistry, *see* Institute of Chemistry, 289
Royal Institution, 10–11, 14, 23, 52, 226, 293, 386, 390–391, 393–394
Royal Institution Laboratory, Manchester, 548
Royal Microscopical Society, 464
Royal Naval College, Greenwich, 504
Royal Pharmaceutical Society, 36
Royal Photographic Society, 546
Royal School for Daughters of Officers, Bath, 164
Royal School of Mines, London, 123
Royal Society, 4, 119, 132, 376, 395, 410, 490, 543
Royal Society, Admission of Women, 371–373
Royal Society of Chemistry, 144, 546
Royal Society of Edinburgh, 248
Royal Statistical Society, 531
Royal Technical College, Glasgow, 226, 232–233
Rutherford, Ernest, 71, 189, 386

S

Sadler, Eileen (Mrs. Doran), 182–183
Saltburn High School, 147
Sanderson, Phyllis, 165–168
Schofield, Charlotte (Mrs. Cole), 210–211

School of Technical Chemistry, Liverpool, 179
School Science Libraries, 58
School to University, 59
School–University Links, 63
School Visits to Local Gas Works, 54
School Visits to Soap Manufacturing Plants, 56
School Visits to Steel Works, 56
School Visits to Sulphuric Acid Plants, 55
Schuster, Arthur, 490
Science Club Magazines, 58
Science Clubs and Societies, 47–59
Science for Peace Organization, 412
Scientific Lady, 10
Scientific Librarian, 549
Scottish Institution for the Education of Young Ladies, 34
Scottish Seaweed Research Association, 248
Scottish Universities, Admission Battle Finally Won, 220
Scottish Universities, Entry of Women, 218
Scottish Universities, Life of Women Students, 221
Searle, Rev. H., 316
Sebright, Caroline, 16
Sebright, Frederica, 16
Sedgwick, Anne (Lady Walker), 469–470
Seward, Margaret (Mrs. McKillop), 94, 118–120, 154, 158, 267, 272, 469
Shaen, William, 109, 160
Sheffield High School, 56, 195, 329
Shirley Institute, Manchester, 178
Sidgwick, Arthur, 88

576   *Index*

Sidgwick, Henry, 67–68, 88
Sidgwick, William, 90
Simpson, Beatrice, 241
Simpson, Delia Margaret
  (Mrs. Agar), 83–84, 150, 418
Sinnatt, Frank, 467
Slater, Jesse (Mrs. Baily), 507–508
Smedley, Ida (Mrs. Smedley
  Maclean), 52, 126, 176, 269,
  292–295, 299, 307, 312, 349, 365,
  478, 506, 520–521
Smith, Alice, 256, 494
Smithells, Arthur, 185–186, 188, 261,
  267, 270
Smith, Isobel, 250, 253
Soar, Marion, 272, 280–281
Society for Analytical Chemistry,
  136, 287–288, 533
Society for Chemical Industry, 184,
  309
Society for Promoting the
  Employment of Women, 432–433
Society for Public Analysts, 533
Society of Apothecaries, 426, 430,
  432, 441–442
Society of Chemical Industry, 533,
  535
Society of Dyers and Colourists, 309
Society of Indexers, 554
Society of Public Analysts, 536
Soddy, Frederick, 97, 231, 237–239,
  241, 462–463
Somerville College, Oxford, 70, 89–90,
  93–95, 119, 385, 405–407, 489
Somerville, Mary, 5, 90
Soulsby, Lucy, 37
Southampton University College, *see*
  University of Southampton, 213,
  504

South Hampstead High School, 63,
  116, 507, 552
South London School of Chemistry
  and Pharmacy, *see* South London
  School of Pharmacy, 436, 445
South London School of Pharmacy,
  447
South-Western Polytechnic, 265–266
South-West London Chemists'
  Association, 452
Spencer, James, 102, 140, 147–148,
  152, 532, 541, 552
Stammwitz, Louisa, 432–433,
  435–438, 447–448
Stanford University, California, 474
Stedman, Edgar, 471–473
Steele, Jannette (Ettie), 223–225, 557
Stephenson, Marjory, 368–374, 376,
  395, 399, 507, 521, 553–554
Stern, Rose, 58, 198, 290, 293,
  346–347, 349–350
St. Gabriel's Training College,
  Camberwell, 126, 494
St. George's School, Edinburgh, 34,
  340
St. Hilda's College, Oxford, 89, 91,
  96, 98–99, 102–104, 338, 514
Stirling-Taylor, Effie Isobel (Mrs.),
  169
St. John's College, Cambridge, 74
St. John's Wood High School, 94
St. Leonard's School, St. Andrews,
  34, 38, 45, 110
St. Martin-in-the-Fields High School,
  53, 151, 318
St. Mary's Training College, London,
  82
St. Paul's Girls' School, 39, 43, 50,
  57, 340, 347–348, 414

Streatham Hill High School, 51, 328–329, 454, 497

St. Swithun's School, 32, 40, 45, 50, 98, 102, 316

Sutherland, Maggie, 232–233

Sutton High School, 49, 128, 328, 331

Swanley Horticultural College, 129–130, 214, 240

Swansea High School, 33

Swansea Municipal School, 132

Sydenham High School, 348, 541

Synesthesia, 103

T

Taylor, Clara, 339–340, 349–350

Taylor, Millicent, 207–208, 474, 490, 503, 557

Tchaykovsky, Barbara (Ally), 145–146

Teachers' Training Colleges, 318

Teachers' Training & Registration Society College, *see* Maria Grey College, 319

Teaching of Domestic Chemistry, 263

The Metabolism of Fat, 294

The Pavement, 439

Thomas, Beatrice, 78, 96, 293, 349

Thompson, Gartha, 310–311

Thompson, J. J., 507

Thompson, Mary (Mrs. Clayton), 273–275

Thorne, Isabel, 218–219, 245

Thorpe, Edward, 128, 301, 317, 551–553

Thorpe, Jocelyn, 128, 131, 306

Tideman, Catherine (Mrs. LeFevre), 476–477

Tilden, William, 126, 203, 205, 298, 300–301, 306, 334

Tinkler, Charles Kenneth, 272, 274, 281

Tohoku Imperial University, Japan, 100

Tokyo Women's Higher Normal School, Japan, 100

Tomlinson, Muriel, 96, 101–102, 104

Toynbee, Grace (Mrs. Frankland), 460, 463–465

Trew, Violet, 148, 150–152

Trinity College, Dublin, 69–70, 116, 334, 362

Turner, Emily, 193–194, 491

Turner, Eustace E., 140, 149, 274–275, 555

Turner, Margaret K., 492

U

University College, Aberystwyth, *see* University of Wales, 254, 290, 483, 492–493

University College, Bristol, *see* University of Bristol, 201, 203, 205, 207

University College, Dundee, 148, 249–253, 304, 470

University College, Exeter, *see* University of Exeter, 553

University College in Bangor, Wales, 189

University College, Liverpool, *see* University of Liverpool, 172, 179

University College, London, 78, 98, 108, 110–111, 114–117, 123, 273, 295, 305, 348, 369, 386, 392–393, 395–397, 399, 470, 476–477, 483–484, 494, 503, 507, 535, 540

578   *Index*

University College, London,
Admission of Women Chemistry
Students, 112
University College of North Wales,
Bangor, *see* University of Wales,
254–255, 488, 493–494
University College of South West
England, *see* University of Exeter,
495
University College, Reading, 525
University College, Sheffield, *see*
University of Sheffield, 191
University College, Swansea, *see*
University of Wales, 254
University of Aberdeen, 235–238
University of Aberdeen, Life of
Women Students, 235
University of Aberdeen, Women
Chemistry Students, 236
University of Birmingham, 159, 196,
248, 322, 373
University of Birmingham, Chemical
Society, 198
University of Birmingham, Women
Chemistry Students, 197
University of Bristol, 201, 208, 219,
263, 365, 474, 524–525
University of Bristol, Chemical
Society, 203
University of Bristol, Women
Chemistry Students, 202
University of Bristol, Women
Students, 202
University of British Columbia, 495
University of Cambridge, 386,
399–400, 407, 414, 418, 420, 537,
540, 542
University of Cambridge, Balfour
Biological Laboratory, 110, 367

University of Cambridge, Cavendish
Laboratory, 401, 414, 507
University of Cambridge, Chemical
Club, 95
University of Cambridge, Chemistry
Laboratories, 75
University of Cambridge, Chemistry
Research Laboratory, 79, 81, 83
University of Cambridge, Life of
Women Students, 73
University of Cambridge,
Mineralogical Laboratory, 389
University of Cambridge, Women
Chemistry Students, 74
University of Delhi, India, 237
University of Dundee, *see* University
College, Dundee, 464, 490
University of Dundee, Women
Chemistry Students, 250–251
University of Durham, 209–210
University of Edinburgh, 121, 242,
246, 248, 340, 472
University of Edinburgh, Life of
Women Students, 242
University of Geneva, 113, 115
University of Glasgow, 226–228,
230, 233, 238, 335, 490
University of Glasgow, Alchemists
Club, 229
University of Glasgow, Women
Chemistry Students, 229
University of Leeds, 185, 261, 386,
393–394, 482, 490
University of Leeds, Cavendish
Society, 186–187
University of Leeds, Organic Lab
Club, 187
University of Leeds, Women
Chemistry Students, 186

University of Leicester, 481
University of Liverpool, 179, 490
University of Liverpool, Chemical Society, 181
University of Liverpool, Women Analytical Chemists, 183
University of Liverpool, Women Chemistry Students, 179
University of Liverpool, Women's Chemical Society, 181
University of London, 108
University of London, Admission of Women to Examinations, 108
University of London, Convocation, 109–110
University of London, Regulations, 111
University of London, Senate, 109–110
University of Natal, S.A., 497
University of Newcastle, 209
University of Newcastle, Women Science Students, 210
University of Oxford, 36, 239, 241, 384, 407, 479, 482
University of Oxford, Alembic Club, 95–97, 406–407, 481
University of Oxford, Clarendon Laboratory, 401
University of Oxford, Degrees for Women, 91
University of Oxford, Dyson Perrin Laboratory, 97–100, 102–104, 338, 480
University of Oxford, Earliest Women Chemistry Staff, 101
University of Oxford, Founding of the Women's Colleges, 88

University of Oxford, Junior Scientific Club, 96
University of Oxford, Proportion of Women Science Students, 93
University of Oxford, Quota for Women, 92
University of Oxford, Women Chemists at the DP Laboratory, 97
University of Oxford, Women Chemistry Students, 92
University of Oxford, Women's Scientific and Philosophical Society, 93
University of Sheffield, 86, 121, 191, 490
University of Sheffield, Chemical Society, 192
University of Sheffield, Women Chemistry Students, 191
University of Southampton, 214
University of St. Andrews, 211, 218, 220, 222, 224–225, 249, 309, 490–491, 496
University of St. Andrews, Life of Women Students, 222
University of St. Andrews, Women Chemistry Students, 223
University of Strathclyde, *see* Royal Technical College, Glasgow, 226
University of Sydney, Australia, 479
University of Toronto, 502
University of Wales, 254
University of Zürich, 143
Usherwood, Edith Hilda (Mrs. Ingold), 129, 477–478, 482–484

V

Vanderstichele, Paule Laure, 146–147
Vickery, Alice, 429, 445

580 *Index*

Victoria University of Manchester, *see* Manchester University, 173, 293, 386, 461, 466–467, 478, 505, 538

Virgo, M., 59

W

Wakefield, Priscilla, 22

Walker, James, 113, 246–247, 304, 470

Walker, Nellie (Mrs. Wishart), 250, 252–253

Walsh, Gertrude (Mrs. Robinson), 97, 469, 477–482

Walter, Edna, 320, 344–345

Warden of Women's Residence, Mary Crewdson, 147

Warden of Women's Residence, Mary Lesslie, 150, 190

Warden of Women's Residences, Dorothy Bennett (Mrs. Leighton), 195

Warden Women's Residences, Ettie Steele, 225

Warden Women's Residences, Grace Leitch, 212

Warden Women's Residences, Helen Masters, 282

Warden Women's Residences, Millicent Taylor, 208

Warden Women's Residences, Ruth King, 495

Warden Women's Residences, Ruth Pirret, 231–232

Watson, James, 418–419, 421–422

Watson, Mary, 94

Werner, Alfred, 144

Westfield College, London, 494

Westminster College of Chemistry, Pharmacy and Dispensing, *see* Westminster College of Pharmacy, 441

Westminster College of Pharmacy, 122, 442

Weymouth High School, 331

Wheelwright Grammar School, 334, 336

Wheldale, Muriel (Mrs. Onslow), 131, 239, 292, 363–367, 373, 469, 479

Whetham, Margaret Dampier (Mrs. Anderson), 553–554

White, Jessie Meriton, 108–109

Whiteley, Martha, 48, 126–130, 153–154, 292, 299, 306, 312, 314, 342, 482, 491, 496, 504, 521

Widdowson, Elsie, 131, 541–543

Widdows, Sibyl, 157–158, 164–166, 471, 491

Wilkins, Maurice, 416, 418–419, 421

Willcock, Edith (Mrs. Gardiner), 361–363, 373

William Perkin Jr., 256

Williams, Katherine, 113, 203–204

Williams, May, 169

Willis, Olive, 324

Wilson, Charlotte, 88

Wilson, Forsyth James, 233

Wilson, Robert, 219

Wilson-Smith, Marjory (Mrs. Farmer), 468

Wimbledon High School, 48, 51, 57, 126, 325, 330

Wimbledon House School, *see* Roedean School, 32

Winchester High School, 32

Winch, Hope, 456
Woman Chemist, Amateur Assistants, 460
Woman Chemist, Collaborative Couples, 469
Woman Chemist, Contrasting Paths, 477
Woman Chemist, Professor's Wife, 465
Women and Biochemistry, 356
Women at University, First Generation, 60
Women at University, Second Generation, 61
Women Biochemists at Cambridge, 360
Women Chemistry Teachers, Pioneers, 325
Women Chemists' Dining Club, 311–313, 532
Women Crystallographers, Early History, 382
Women Pharmacists, Employment, 440
Women Pharmacists, Pioneers, 445
Women Science Teachers, Shortage, 326

Women's Department of King's College, *see* King's College for Women, London, 117
Women's Educational Union, 30
Women's Magazines, 12
Women Teachers, Life of, 322
Wood, Florence, 311
Wordsworth, Elizabeth, 88–89
Workman, Olive, 310
Wren, Gertrude, 438
Wright, Albert, John, 497
Wright, Winifred (Freda), 539–540
Wycombe Abbey School, 32, 38, 49
Wynne, William Palmer, 192–193, 491
Wyss, Clotilde van, 321

Y
Yardley, Kathleen (Mrs. Lonsdale), 312, 373, 387, 391–397, 401, 420, 469
Yorkshire College, *see* University of Leeds, 172, 185–186, 305, 317
Yorkshire Ladies Education Association, 185
Young, Sydney, 205
Younie, Elinor, 340–341

CPSIA information can be obtained
at www.ICGtesting.com
Printed in the USA
JSHW011434090120
3449JS00001B/4